Mechanical Desktop Instructor

Sham Tickoo
Professor
Department of Mechanical Engineering Technology
Purdue University Calumet
Hammond, Indiana
U.S.A.

CADCIM Technologies
(www.cadcim.com)
U.S.A.

Boston Burr Ridge, IL Dubuque, IA Madison, WI New York San Francisco St. Louis
Bangkok Bogotá Caracas Kuala Lumpur Lisbon London Madrid Mexico City
Milan Montreal New Delhi Santiago Seoul Singapore Sydney Taipei Toronto

McGraw-Hill Higher Education 

A Division of The **McGraw-Hill** *Companies*

MECHANICAL DESKTOP INSTRUCTOR

Published by McGraw-Hill, a business unit of The McGraw-Hill Companies, Inc., 1221 Avenue of the Americas, New York, NY 10020.

 This book is printed on recycled, acid-free paper containing 10% postconsumer waste.

1 2 3 4 5 6 7 8 9 0 QPD/QPD 0 9 8 7 6 5 4 3 2 1

ISBN 0–07–249941–9

General manager: *Thomas E. Casson*
Publisher: *Elizabeth A. Jones*
Sponsoring editor: *Kelly Lowery*
Senior developmental editor: *Kelley Butcher*
Executive marketing manager: *John Wannemacher*
Project manager: *Mary Lee Harms*
Senior production supervisor: *Sandy Ludovissy*
Coordinator of freelance design: *Michelle D. Whitaker*
Cover designer: *Swimmer Design Associates*
Cover illustration: *Created by Deepak Maini and Puja Bahl, Cadsoft Technologies, using MDT and MAX software*
Senior supplement producer: *Stacy A. Patch*
Media technology senior producer: *Phillip Meek*
Compositor: *Interactive Composition Corporation*
Typeface: *10/12 New Baskerville Bt*
Printer: *Quebecor World Dubuque, IA*

Library of Congress Cataloging-in-Publication Data

Tickoo, Sham.
Mechanical Desktop instructor / Sham Tickoo. — 1st ed.
p. cm.
Includes index.
ISBN 0–07–249941–9
1. Engineering graphics. 2. Mechanical desktop. 3. Engineering design—Data processing. I. Title.

T353 .T55 2002
620'.0042'02855369—dc21 2001052202
CIP

www.mhhe.com

DEDICATION

To teachers, who make it possible to disseminate knowledge
to enlighten the young and curious minds
of our future generations

To students, who are dedicated to learning new technologies
and making the world a better place to live

Thanks

To the faculty and students of the MET department of
Purdue University Calumet for their cooperation

To Deepak Maini, Senior CADD Engineer, CADCIM Technologies
for his valuable help

Table of Contents

Chapter 3: Sketch Planes, Work Features, and Other Extrusion and Revolution Options

Chapter 4: Advanced Dimensioning Techniques, Design Variables, and Visibility Options

Chapter 5: Placed Features I

Chapter 8: 2D Path, 3D Path, Sweep, and Loft

Chapter 9: Creating New Parts, Activating the Part, and Mirroring and Combining the Parts

Chapter 10: Assembly Modeling I

Chapter 11: Assembly Modeling II

Chapter 12: Creating and Modifying the Drawing Views

Chapter 13: Dimensioning the Drawing Views

Chapter 14: Surface Modeling

Author's Web Sites

For Faculty: Please contact the author at **stickoo@calumet.purdue.edu** or **tickoo@cadcim.com** to access the Web site that contains the PowerPoint presentations, drawings used in this textbook, and student projects.

For Students:You can download drawing-exercises, tutorials, and special topics by accessing the author's web site at **www.cadcim.com** or **www.calumet.purdue.edu/public/mets/tickoo/mdt/index.htm**.

Preface

MECHANICAL DESKTOP 5

Mechanical Desktop, developed by Autodesk Inc., is one of the the world's best selling solid modeling softwares. It is a parametric feature-based solid modeling tool and it not only unites the 3D parametric features with 2D tools but also addresses every design-through-manufacturing process. Thus, it provides a link between the 2D and 3D drafting techniques leading to increased productivity. This software is completely focused on designing and you are allowed to access and edit multiple files at a time using Multiple Design Environment. Mechanical Desktop 5 creates the complex models with blends, lofts and sweeps using the ACIS 6 kernal. It allows you to use polylines, splines or an existing edge along with the simple helical paths for sweeping.

You can view the thumbnail of the external files used in the assemblies without actually opening them. You can use the visual feedback to instantly determine the current editing state. This software allows you to exchange the data in the formats like BMP, EPS, DWF, IGES (Version 5.3), VDA-FS, VRML, SAT and so on.

The 2D drawing views of the components are automatically generated in the layouts. The drawing views that can be generated include detailed, aligned, orthographic, isometric, auxiliary, section, partial section and so on. The dimensions in these views are automatically arranged and the bidirectional associative nature of this software ensures that any modification made in the model is automatically reflected in the drawing views and any modification made in the dimensions in drawing views automatically updates the model.

Mechanical Desktop Instructor is a book that is written with an intent of helping the people who are into 3D design. This book is written with the tutorial point of view with learn-by-doing as the theme. However, it is presumed that the user of this book has the basic knowledge of AutoCAD. This book includes the following features:

- AutoSurf Module.

  This is one of the very few books that includes the AutoSurf module along with the AutoCAD Designer and the Assembly Modeling modules. This will provide the students with a thorough knowledge of the software. This will also prove helpful for the students who are considering a future in Surface Modeling industry.

- Learning objectives and commands covered at the beginning of the chapter.
  Each chapter of this book starts with learning objectives and the commands covered in the chapter. This will be an easy reference tool for the students as well as instructors.

- Command section at the beginning of every chapter.
  Every chapter has a command section preceding the tutorials. It covers the commands related to that particular chapter. The detailed explanation of the commands allow easy understanding of the concepts for the student. The student can refer to these commands while working with the tutorials to ensure proper understanding.

- Use of the desktop browser.
  Apart from accessing the commands using the toolbars and the command line, this book covers the explanation of executing the commands using the desktop browser. This is a very easy and convenient way of accessing the commands.

- Live day-to-day projects as tutorials.
  The book consists of the tutorials that are live day-to-day projects used in the mechanical engineering industry. For example, this book includes the assembly of the Screw Jack starting from creating the components of Screw Jack to generating the drawing views of the assembly and adding Bill Of Material and balloons to the drawing views. This helps the student to learn the actual practical use of the commands.

- Detailed step-by-step explanation of the tutorials.
  The projects included as tutorials are explained in a detailed step-by-step procedure that includes the commands as well as the command responses. This will allow the student to learn Mechanical Desktop without the help of an instructor. This book will also be a helpful tool for the instructors.

- Easily understandable sequence of the tutorials.
  The tutorials discussed in this book have been written in an easily understandable sequence. This allows the student to easily build on the knowledge as he progressed from one chapter to the other.

- Extensive use of figures supporting the commands and tutorials.
  All the commands and tutorials are supported with the extensive use of figures for easy understanding of the concept.

- Additional informations in the form of tips and notes.
  Additional information related to the topics has been provided in the form of a tip or a note. This enhances the knowledge of the student and allows the student to use the software effectively.

- Includes Review Questions and Exercises.
  Every chapter ends with the Review Questions and Exercises for the student so that they can assess their performance.

Introduction

MECHANICAL DESKTOP 5

Welcome to the world of Mechanical Desktop! In case you are new to Mechanical Desktop or 3D design, you have just joined thousands of people worldwide who are already into 3D design using Mechanical Desktop. Those who have been using Mechanical Desktop will find major enhancement in Mechanical Desktop 5 over the previous releases. For the ease of finding new and enhanced commands, an asterisk has been put on the new or the enhanced commands.

Mechanical Desktop is a 3D feature-based solid modeling tool which will allow you to create very complex 3D models with lots of ease and then create 2D views from those 3D models. Thus it provides a link between 2D and 3D drafting techniques leading to increased productivity. It is not only capable of parametric 3D solid output but also sophisticated surface modeling and bi-directional associative drafting. The engine for Mechanical Desktop 5 is AutoCAD 2000i which means that Mechanical Desktop 5 works on top of AutoCAD 2000i. Hence, in addition to its own commands, most of the commands of AutoCAD 2000i will also work in Mechanical Desktop 5.

Apart from AutoCAD, Mechanical Desktop consists of 3 other modules:

AutoCAD Designer. It is a part modeling or a feature-based parametric modeling technique. It uses the basic drawing and editing commands of AutoCAD for creating the sketches. However, this does not mean that you should have a detailed knowledge of AutoCAD. The reason for this is that the 3D designer models are created using the commands of Mechanical Desktop.

AutoSURF. It is a surface modeling technique in which surfaces are created using the **Non Uniform Rational Bezier Splines** (**NURBS**) surfaces.

Assembly Modeling. Although some designs consist of a single part, the vast majority of designs are created by assembling multiple components. In this module, the individual parts created in the **AutoCAD Designer** are assembled and arranged. Also in this module you will learn to create and dimension the 2D views from the designer models. You will also learn to add a Bill of Material to the views.

WHY MECHANICAL DESKTOP

Autodesk Mechanical Desktop is a software package comprising of **AutoCAD Designer**, **Assembly Modeler**, and **AutoSURF**. It not only increases the productivity, but also minimizes the drafting and designing process by providing a link between 2D and 3D drafting techniques.

Mechanical Desktop's basic solids are created with the easy to use tools like extrude, revolve, sweep, path, loft, bend and so on. Mechanical Desktop has intelligent parametric features like holes, fillets, chamfers, ribs etc. to easily and quickly convert a rough 2D profile into a finished designer model. All of these parametric features easily understand their fit and function and thus can be easily edited by simply changing the dimension. For instance, a through hole in a rectangular plate will automatically adjust its depth if the thickness of the plate is increased. Another important advantage of Mechanical Desktop is bi-directional associativity. It is the property that ensures that if any editing is done in the 3D model, the drawing views are updated automatically. Also if any changes are made in the dimensions of drawing views, the model updates automatically.

Non-Uniform Rational Bezier Spline (NURBS) based surface modeling is another important feature that makes Mechanical Desktop a very useful software package. As you take a look around yourself, you will find that most of the objects, from a computer to a simple pen stand, are created using the surfaces. To fulfill these design requirements, Mechanical Desktop provides one of the most advanced surface modeling tools available on the PC in the form of the **NURBS** based surface modeling.

Assembly Modeling Technique of Mechanical Desktop is an important feature that a mechanical engineer will find very useful in his work. Majority of designs consists of multiple components assembled together. This could be done very easily in Mechanical Desktop using intelligent constraints like angle, flush, insert and mate. The parts can be created in the same drawing or can be added from an external drawing.

SYSTEM REQUIREMENTS

The minimum system requirements for your computer to enable it to run Mechanical Desktop are:

- Windows NT 4.0, Windows 98 or Windows 2000.
- 300 MHz Pentium II Processor. 800x600x64K graphics card.
- 96 MB RAM for training, 128 MB RAM for Part modeling and 256 MB RAM for Assembly modeling.
- 680 MB of hard disk space.
- 60 MB of free disk space in the system folder.

GETTING STARTED

Once you have loaded Mechanical Desktop onto your system, you can start it by double-clicking on the icon of Mechanical Desktop 5 on the desktop of your computer. You can also use the taskbar shortcuts by choosing the **Start** button at the lower left corner of the screen. Choose **Programs** to display the program folders. Now, choose **Mechanical Desktop 5** to display the Mechanical desktop programs and then choose **Mechanical Desktop 5** (Figure 1). Since the engine for Mechanical Desktop 5 is AutoCAD 2000i, therefore, the computer will first start AutoCAD 2000i and then it will start Mechanical Desktop by loading Mechanical modules. The **Mechanical Desktop Today** window (Figure 2) will be displayed once the loading is completed. In this dialog box, choose the **Create Drawings** tab. Select **Start from Scratch** from the **Select how to begin** drop-down list and then select **English (feet and inches)**. The screen of Mechanical Desktop is quite different from that of AutoCAD. Apart from toolbars and menubar, you will also find a **desktop browser** on left of the screen (Figure 3).

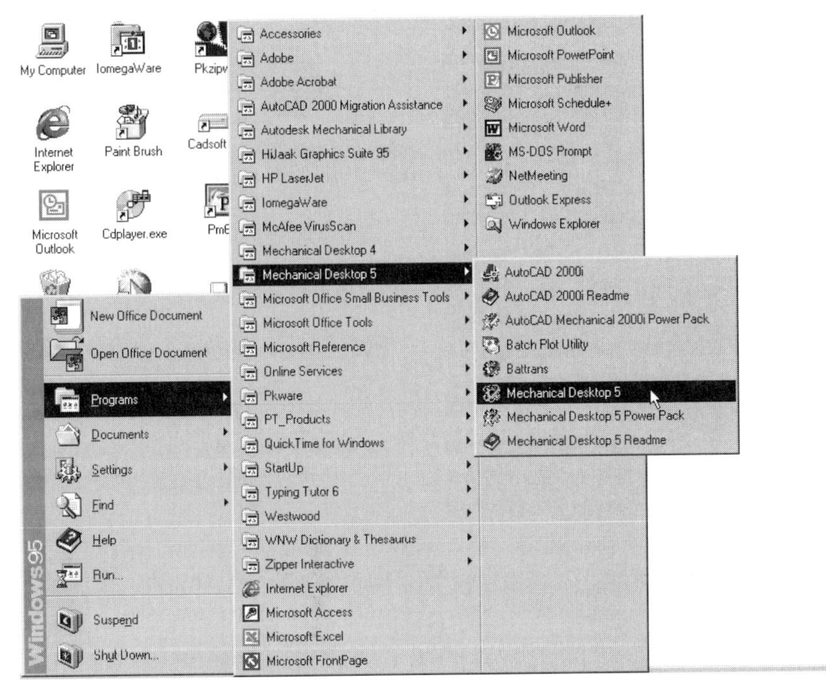

Figure 1 *Windows screen with task bar and application icons*

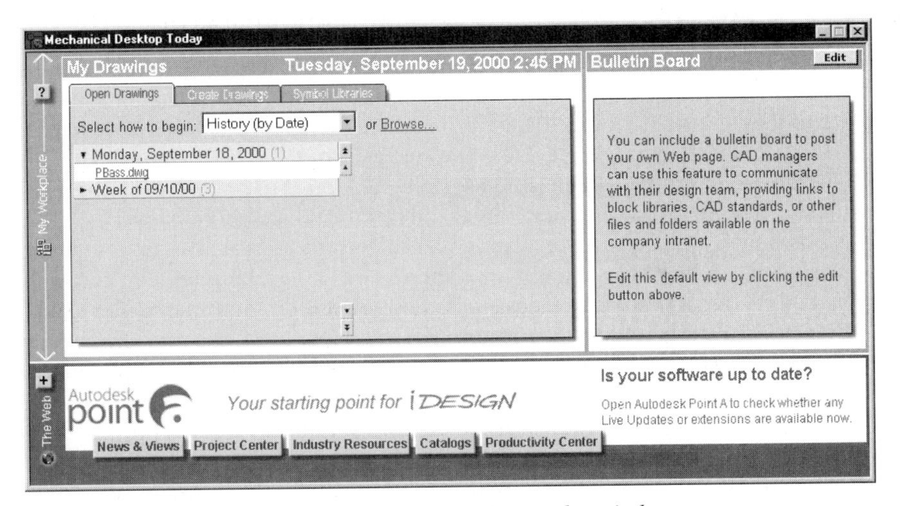

Figure 2 *Mechanical Desktop Today window*

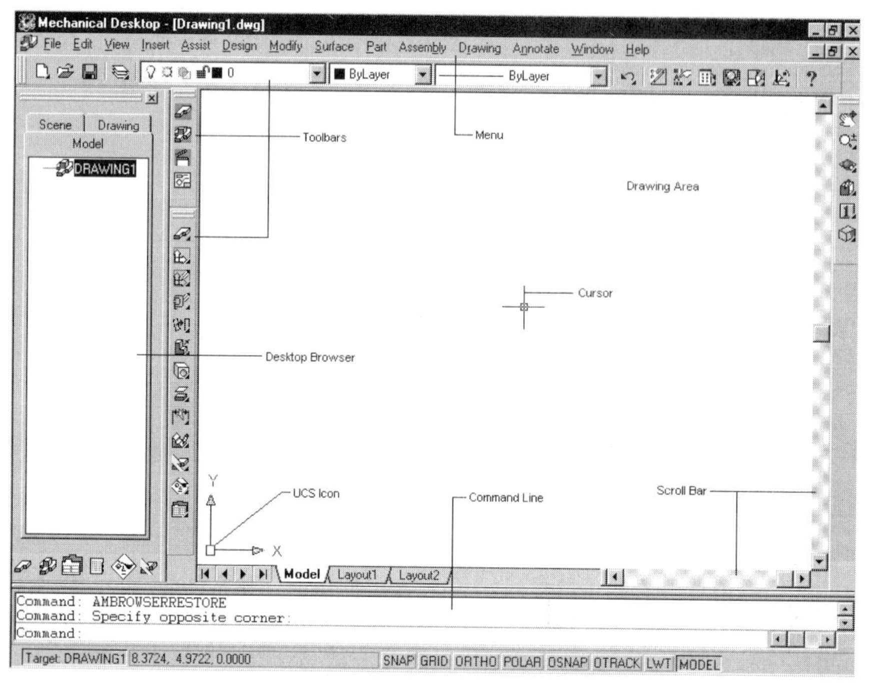

Figure 3 *Mechanical Desktop screen showing toolbars, menubar and desktop browser*

DESKTOP BROWSER

The desktop browser is a very important tool provided in Mechanical Desktop. It reduces the use of toolbars and menubar appreciably for creating and editing the designer models or the assemblies. In the desktop browser, the operations that are performed on the sketch, starting from the basic sketch itself, are displayed as the tree view in the sequence in which they were performed. You can select any operation and right-click on it to display the shortcut menu. Now in this shortcut menu, choose the editing operation you want to perform on the sketch. You can choose any kind of editing operation like changing the basic sketch of the model, or changing the diameter of a hole in the model and so on. The desktop browser can be hidden or displayed by choosing **Display > Desktop Browser** from the **View** menu. You can also use the **AMBROWSER** command for this purpose.

The desktop browser has three tabs:

Model. Choosing this tab allows you to switch to the model space. It is similar to setting the value of the **TILEMODE** variable to 1. It is the default tab and generally, all the drawing work is done in the **Model** tab.

Scene. This tab is similar to the **Model** tab except that it allows you to explode the assemblies created in the **Model** tab as shown in Figure 4. You can also add, delete or edit tweaks and trails in the assembly in this tab.

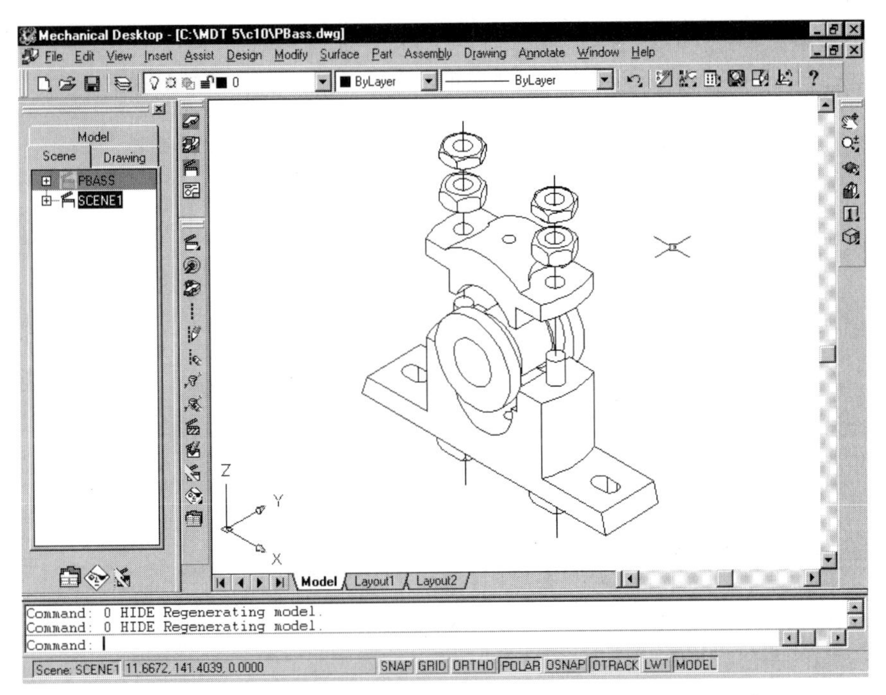

***Figure 4** Figure showing exploded assembly under the **Scene** tab*

Drawing. This tab is similar to the **Layout** in AutoCAD. It allows you to generate the 2D views of the designer model created in the **Model** tab or of the assemblies exploded in the **Scene** tab. These views can be dimensioned, modified or deleted in this tab. You can also add Bill of Material to these views. To edit any view or the Bill Of Material, select it in the hierarchy of the **Scene** and then right-click on it.

Apart from these tabs the desktop browser also has seven buttons, see Figure 5.

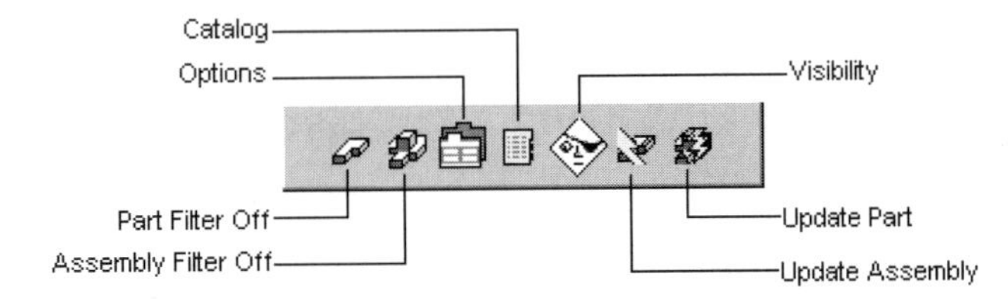

***Figure 5** Tool buttons available in the Desktop Browser*

Tip: *You should try to use the desktop browser to execute the commands wherever possible. It not only reduces the use of the toolbars and menus for invoking the commands but makes the designing faster.*

TOOLBARS

Mechanical Desktop provides different toolbars while working with different tabs of the desktop browser. The **Model**, **Scene**, and the **Drawing** tabs have different toolbars because of the different commands used in these tabs. There are three toolbars that are common to all the tabs of the desktop browser. They are **:**

Mechanical Main (Figure 6), **Desktop Main** (Figure 7) and **Mechanical View** (Figure 8).

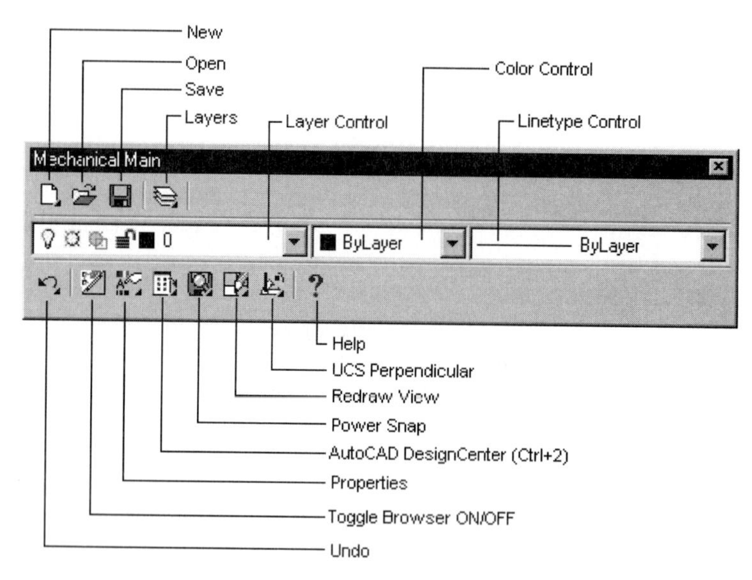

Figure 6 *Mechanical Main toolbar*

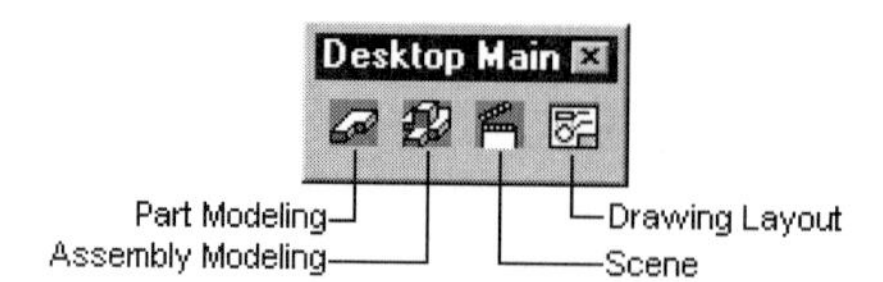

Figure 7 *Desktop Main toolbar*

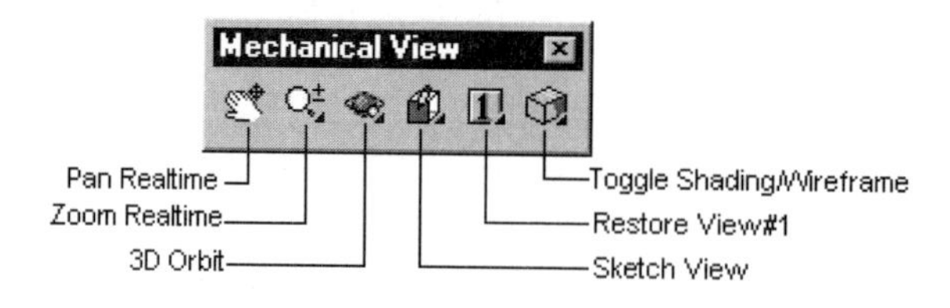

Figure 8 *Mechanical View toolbar*

Apart from these three toolbars, the **Model** tab has the **Part Modeling** toolbar (Figure 9).

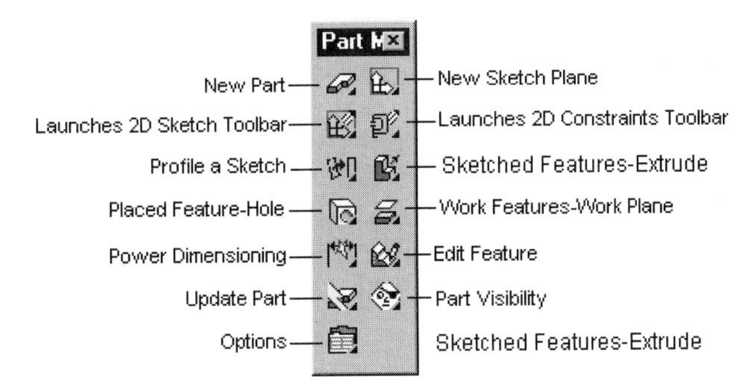

Figure 9 *Part Modeling toolbar*

Instead of the **Part Modeling** toolbar, the **Scene** tab has the **Scene** toolbar (Figure 10).

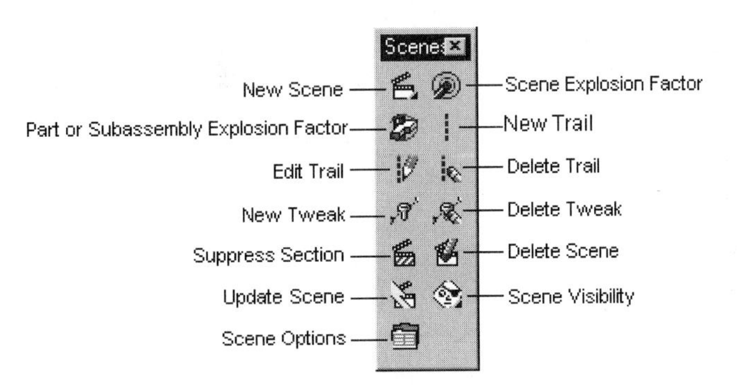

Figure 10 *Scene toolbar*

The **Drawing** tab has the **Drawing Layout** toolbar (Figure 11) apart from the **Standard**, **Desktop Main**, and the **Mechanical View** toolbars.

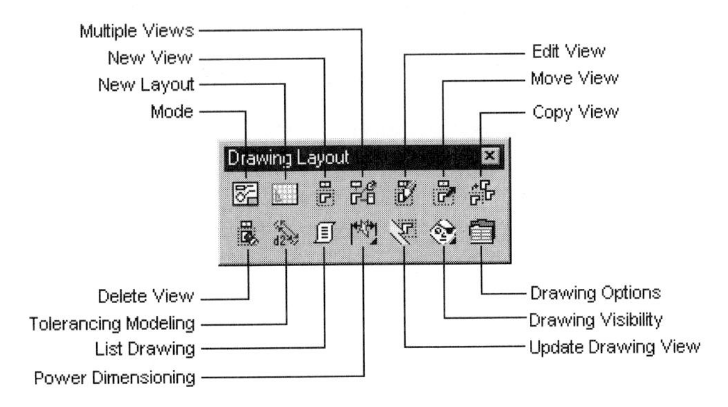

Figure 11 *Drawing Layout toolbar*

Some of the buttons in these toolbars have a small arrow on the lower right corner (Figure 12). These buttons with the arrow are called flyouts. To display the flyout, press on button with an arrow using the pick button of the mouse (or other pointing device). The flyout is displayed showing the other related buttons as shown in Figure 13.

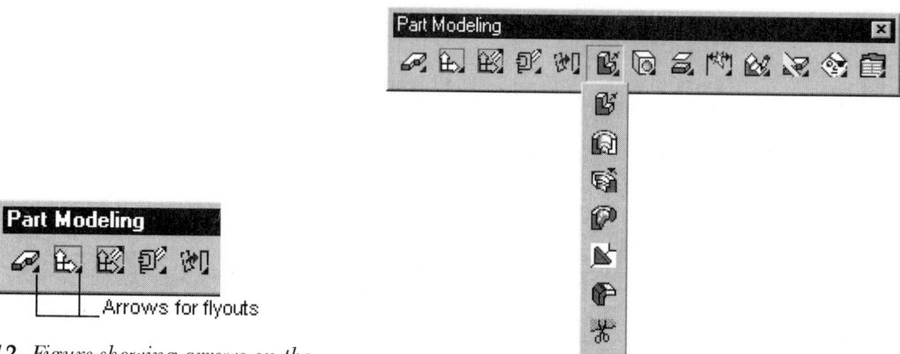

Figure 12 *Figure showing arrows on the lower-right corner of the buttons*

Figure 13 *Toolbar showing flyout*

You will also find some buttons in the **Part Modeling** toolbar having an arrow on the upper right corner (Figure 14). When you choose any of these buttons with an arrow on the upper right corner, they launch another toolbar.

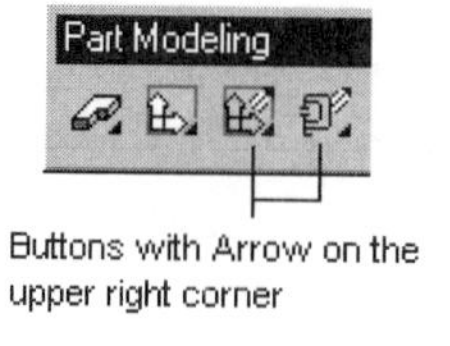

Figure 14 *Buttons showing the arrow on the upper right corner*

IMPORTANT TERMS AND THEIR DEFINITIONS

Parametric Modeling

The parametric is defined as the ability of the software package to use some standard parameters or properties of the geometry to define it. In AutoCAD, the sketch has to be drawn very accurately and to its actual dimensions. On the other hand, in Mechanical Desktop all you have to do is to draw the sketch to relative shape and then simply add the actual dimensions which will drive the sketch to its actual size. For instance, you have created a circle of some relative diameter. Now you want that the circle should have a diameter of 20 units. This can be easily done by invoking the command then selecting the circle and entering the desired value of the diameter. The diameter of the circle will be updated automatically. It is very important to mention here that if you stretch the sketch using the **GRIPS** or the **STRETCH** command, the dimensions will be updated automatically.

Feature-based Modeling

The feature-based modeling is the property of the software package that ensures that the features like hole, loft, 3D path, fillet and so on, in the designer model are all independent features and that they understand their fit and function properly so that they can be edited independently. For instance, a designer model comprising of the counterclock threading can be converted into the clockwise threading just by editing some values because of it being an independent feature.

Bi-directional Associativity

It is the property that provides a link between the 2D and the 3D modeling techniques which leads to reduced drafting and designing process and increased productivity. As a result of this property if any modifications are made in the designer model, its drawing views get automatically updated. Also, the changes in the dimensions of the drawing views are reflected in the designer model.

Constraints

These are the logical operations performed on the sketch to make them more accurate. For example, when you apply the constraints, the lines that are almost horizontal become horizontal, the lines that are almost vertical become vertical. AutoCAD Designer has provided three types of logical constraints. All three types are discussed next.

Geometric Constraints

These are the logical operations performed on the sketch that determines and evaluates the basic entities in the sketch and relates them with the properties like concentricity, tangency, collinearity and so on in the sketch. When you create the designer model, these geometric constraints are applied to it automatically. Mechanical Desktop has provided fifteen types of geometric constraints which are as follows:

Horizontal

As you apply this constraint, the selected line segment is made parallel to X axis of the sketch plane.

Vertical

This constraint is applied to make the selected line segment parallel to Y axis of the sketch plane.

Perpendicular

This constraint is applied to make two line segments mutually perpendicular.

Parallel

Applying this constraint makes the selected line segments parallel to each other.

XValue

This constraint is applied on two different entities to make their X placement value the same.

YValue
This constraint is applied on two different entities to make their Y placement value the same.

Collinear
Applying this constraint allows two different line segments to be placed in same line.

Concentric
Applying this constraint allows two different arcs or circles to share the X and Y values of their center points. It can also be used for a work point and center of an arc or circle.

Project
This constraint is applied to join any vertex with a point, line, circle or an arc.

Join
This constraint is applied to close the gap between two points in a geometry.

Radius
Applying this constraint allows two or more arcs or circles to share same radius. Before applying this constraint, one of the circles or arcs has to be allotted some radius value.

Tangent
This constraint when applied makes a line, arc, or a circle tangent to another arc or circle.

Length
Applying this constraint forces the selected line segments to have same length.

Mirror
This constraint forces the selected entity to become the mirror image of the other selected entity.

Fix
This constraint is applied to fix the selected point or entity at its place.

Numeric Constraints

This type of constraints are very similar to geometric constraints except that they are specified to define the size of a feature like hole, fillet and so on in the designer model.

Assembly Constraints

These constraints are used in the assembly modeling to assemble different designer parts together. Mechanical Desktop has provided four types of assembly constraints. They are:

Angle
This constraint is used to assemble two different parts at certain angle to each other. The angle can be defined between two planes, vectors, or a combination of both.

Flush
This constraint defines coplanar position between two different components. This constraint can assemble two different parts parallel to each other. You can also define the offset distance between these parts.

Insert
This constraint is defined in terms of a central axis between two different circular parts. Applying this constraint allows two different designer parts to share the same orientation of central axis and at the same time makes the faces coplanar. However you can define an offset distance between the two faces.

Mate
Applying this constraint makes a point, line, or a plane of a designer model coincidental with that of another designer model.

CREATING THE DESIGNER MODEL

The advantage of working with Mechanical Desktop is that you do not have to draw the sketch to the proper dimensions. You just have to create a rough sketch in which even the lines may not be exactly horizontal, vertical, or even tangent to any arc. With the help of just one command the rough sketch gets converted into a cleaned-up sketch. This means that the lines that are almost horizontal become horizontal, the lines that are almost vertical become vertical, the lines that are almost tangent to an arc become properly tangent. This is because the software applies the logical constraints to the sketch depending upon the shape of initial sketch. Thus creating a designer model has been divided into four steps as given below:

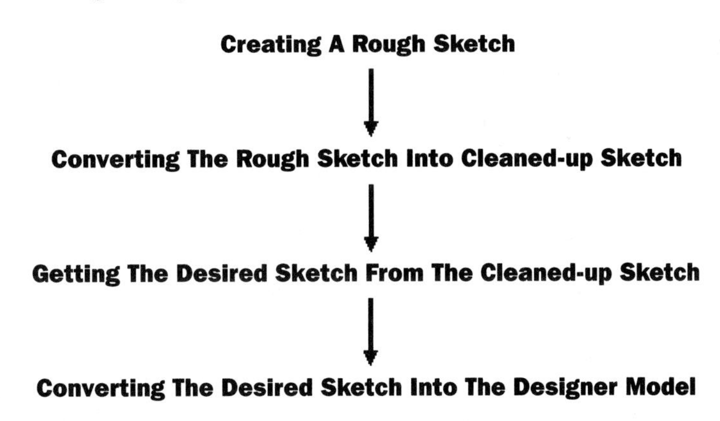

Creating A Rough Sketch

In this step you have to draw a rough outline of the designer model as shown in Figure 15. It is clear in this figure that the lines are not properly horizontal or properly vertical. Also the lines are not properly tangent to the arc. The rough sketch can be created using the **PLINE** command or a combination of the **LINE** and **ARC** commands, see Figure 15. However, it has to be kept in mind that the rough sketch should be drawn in proportion with the actual model. Therefore, if the actual model has large dimensions, then the limits of the drawing should be increased before starting the sketch.

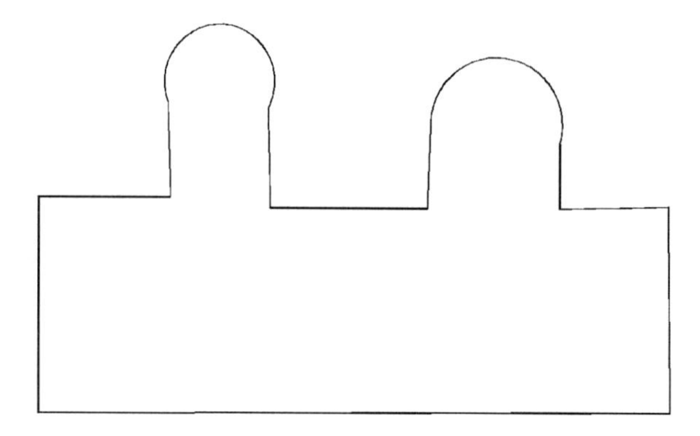

Figure 15 *Rough sketch*

Converting The Rough Sketch Into Cleaned-up Sketch

The next step after creating a rough sketch is to convert it into a cleaned-up sketch. This is done using the **AMPROFILE** command. You can invoke this command by choosing the **Profile a Sketch** button from the **Part Modeling** toolbar or by entering **AMPROFILE** at the Command prompt. The main function of this command is to apply logical constraints to the rough sketch thus converting it into a cleaned-up sketch. For instance, in Figure 15, the lines that were not properly horizontal, or vertical are forced into one, automatically, using the **AMPROFILE** command (Figure 16).

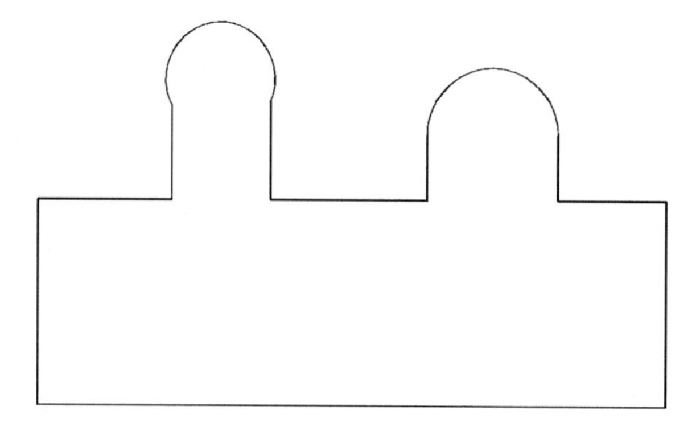

Figure 16 *Partially cleaned-up sketch*

However, if the orientation difference of these lines from the horizontal or vertical axis is more than four degrees on either side, then you have to apply the geometric constraints manually. If you want to view all the constraints that are applied to the sketch, use the

AMSHOWCON command. You can invoke this command by choosing the **Show Constraint** button from the **2D Constraint** toolbar. This command displays all the constraints applied to the sketch as shown in Figure 17. It is clear in Figure 17, the upper left vertical lines are not tangent to the arc and the lengths of the upper vertical lines are different. Therefore, you will have to make them tangent to the arc and add the equal length constraint manually with the help of the **AMADDCON** command. This command is used to apply the geometric constraints like horizontal, vertical, tangent, concentric, collinear, equal length and so on to the entities of the sketch. You can also directly choose the required constraint button from the **2D Constraint** toolbar to apply the required constraint. For the given sketch, choose the **Tangent** button and select the arc and then the line to make it tangent to the arc. Now choose the **Equal Length** button and apply it to the upper vertical lines and upper horizontal lines. This applies the equal length constraints to the sketch and finally converts the rough sketch into a cleaned-up sketch as shown in Figure 18.

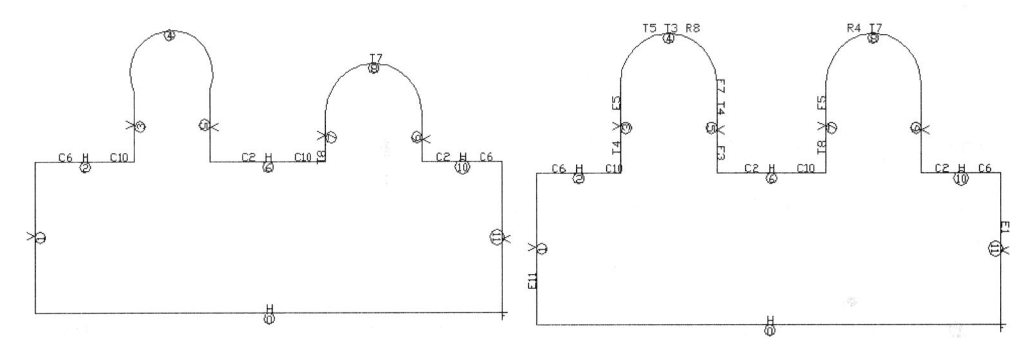

Figure 17 *Figure showing different constraints*

Figure 18 *Fully cleaned-up sketch displaying all the constraints applied on it*

Getting The Desired Sketch From The Cleaned-up Sketch

Once you have cleaned-up the rough sketch using various command, almost half of the job is done. Now you will have to convert this cleaned-up sketch into the desired sketch. This is because the cleaned-up sketch has the dimensions different than that are required. To get the desired sketch, you will use the **AMPARDIM** command. Since the dimensioning in Mechanical Desktop is parametric, therefore, you simply have to select the entity you want to dimension and then enter the dimension value for that entity in the **AMPARDIM** command. This will automatically drive the entity to the dimensions entered as shown in Figure 19.

Converting The Desired Sketch Into The Designer Model

The final step is to convert the desired sketch into a 3D designer model. This can be done using various commands like the **AMEXTRUDE** command, **AMREVOLVE** command, **AMSWEEP** command and so on depending upon the final model you want. All these commands are used to convert the 2D sketches into the 3D designer models. The **AMEXTRUDE** command extrudes the sketch in Z direction of current sketch plane, the **AMREVOLVE** command revolves the 2D sketch about a revolution axis, and the **AMSWEEP** command sweeps the 2D sketch

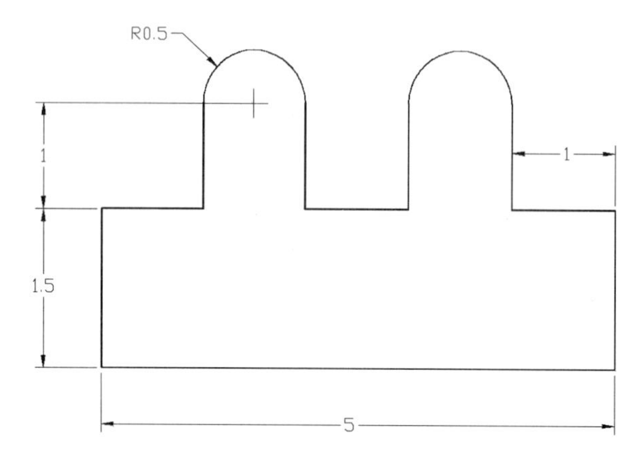

Figure 19 *Desired Sketch*

about a path defined by you. For the given sketch you will use the **AMEXTRUDE** command to convert it into a 3D model as shown in Figure 20.

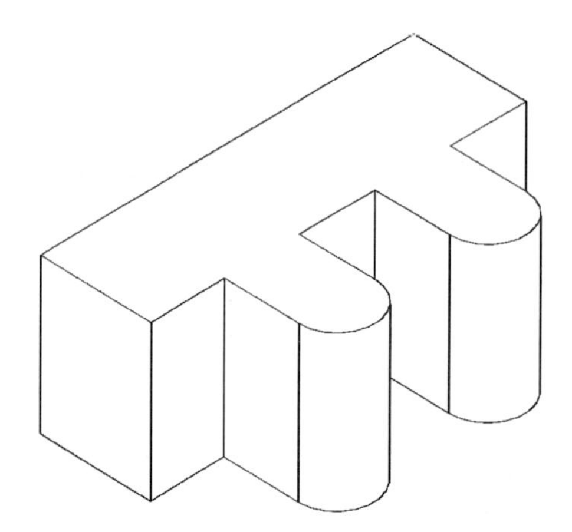

Figure 20 *3D Designer model created from desired sketch using the* ***AMEXTRUDE*** *command*

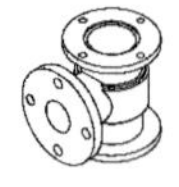

Note

The commands like ***AMPROFILE, AMADDCON, AMPARDIM, AMEXTRUDE*** *are discussed in detail in the later chapters.*

ACCELERATOR KEYS

The number keys have a very significant importance in Mechanical Desktop. When entered alone these keys are used to split the screen into various viewports. Some of these keys are also used to view the model from various viewpoints. These keys are generally divided into two sets.

The first set of keys is used to split the screen into various viewports. These keys, along with the number of viewports they split the screen into, and the view they display are given below:

Key	Viewport(s)
1	One (the last active viewport)
2	Two (showing Plan, and SE Isometric views)
3	Three (showing Plan, SE Isometric, and Front views)
4	Four (showing Plan, SE Isometric, Front, and Side views)

The second set of keys is used to view the model from the various viewpoints. This is the reason the **Viewpoints** toolbar is not present in the Mechanical Desktop. These keys, along with the view they display are shown next.

Key	View
5	Top view of the World XY plane
55	Bottom view of the World XY plane
6	Front view
66	Back View
7	Right side view
77	Left side view
8	Southeast Isometric view of the World XY plane
88	Southwest Isometric view of the World XY plane
9	Plan view of the current sketch plane

Apart from the above mentioned keys, there are some other keys that are extensively used in Mechanical Desktop. These keys, along with their functions are given below:

Key	Function
0	Hides the hidden lines in the designer models
=	Rotates the current view upwards
-	Rotates the current view downwards
[	Rotates the current view towards left
]	Rotates the current view towards right

PERSONALIZED SNAP SETTINGS

One of the major enhancement of Mechanical Desktop 5 over the previous releases is the **Power Snap Settings**. The **Power Snap Settings** ensures that apart from the default snap setting, you can also create four more personalized snap settings. These personalized snap settings can be used globally at any point of time to snap on to a point. These settings can be made using the **Power Snap** button in the **Mechanical Main** toolbar. When you choose this button, the **Power Snap Settings** dialog box is displayed as shown in Figure 21. This dialog box has four more tabs in addition to the **Current Settings** tab. They are **Settings 1**, **Settings**

2, **Settings 3** and **Settings 4**. You can select the desired snap settings in each of the tabs. You can also set different **Polar Snap Settings** for each tab using the **Polar Snap Settings** button.

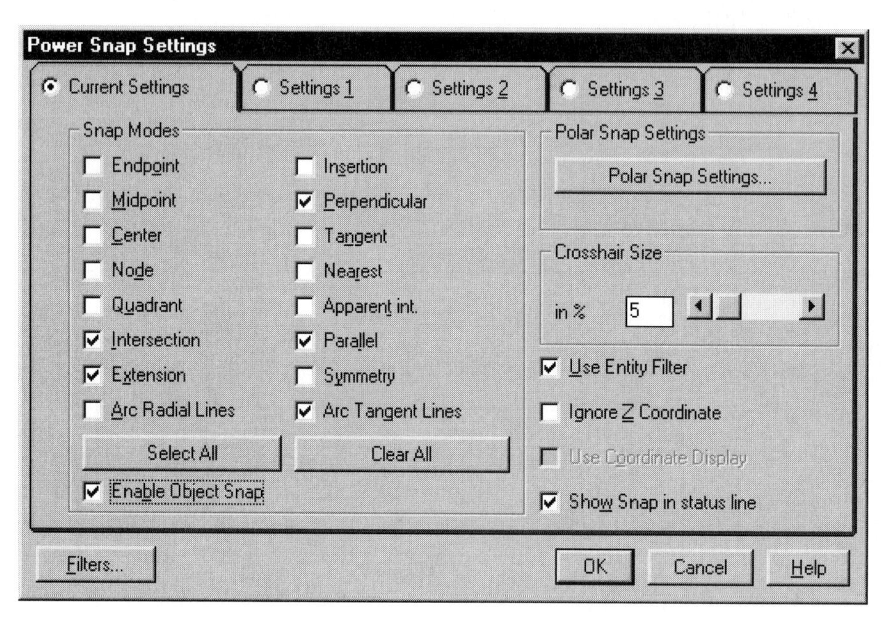

Figure 21 *Power Snap Settings dialog box*

The four personalized power snap settings can be loaded using the **AMPSNAP1**, **AMPSNAP2**, **AMPSNAP3** and **AMPSNAP4** commands respectively. When you enter any of these commands at the command line, the snap settings assigned to it will be displayed in the command line. This snap setting will be made current now. All these commands are transparent commands. Therefore any of these personalized snap settings can be loaded when you are inside any other command.

Chapter 1

Creating, Profiling, Constraining, and Dimensioning the Basic Sketch

Learning Objectives

After completing this chapter, you will be able to:

- *Draw the basic outline (sketch) of designer model.*
- *Profile the basic sketch using the* ***AMPROFILE*** *command.*
- *Add constraints to the sketch using the* ***AMADDCON*** *command.*
- *View the constraint applied on the sketch with the help of the* ***AMSHOWCON*** *command.*
- *Delete the constraint applied on the sketch using the* ***AMDELCON*** *command.*
- *Dimension the basic sketch using the* ***AMPARDIM*** *command.*
- *Resolve the sketches and add additional geometries to the profiled sketch using the* ***AMRSOLVESK*** *command.*
- *Create cut lines.*
- *Create split lines.*
- *Create break lines.*

Commands Covered

- ***AMPROFILE***
- ***AMADDCON***
- ***AMSHOWCON***
- ***AMDELCON***
- ***AMPARDIM***
- ***AMRSOLVESK***
- ***AMCUTLINE***
- ***AMSPLITLINE***
- ***AMBREAKLINE***

GETTING STARTED WITH MECHANICAL DESKTOP

It has already been discussed that creating the designer model requires that you start with a basic sketch called the rough sketch. This rough sketch should be in proportion with the designer model. In the second step, this rough sketch is converted into cleaned-up sketch. The third step is to get the desired sketch out of this cleaned-up sketch. In the fourth and the final step, this desired sketch is converted into the designer model. The first three steps are common for almost all of the drawings.

You will create the designer models by using these four steps only. However, in this chapter only the first three steps will be discussed and the fourth step will be discussed in the next chapter. The commands required to create various sketches in steps 1, 2, and 3 are discussed below.

To proceed, you first need to start Mechanical Desktop by double-clicking on the Mechanical Desktop icon at the desktop of your computer. You can also start Mechanical Desktop using the taskbar shortcut. Choose the **Start** button at the bottom left corner of the screen to display the menu. Choose **Program** to display the program folder. Now choose **Mechanical Desktop 5** to display the Mechanical Desktop programs and then choose **Mechanical Desktop 5** to start Mechanical Desktop.

CREATING A ROUGH SKETCH*

One of the major enhancements of Mechanical Desktop 5 over the previous releases is the scheme of creating the rough sketches for the designer models. In the previous releases you had to make sure that the sketches created were closed. For example, consider the designer model shown in Figure 1-1. The basic sketch you had to draw for this model in the previous releases is shown in Figure 1-2.

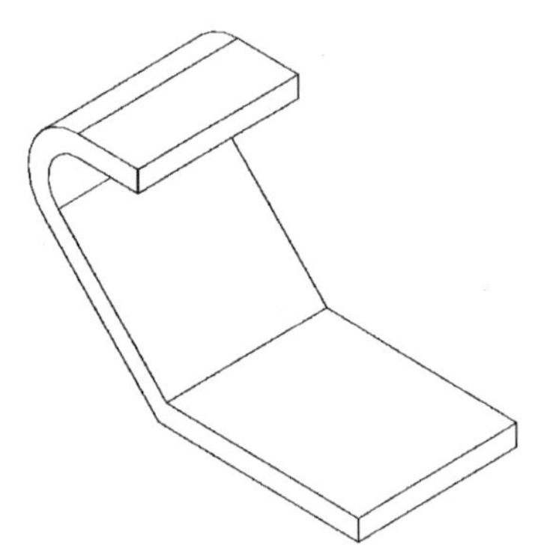

Figure 1-1 *Figure showing a designer model*

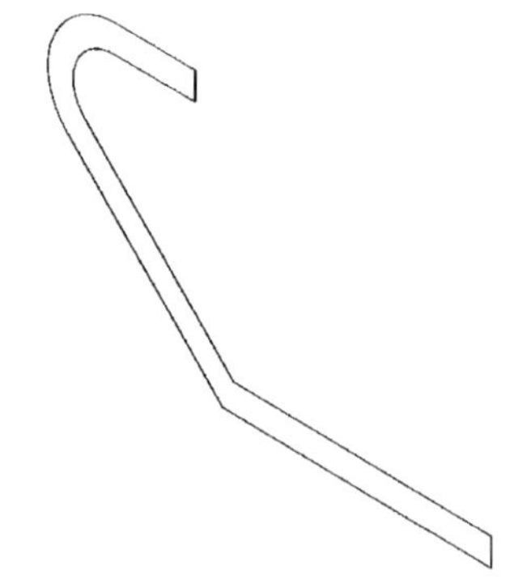

Figure 1-2 *Sketch created in previous releases*

In Mechanical Desktop 5, concepts called the **open profile** and **thin extrusion** has been incorporated. These concepts ensure that to create a designer model similar to the one shown in Figure 1-1, you have to draw an open sketch as shown in Figure 1-3. When you extrude it, you will be prompted to specify the extrusion height and the thickness of the designer model.

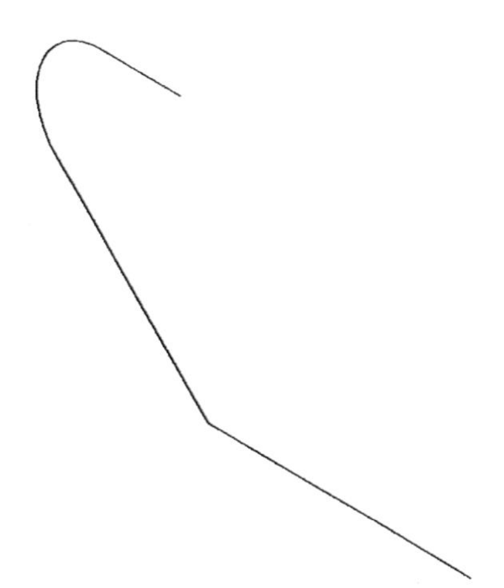

Figure 1-3 *Basic sketch for the model*

Therefore, now you can create the designer model by **drawing either an open profile** or **a closed profile**.

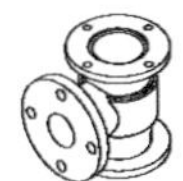

Note

The method used for converting the open profile into a designer model is discussed in the next chapter.

Before drawing the basic sketch of the designer model it should be kept in mind that the sketch is drawn in proportion to the dimensions of the designer model. If the dimensions of the designer model are very large, it is recommended that you first increase the limits in proportion with the dimensions of designer model, using the **LIMITS** command. Once the limits have been increased, you will also have to increase different scale factors like **dimension scale factor**, **linetype scale factor** and so on. Instead, you can simply increase the value in the **Use overall scale of** edit box in the **Fit** tab of the **Modify Dimension Style** dialog box. This dialog box can be invoked by choosing **Edit Dimensions > Dimension Style** from the **Annotate** menu.

Rough sketches for the figures can be created using the **PLINE** command or by using a combination of the **LINE** and **ARC** commands.

CONVERTING THE ROUGH SKETCH INTO A CLEANED-UP SKETCH

Once you have created the rough sketch, the second step is to convert this rough sketch into a cleaned-up sketch. A cleaned-up sketch is one in which all the entities comprising the sketch are in proper order. This is done by using the **AMPROFILE** and **AMADDCON** commands. The main function of these commands is to add geometric constraints to the sketch. To proceed, you have to first use the **AMPROFILE** command and then the **AMADDCON** command. Both of these commands are discussed next.

AMPROFILE Command

Toolbar: Part Modeling > Profile a Sketch
Menu: Part > Sketch Solving > Profile
Context Menu: Sketch Solving > Profile
Command: AMPROFILE

This is the basic command in Mechanical Desktop without which you cannot proceed. Figure 1-4 shows the **Profile a Sketch** button in the **Part Modeling** toolbar. This command can also be invoked from the context menu that is displayed by right-clicking in the drawing area, and choosing **Sketch Solving > Profile**. This command is used to solve (profile) the rough 2D sketches and apply geometric constraints on them. If the selected sketch is closed, it will be converted into a closed profile (Figure 1-5) and if the selected sketch is open, it will be converted into an open profile (Figure 1-6).

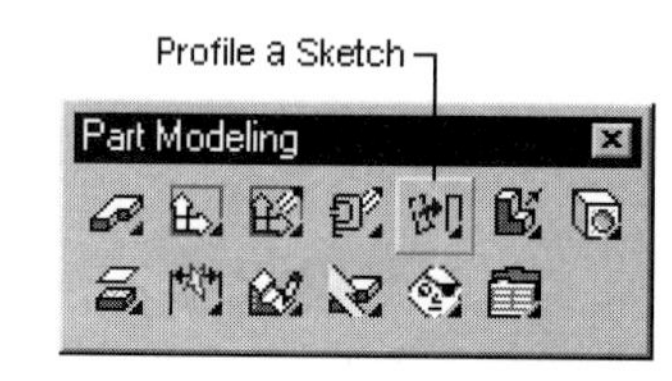

Figure 1-4 *Invoking the AMPROFILE command from the Part Modeling toolbar*

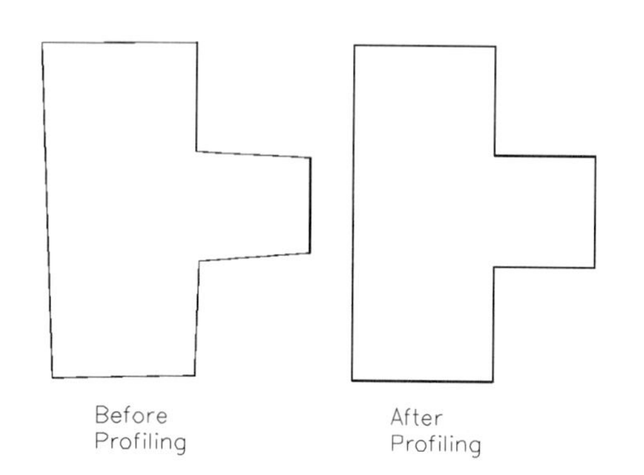

Figure 1-5 *Figure showing a closed sketch before and after profiling*

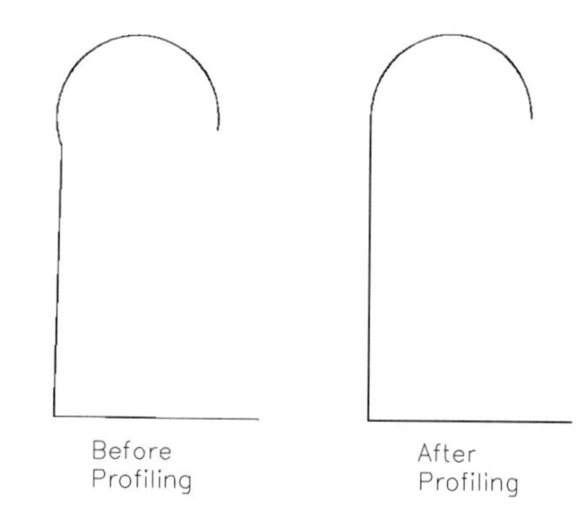

Figure 1-6 *Figure showing an open sketch before and after profiling*

Profiling A Single Entity

Toolbar: Part Modeling > Profile a Sketch > Single Profile
Menu: Part > Sketch Solving > Single Profile
Context Menu: Sketch Solving > Single Profile
Command: AMPROFILE

If the sketch you want to profile consists only of a single entity like a single line segment, a circle, an arc, and so on, you can profile it using the **Single Profile** button. You can choose this button from the **Profile a Sketch** flyout in the **Part Modeling** toolbar. You will not be prompted to select the entity to profile. The last drawn entity will be automatically selected and profiled.

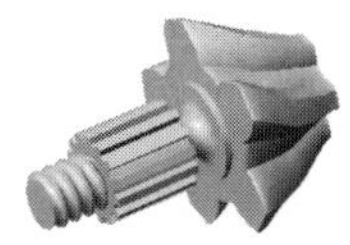

Tip: *The sketch drawn using the **PLINE** command is considered as a single entity. When you choose the **Single Profile** button, the complete sketch created using the **PLINE** command is selected and profiled automatically.*

AMADDCON Command

Toolbar:	2D Constraints
Menu:	Part > 2D Constraints
Context Menu:	2D Constraints
Command:	AMADDCON

This command is used to add the missing or the additional constraints to the profiled sketch. The default angular tolerance is 4 degrees, which means that if the angular deviation of the horizontal or vertical lines of the sketch is more than 4 degrees on either side then you will have to add the constraints manually using the **AMADDCON** command. The name of the command itself implies that it is used to add 2D constraints on the sketch. There are fifteen types of 2D constraints that can be added using this command. You can also add any of these constraints by choosing the button of the required constraint from the **2D Constraints** toolbar as shown in Figure 1-7.

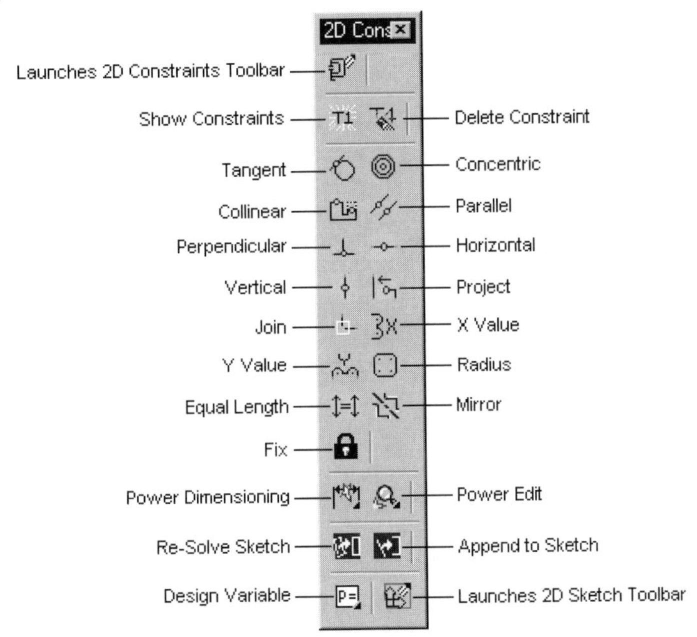

Figure 1-7 *2D Constraints toolbar*

AMSHOWCON Command

Toolbar:	2D Constraints > Show Constraints
Menu:	Part > 2D Constraints > Show Constraints
Context Menu:	2D Constraints > Show Constraints
Command:	AMSHOWCON

The **AMSHOWCON** command is used to find out the constraints that are applied on the sketch. The main purpose of this command is to show the constraints applied on the sketch. The options provided under this command are:

All

The **All** option is used to display all the constraints applied on the sketch.

Select

The **Select** option is used to view some selected constraints. You will be prompted to select the segment in the current sketch whose constraint you want to view.

Next

The **Next** option is used to cycle through the selected segments in the sketch.

Exit

The **Exit** option is used to exit the **AMSHOWCON** command.

If you want to delete a constraint from the sketch, use the **AMDELCON** command.

AMDELCON Command

Toolbar:	2D Constraints >Delete Constraints
Menu:	Part > 2D Constraints > Delete Constraints
Context Menu:	2D Constraints > Delete Constraints
Command:	AMDELCON

There might arise a case where you will need to delete some of the constraints that are applied on the sketch. This can be done with the help of the **AMDELCON** command. As the name of command indicates, it is used to delete a constraint from the current sketch. You have an option of either selecting the constraint you want to delete or you can even delete all of the constraints at a time. The options provided under this command are:

All

The **All** option is used to delete all the constraints applied to the sketch.

Size

The **Size** option is used to resize the symbols of the constraints. When you enter **SIZE** at the Command prompt, the **Constraint Display Size** (Figure 1-8) dialog is displayed. This dialog box has a slider bar that you can move to resize the symbols of constraints.

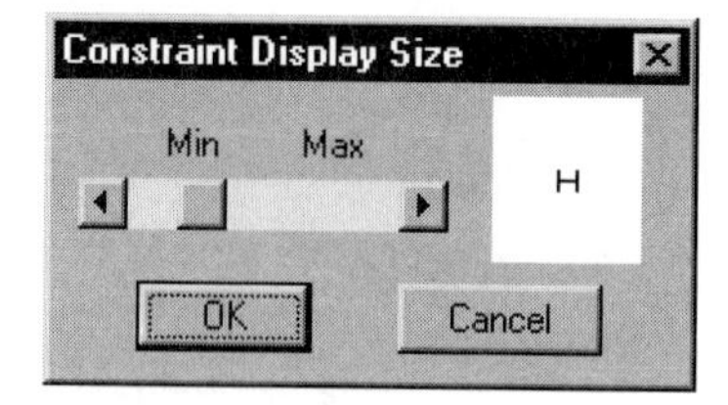

Figure 1-8 *Constraint Display Size dialog box to resize constraint symbols*

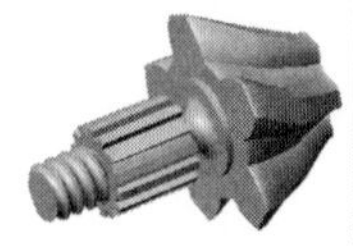

Tip: *The size of the symbols of constraints as they appear on the sketch can also be changed using the AMCONDSPSZ system variable. Default value for this variable is 5 and it can vary from 1 to 19.*

GETTING THE DESIRED SKETCH FROM THE CLEANED-UP SKETCH

Once you have cleaned-up the rough sketch, you have to add the required dimensions. The reason for this is that the rough sketch was drawn to some arbitrary dimensions and so the cleaned-up sketch still has the same arbitrary dimensions. As you know the dimensioning in Mechanical Desktop is parametric, so to dimension the cleaned-up sketch all you have to do is select the segment you want to dimension and enter the value you require. The parametric property will drive (adjust) the selected segment to the dimension you have entered. The dimensions to the cleaned-up sketch are added using the **AMPARDIM** command.

AMPARDIM Command

Toolbar:	Part Modeling > Power Dimensioning >New Dimension
Menu:	Part > Dimensioning > New Dimension
Context Menu:	Dimensioning > New Dimension
Command:	AMPARDIM

As the name implies, this command is used to dimension the parts of sketch. Since the dimensioning in Mechanical Desktop is parametric, therefore, irrespective of the actual dimensions, the object selected in the sketch is driven to the dimension value you have entered.

Depending upon the object selected to dimension, the options provided by this command change. For example, if the object selected to dimension is an arc or a circle, the options provided by this command are:

Undo

The **Undo** option is used to clear the current selection set so that you can select another object to dimension.

Diameter or Radius

These options are used to toggle the dimensions displayed between diameter or radius.

Ordinate

The **Ordinate** option is used to place ordinate dimension of center of arc or circle. Generally value of ordinate dimension of center point of arc or circle is placed as zero.

Placement point

The **Placement point** option is used to change the location of the dimension that you placed earlier.

Similarly, if the selected object is a line segment, the options are as follows:

Hor

The **Hor** option is used to place the horizontal dimension as shown in Figure 1-9.

Ver

The **Ver** option is used to place the vertical dimension, see Figure 1-9.

Align

The **Align** option is used to apply the aligned dimension. It is generally used for dimensioning inclined lines.

Par

The **Par** option is used to define parallel distance between any two selected segments. To use this option, you need to select two segments.

aNgle

The **aNgle** option is used to define angle between any two selected segments. This option is valid only if two objects are selected.

Diameter

The **Diameter** option is used to place the distance between two selected objects in the terms of diameter. It is placed as double of the actual distance between two selected objects. For instance, in Figure 1-10, the actual distance is between top line and the bottom line (shown as extended in dashed line) is 3 units, but the dimension is placed as the double of this distance as 6 units, with a symbol of **Ø** preceding the dimension value. The diameter dimensions are used to dimension the sketches of the revolved features.

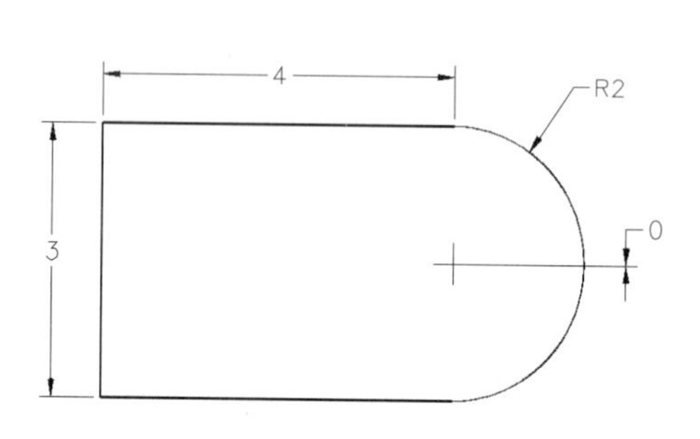

Figure 1-9 *Figure showing horizontal, vertical, radius, and ordinate dimensions*

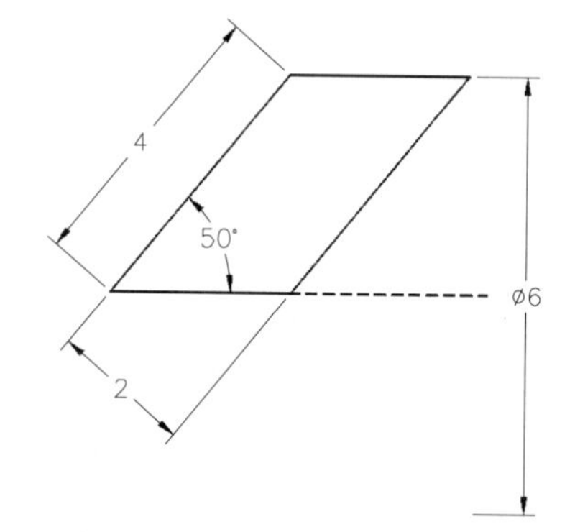

Figure 1-10 *Figure showing parallel, angular, aligned, and diameter (for lines) dimensions*

The **Undo**, **pLace**, and **Ord** options are same as the **Undo**, **Placement point**, and **Ordinate** options discussed earlier for arcs or circles.

The sketches can also be dimensioned using the **AMPOWERDIM** and the **AMAUTODIM** commands that are discussed in later chapters.

Resolving The Sketches (AMRSOLVESK Command)

Toolbar: Part Modeling > Profile a Sketch > Re-Solve Sketch
Menu: Part > Sketch Solving > Re-Solve
Context Menu: Sketch Solving > Re-Solve
Command: AMRSOLVESK

This command is used to resolve the sketch and display the remaining constraints required to fully constrain the sketch. This command is generally used to find the number of constraints further required to fully constrain the sketch.

Adding Additional Geometries To The Sketch (AMRSOLVESK Command)

Toolbar: Part Modeling > Profile a Sketch > Append to Sketch
Menu: Part > Sketch Solving > Append
Context Menu: Sketch Solving > Append
Command: AMRSOLVESK

When entered at the Command prompt, this command will provide the **Append** option to add the additional geometries to the sketch. This command is generally used to change the basic contour of the profiled sketch. When you invoke this command, the part of the sketch that is already profiled is highlighted and you will be asked to select the additional geometry to the sketch.

AMCUTLINE Command

Toolbar: Part Modeling > Profile a Sketch > Cut Line
Menu: Part > Sketch Solving > Cut Line
Context Menu: Sketch Solving > Cut Line
Command: AMCUTLINE

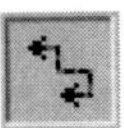

This command is used to create parametric cut lines to be used for creating the offset or aligned section views in the **Drawing** mode. Mechanical Desktop allows you to create two types of cut lines; the offset cut lines and the aligned cut lines.

The cut lines used to generate the offset section views can have as many number of segments. However, it is very important to mention here that the first and the last segment of the cut line should be codirectional and the valid angle variation between any of the segments is 90 degrees. The cut lines used to generate the aligned section views can be aligned but should consist of only two segments.

AMSPLITLINE Command

Toolbar: Part Modeling > Profile a Sketch > Split Line
Menu: Part > Sketch Solving > Split Line
Context Menu: Sketch Solving > Split Line
Command: AMSPLITLINE

The **AMSPLITLINE** command is used to create the parametric split lines that can be later used to split the entire designer model or the faces of the designer model.

AMBREAKLINE Command

Toolbar:	Part Modeling > Profile a Sketch > Break Line
Menu:	Part > Sketch Solving > Break Line
Context Menu:	Sketch Solving > Break Line
Command:	AMBREAKLINE

This command is used to generate a parametric break line, to be used for generating the breakout section views. The sketch used for the breakline should be a closed entity.

TUTORIALS

Tutorial 1

In this tutorial you will create the basic sketch of the drawing shown in Figure 1-11, then apply the geometric constraints and finally dimension it to obtain the required sketch. The dimensions for the drawing are given in Figure 1-12.

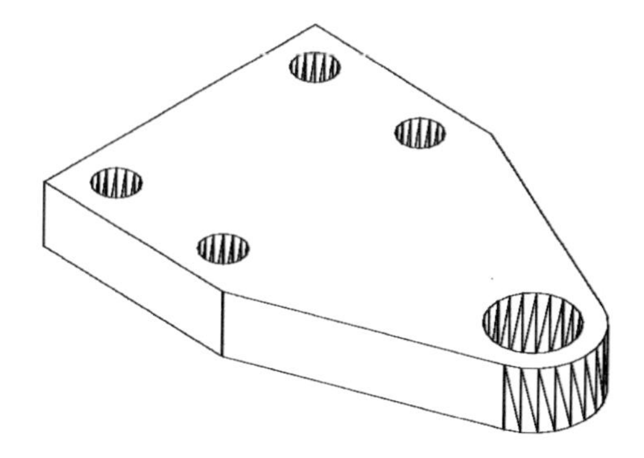

Figure 1-11 *Drawing for tutorial 1*

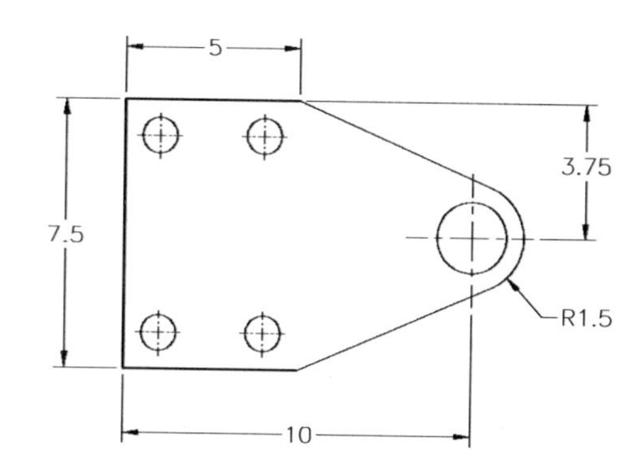

Figure 1-12 *Dimensions for tutorial 1*

Step 1: Creating A Rough Sketch

1. Create the rough sketch of the drawing shown in Figure 1-11 using the **PLINE** command or a combination of the **LINE** and **ARC** commands as shown in the Figure 1-13.

Step 2: Converting The Rough Sketch Into A Cleaned-up Sketch

2. Once you have created the rough sketch, the next step is to convert this rough sketch into a cleaned-up sketch. This is done using the **AMPROFILE** command. To invoke this command, choose the **Profile a Sketch** button from the **Part Modeling** toolbar. You can also invoke this command by entering **AMPROFILE** at the Command prompt.

   The prompt sequence is as follows:

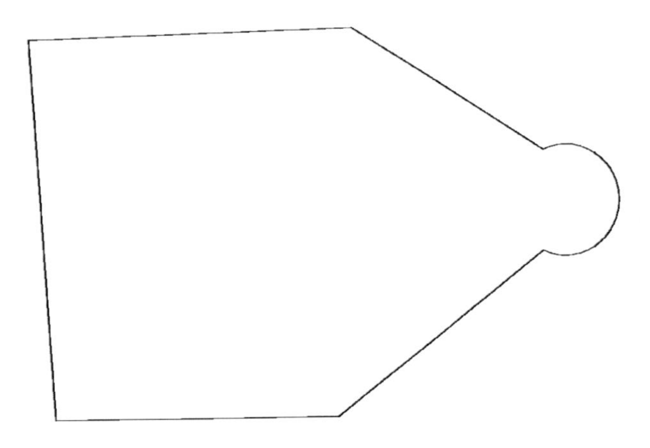

Figure 1-13 *Rough sketch for Tutorial 1*

Select object for sketch: *Select the sketch using any one of the object selection methods.*
Select object for sketch:

This operation applies the geometric constraints to the sketch as shown in Figure 1-14. When you press ENTER at the ***Select object for sketch*** *prompt, the total number of constraints require to fully constrain the sketch will be displayed in the Command prompt.*

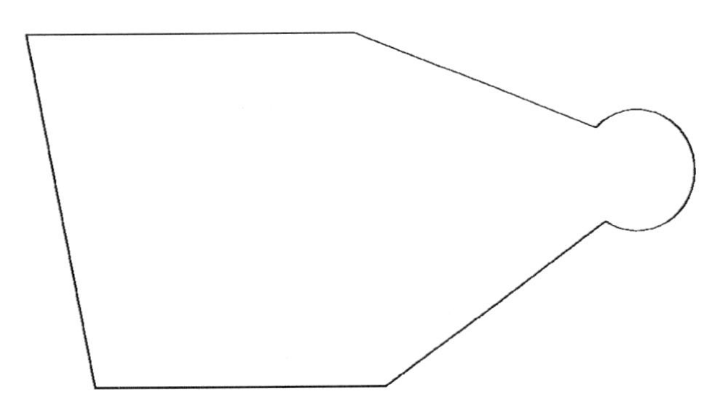

Figure 1-14 *Sketch after profiling*

3. If the angular tolerance of the lines drawn is within 4 degrees, the horizontal and vertical constraints will be applied automatically on them. Check whether all the geometric constraints are applied to the sketch using the **AMSHOWCON** command. This command can be invoked by choosing the **Show Constraints** button from the **2D Constraints** toolbar. The prompt sequence is as follows:

Enter an option [All/Select/Next/eXit] <eXit>: **A**

All the constraints applied on the sketch will be displayed as shown in Figure 1-15.

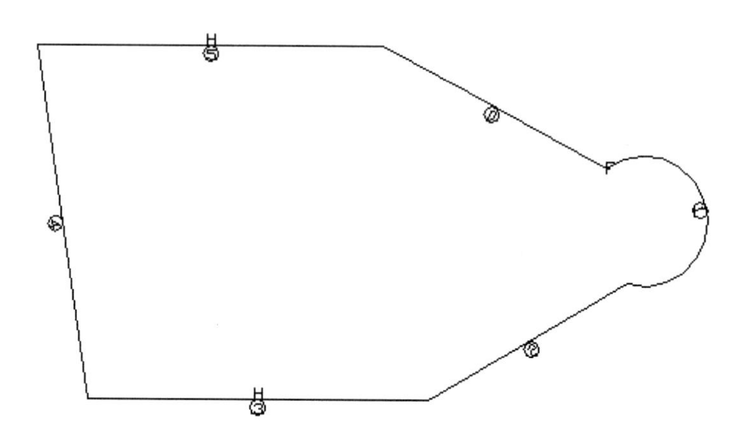

Figure 1-15 *Sketch showing all the constraints applied*

Tip: *When you choose the* ***Show Constraints*** *button, you do not have to enter* ***ALL*** *at the* ***Enter an option [All/Select/Next/eXit] <eXit>*** *prompt, as it will automatically display all the constraints that are applied on the sketch.*

It is clear from Figure 1-15 that neither the vertical nor the tangent constraint is applied to the sketch. Therefore, you need to apply these constraints manually. Both of these constraints are applied, one by one, in the following steps.

4. Choose the **Vertical** button from the **2D Constraints** toolbar. You can also enter **AMADDCON** at the Command prompt and use the **Ver** option. The prompt sequence is as follows:

   Valid selections: line, ellipse or spline segment
   Select object to be reoriented: *Select the vertical line.*

   Solved under constrained sketch requiring 8 dimensions or constraints.
   Valid selections: line, ellipse or spline segment
   Select object to be reoriented: *Press ESC.*

5. Choose the **Tangent** button from the **2D Constraints** toolbar. You can also enter **AMADDCON** at the Command prompt and use the **Tan** option. The prompt sequence is as follows:

   Valid selections: line, circle, arc, ellipse or spline segment
   Select object to be reoriented: *Select the arc.*
   Valid selections: line, circle, arc, ellipse or spline segment
   Select object to be made tangent to: *Select the lower inclined line.*

Solved under constrained sketch requiring 7 dimensions or constraints.
Valid selections: line, circle, arc, ellipse or spline segment
Select object to be reoriented: *Select the arc.*
Valid selections: line, circle, arc, ellipse or spline segment
Select object to be made tangent to: *Select the upper inclined line.*

Solved under constrained sketch requiring 6 dimensions or constraints.
Valid selections: line, circle, arc, ellipse or spline segment
Select object to be reoriented: *Press ESC.*

This applies all the required geometric constraints to the sketch. You can also add some additional geometric constraints to reduce the number of dimensions that have to be applied to the sketch. For example, in this sketch both the horizontal lines have to be of same length. Therefore, if you add the equal length constraint to these lines then you will have to apply the dimension only to one of the lines and the other line will be driven to same dimension value automatically by the equal length constraint.

6. Choose the **Equal Length** button from the **2D Constraints** toolbar. The prompt sequence is as follows:

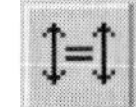

Valid selections: line or spline segment
Select first object: *Select one of the horizontal lines.*

Valid selections: line or spline segment
Select second object: *Select the other horizontal line.*
Solved under constrained sketch requiring 5 dimensions or constraints.
Valid selections: line or spline segment
Select first object: *Press ESC.*

After applying all the constraints, the sketch should look similar to the one shown in Figure 1-16.

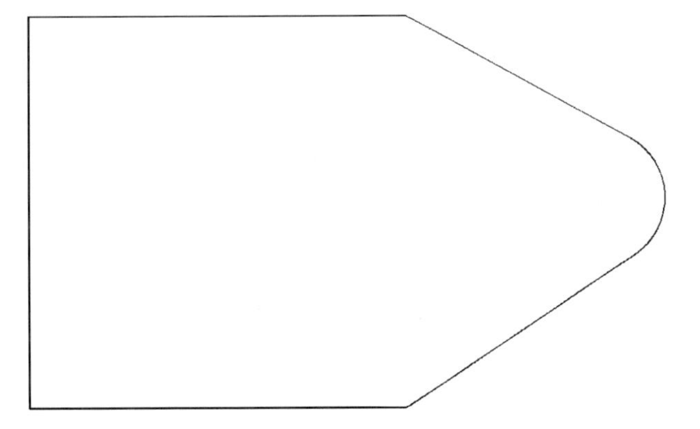

Figure 1-16 *Sketch after applying all the geometric constraints*

Step 3: Getting The Desired Sketch From The Cleaned-up Sketch

7. Convert this fully cleaned-up sketch into the desired sketch using the **AMPARDIM** command. This command can be invoked by choosing the **New Dimension** button from the **Power Dimensioning** flyout in the **Part Modeling** toolbar. You can also invoke this command by entering **AMPARDIM** at the Command prompt.

The prompt sequence is as follows:

Select first object: *Select the vertical line.*
Select second object or place dimension: *Place the dimension.*
Enter dimension value or [Undo/Hor/Ver/Align/Par/aNgle/Ord/Diameter/pLace] <default value>: **7.5**

Solved under constrained sketch requiring 4 dimensions or constraints.
Select first object: *Select the upper horizontal line.*
Select second object or place dimension: *Place the dimension.*
Enter dimension value or [Undo/Hor/Ver/Align/Par/aNgle/Ord/Diameter/pLace] <default value>: **5**

Solved under constrained sketch requiring 3 dimensions or constraints.
Select first object: *Select the arc.*
Select second object or place dimension: *Place the dimension.*
Enter dimension value or [Undo/Diameter/Ordinate/Placement point]: <default value>: **1.5**

Solved under constrained sketch requiring 2 dimensions or constraints.
Select first object: *Select the arc*
Select second object or place dimension: *Select the bottom horizontal line.*
Specify dimension placement: *Place the dimension.*
Enter dimension value or [Undo/Hor/Ver/Align/Par/aNgle/Ord/Diameter/pLace] <default value>: **V**
Enter dimension value or [Undo/Hor/Ver/Align/Par/aNgle/Ord/Diameter/pLace] <default value>: **3.75**

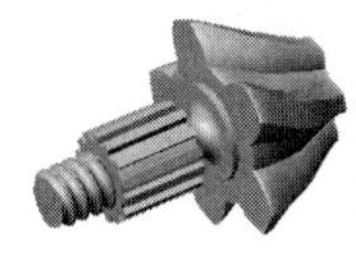

Tip: *If you apply the equal length constraint to the aligned lines then even this vertical dimension between the horizontal line and the arc will not be required to fully constrain the sketch.*

Solved under constrained sketch requiring 1 dimension or constraint.
Select first object: *Select the arc.*
Select second object or place dimension: *Select the vertical line.*
Specify dimension placement: *Place the dimension.*
Enter dimension value or [Undo/Hor/Ver/Align/Par/aNgle/Ord/Diameter/pLace] <default value>: **H**
Enter dimension value or [Undo/Hor/Ver/Align/Par/aNgle/Ord/Diameter/pLace] <default value>: **10**
Solved fully constrained sketch.

Select first object: Enter

8. Enter **FF** at the Command prompt. This is the keyboard shortcut for the **Extents** option of the **ZOOM** command. The final sketch should be similar to the sketch shown in Figure 1-17.

9. Save this drawing with the name given below:

   \MDT Tut\Ch-1**Tut1.dwg**

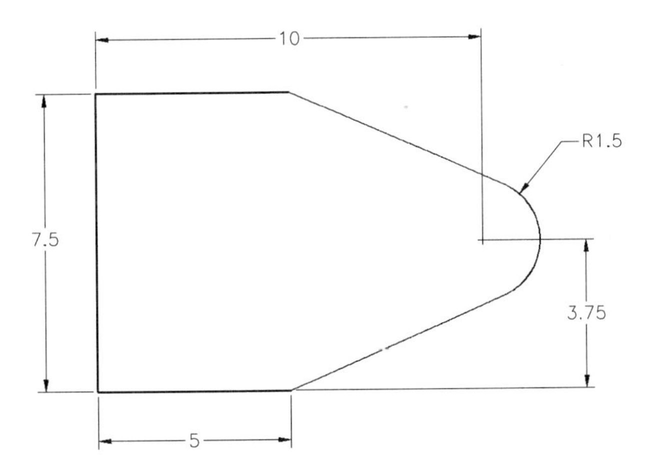

Figure 1-17 *Desired sketch for Tutorial 1*

Tip: *The reason for fully constraining the sketch is that you can edit the sketch even after you have created the designer model from that sketch. The reason for this is that all the dimension values that you assign to the sketch using the* ***AMPARDIM*** *command are displayed when you edit the designer model.*

Tutorial 2

In this tutorial you will create the fully dimensioned sketch for the object shown in Figure 1-18. The dimensions to be used are given in Figure 1-19. The vertical distance between the center of arcs is 10 units.

Step 1: Creating A Rough Sketch

1. It is clear from Figure 1-18 that the limits required for the drawing are more than the default limits. Therefore, before proceeding with the drawing you first need to increase the limits. Use the **LIMITS** command for this purpose. To invoke this command choose **Format > Drawing Limits** from the **Assist** menu. The prompt sequence is as follows:

   Specify lower left corner or [ON/OFF] <0.0000,0.0000>: Enter
   Specify upper right corner <12.0000,9.0000>: **60,60**

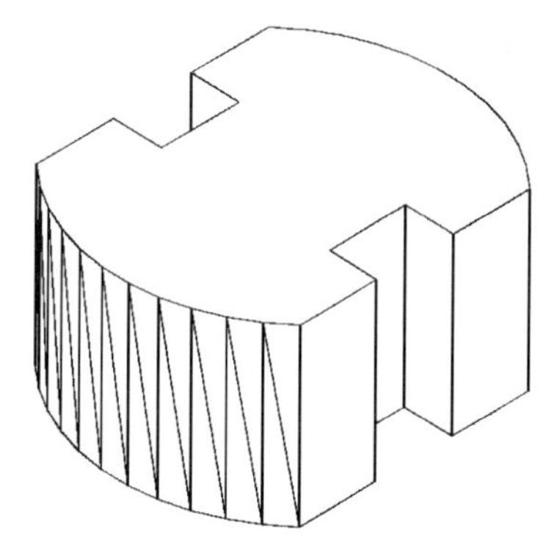

Figure 1-18 Drawing for Tutorial 2

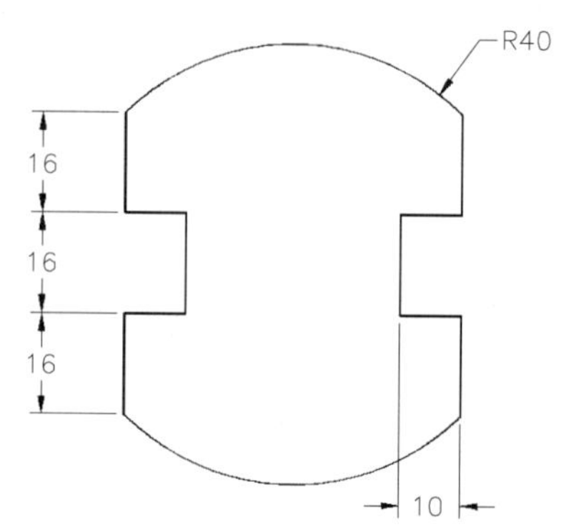

Figure 1-19 Figure showing dimensions for Tutorial 2

2. The **LIMITS** command does not increase the drawing display area. Use the **ALL** option of the **ZOOM** command to increase the drawing display area. This command can be invoked by choosing the **Zoom All** button from the **Zoom Realtime** flyout in the **Mechanical View** toolbar.

3. Now as the limits have been increased so you will also have to increase various scale factors. Instead, you can simply increase the overall scale factor with the help of the **DIMSTYLE** command. To invoke this command choose **Edit Dimensions > Dimension Style** from the **Annotate** menu. In the **Dimension Style Manager** dialog box, choose the **Modify** button to display the **Modify Dimension Style** dialog box.

4. Choose the **Fit** tab. In this tab increase the overall scale factor by setting the value of the **Use overall scale of** spinner to **6**.

5. Now draw the rough sketch for the object using the **PLINE** command or a combination of the **ARC** and **LINE** command as shown in Figure 1-20.

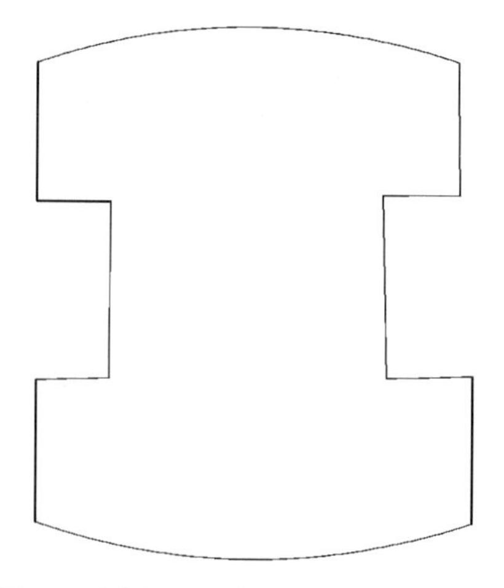

Figure 1-20 Rough sketch for Tutorial 2

Step 2: Converting The Rough Sketch Into The Cleaned-up Sketch

6. Convert the rough sketch shown in Figure 1-19 into a cleaned-up sketch by choosing the **Single Profile** button from the **Part Modeling** toolbar. This converts the rough sketch into cleaned-up sketch, see Figure 1-20.

Tip: *Profile the sketch using the* ***Single Profile*** *command only if the sketch is drawn as a single entity using the* ***PLINE*** *command. If it is not a single entity, profile the complete sketch using the* ***Sketch a Profile*** *button.*

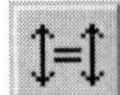

7. Choose the **Equal Length** button from the **2D Constraints** toolbar to apply the additional geometric constraints to the sketch. The prompt sequence is as follows:

Valid selections: line or spline segment
Select first object: *Select the upper left vertical line.*

Valid selections: line or spline segment
Select second object: *Select the middle left vertical line.*
Solved under constrained sketch requiring 9 dimensions or constraints.
Valid selections: line or spline segment
Select first object: *Select the middle left vertical line.*

Valid selections: line or spline segment
Select second object: *Select the lower left vertical line.*
Solved under constrained sketch requiring 8 dimensions or constraints.
Valid selections: line or spline segment
Select first object: *Select the lower left vertical line.*

Valid selections: line or spline segment
Select second object: *Select the upper right vertical line.*
Solved under constrained sketch requiring 7 dimensions or constraints.
Valid selections: line or spline segment
Select first object: *Select the upper right vertical line.*

Valid selections: line or spline segment
Select second object: *Select the middle right vertical line.*
Solved under constrained sketch requiring 7 dimensions or constraints.
Valid selections: line or spline segment
Select first object: *Select the middle right vertical line.*

Valid selections: line or spline segment
Select second object: *Select the lower right vertical line.*
Solved under constrained sketch requiring 6 dimensions or constraints.
Valid selections: line or spline segment
Select first object: *Select the upper left horizontal line.*

Valid selections: line or spline segment
Select second object: *Select the lower left horizontal line.*

Solved under constrained sketch requiring 6 dimensions or constraints.
Valid selections: line or spline segment
Select first object: *Select the lower left horizontal line.*

Valid selections: line or spline segment
Select second object: *Select the lower right horizontal line.*
Solved under constrained sketch requiring 5 dimensions or constraints.
Valid selections: line or spline segment
Select first object: *Select the lower right horizontal line.*

Valid selections: line or spline segment
Select second object: *Select the upper right horizontal line.*
Solved under constrained sketch requiring 5 dimensions or constraints.
Valid selections: line or spline segment
Select first object: *Press ESC.*

The sketch after applying the **Equal Length** constraint should look similar to the one shown in Figure 1-21.

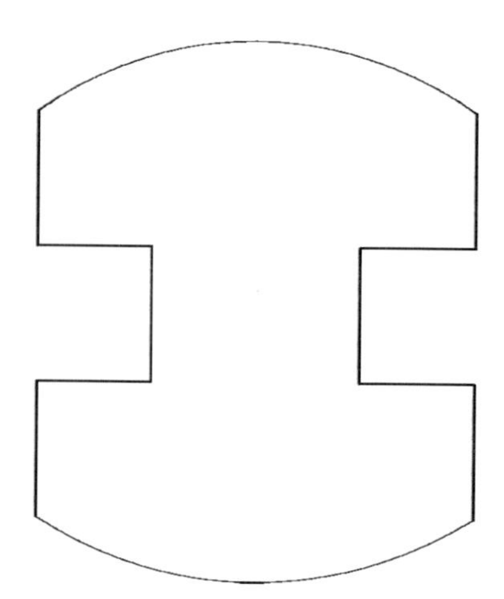

Figure 1-21 *Sketch after applying the equal length constraint*

Step 3: Getting The Desired Sketch From The Cleaned-up Sketch

8. Once you have added all the constraints to the sketch, you need to add the required dimensions. Add the dimensions using the **AMPARDIM** command. To invoke this command, choose the **New Dimension** button from the **Power Dimensioning** flyout in the **Part Modeling** toolbar. You can also invoke this command by entering **AMPARDIM** at the Command prompt. The prompt sequence is as follows:

   Select first object: *Select one of the vertical lines.*
   Select second object or place dimension: *Place the dimension.*
   Enter dimension value or [Undo/Hor/Ver/Align/Par/aNgle/Ord/Diameter/pLace] <default value>: **16**

Solved under constrained sketch requiring 4 dimensions or constraints.
Select first object: *Select one of the horizontal lines.*
Select second object or place dimension: *Place the dimension.*
Enter dimension value or [Undo/Hor/Ver/Align/Par/aNgle/Ord/Diameter/pLace] <default value>: **10**

Solved under constrained sketch requiring 3 dimensions or constraints.
Select first object: *Select the upper arc.*
Select second object or place dimension: *Place the dimension.*
Enter dimension value or [Undo/Diameter/Ordinate/Placement point] <default value>: **40**

Solved under constrained sketch requiring 2 dimensions or constraints.
Select first object: *Select the lower arc.*
Select second object or place dimension: *Place the dimension.*
Enter dimension value or [Undo/Diameter/Ordinate/Placement point] <default value>: **40**

Tip: *Instead of adding the dimension to the second arc you can also add the* ***Radius*** *constraint. To add this constraint you must have an arc that has been assigned some dimension value so that it can be used for the other arcs or circles.*

Solved under constrained sketch requiring 1 dimension or constraint.
Select first object: *Select the upper arc.*
Select second object or place dimension: *Select the lower arc.*
Specify dimension placement: *Place the dimension on the left of the sketch.*
Enter dimension value or [Undo/Hor/Ver/Align/Par/aNgle/Ord/Diameter/pLace] <default value>: **V**
Enter dimension value or [Undo/Hor/Ver/Align/Par/aNgle/Ord/Diameter/pLace] <default value>: **10**
Solved fully constrained sketch.
Select first object: Enter

Tip: *If the sketch is not fully constrained still and you are prompted to add another dimension or constraint to fully constrain the sketch, make the horizontal distance between the two arcs as zero.*

9. Enter **FF** at the Command prompt. The final sketch for Tutorial 2 should look similar to the one shown in Figure 1-22.

10. Save this drawing with the name given below:

    \MDT Tut\Ch-1**Tut2.dwg**

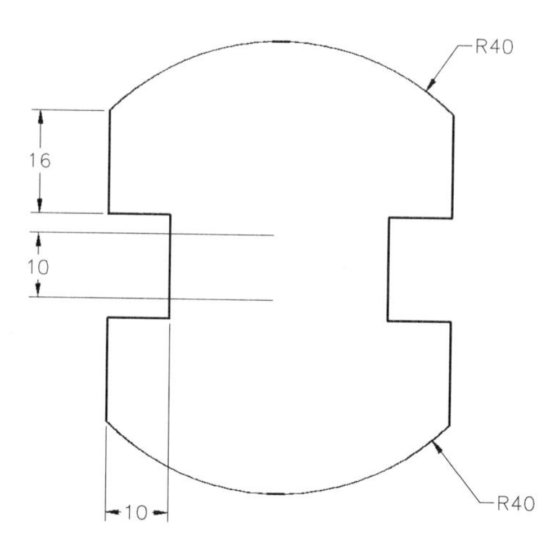

Figure 1-22 Desired sketch for Tutorial 2

Tutorial 3

In this tutorial you will create a proper dimensioned sketch for the object shown in Figure 1-23. The dimensions to be used are given in Figure 1-24.

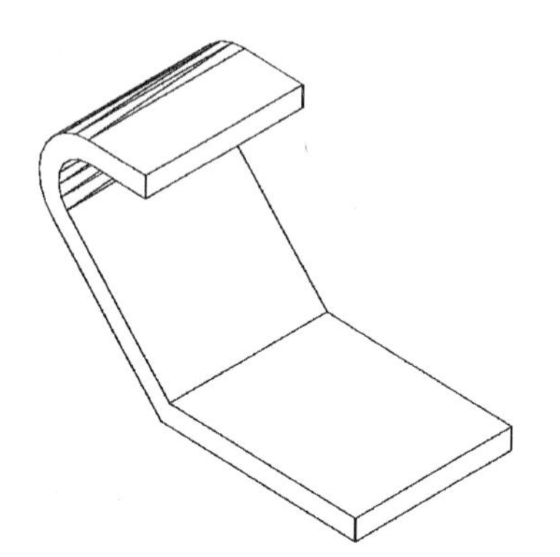

Figure 1-23 Dimensions for Tutorial 3

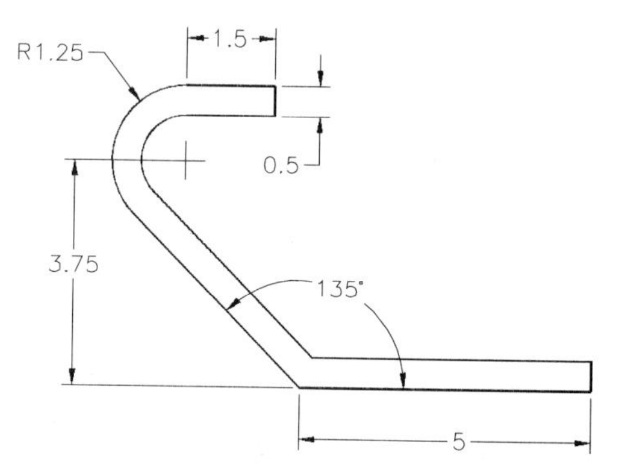

Figure 1-24 Dimensions for Tutorial 3

Step 1: Creating A Rough Sketch

1. It is clear from the figure that the sketch for this model should be an open sketch. It is also clear from the figure that the face of the model is not parallel to the world XY plane. Therefore, either you will have to change the UCS before drawing the sketch of this model or rotate the designer model after creating. In this case the UCS is changed before drawing the sketch.

2. Choose **New UCS > X** from the **Assist** menu. The prompt sequence is as follows:

   Specify rotation angle about X axis <90>: Enter

3. Enter **9** at the Command prompt. It is the keyboard shortcut to switch to the Plan view of the current UCS.

4. Enter **FF** at the Command prompt.

5. Create the rough sketch for the designer model using the **PLINE** command, see Figure 1-25.

Step 2: Converting The Rough Sketch Into A Cleaned-up Sketch

6. Choose the **Single Profile** button from the **Profile a Sketch** flyout in the **Part Modeling** toolbar. The open sketch is converted into an open profile, see Figure 1-26.

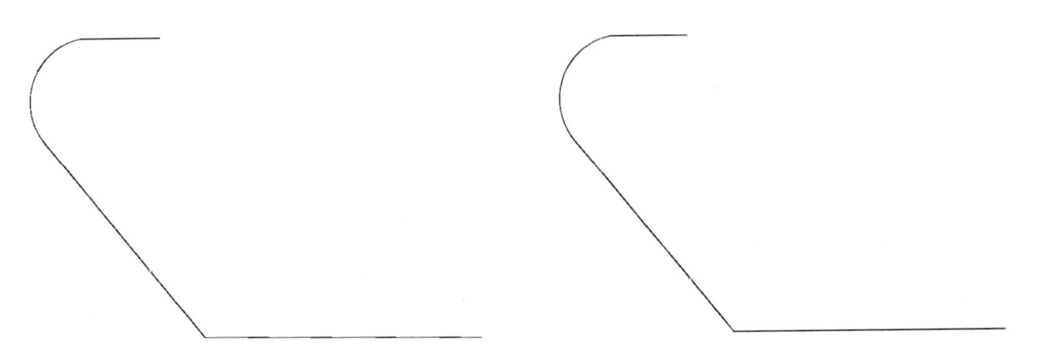

Figure 1-25 *Rough sketch for Tutorial 3*

Figure 1-26 *Profiled sketch for Tutorial 3*

Step 3: Getting The Desired Sketch From The Cleaned-up Sketch

7. Choose the **New Dimensions** button from the **Power Dimensioning** flyout in the **Part Modeling** toolbar. The prompt sequence is as follows:

Select first object: *Select the lower horizontal line.*

Select second object or place dimension: *Place the dimension.*
Enter dimension value or [Undo/Hor/Ver/Align/Par/aNgle/Ord/Diameter/pLace] <default value>: **5**
Solved under constrained sketch requiring 4 dimensions or constraints.
Select first object: *Select the upper horizontal line.*

Select second object or place dimension: *Place the dimension.*
Enter dimension value or [Undo/Hor/Ver/Align/Par/aNgle/Ord/Diameter/pLace] <default value>: **1.5**
Solved under constrained sketch requiring 3 dimensions or constraints.
Select first object: *Select the arc.*

Select second object or place dimension: *Place the dimension.*
Enter dimension value or [Undo/Diameter/Ordinate/Placement point] <default value>: **1.25**

Solved under constrained sketch requiring 2 dimensions or constraints.
Select first object: *Select the lower horizontal line.*

Select second object or place dimension: *Select the arc.*

Specify dimension placement: *Place the dimension.*
Enter dimension value or [Undo/Hor/Ver/Align/Par/aNgle/Ord/Diameter/pLace] <default value>: **V**
Enter dimension value or [Undo/Hor/Ver/Align/Par/aNgle/Ord/Diameter/pLace] <default value>: **3.75**
Solved under constrained sketch requiring 1 dimension or constraint.
Select first object: *Select the inclined line.*

Select second object or place dimension: *Select the lower horizontal line.*

Specify dimension placement: *Place the dimension.*
Enter dimension value or [Undo/Hor/Ver/Align/Par/aNgle/Ord/Diameter/pLace] <0.0000>: **N**

Enter dimension value or [Undo/Placement point] <133>: **135**

Solved fully constrained sketch.
Select first object: Enter

8. Enter **FF** at the Command prompt. The final sketch should look similar to the one shown in Figure 1-27.

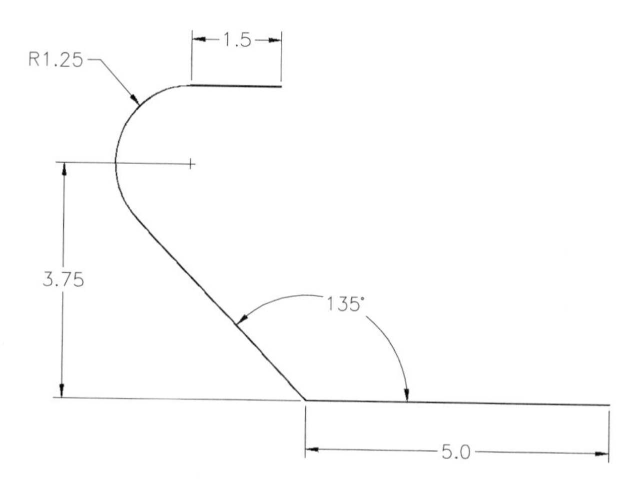

Figure 1-27 *Final sketch for Tutorial 3*

9. Save this drawing with the name given below:

\MDT Tut\Ch-1**Tut3.dwg**

Tutorial 4

In this tutorial you will create the basic sketch for the drawing shown in Figure 1-28. The dimensions to be used are given in Figure 1-29.

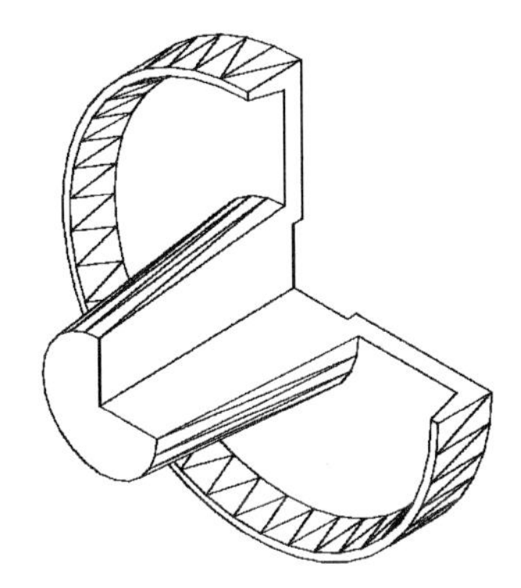

Figure 1-28 Drawing for Tutorial 4

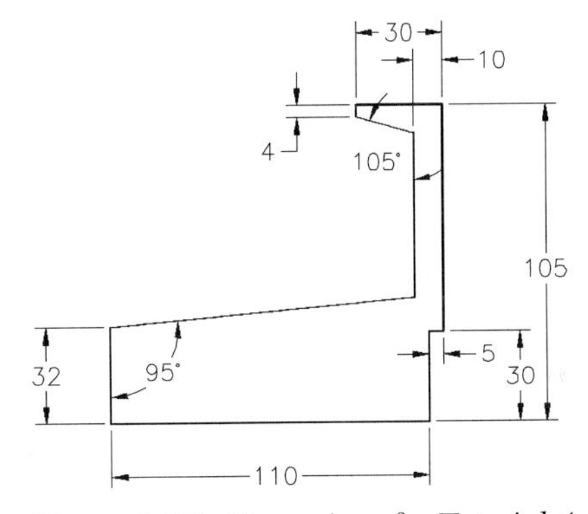

Figure 1-29 Dimensions for Tutorial 4

Step 1: Creating A Rough Sketch

1. Choose **Format > Drawing Limits** from the **Assist** menu. The prompt sequence is:

   Specify lower left corner or [ON/OFF] <0.0000,0.0000>: Enter
   Specify upper right corner <9.0000,12.0000>: **150,150**

2. Choose the **Zoom All** button from the **Zoom Realtime** flyout in the **Mechanical View** toolbar.

3. Choose **Edit Dimensions > Dimension Style** from the **Annotate** menu.

4. Choose the **Modify** button in the **Dimension Style Manager** dialog box.

5. Choose the **Fit** tab. Now set the value of the **Use overall scale of** spinner to **15**.

6. Draw the rough sketch using the **PLINE** command as shown in Figure 1-30.

Step 2: Converting The Rough Sketch Into A Cleaned-Up Sketch

7. Choose the **Single Profile** button from the **Part Modeling** toolbar to profile the rough sketch.

8. If any constraint is missing, apply it by choosing the button of that constraint from the **2D Constraints** toolbar and convert it into a cleaned-up sketch, see Figure 1-31.

Step 3: Getting The Desired Sketch From The Cleaned-up Sketch

9. Since it is difficult to understand which line is selected to dimension, therefore, all the lines have been numbered (Figure 1-31). This makes it easier to understand which line is selected to dimension. Now apply the required dimensions to the

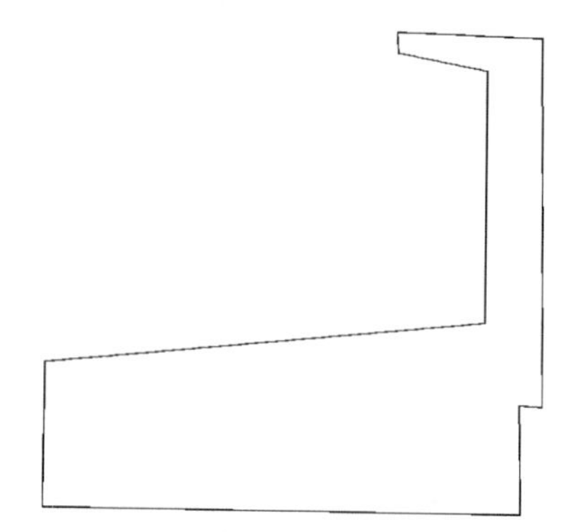

Figure 1-30 *Rough sketch for Tutorial 4*

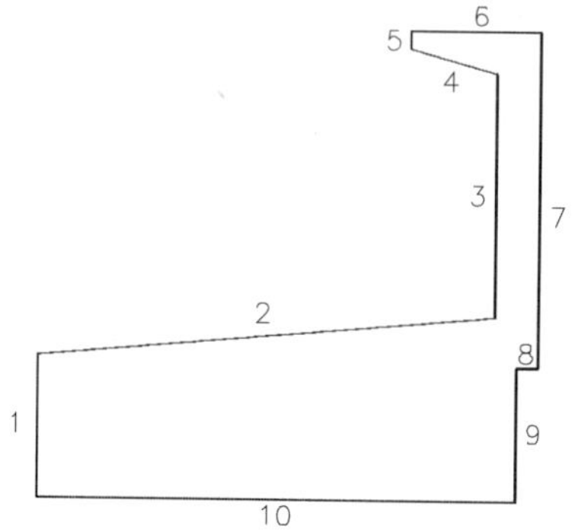

Figure 1-31 *Fully cleaned-up sketch for Tutorial 4*

sketch by choosing the **New Dimension** button from the **Power Dimensioning** flyout in the **Part Modeling** toolbar. The prompt sequence is as follows:

Select first object: *Select line 1.*
Select second object or place dimension: *Place the dimension.*
Enter dimension value or [Undo/Hor/Ver/Align/Par/aNgle/Ord/Diameter/pLace] <default value>: **32**

Solved under constrained sketch requiring 9 dimensions or constraints.
Select first object: *Select line 10.*
Select second object or place dimension: *Place the dimension.*
Enter dimension value or [Undo/Hor/Ver/Align/Par/aNgle/Ord/Diameter/pLace] <default value>: **110**

Solved under constrained sketch requiring 8 dimensions or constraints.
Select first object: *Select line 9.*
Select second object or place dimension: *Place the dimension.*
Enter dimension value or [Undo/Hor/Ver/Align/Par/aNgle/Ord/Diameter/pLace] <default value>: **30**

Solved under constrained sketch requiring 7 dimensions or constraints.
Select first object: *Select line 8.*
Select second object or place dimension: *Place the dimension.*
Enter dimension value or [Undo/Hor/Ver/Align/Par/aNgle/Ord/Diameter/pLace] <default value>: **30**

Solved under constrained sketch requiring 6 dimensions or constraints.
Select first object: *Select line 7.*
Select second object or place dimension: *Select line 3.*
Specify dimension placement: *Place the dimension.*
Enter dimension value or [Undo/Hor/Ver/Align/Par/aNgle/Ord/Diameter/pLace] <default value>: **P**

Enter dimension value or [Undo/Hor/Ver/Align/Par/aNgle/Ord/Diameter/pLace] <default value>: **10**

Solved under constrained sketch requiring 5 dimensions or constraints.
Select first object: *Select line 6.*
Select second object or place dimension: *Place the dimension.*
Enter dimension value or [Undo/Hor/Ver/Align/Par/aNgle/Ord/Diameter/pLace] <default value>: **30**

Solved under constrained sketch requiring 4 dimensions or constraints.
Select first object: *Select line 5.*
Select second object or place dimension: *Place the dimension.*
Enter dimension value or [Undo/Hor/Ver/Align/Par/aNgle/Ord/Diameter/pLace] <default value>: **4**

Solved under constrained sketch requiring 3 dimensions or constraints.
Select first object: *Select line 6.*
Select second object or place dimension: *Select line 10.*
Specify dimension placement: *Place the dimension.*
Enter dimension value or [Undo/Hor/Ver/Align/Par/aNgle/Ord/Diameter/pLace] <default value>: **V**
Enter dimension value or [Undo/Hor/Ver/Align/Par/aNgle/Ord/Diameter/pLace] <default value>: **105**

Solved under constrained sketch requiring 2 dimensions or constraints.
Select first object: *Select line 4.*
Select second object or place dimension: *Select line 3.*
Specify dimension placement: *Place the dimension.*
Enter dimension value or [Undo/Hor/Ver/Align/Par/aNgle/Ord/Diameter/pLace] <default value>: **N**
Enter dimension value or [Undo/Placement point] <default value>: **105**

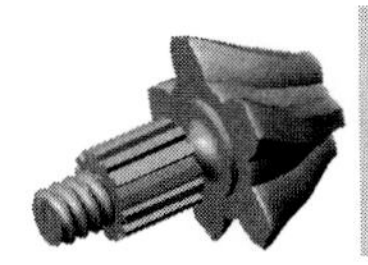

Tip: *Sometimes, when you select two lines to assign the angular dimension and place the dimension, you will be prompted to enter the angle value. In such cases do not enter* ***N*** *at the Command prompt. Instead, enter the angle value directly.*

Solved under constrained sketch requiring 1 dimensions or constraints.
Select first object: *Select line 1.*
Select second object or place dimension: *Select line 2.*
Specify dimension placement: *Place the dimension.*
Enter dimension value or [Undo/Hor/Ver/Align/Par/aNgle/Ord/Diameter/pLace] <default value>: **N**
Enter dimension value or [Undo/Placement point] <default value>: **105**
Solved fully constrained sketch.
Select first object: [Enter]

10. Enter **FF** at the Command prompt. The final sketch for Tutorial 4 should look similar to

the one shown in Figure 1-32.

11. Save this drawing with the name given below:

    \MDT Tut\Ch-1**Tut4.dwg**

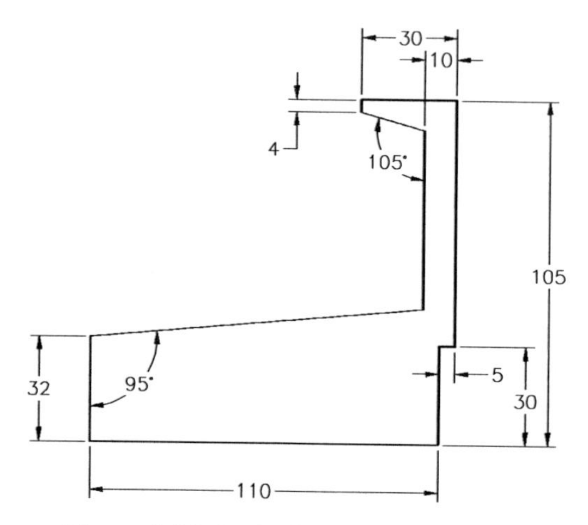

***Figure 1-32** Desired sketch for Tutorial 4*

Review Questions

Answer the following questions.

1. List various geometric constraints available in Mechanical Desktop.
2. What are the four steps for creating a designer model?
3. Which command is used to view the constraints applied to the sketch?
4. Which command is used to add the geometric constraints to the sketch?
5. Which constraint is applied to place two different segment in the same line?
6. Which command is used to delete the constraints from the sketch?
7. You can dimension the sketch using the **AMPARDIM** command without profiling the sketch using the **AMPROFILE** command. (T/F)
8. What is the reason for fully constraining the sketch?

Exercises

Exercise 1

Draw the sketch for the solid model shown in Figure 1-33. After drawing the sketch, profile it and add the required constraints and dimensions so that the sketch is fully constrained. The dimensioned sketch is shown in Figure 1-34. Save the sketch with the name given below:

\MDT Tut\Ch-1**Exr1.dwg**

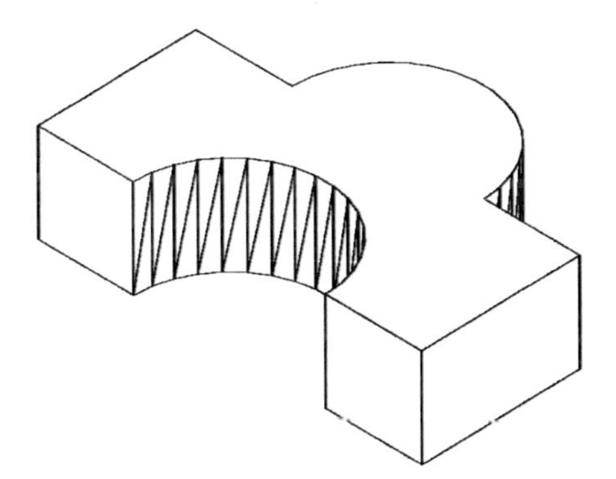

Figure 1-33 Drawing for Exercise 1

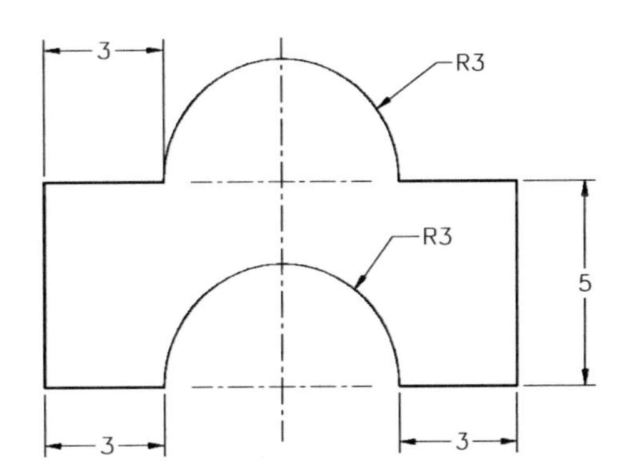

Figure 1-34 Dimensions for Exercise 1

Exercise 2

Draw the sketch for the solid model shown in Figure 1-35. After drawing the sketch, profile it and add the required constraints and dimensions so that the sketch is fully constrained. The dimensioned sketch is shown in Figure 1-36. Save the sketch with the name given below:

\MDT Tut\Ch-1**Exr2.dwg**

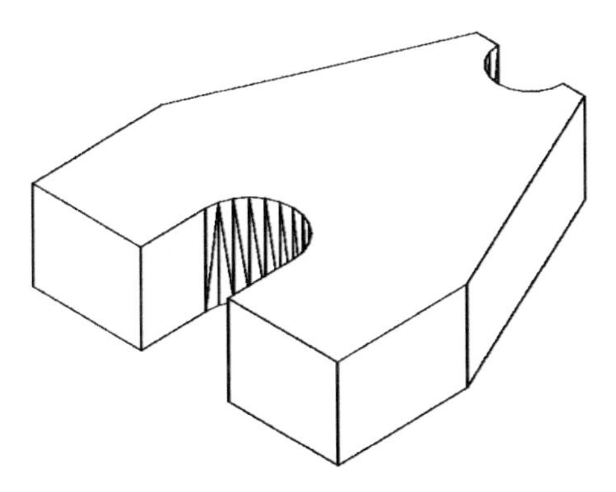

Figure 1-35 Drawing for Exercise 2

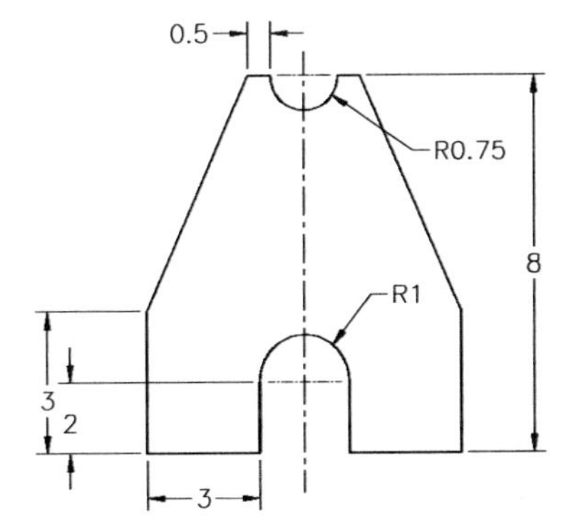

Figure 1-36 Dimensions for Exercise 2

Exercise 3

Draw the sketch for the solid model shown in Figure 1-37. After drawing the sketch, profile it and add the required constraints and dimensions so that the sketch is fully constrained. The dimensioned sketch is shown in Figure 1-38. Save the sketch with the name given below:

\MDT Tut\Ch-1**Exr3.dwg**

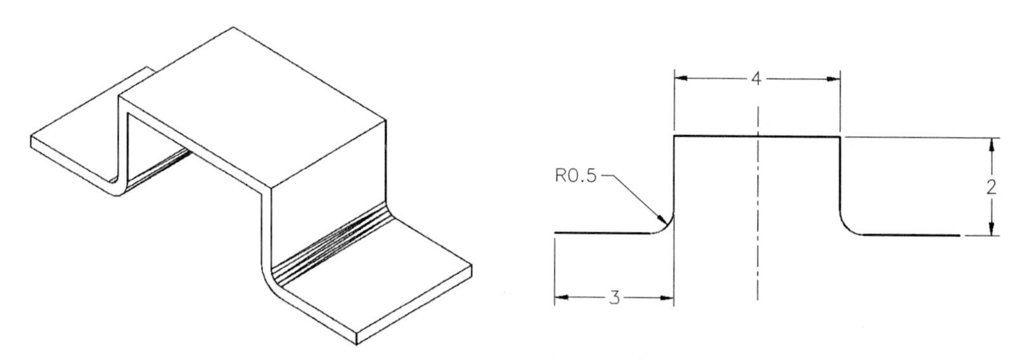

Figure 1-37 Drawing for Exercise 3

Figure 1-38 Dimensions for Exercise 3

Exercise 4

Draw the sketch for the solid model shown in Figure 1-39. After drawing the sketch, profile it and add the required constraints and dimensions so that the sketch is fully constrained. The dimensioned sketch is shown in Figure 1-40. Save the sketch with the name given below:

\MDT Tut\Ch-1**Exr4.dwg**

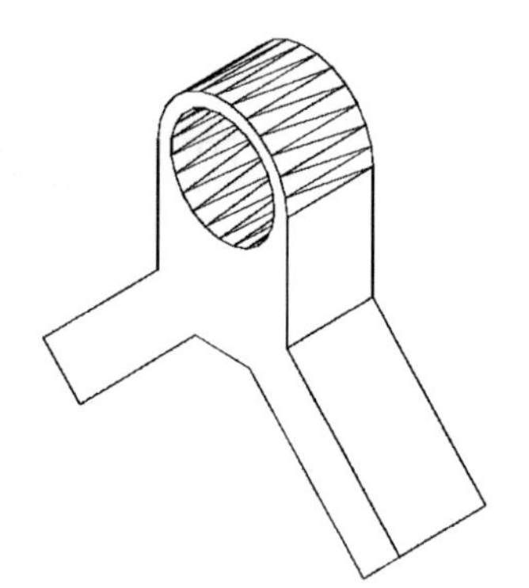

Figure 1-39 Drawing for Exercise 4

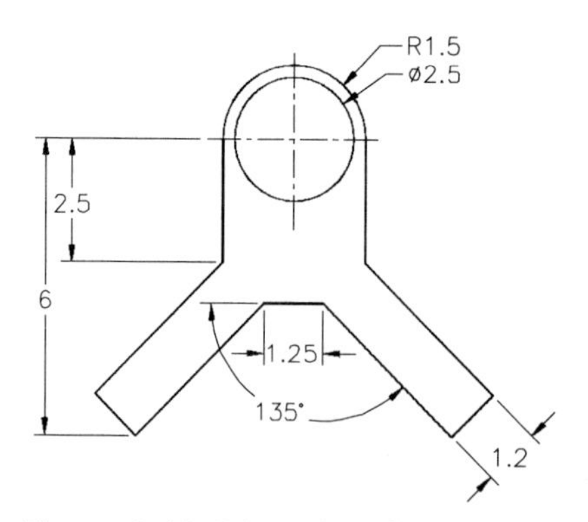

Figure 1-40 Dimensions for Exercise 4

Tip: *To assign the parallel dimension between two lines, select the first line, then select the second line, place the dimension and then enter **P** at the Command prompt. The dimension will be placed in terms of parallel distance between two lines. You can modify the dimension value in the Command line.*

Chapter 2

Modifying, Extruding and Revolving the Sketches

Learning Objectives

After completing this chapter, you will be able to:

- *Modify the desired sketch using the **AMMODDIM** command.*
- *Extrude the desired sketch using the **AMEXTRUDE** command.*
- *Revolve the desired sketch using the command **AMREVOLVE**.*

Commands Covered

- *AMMODDIM*
- *AMEXTRUDE*
- *AMREVOLVE*

MODIFYING THE SKETCHES

Sometimes, after creating the desired sketch you may want to modify the dimensions assigned to the sketch. This is done with the help of the **AMMODDIM** command. This command is discussed below.

AMMODDIM Command

Toolbar:	Part Modeling > Power Dimensioning > Edit Dimension
Menu:	Part > Dimensioning > Edit Dimension
Context Menu:	Dimensioning > Edit Dimension
Command:	AMMODDIM

One of the methods of modifying the dimensions assigned to a sketch is to erase or delete the dimension and then assign a new dimension in its place. This is a very tedious and time consuming job as you have to first erase the dimensions and then enter new dimension. To overcome this problem, Mechanical Desktop has provided you with the **AMMODDIM** command. Choose the **Power Dimensioning** flyout in the **Part Modeling** toolbar (Figure 2-1) and choose **Edit Dimension** to invoke this command. This command not only modifies the dimension, but also drives the selected entity to the new dimension value you have entered. For example, consider the sketch shown in Figure 2-2a. It shows a fully dimensioned sketch. Choose the **Edit Dimension** button and select the dimension of the bottom horizontal line which is 8.0 units. You will be prompted to specify the new dimension value. Specify the new dimension value as 6.0 units. Now, the new dimension value will be displayed and the line segment will be driven to the new length as shown in Figure 2-2b.

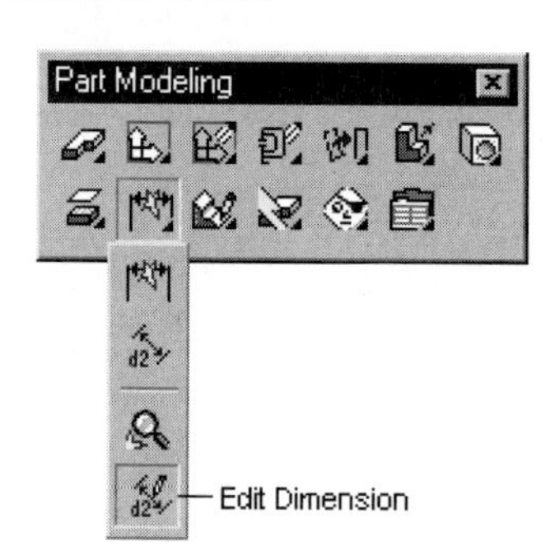

Figure 2-1 *Choosing Edit Dimension button from Power Dimensioning flyout*

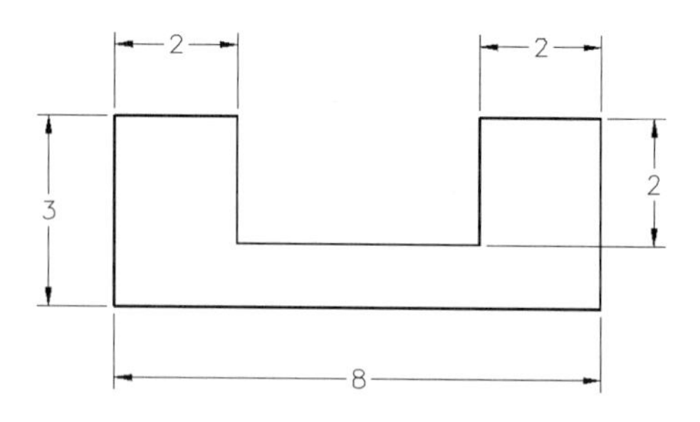

Figure 2-2a *Figure showing sketches before*

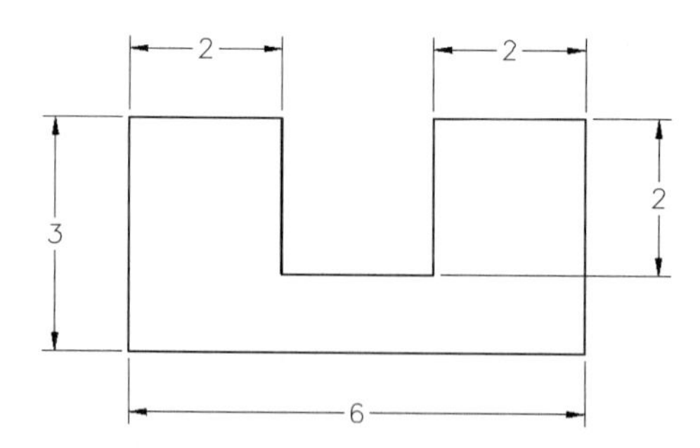

Figure 2-2b *Figure showing sketches after*

EXTRUDING THE SKETCHES (AMEXTRUDE COMMAND*)

Toolbar:	Part Modeling > Sketched Features - Extrude
Menu:	Part > Sketched Features > Extrude
Context Menu:	Sketched & Worked Features > Extrude
Command:	AMEXTRUDE

The first three steps of creating the designer model have already been discussed in chapter 1. In this chapter the fourth step of creating the designer model will be discussed. In this step you will create the designer model from the desired sketch. This can be done with the help of various commands depending upon the final designer model required. You will start with extruding the desired sketch with the help of the **AMEXTRUDE** command. This command is used to extrude the profiled sketch in the Z direction of the current sketch plane. Depending upon whether the sketch selected for extruding is an open profile or a closed profile, this dialog box changes. Therefore, the **AMEXTRUDE** command is discussed separately for both open and closed profiles.

Extruding The Closed Profiles

When you select a closed profile for extruding, the **Extrusion** dialog box that will be displayed is shown in Figure 2-3.

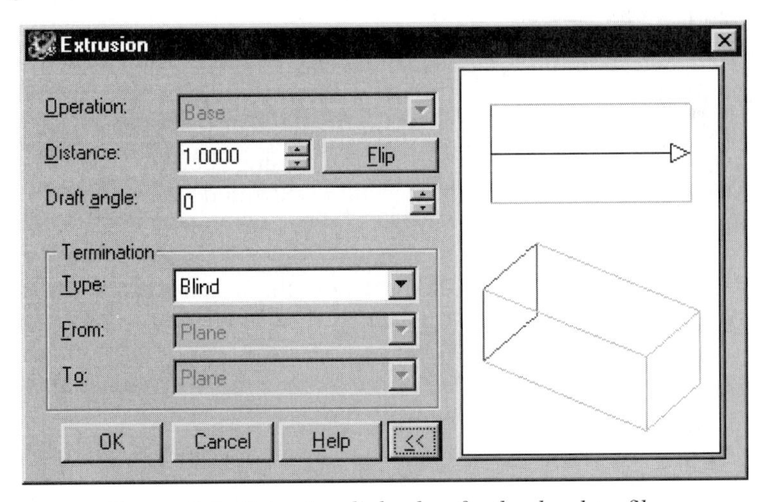

Figure 2-3 *Extrusion dialog box for the closed profiles*

Operation

The first feature of every designer model is the base feature. Since, no base feature is created as yet, therefore, the options under this drop-down list are not available. Once you have created the base feature, this drop-down list gets activated providing you with various other options. The remaining options of this drop-down list will be discussed later.

Distance

This spinner is used to specify the height of the extrusion. You can either enter the value or use the spinner for specifying the extrusion height.

Flip

This button is chosen to reverse the direction of extrusion. The direction of extrusion is displayed by an arrow that appears on the sketch as soon as the **AMEXTRUDE** command is invoked. When you choose this button, the direction of the arrow changes to the opposite direction.

Draft angle

This spinner is used to specify the taper angle by which you want the sketch to taper. If the draft angle is negative, the sketch is tapered inwards thus creating a negative draft. If the draft angle is positive, the sketch is expanded outwards thus creating a positive draft. The **Draft Angle** edit box can have a value that lies between -89 and +89 degrees.

Termination Area

The options under this area are used to specify the termination of the selected sketch.

Type

The options under this drop-down list are used to specify the type of the termination of the sketch. Since this is the base feature, therefore, only the following four options are available. The remaining options of this drop-down list are discussed in the later chapters.

Blind. Selecting this option extrudes the sketch in only one direction. As soon as the **AMEXTRUDE** command is invoked, an arrow appears on the sketch that displays the direction in which the sketch will be extruded. However, you can reverse the direction of extrusion by choosing the **Flip** button in the **Extrusion** dialog box. Figure 2-4 shows a rectangle extruded with negative draft angle using the **Blind** option.

MidPlane. Selecting this option extrudes the sketch equally in both the directions above and below the plane in which the sketch is created. When you use this option no arrow is displayed as the sketch is extruded in both the directions. Figure 2-5 shows a rectangle extruded using the **MidPlane** option.

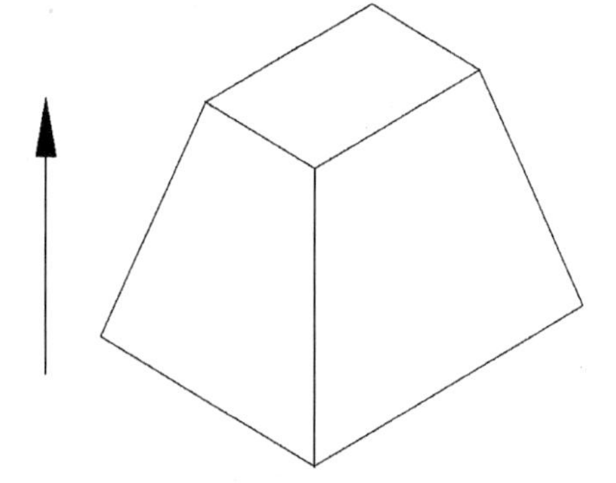

Figure 2-4 *Extruding a rectangle using the Blind options with a negative draft angle*

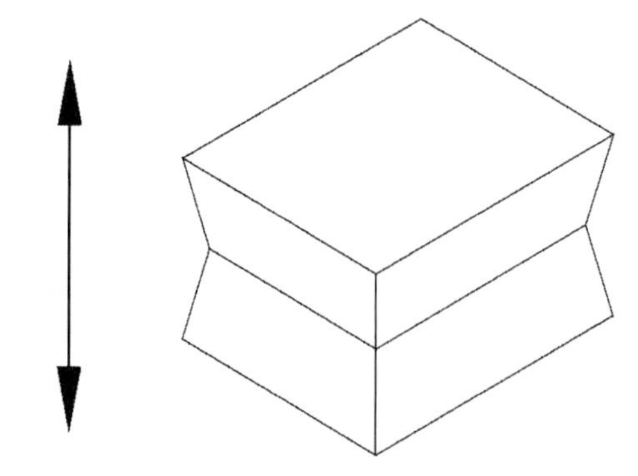

Figure 2-5 *Extruding a rectangle using the MidPlane option and with a positive draft angle*

Tip: *If the height of extrusion is 4 units, then selecting the **MidPlane** option from the **Type** drop-down list will extrude the sketch 2 units above and 2 units below the plane of the sketch, thus making it a total of 4 units.*

Plane. This option is used to extrude the selected sketch to a specified work plane.

From-To. This option is used to extrude the selected sketch using two planes. The extrusion will start from one specified work plane and end at the other specified work plane. When you select this option, the **From** and **To** drop-down lists will get activated. However, both the drop-down lists will provide only the **Plane** option.

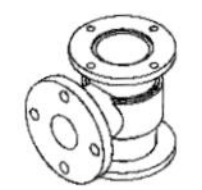

Note

The different methods for creating the work planes will be discussed in the later chapters.

Extruding The Open Profiles (Thin Extrusions)*

When you select an open profile to extrude, the **Extrusion** dialog box that will be displayed is shown in Figure 2-6. This dialog box has the **Thickness Area** in addition to the other options that are same as those in the **Extrusion** dialog box for the closed profiles.

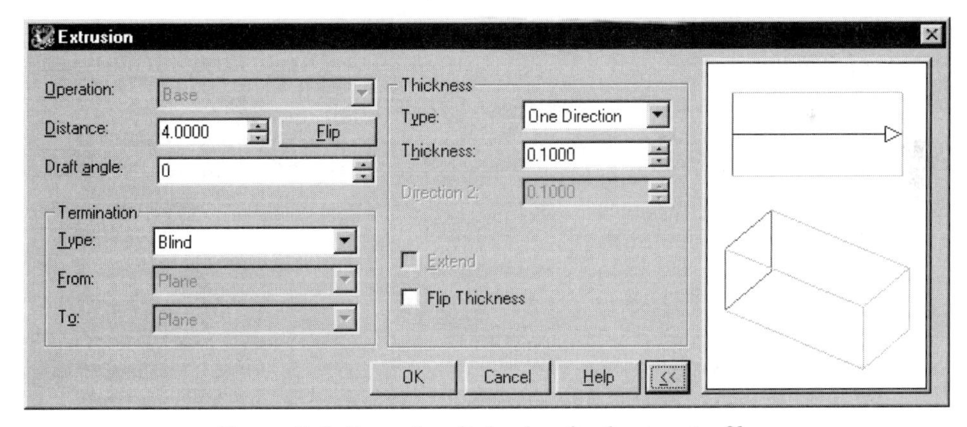

***Figure 2-6** Extrusion dialog box for the open profiles*

Thickness Area

The options under this area are used to specify the thickness of the open profiles.

Type

This drop-down list provides the methods of specifying the thickness of the open profile. You can specify the thickness using the following three methods:

One Direction. This method is used when you want to specify the thickness in only one direction. This direction can be either below or above the open profile. The value of the thickness can be specified using the **Thickness** spinner.

Two Directions. This method is used to specify the thickness in two directions;

above and below the open profile. You can specify different values of thickness in both directions. When you select this method from the **Type** drop-down list, the **Distance 1** spinner appears in place of the **Thickness** spinner and the **Distance 2** spinner is also activated. Using these two spinners you can specify the values of both the distances. The value of first distance will be taken above the sketch and the value of second distance will be taken below the sketch.

MidPlane. This method is used to specify the thickness value equally on the either side of the open profile. The thickness value can be specified using the **Thickness** spinner.

Extend

This check box is not available as there is no base feature. This check box is used when you want to extend the open profile to the next face available.

Flip Thickness

This check box is selected to reverse the direction in which the thickness is applied to the open profile. This check box will be available only when you select **One Direction** from the **Type** drop-down list.

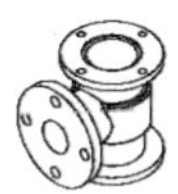

Note

The remaining options of this dialog box are similar to those of the ***Extrusion*** *dialog box for the closed profiles.*

REVOLVING THE SKETCHES (AMREVOLVE COMMAND*)

Toolbar:	Part Modeling > Sketched Feature -Extrude > Revolve
Menu:	Part > Sketched Feature > Revolve
Context Menu:	Sketched & Worked Features > Revolve
Command:	AMREVOLVE

Revolving the sketches is also one of the methods of creating a designer model. It is generally used to create circular models like couplings, shafts and so on or for creating the circular cavities. You can revolve the sketch using the **AMREVOLVE** command. This command is used to create a 3D model by revolving the sketch about a straight edge. However, it has to be kept in mind that the edge about which you are revolving the sketch should be a part of the sketch or should be a work axis. When you invoke this command, you will be prompted to define the revolution axis about which you want to revolve the sketch. On specifying the revolution axis, the **Revolution** dialog box is displayed as shown in Figure 2-7.

Revolution Dialog Box Options

The various options provided under the **Revolution** dialog box are discussed next.

Operation

The feature you are going to create is the base or the first feature. This is the reason that the options under this drop-down list will not be available. Once you have created the base, this drop-down list will be activated and will provide you with various other options.

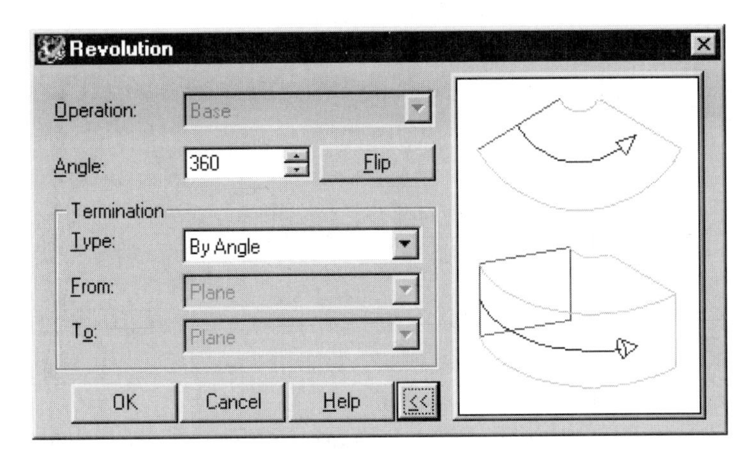

Figure 2-7 *Revolution dialog box*

The remaining options of this drop-down list will be discussed in the later chapters.

Angle

This spinner is used to specify the angle by which you want to revolve the sketch. You can enter the value of the angle in this edit box or specify it with the help of the spinners available on the right side of this edit box. This value can vary from 0 to 360 degrees.

Flip

This button is used to reverse the direction of revolution. As you define the axis of revolution in the sketch, the **Revolution** dialog box is displayed and an arrow appears on the sketch showing the direction in which the sketch will be revolved. When you choose this button, the direction of the revolution is reversed and thus the direction of the arrow is also reversed.

Termination Area

The options provided under this area are used to define the termination of the revolution.

Type

This drop-down list is used to specify the method of revolving the selected sketch. As this is the base feature, therefore, this drop-down list will provide only the following four methods:

By Angle. This option is used to revolve the selected sketch through a desired angle and in a particular direction as shown in Figures 2-8 and 2-9. The direction in which the sketch is revolved is shown by an arrow displayed on the sketch. You can reverse this direction of revolution by choosing the **Flip** button.

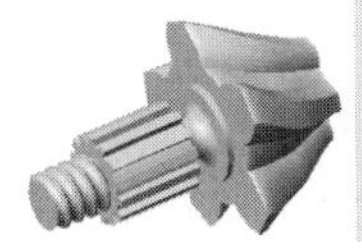

Tip: *To extrude or revolve the sketch using the **AMEXTRUDE** or **AMREVOLVE** command, you first need to profile it using the **AMPROFILE** command. If the sketch is not profiled, these commands will not work.*

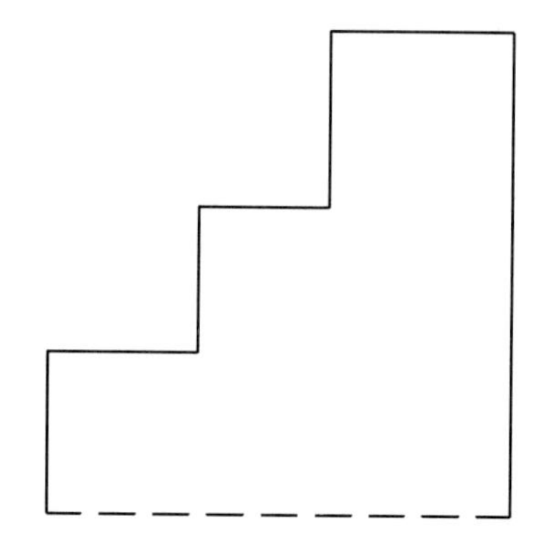

Figure 2-8 *Figure showing basic sketch before revolving*

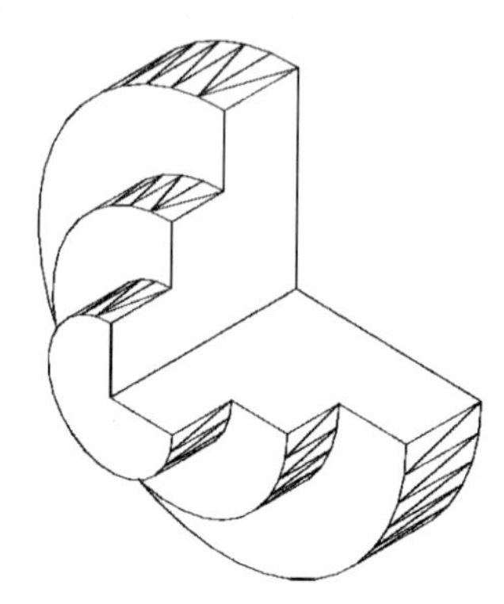

Figure 2-9 *Designer model created after revolving the sketch through an angle of 270 degrees about the axis shown as a dotted line in Figure 2-8*

MidPlane. This option is used to revolve the sketch equally in the either directions of the sketch. The angle by which you want to revolve the sketch can be specified using the **Angle** spinner.

Plane. This option is used to revolve the selected sketch up to a specified work plane.

From-To. This option is used to revolve the selected sketch using two planes. The revolution will start from the first specified plane and will end at the other specified plane.

TUTORIALS

Tutorial 1

In this tutorial you will create the object shown in Figure 2-10. The dimensions for the object are given in Figure 2-11. The extrusion height for the object is 5 units.

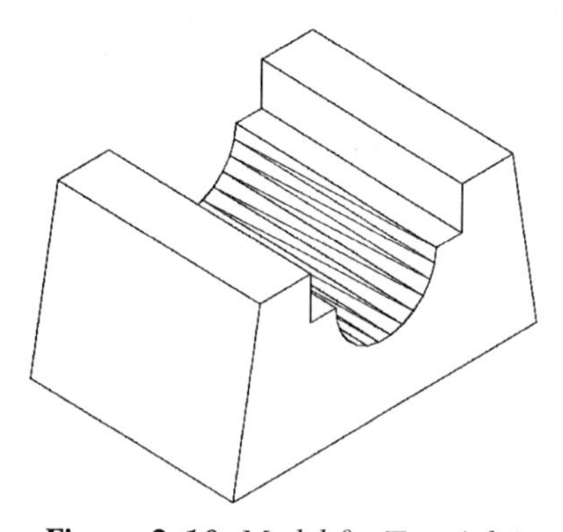

Figure 2-10 *Model for Tutorial 1*

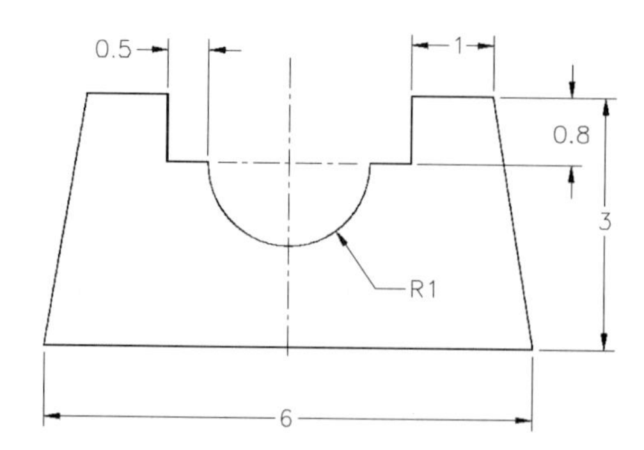

Figure 2-11 *Dimensions for Tutorial 1*

Step 1: Creating The Rough Sketch

1. Rotate the UCS about the X axis through an angle of 90 degrees and then create the rough sketch in proportion with the given figure.

Step 2: Converting The Rough Sketch Into Cleaned-up Sketch

2. Convert this rough sketch into a cleaned-up sketch using the **AMPROFILE** and **AMADDCON** commands.

Step 3: Getting The Desired Sketch From Cleaned-up Sketch

3. Dimension the cleaned-up sketch using the **AMPARDIM** command. The final desired sketch should be similar to the sketch shown in the Figure 2-12.

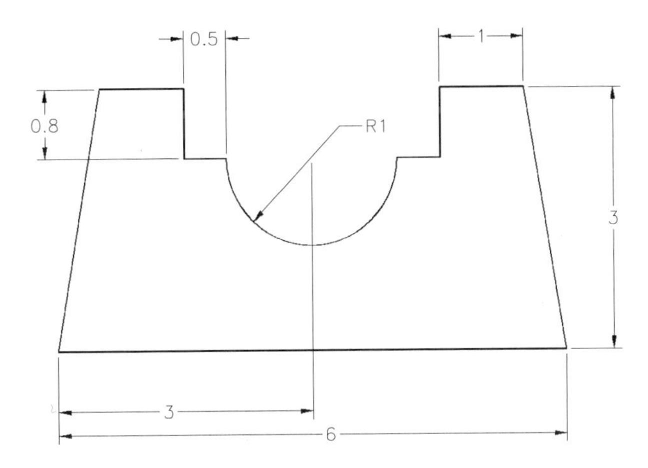

Figure 2-12 *Desired sketch for Tutorial 1*

Step 4: Converting The Desired Sketch Into Designer Model

4. Enter **88** at the Command prompt to shift to the Southwest Isometric view. Here you can get a better view of the object and you can extrude it in any direction you want.

5. Invoke the **AMEXTRUDE** command by choosing the **Sketched Feature-Extrude** button from the **Part Modeling** toolbar. When you choose this button, the **Extrusion** dialog box will be displayed as shown in Figure 2-13.

6. Select the **Blind** option from the **Type** drop-down list from the **Termination** area.

7. Set the value of extrusion in the **Distance** spinner to **5**.

8. Set the value of draft angle to **0** in the **Draft Angle** spinner. Now choose **OK** button.

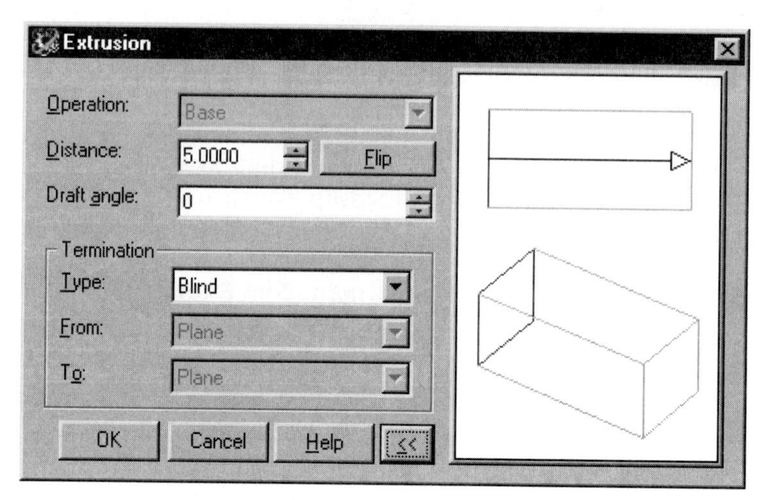

Figure 2-13 *Extrusion dialog box*

9. Enter **FF** at the Command prompt to zoom to the extents of the drawing. Enter **0** (zero) at the Command prompt. This will hide the hidden lines in the designer model as shown in Figure 2-14.

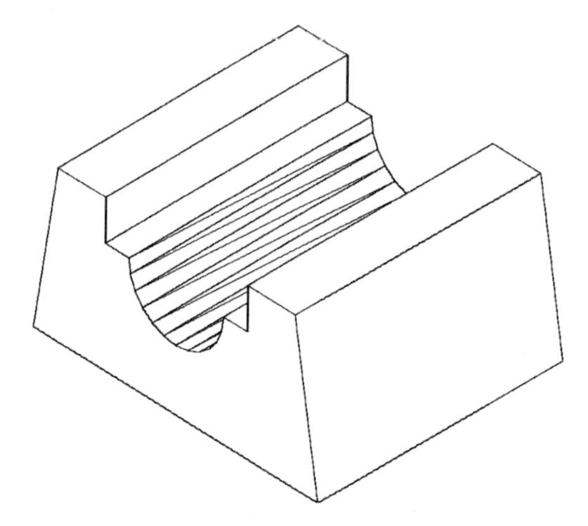

Figure 2-14 *The final designer model after hiding the hidden lines*

10. One of the main advantages of using the Mechanical Desktop is that you can rotate the designer model with the help of your pointing device and view it from various directions. This is done with the help of the **3DORBIT** command. Invoke this command by choosing the **3D Orbit** button from the **Mechanical View** toolbar.

When you choose this button, an arcball is displayed. This arcball is divided into four quadrants with the help of four small circles placed at the quadrants of the arcball.

11. Right-click to display the shortcut menu and choose **Shading Modes > Gouraud Shaded**

to display the shaded object as shown in Figure 2-15.

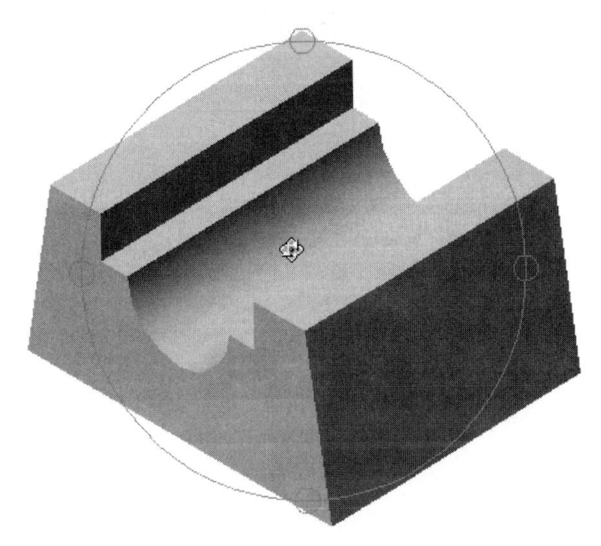

Figure 2-15 *Designer model with Gouraud shaded shading mode displayed along with arcball in 3D Orbit command*

12. Taking the reference of the arcball, rotate the designer model by pressing the pick button of your pointing device and dragging it around on the screen.

13. Right-click to display the context menu. Choose **Exit** to exit this command.

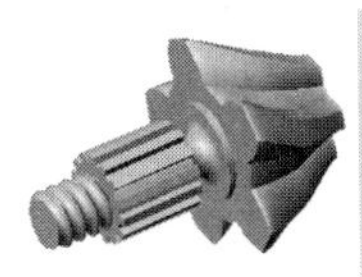

Tip: *Right-click when you are inside the **3D Orbit** command to display the shortcut menu and choose **More > Continuous Orbit**. Now press the pick button of your pointing device, rotate it once and then release the button. You will see that the designer model rotates continuously without your help.*

14. Enter **8** at the Command prompt to return to the Southeast Isometric view.

15. Choose the **Toggle Shading/Wireframe** button from the **Mechanical View** toolbar to display the wireframe model.

16. Save this drawing with the name given below:

\MDT Tut\Ch-2**Tut1.dwg**

Tutorial 2

In this tutorial you will create the object shown in Figure 2-16. The dimensions for the designer model are shown in Figure 2-17. The extrusion height is 100 units.

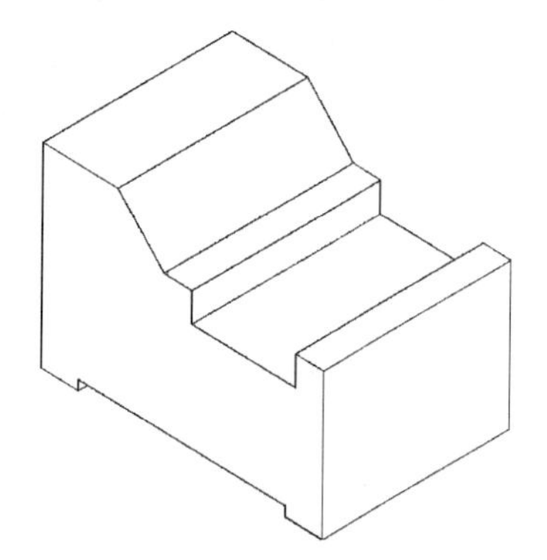

Figure 2-16 Drawing for Tutorial 2

Figure 2-17 Dimensions for Tutorial 2

Step 1: Creating A Rough Sketch

1. Increase the limits, change the UCS and then create a rough sketch in proportion with the actual object.

Step 2: Converting The Rough Sketch Into Cleaned-up Sketch

2. Convert this rough sketch into a cleaned-up sketch with the help of the **AMPROFILE** and **AMADDCON** commands.

Step 3: Getting The Desired Sketch From Cleaned-up Sketch

3. Add the required dimensions to the cleaned-up sketch with the help of the **AMPARDIM** command. The final desired sketch should be similar to the sketch shown in Figure 2-18.

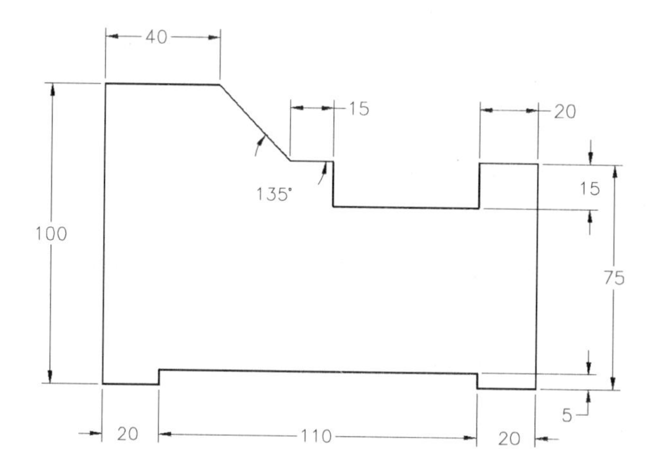

Figure 2-18 Desired sketch for Tutorial 2

Step 4: Converting The Desired Sketch Into Designer Model

4. Enter **88** at the Command prompt.

5. Choose the **Sketched Feature-Extrude** button from the **Part Modeling** toolbar to display the **Extrusion** dialog box. Select the **Blind** option from the **Termination** drop down list.

6. Set the value of the distance in the **Distance** spinner to **100**.

7. In the **Draft Angle** spinner, set the value of draft angle to **0**. Choose **OK**.

8. Enter **FF** at the Command prompt to zoom to the extents of the drawing.

9. Enter **0** at the Command prompt to hide the lines. You will get the model similar to the one shown in Figure 2-19.

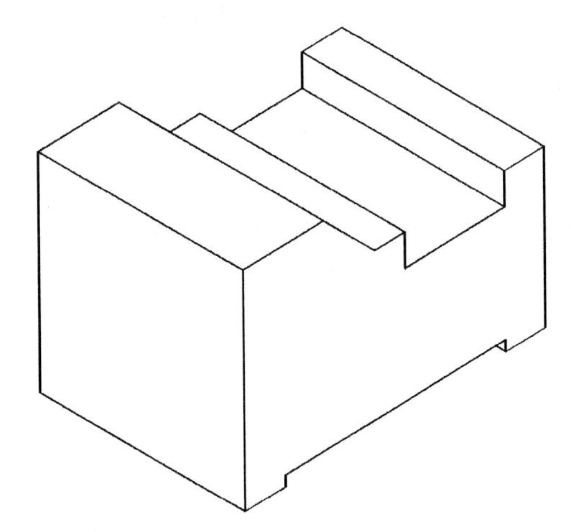

Figure 2-19 *Designer model after hiding the hidden line*

10. Choose the **3D Orbit** button from the **Mechanical View** toolbar.

11. Right-click to display the shortcut menu and choose **Shaded Modes > Gouraud Shading** to shade the model as shown in Figure 2-20.

12. Press the pick button of your pointing device and drag it around on the screen to rotate the designer model.

13. Right-click to display the shortcut menu and choose **Exit**.

14. Enter **8** at the Command prompt.

15. Choose the **Toggle Shading/Wireframe** button from the **Mechanical View** toolbar to toggle the shading to wireframe.

16. Save this drawing with the name given below:

\MDT Tut\Ch-2**Tut2.dwg**

Figure 2-20 Gouraud Shaded model in the 3DORBIT command

Tutorial 3

In this tutorial you will convert the desired sketch created in Tutorial 3 of Chapter 1 into the designer model. The thickness of the model is 0.5 and the extrusion height is 4.

1. Choose the **Open** button from the **Mechanical Main** toolbar to display the **Select File** dialog box. Open the drawing \MDT Tut\Ch-1**Tut3.dwg**, see Figure 2-21.

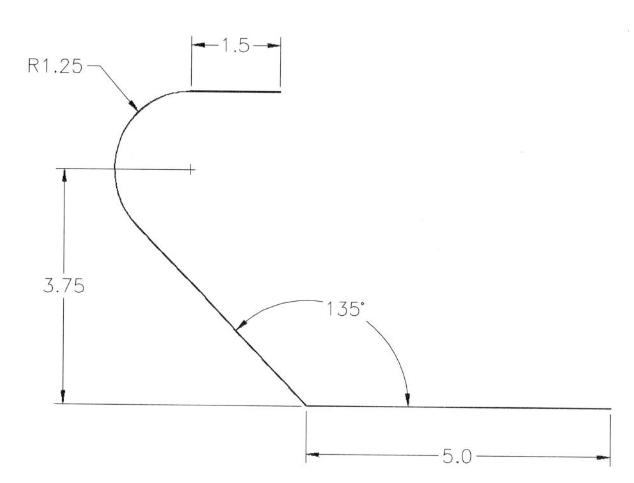

Figure 2-21 *Fully dimensioned sketch for Tutorial 3*

2. As the first three steps for this sketch have been discussed in Chapter 1, therefore, you have to start with Step 4.

Step 4: Converting The Desired Sketch Into The Designer Model

3. Enter **8** at Command prompt.

4. Choose the **Sketched Feature-Extrude** button from the **Part Modeling** toolbar to display the **Extrusion** dialog box for the open profiles, see Figure 2-22.

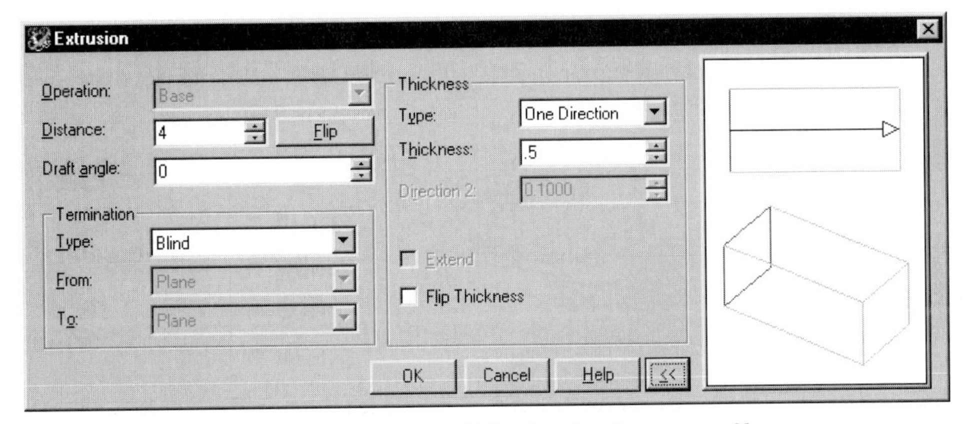

Figure 2-22 *Extrusion dialog box for the open profiles*

5. Set the value of distance to **4** in the **Distance** spinner. In the **Draft Angle** spinner, set the value of draft angle to **0**.

6. Select **One Direction** from the **Type** drop-down list in the **Thickness** area. Set the value of the **Thickness** spinner to **0.5**. Choose **OK**.

7. Enter **FF** at the Command prompt.

8. Enter **0** at the Command prompt. The designer model should look similar to the one shown in Figure 2-23.

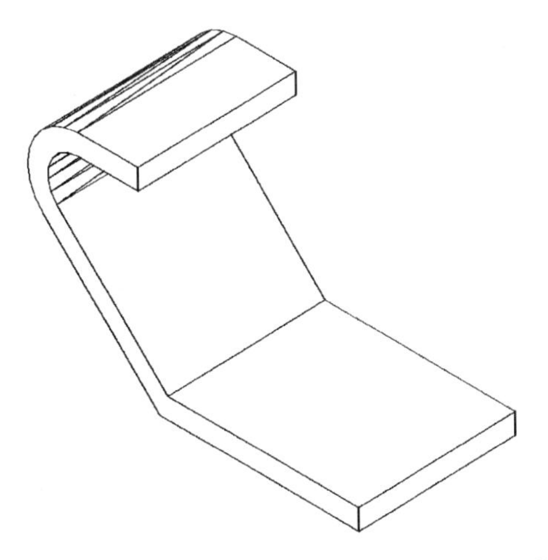

Figure 2-23 *Final designer model for Tutorial 3*

9. Choose the **3D Orbit** button from the **Mechanical View** toolbar.

10. Right-click to display the shortcut menu and choose **Shading Modes > Gouraud Shaded**.

11. Again right-click to display the shortcut menu choose **More > Continuous Orbit**.

12. Press the pick button of the mouse, drag it once on the screen and then release the pick button. By doing this, the designer model will continuously rotate on the screen without your help, see Figure 2-24.

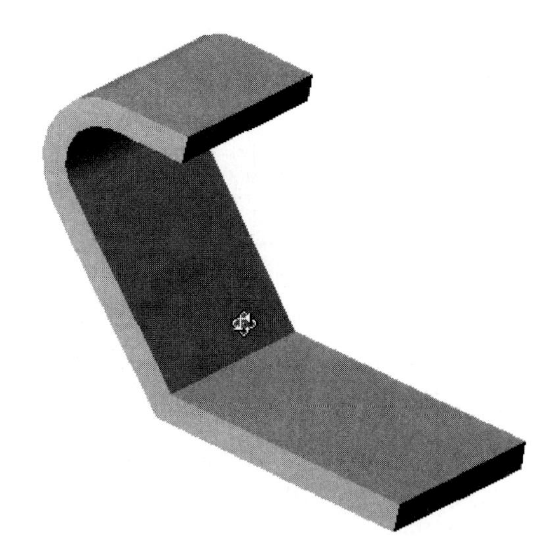

Figure 2-24 *Rotating the model using continuous orbit*

13. Press ESCAPE.

14. Enter **8** at the Command prompt.

15. Choose the **Toggle Shading/Wireframe** button from the **Mechanical View** toolbar to toggle the shading to wireframe model.

16. Save this drawing with the name given below:

    \MDT Tut\Ch-2**Tut3.dwg**

Tutorial 4

In this tutorial you will create a designer model by revolving the sketch you had drawn in Chapter 1 and saved it with the name \MDT Tut\Ch-1**Tut4.dwg**. You will start with step 4 as the first three steps have already been discussed while drawing the sketch in **Tut4.dwg**.

1. Choose the **Open** button from the **Standard** toolbar and open the drawing \MDT Tut\Ch-1**Tut4.dwg**. The sketch should be similar to the one in Figure 2-25.

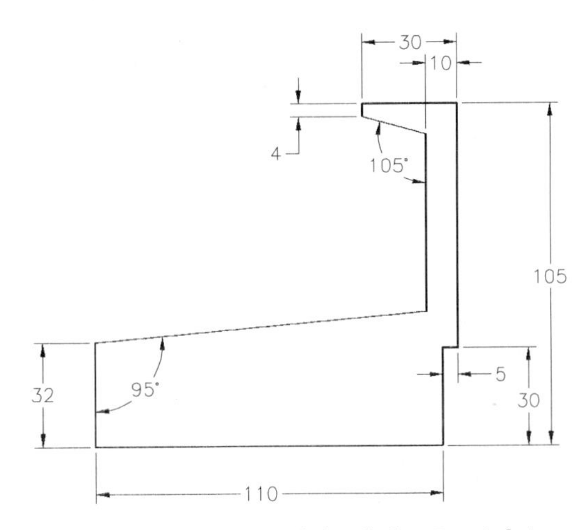

Figure 2-25 *Desired sketch for Tutorial 4*

Step 4: Converting The Desired Sketch Into The Designer Model

2. Enter **8** at the Command prompt.

3. Choose the **Revolve** button from the **Sketched Feature-Extrude** flyout in the **Part Modeling** toolbar. The prompt sequence is as follows:

Select revolution axis: *Select line 10. (See Tutorial 4 of Chapter 1 for line numbers.)*

The revolution dialog box will be displayed when you select the revolution axis, see Figure 2-26.

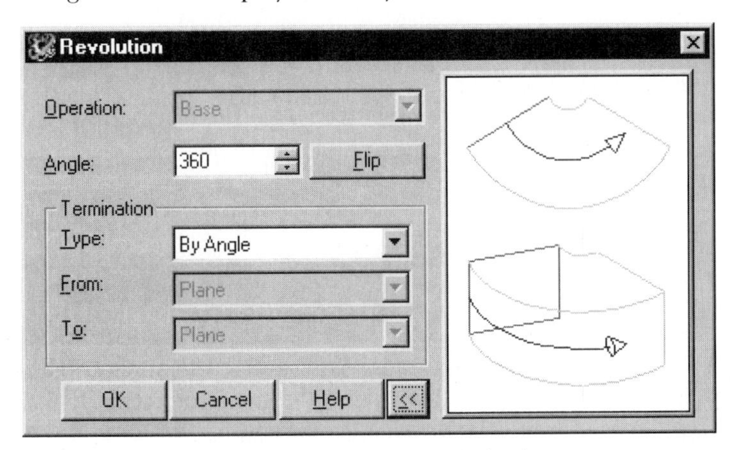

Figure 2-26 *Revolution dialog box*

4. Set the value of the **Angle** spinner to **360**.

5. Select **By Angle** from the **Type** drop-down list in the **Termination** area. Choose **OK**.

6. Enter **FF** at the Command prompt. The designer model should look similar to the one shown in Figure 2-27.

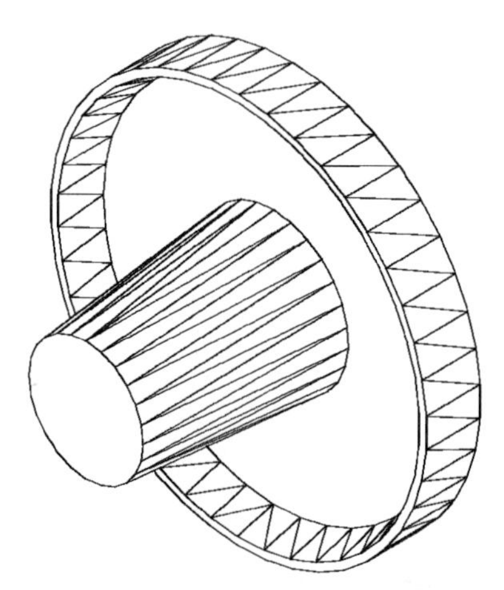

Figure 2-27 *Final designer model for Tutorial 4*

7. Choose the **3D Orbit** button from the **Mechanical View** toolbar to dynamically rotate and view the revolved model.

8. Right-click to display the shortcut menu. Choose **Shading Modes > Gouraud Shading**.

9. Press the pick button of your pointing device and then drag it around on the screen to rotate the designer model.

10. Enter **8** at the Command prompt.

11. Choose the **Toggle Shading/Wireframe** button from the **Mechanical View** toolbar.

12. Save this drawing with the name given below:

    \MDT Tut\Ch-2**Tut4.dwg**

Tutorial 5

In this Tutorial you will open the sketch you had drawn in Chapter 1 and saved with the name \MDT Tut\Ch-1\Exr4.dwg. The extrusion height is 2. The first three steps have been discussed in Chapter 1, so start with step 4.

Step 4: Converting The Desired Sketch Into The Designer Model

1. Choose the **Open** button from the **Mechanical Main** toolbar and open the file with the name given on the next page:

\MDT Tut\Ch-1\Exr4.dwg

The sketch should be similar to the one shown in Figure 2-28.

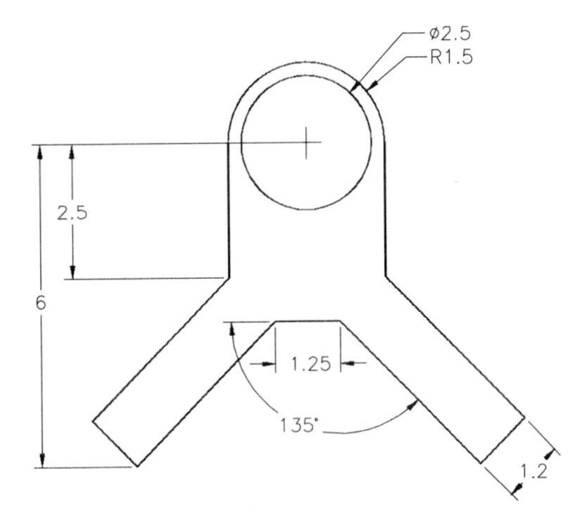

Figure 2-28 *Desired sketch for Tutorial 4*

2. Choose the **Sketched Feature-Extrude** button from the **Part Modeling** toolbar. Select the **Blind** option from the **Termination** drop down list.

3. Set the value of the distance in the **Distance** spinner to **2**.

4. Select **Blind** from the **Type** drop-down list in the **Termination** area. Choose **OK**.

5. Enter **FF** at the Command prompt. The final model should be similar to the one shown in Figure 2-29.

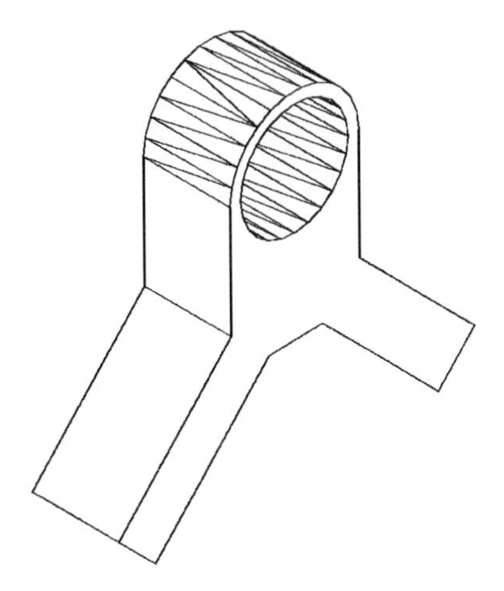

Figure 2-29 *Final designer model for Tutorial 4*

6. Now you can dynamically rotate and view the model using the **3D Orbit** button in the **Desktop View** toolbar.

7. Save this drawing with the name given below:

   \MDT Tut\Ch-2**Tut5.dwg**.

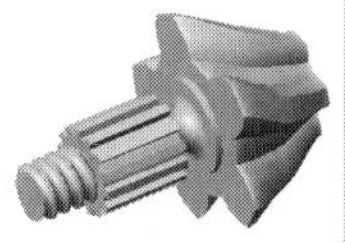

Tip: *While profiling the sketch if you select two closed loops, one inside the other, both the closed loops will be considered as a single profile. Therefore, the inner loop will be subtracted from the outer loop during extrusion or revolution.*

Review Questions

Answer the following questions.

1. Which command is used to modify the dimensions of the desired sketch?

2. What are the options provided in the **Type** drop-down list in the **Termination** area of the **Extrusion** dialog box?

3. What are the options provided in the **Type** drop-down list in the **Termination** area of the **Revolution** dialog box?

4. If you want to extrude the sketch equally in both the directions, which option will you use?

5. Which command is used to rotate the designer model and view it from various directions?

6. When you select an open profile for extrusion, the options related to the __________ are displayed along with the options related to the extrusion of the open profile.

7. When you enter 88 at the Command prompt, which view of the model is displayed?

8. Which view of the model is displayed when you enter 7 at the Command prompt?

Exercises

Exercise 1

Create the designer model shown in Figure 2-30. The dimensions for the designer models are given in Figure 2-31. The extrusion thickness is 30 units. Save the drawing after creating with the names given below:

\MDT Tut\Ch-2**Exr1.dwg**

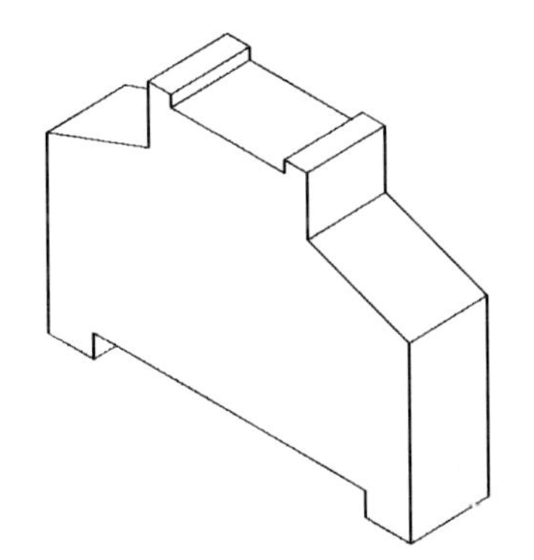

Figure 2-30 Drawing for Exercise 1

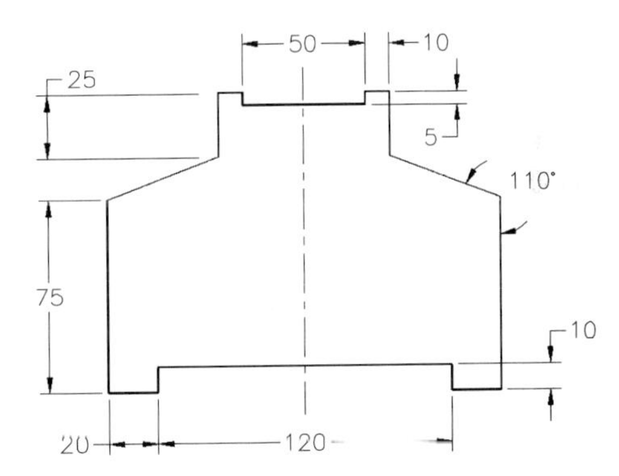

Figure 2-31 Dimensions for Exercise 1

Exercise 2

Create the designer model shown in Figure 2-32. The dimensions for the designer models are given in Figure 2-33. After drawing the sketch revolve it through an angle of 270°. Save the drawing after creating with the names given below:

\MDT Tut\Ch-2**Exr2.dwg**

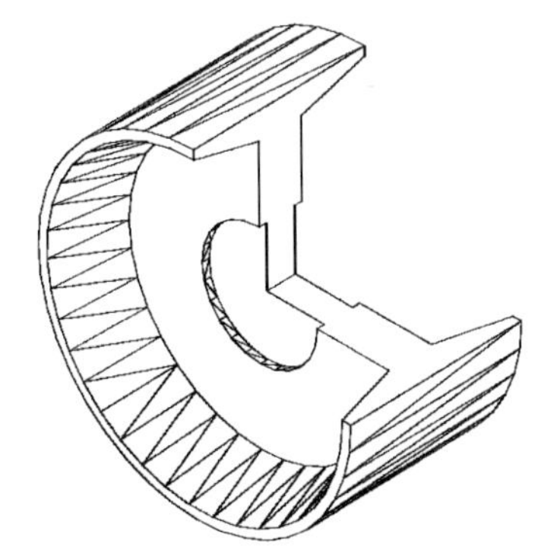

Figure 2-32 Drawing for Exercise 2

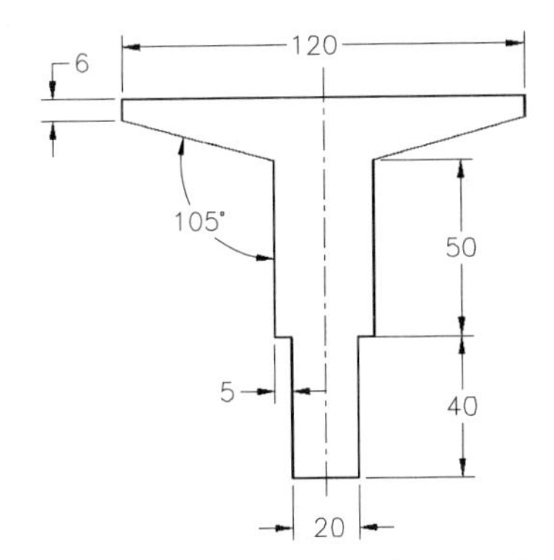

Figure 2-33 Dimensions for Exercise 2

Exercise 3

Create the designer model shown in Figure 2-34. The dimensions for the designer models are given in Figure 2-35. The extrusion height of the model is 60 units. Save the drawing after creating with the names given below:

\MDT Tut\Ch-2**Exr3.dwg**

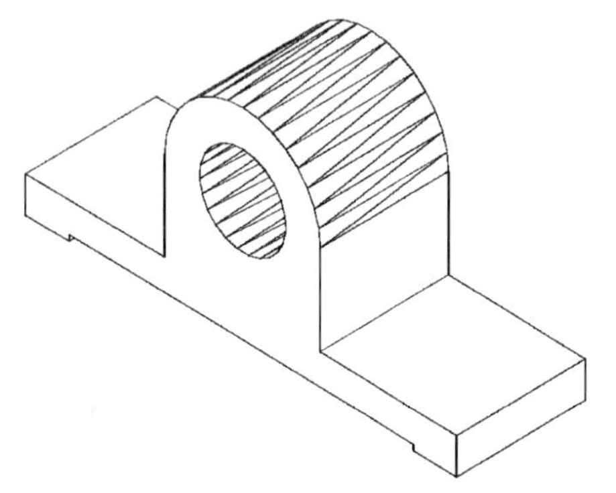

Figure 2-34 Drawing for Exercise 3

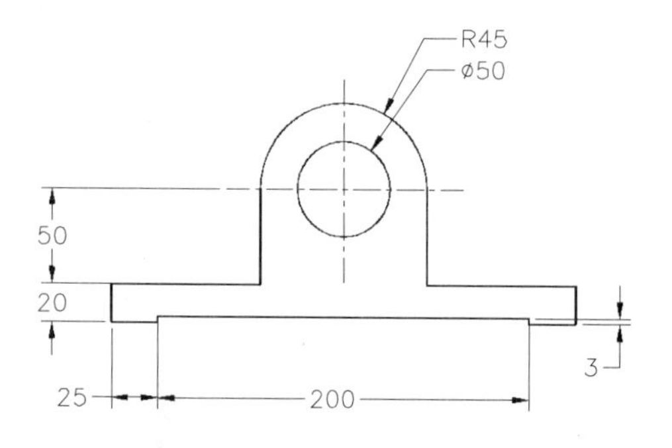

Figure 2-35 Dimensions for Exercise 3

Chapter 3

Sketch Planes, Work Features, and Other Extrusion and Revolution Options

Learning Objectives

After completing this chapter, you will be able to:

- *Define the location and orientation of the sketch plane using the* ***AMSKPLN*** *command.*
- *Place the work features like work point, work axis, and work plane.*
- *Place all three basic planes using the* ***AMBASICPLANES command.***
- *Use the rest of the options of the* ***Extrusion*** *and* ***Revolution*** *dialog boxes.*

Commands Covered

- ***AMSKPLN***
- ***AMWORKPOINT***
- ***AMWORKAXIS***
- ***AMWORKPLN***
- ***AMBASICPLANES***
- ***AMEXTRUDE***
- ***AMREVOLVE***

CREATING THE SKETCH PLANES

In the previous chapters you have created the basic designer models by extruding or revolving the sketches. All of these models were created in a single plane; the World XY plane. But a complex designer model consist of the features lying in different planes, see Figure 3-1. The designer model shown in Figure 3-1 consists of four holes in the front face and a hollow cylindrical feature on the top face. To define these features you first need to define the sketch planes separately on these faces. The sketch planes are the active infinite planes used to define the next feature. The location and orientation of the new sketch plane can be defined depending upon your requirement. You can define the sketch plane using the **AMSKPLN** command discussed below.

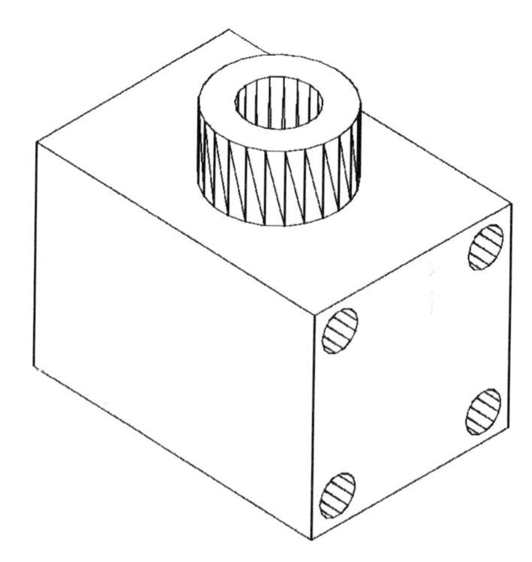

Figure 3-1 *Designer model showing features on different faces*

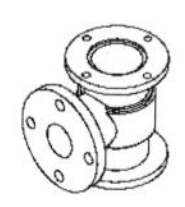

Note

The new drawing planes cannot be defined by simply changing the UCS as it is done in AutoCAD. The reason for this is that Mechanical Desktop follows the sketch planes and profile the sketches drawn only on the current sketch planes.

AMSKPLN Command*

Toolbar:	Part Modeling > New Sketch Plane
Menu:	Part > New Sketch Plane
Context Menu:	New Sketch Plane
Command:	AMSKPLN

Figure 3-2 *New Sketch Plane button in the Part Modeling toolbar*

The sketch planes can only be defined on the designer models created using the Mechanical Desktop commands, see Figure 3-2. The designer model must also contain a planar face or a plane on which the sketch plane will be defined. However, the planar face may or may not have straight edges. For example, you can define the sketch plane on either the top or the bottom face of the cylinder. When you select any face or a plane to define the new sketch plane, a temporary sketch plane appears on that plane showing the orientation of the new sketch plane. You can define the orientation of the X and Y axes and the direction of the Z axis of the new sketch plane depending upon your requirements. It has to be kept in mind that at a time you can have only one active sketch plane. The options provided in this command are:

Select work plane

The **Select work plane** option is used when you have defined more than one work plane and want to make any one of them active.

Planar face

The **planar face** option is used to define a sketch plane on a planar face selected in the designer model. As soon as the cursor is moved close to the planar face, the face is highlighted for the ease of selecting as shown in Figure 3-3.

Tip: *You can cycle through to the next face using the* **Next** *option. This can also be done by pressing the pick button of your mouse.*

World XY

The **World XY** option is used when you want to define a sketch plane on the World XY plane. You can define any orientation of the X and Y axes using the **Rotate** option.

World YZ

The **World YZ** option is used to define a sketch plane on the World YZ plane.

World ZX

The **World ZX** option is used to define a sketch plane on the World ZX plane.

UCS

The **UCS** option is used to define a sketch plane on the UCS that is defined using the **UCS** command. Figure 3-4 defines the sketch plane tangent to a cylindrical face.

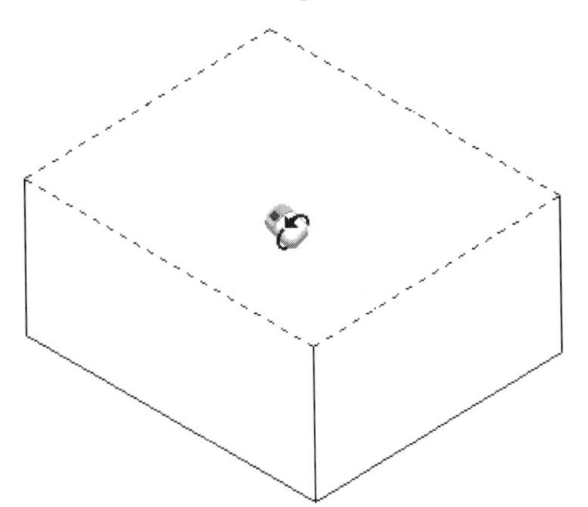

Figure 3-3 *Selecting the top face of the designer model to define the sketch plane*

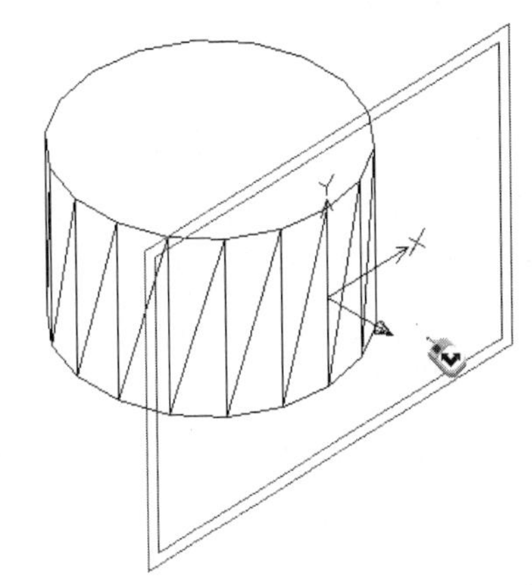

Figure 3-4 *Defining the sketch plane tangent to a cylinder using the UCS option*

Tip: *If you want to define a sketch plane tangent to the cylinder, you first have to define a UCS tangent to the cylinder and then use the* ***Ucs*** *option of the* ***AMSKPLN*** *command to place the sketch plane.*

Flip

The **Flip** option is used to reverse the direction of the Z axis of the new sketch plane.

Rotate

The **Rotate** option is used to rotate the X and Y axes of the new sketch plane you are defining.

Origin

The **Origin** option is used to define the origin point of the new sketch plane.

WORK FEATURES

The work features are the special features associated parametrically with the designer part that was active at the time of their creation. These work features are used to place the geometric features on the active part. There are three types of work features. They are:

1. Work Point
2. Work Axis
3. Work Plane

All these work features can be placed using the following commands:

AMWORKPT Command

Toolbar:	Part Modeling > Work Point
Menu:	Part > Work Features > Work Point
Context Menu:	Sketched & Worked Features > Work Point
Command:	AMWORKPT

This command can be invoked by choosing the **Work Point** button from the **Work Features - Work Plane** flyout in the **Part Modeling** toolbar, see Figure 3-5. This command is used to place a parametric point any where on the active sketch plane. The work point is used to place holes, polar array, work plane and so on. When you invoke this command, you will be asked to specify the location of the work point. You can specify the location of the work point using your pointing device, see Figure 3-6.

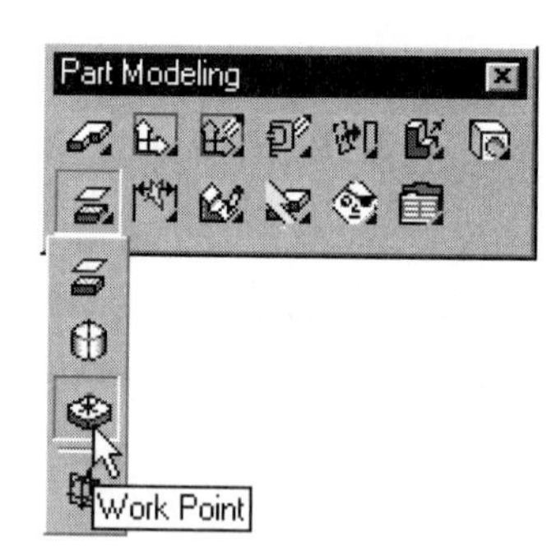

Figure 3-5 *Invoking the AMWORKPT command from the Part Modeling toolbar*

Tip: *The work points are the only non-cylindrical features that can be constrained with the help of the* ***Concentric*** *constraint.*

AMWORKAXIS Command

Toolbar: Part Modeling > Work Axis
Menu: Part > Work Features > Work Axis
Context Menu: Sketched & Worked Features > Work Axis
Command: AMWORKAXIS

The work axis is a parametric axis passing through the model. To invoke this command, choose the **Work Axis** button from the **Work Features - Work Plane** flyout in the **Part Modeling** toolbar. The work axis passing through the center of a circular part can be defined just by selecting the circular edge of that part. You can also create the work axis by sketching its location in the current sketch plane. The work axis is used as reference for dimensioning the sketches, to revolve the sketches, to place the work planes passing through the center of a cylindrical part, to place polar arrays and so on, see Figure 3-7.

Tip: *The work axis created using the **Sketch** option is placed on the current sketch plane. However, when you select a circular part, the work axis is placed passing through its center even if the current sketch plane does not lie at the center of the circular part.*

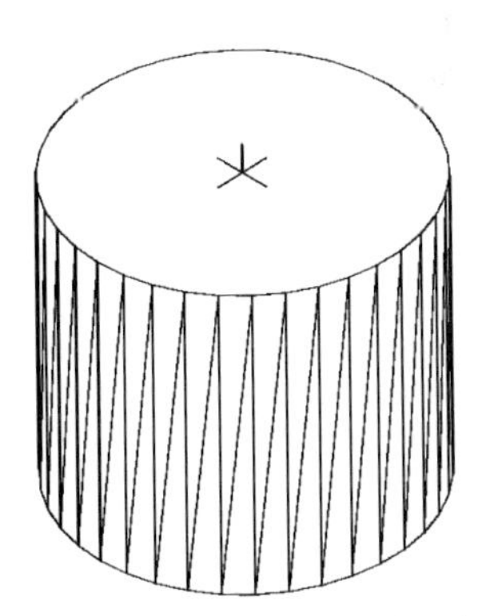

Figure 3-6 *Figure showing a work point placed at the center of the top face of the cylinder*

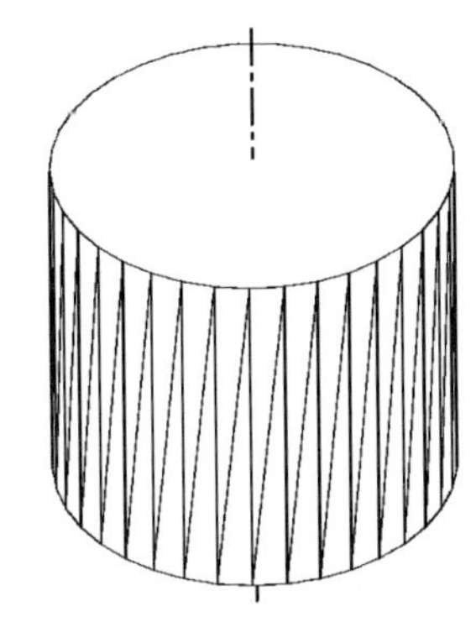

Figure 3-7 *Figure showing a work axis passing through the center of cylinder*

AMWORKPLN Command

Toolbar: Part Modeling > Work Feature-Work Plane
Menu: Part > Work Features > Work Plane
Context Menu: Sketched & Worked Features > Work Plane
Command: AMWORKPLN

This command is used to place parametric work planes on the designer models. The work planes are similar to the sketch planes and are defined when you want to draw the features on different planes. However, the **AMWORKPLN** command is more versatile than the **AMSKPLN** command as it provides more options for defining the work planes compared to the **AMSKPLN** command. Using the **AMWORKPLN** command you can define the work planes even on the models that do not have a planar face. Unlike the sketch planes, the work planes are visible on the screen. When you invoke this command, the **Work Plane** dialog box (Figure 3-8) is displayed. This dialog box helps you in defining the work

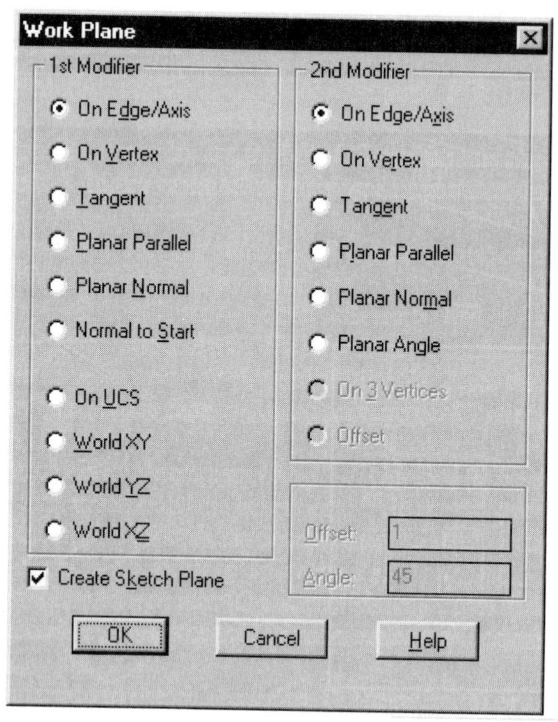

Figure 3-8 *Work Plane dialog box displayed upon invoking the AMWORKPLN command*

planes using two modifiers; the first modifier and the second modifier. When you select the first modifier for defining the work plane, the options of the second modifier that can combine with the first modifier remain activated and the rest are not available. Defining the work planes using different options are discussed below.

On Edge/Axis - On Edge/Axis

This option helps you to create a work plane by selecting two edges, two axes, or one axis and one edge, Figures 3-9a and 3-9b. The axis that has to be selected can be a work axis.

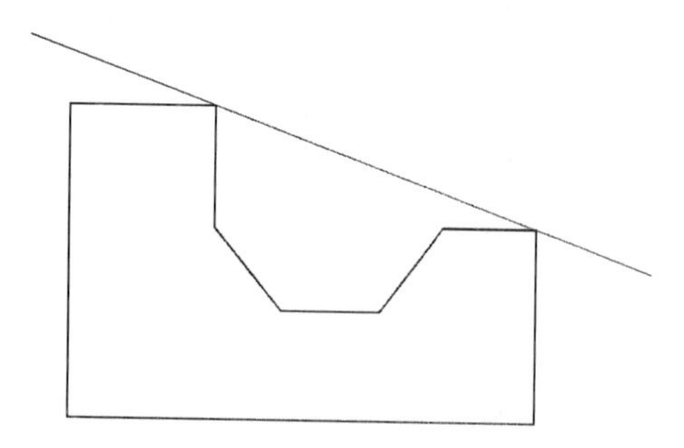

Figure 3-9a *Front view showing the location of the work plane*

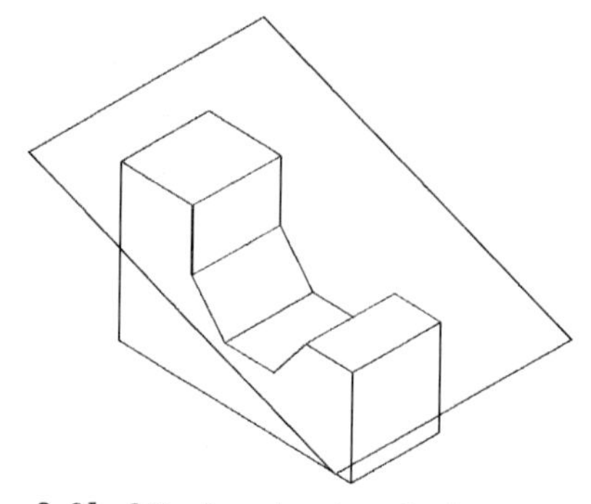

Figure 3-9b *3D view showing the location of the work plane*

Tip: *You can select the first and the second modifier for defining the work plane in any sequence. For example, whether you select the first and second modifiers in the sequence of On Edge/Axis - On Vertex or On Vertex - On Edge/Axis, the creation of the work plane remains the same.*

On Edge/Axis - On Vertex or On Vertex - On Edge/Axis

The **On Edge/Axis - On Vertex** or the **One Vertex - On Edge/Axis** options helps you to create a work plane by selecting one edge or axis and one vertex, see Figures 3-10a and 3-10b. The vertex can be an end point of an edge.

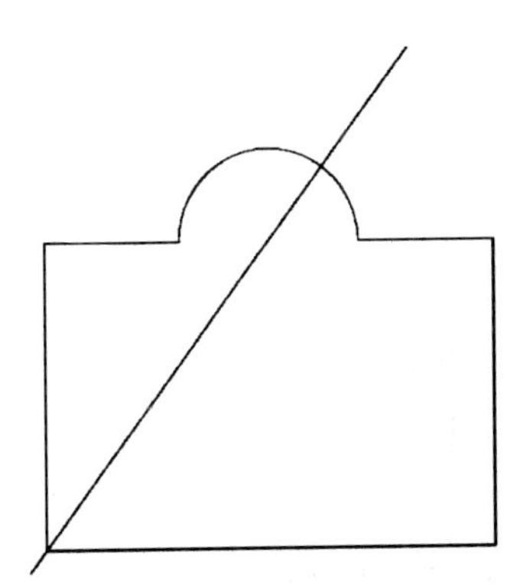

Figure 3-10a *Top view showing the location of the work plane*

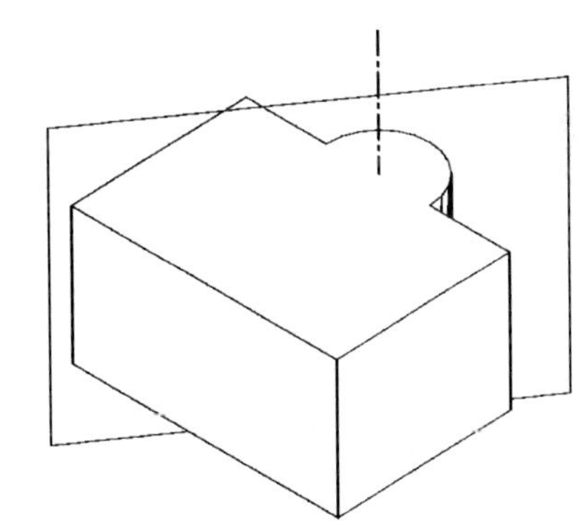

Figure 3-10b *3D view showing the location of the work plane*

On Edge/Axis - Tangent or Tangent - On Edge/Axis

Using the **On Edge/Axis - Tangent** or the **Tangent - On Edge/Axis** options you can create a work plane by selecting an edge or an axis and a cylindrical face, Figures 3-11a and 3-11b. The work plane thus created will be tangent to the cylindrical face and will pass through the edge or the axis you selected.

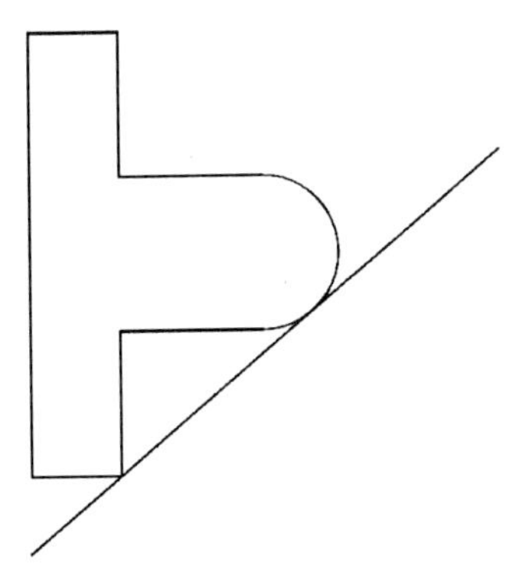

Figure 3-11a *Top view showing the location of the work plane*

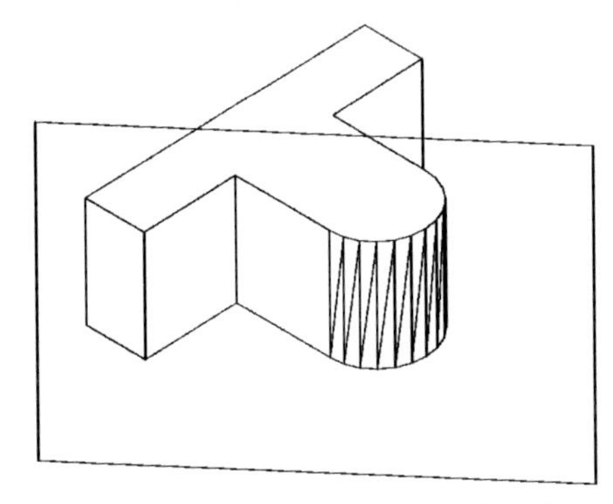

Figure 3-11b *3D view showing the location of the work plane*

On Edge/Axis - Planar Parallel or Planar Parallel - On Edge/Axis

These options allows you to create a work plane parallel to a plane and passing through an edge or an axis, Figures 3-12a and 3-12b. In this figure a work plane is created which passes through the front edge of the model and is parallel to the back face of the model.

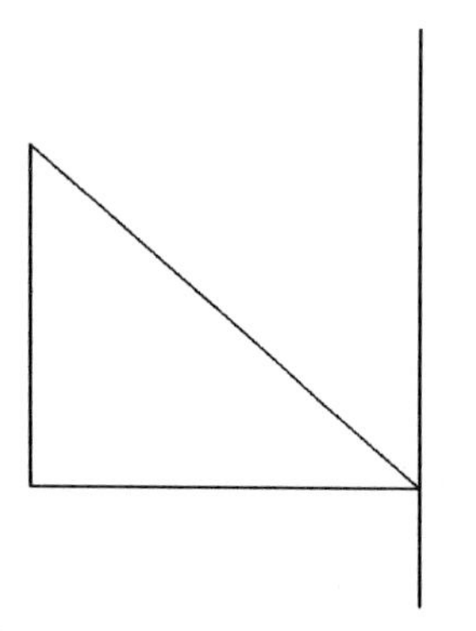

Figure 3-12a Front view showing the location of the work plane

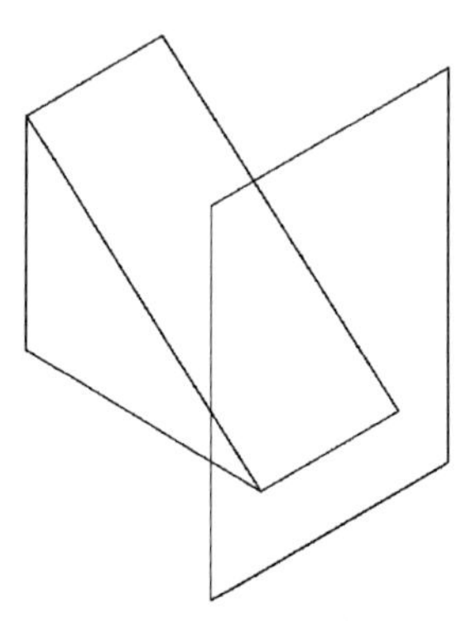

Figure 3-12b 3D view showing the location of the work plane

On Edge/Axis - Planar Normal or Planar normal - On Edge/Axis

These options allows you to create a work plane which is normal to a selected plane and passing through an edge or an axis, Figures 3-13a and 3-13b. This figure shows a work plane created normal to the bottom face of the model and passing through the front edge.

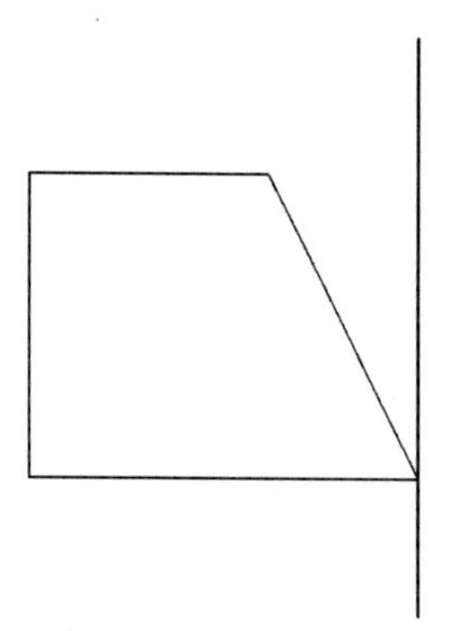

Figure 3-13a Front view showing the location of the work plane

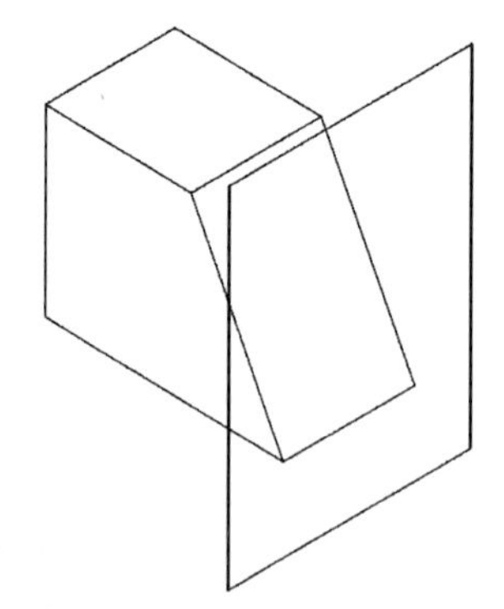

Figure 3-13b 3D view showing the location of the work plane

On Edge/Axis - Planar Angle or Planar Angle - On Edge/Axis

The **On Edge/Axis - Planar Angle** or the **Planar Angle - On Edge/Axis** options allows you to create a work plane passing through an axis or an edge and placed at a certain angle to a selected plane, see Figures 3-14a and 3-14b. The figure shows a work plane passing through an edge and placed at an angle of 45 degrees to the top face.

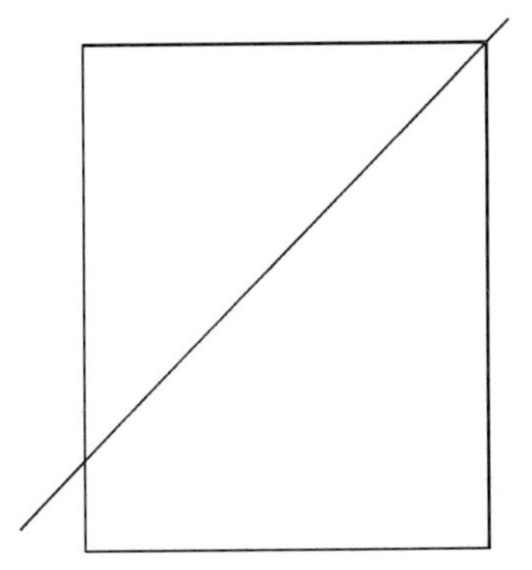

Figure 3-14a Front view showing the location of the work plane

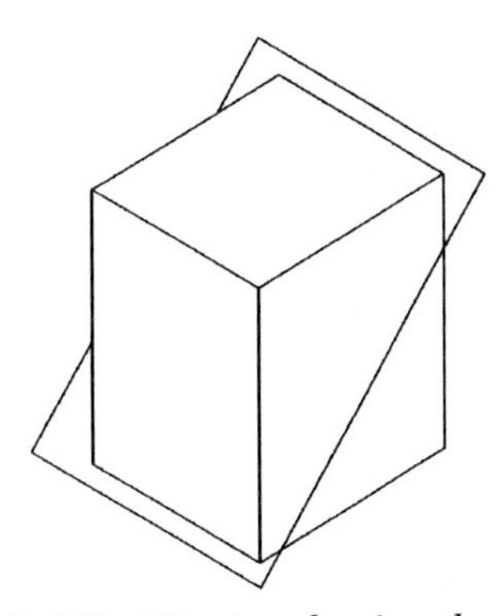

Figure 3-14b 3D view showing the location of the work plane

On Vertex - Planar Parallel or Planar Parallel - On Vertex

Using the **On Vertex - Planar Parallel** or the **Planar Parallel - On Vertex** options you can create a work plane parallel to a selected plane and passing through a selected vertex, see Figures 3-15a and 3-15b.

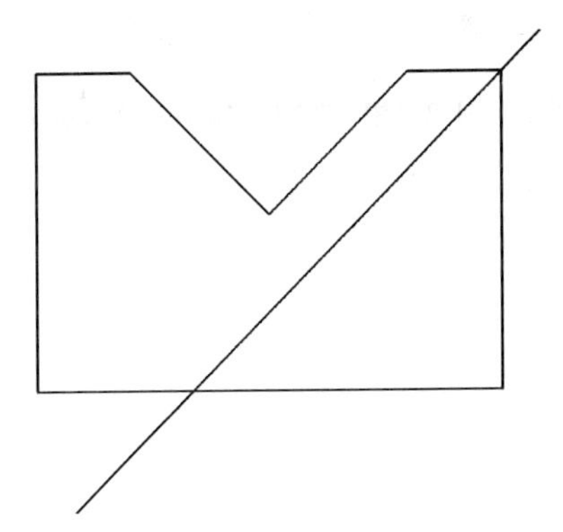

Figure 3-15a Front view showing the location of the work plane

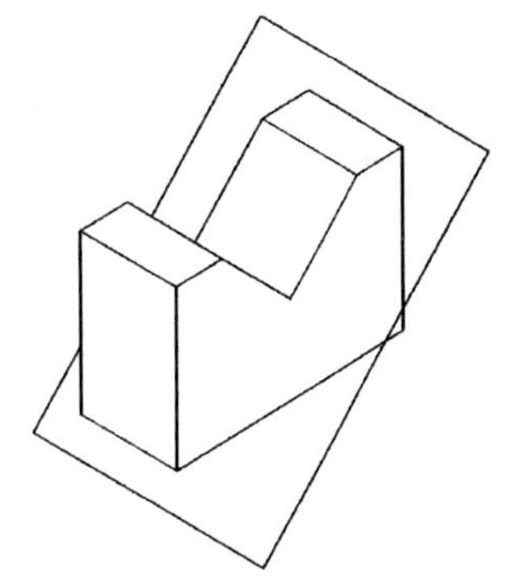

Figure 3-15b 3D view showing the location of the work plane

On Vertex - On 3 Vertices

The **On Vertex - On 3 Vertices** option allows you to create a work plane passing through three vertices, see Figures 3-16a and 3-16b.

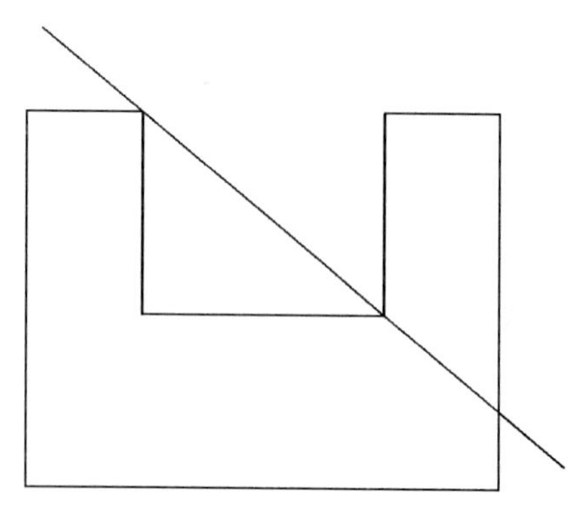

Figure 3-16a *Front view showing the location of the work plane*

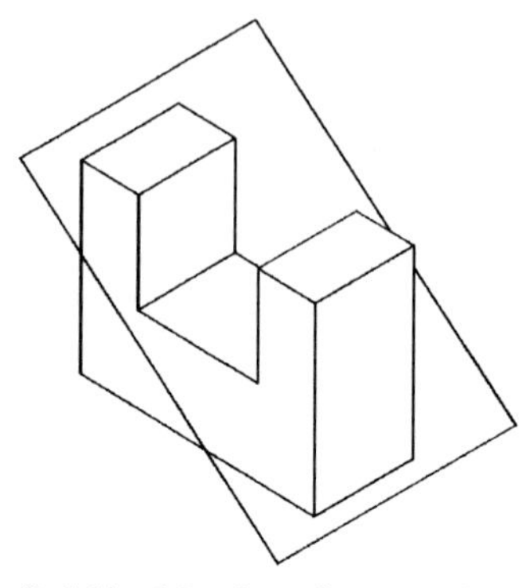

Figure 3-16b *3D view showing the location of the work plane*

Tangent - Tangent

Using the **Tangent - Tangent** option you can create a work plane tangent to two cylindrical faces, see Figures 3-17a and 3-17b.

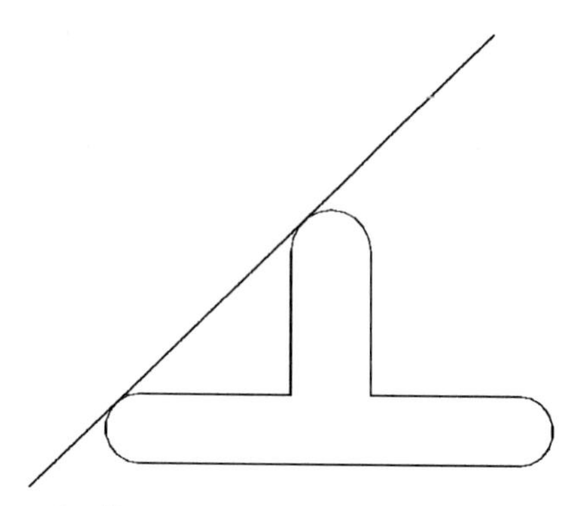

Figure 3-17a *Front view showing the location of the work plane*

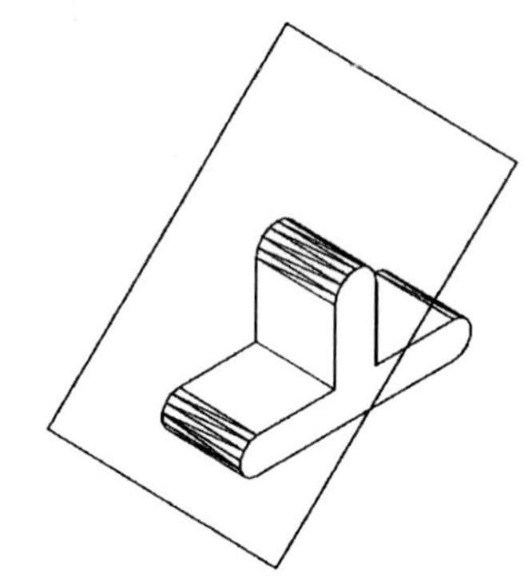

Figure 3-17b *3D view showing the location of the work plane*

Tangent - Planar Parallel or Planar Parallel - Tangent

The **Tangent - Planar Parallel** or the Planar Parallel - Tangent options allow you to create a work plane parallel to a selected plane and tangent to a cylindrical face, see Figures 3-18a and 3-18b.

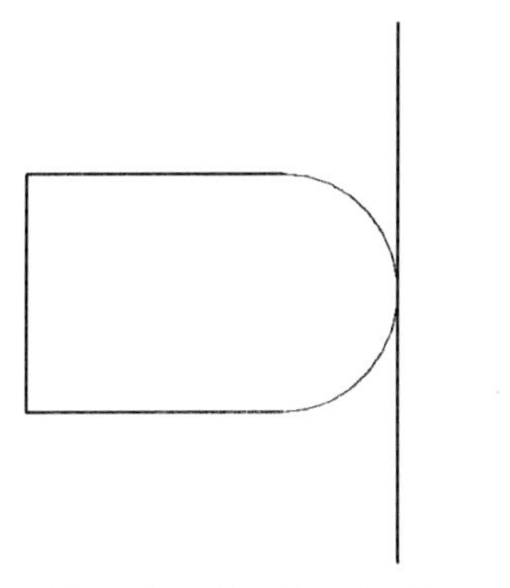

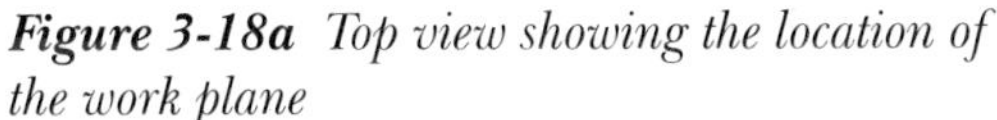

Figure 3-18a *Top view showing the location of the work plane*

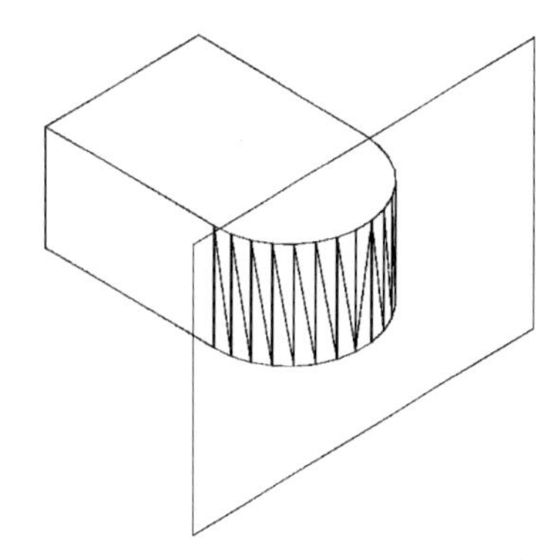

Figure 3-18b *3D view showing the location of the work plane*

Tangent - Planar Normal or Planar Normal - Tangent

The **Tangent - Planar Normal** or the **Planar Normal - Tangent** options allow you to create a work plane tangent to a cylindrical face and normal to another face, see Figures 3-19a and 3-19b. The figure shows a work plane created tangent to a cylindrical face and normal to the side face of the model.

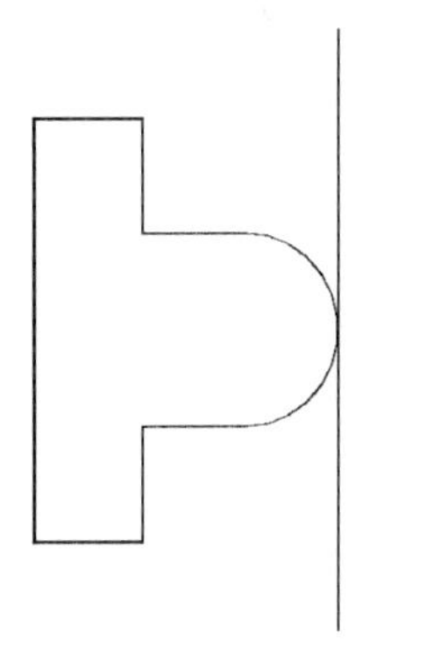

Figure 3-19a *Top view showing the location of the work plane*

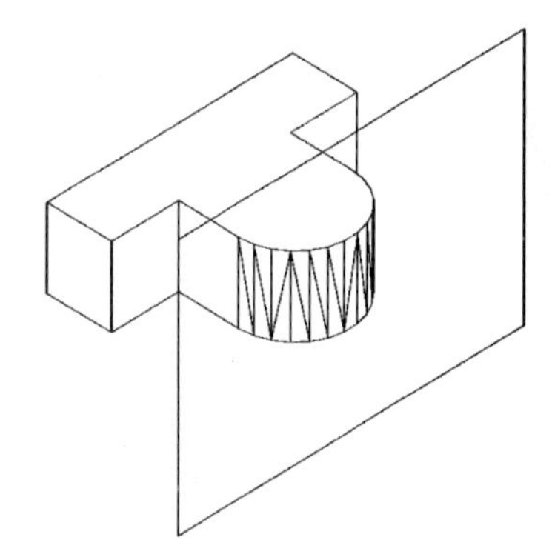

Figure 3-19b *3D view showing the location of the work plane*

Planar Parallel - Offset

Using the **Planar Parallel - Offset** option you can create a work plane parallel to a plane and at a certain distance from that plane, see Figures 3-20a and 3-20b. In this figure a work plane is created that is parallel to the front face of the model and is at a distance of 1 unit from the plane.

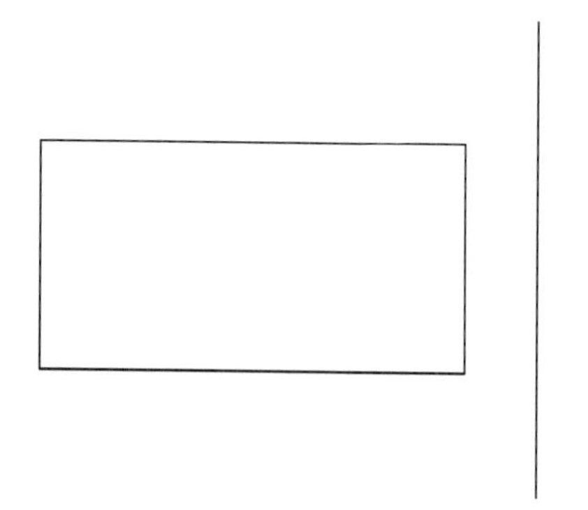

Figure 3-20a Top view showing the location of the work plane

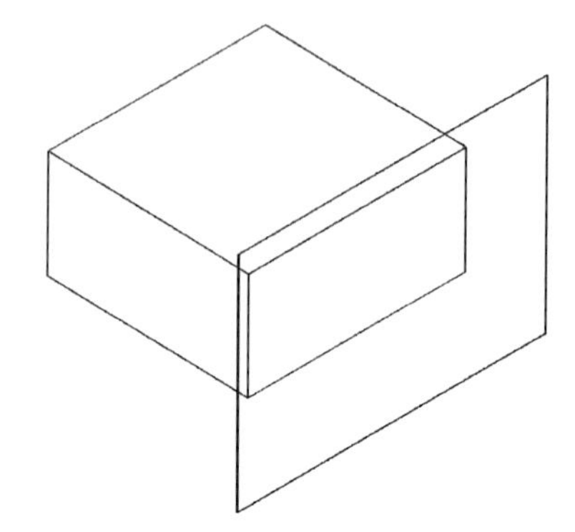

Figure 3-20b 3D view showing the location of the work plane

The **Normal to Start** option will be discussed in the later chapters.

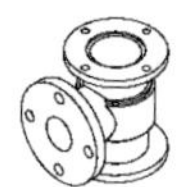

Note

*The **On UCS**, **World XY**, **World YZ**, and the **World XZ** options are similar to the options discussed in the **AMSKPLN** command.*

*If the **Create Sketch Plane** check box is cleared, the current sketch plane will still be the last sketch plane even if you have created a new work plane.*

Tip: *While creating a feature tangent to the cylinder, a new work plane should be created planar parallel to the work plane that is tangent to cylinder, and at a small distance inside the cylinder. The reason for this is that the cylinder has a circular face and the new feature will not start from inside the cylinder if it is created on the work plane tangent to the cylinder thus leaving small gaps on the sides.*

AMBASICPLANES Command

Toolbar:	Part Modeling > Work Features - Work Planes > Create Basic Work Planes
Menu:	Part > Work Features > Basic 3D Work Planes
Command:	AMBASICPLANES

You can also create all the three basic planes in a single effort using the **AMBASICPLANES** command. The planes will be created on the basis of the plan, front elevation, and the side elevation views of a point (which is a work point) you specify. All three of these planes will be placed parallel to the world XY, YZ, and XZ planes. However, the current sketch plane will be the plane parallel to the world XY plane.

OTHER EXTRUSION AND REVOLUTION OPTIONS

In the previous chapters you have created the base of the designer model simply using the **AMEXTRUDE** or the **AMREVOLVE** commands. While creating the base part the **Operation** drop-down list was not activated. When you create the next feature on a different sketch plane or a work plane, the **Operation** drop-down list of the **Extrusion** and the **Revolution** dialog box is activated providing different options. All of these options, along with the remaining options of the **Type** drop-down list in the **Termination** area, are discussed individually for the **Extrusion** and the **Revolution** dialog boxes below:

Other Extrusion Dialog Box Options*

The other options of the **Extrusion** dialog box are discussed next.

Operation

The other options of this drop-down list that will be available once the base feature is created are:

Cut

Selecting the **Cut** option creates a feature by removing the material by creating a cavity or a slot.

Join

The **Join** option creates a feature by adding the material in the direction you select, thus forming a complex solid. However, the new feature created is a part of the designer model and still remains a single entity.

Intersect

The intersection of two objects result in the solid that is common to both of the objects selected. The **Intersect** option in the **Operation** drop-down list also creates a feature by retaining the solid area common to the existing base feature and the new extrusion created.

Split

The split option is used to split the existing part and use the area common to the existing part and the new feature extruded for creating a new part. You can specify the name of the new part in the **Enter name of the new part <default name>** prompt.

Termination Area

Type

Once you have created the base feature, this drop-down list will provide the following options in addition to the options discussed in chapter 2.

Through. The **Through** option is used to create a new feature by cutting the selected sketch right through the existing feature. The **Through** termination option does not combine with the **Join** operation. If you select the **Through** termination option along with the **Join** operation and choose the **OK** button, Mechanical Desktop displays the **Invalid input** dialog box, Figure 3-21. When the **Through** termination option is selected, the **Distance** edit box is not available because the distance value is not required as it has to cut right through the solid.

Mid-Through. The **Mid-Through** option is a combination of the **MidPlane** and

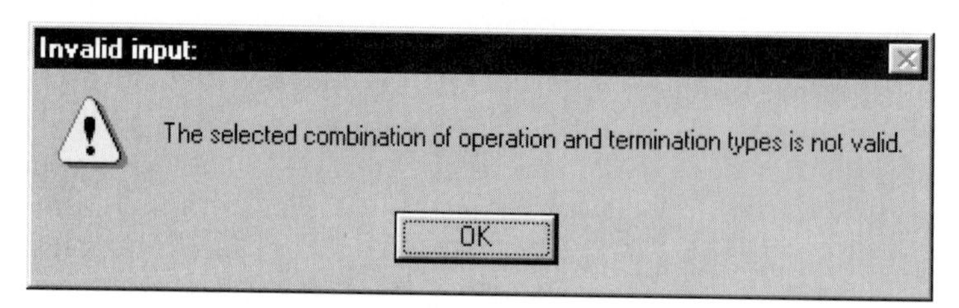

Figure 3-21 Invalid input dialog box

the **Through** options. When selected, this option creates a feature by extruding the selected profile through the existing feature in both the directions taking the current sketch plane as the mid plane. The **Distance** edit box is not available when this option is selected.

Next. The **Next** option is used to extrude the selected profile to the next face that it comes across.

Face. The **Face** option is used when the termination of the profile is defined using a face of the existing feature. You will be prompted to select the termination face when you select this option.

Extended Face. The **Extended Face** option is used to specify the termination of the selected profile using a face that may or may not actually intersect the selected profile. It is generally used when the profile is drawn somewhere above or below the face to be used for termination.

From-To. The **From-To** option is used to define the start and termination of a feature using two faces, planes, extended faces or a combination of any of these. When you select the **From-To** option for defining the termination of the new feature, the **From** and **To** drop-down lists are activated and will provide you with the options of defining the start and termination plane using faces, planes or extended faces.

Other Revolution Dialog Box Options*

Operation

The options that will be provided under this drop-down list after the base feature is created are discussed next.

Cut

The **Cut** option when selected creates a feature by removing the material thus creating a revolved cavity.

Join

The **Join** option is used to create a feature by joining a revolved feature to an existing feature.

Intersect

The **Intersect** option creates a feature by retaining the area common to the existing part and the new revolved feature.

Split

The **Split** option creates a new part from the area common to the existing part and the new feature revolved. You can specify the name of the new part in the **Enter name of the new part <default name>** prompt.

Termination Area

Type

Once you have created the base feature, this drop-down list will provide the following options in addition to the options discussed in chapter 2.

Next. The **Next** option is used to revolve the selected profile to the next face that it comes across.

Face. The **Face** option is used to define the termination of the new revolved feature using a planar face. You will be prompted to select the planar face when you select this option.

Extended Face. The **Extended Face** option is used to specify the termination of the revolution using a face that may or may not actually intersect the selected profile. It is generally used when the profile is drawn somewhere above or below the face to be used for termination.

From-To. When you select the **From-To** option for defining the termination of the new feature, the **From** and **To** drop-down lists are activated and will provide the following options in addition to the option discussed in chapter 2.

Face. The **Face** option is used to specify the termination using a face of the existing feature.

Extended Face. The **Extended Face** option is used to specify the termination using a face that may or may not actually intersect the selected profile

TUTORIALS

Tutorial 1

In this tutorial you will create the designer model shown in Figure 3-22a. The dimensions to be used are given in Figures 3-22b, 3-22c and 3-22d. Assume the missing dimensions.

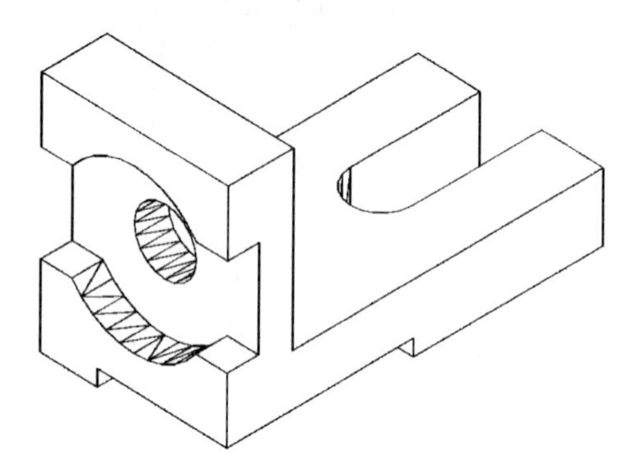

Figure 3-22a Designer model for Tutorial 1

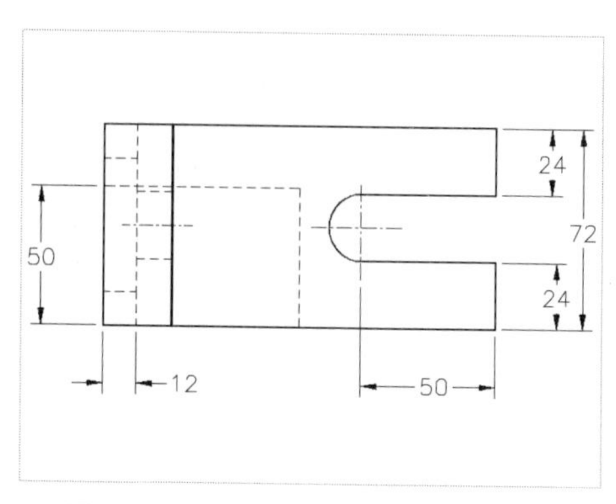

Figure 3-22b Top view of the model

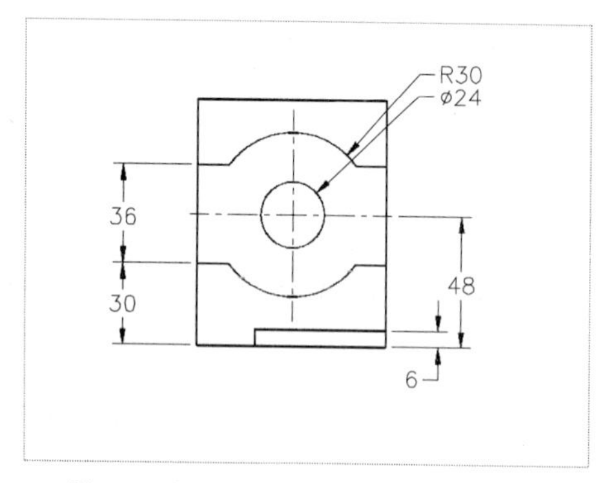

Figure 3-22c Side view of the model

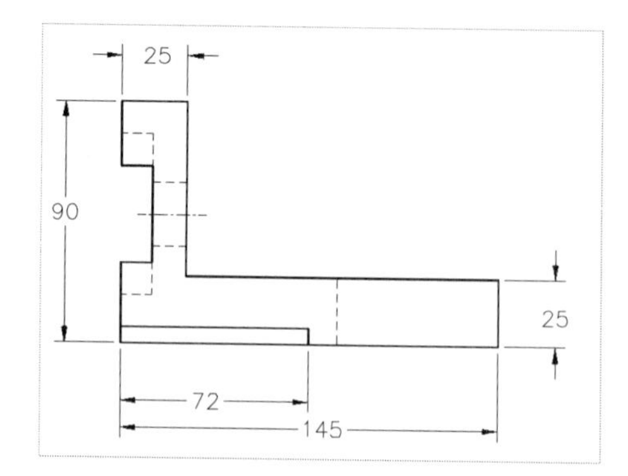

Figure 3-22d Front view of the model

1. Rotate the UCS through an angle of 90 degrees about the X axis, increase the limits, and then create the rough sketch for the base feature as shown in Figure 3-23.

Figure 3-23 Rough sketch for Tutorial 1

2. Convert this rough sketch into an open profile using the required commands.

3. Add the required dimensions to the cleaned-up sketch and convert it into the desired sketch, see Figure 3-24.

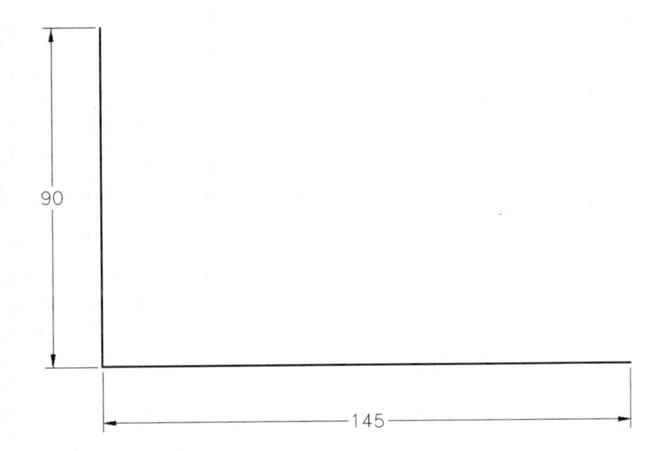

***Figure 3-24** Desired sketch for the base part*

4. Enter **8** at the Command prompt.

5. Choose the **Sketched Features - Extrude** button from the **Part Modeling** toolbar. Set the value of the **Distance** spinner to **90**.

6. Select the **Blind** option from the **Type** drop-down list in the **Termination** area.

7. Select **One Direction** from the **Type** drop-down list in the **Thickness** area. Set the value of the **Thickness** spinner to **25**.

8. Make sure the thickness is applied in such a way that the dimensioned edges are the outer edges of the model. If not, select the **Flip Thickness** radio button. Choose **OK**.

9. Enter **FF** at the Command prompt. The model should be similar to the one shown in Figure 3-25.

10. Now, you have to create the cut feature on the front face of the model. But before drawing the sketch for this feature, you will have to define a new sketch plane on the front face of the model. Choose the **New Sketch Plane** button from the **Part Modeling** toolbar. The prompt sequence is as follows:

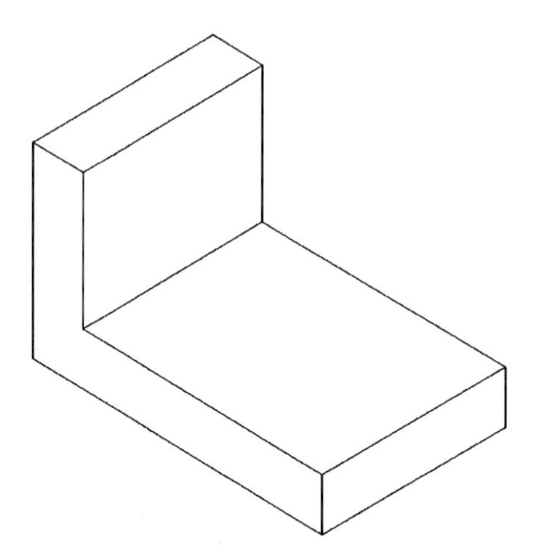

Figure 3-25 *Designer model after extruding the base sketch*

Select work plane, planar face or [worldXy/worldYz/worldZx/Ucs]: *Select the front face of the model as shown in Figure 3-26.*
Enter an option [Next/Accept] <Accept>: *Press ENTER.*
Plane=Parametric
Select edge to align X axis or [Flip/Rotate/Origin] <Accept>: *Align the direction of the X axis using the bottom horizontal edge of the front face.*

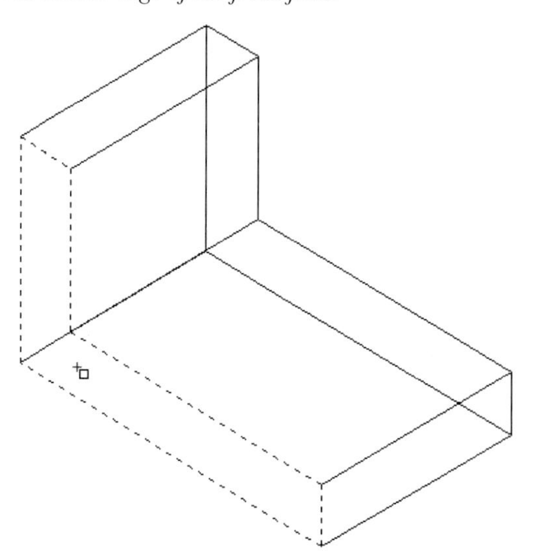

Figure 3-26 *Selecting the front face for defining the sketch plane*

Tip: *If the direction of the X axis is not where you want it to be, continue entering **R** at the **Select edge to align X axis or [Z-flip/Rotate] <Accept>** prompt. The X axis direction will keep on rotating as you enter **R** at the command prompt. Enter **A** when the desired direction is displayed. This can also be done by pressing the pick button of the mouse*

11. Draw the rough sketch for the cut feature. The rough sketch should be a single line segment.

12. Convert this rough sketch into an open profile and add the dimensions, see Figure 3-27.

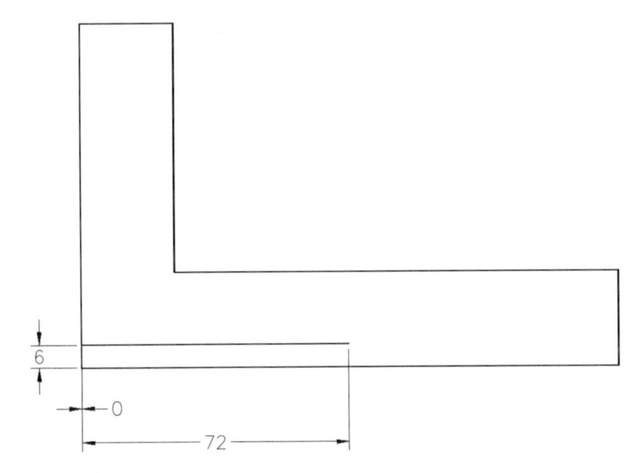

Figure 3-27 *Figure showing the open profile to be used for the cut*

13. Choose the **Sketched Features - Extrude** button from the **Part Modeling** toolbar to display the **Extrusion** dialog box. In this dialog box select the **Cut** option from the **Operation** drop-down list.

14. Set the value of the **Distance** spinner to **50**. Select the **Blind** option from the **Type** drop-down list in the **Termination** area.

15. Set the value of the **Thickness** spinner under the **Thickness** area to **6**. Make sure the thickness value is applied in the downward direction. If not, select the **Flip Thickness** check box. Choose **OK**. The model should look similar to the one shown in Figure 3-28.

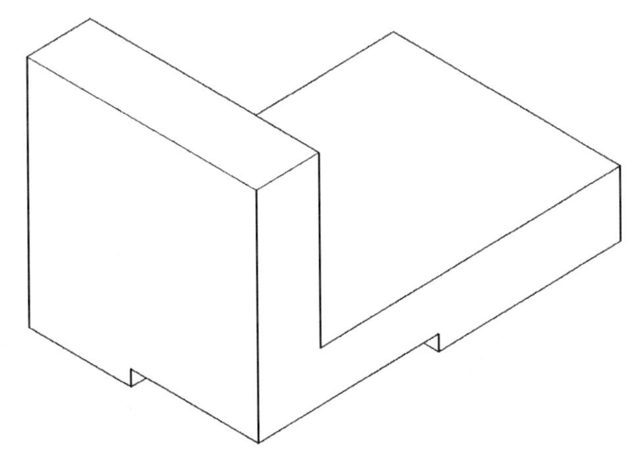

Figure 3-28 *Designer model after creating the cut feature*

16. Define new sketch plane on the left face of the model to draw the next cut feature. Choose the **New Sketch Plane** button from the **Part Modeling** toolbar. The prompt sequence is as follows:

Select work plane, planar face or [worldXy/worldYz/worldZx/Ucs]: *Select the left face of the model as shown in Figure 3-29.*

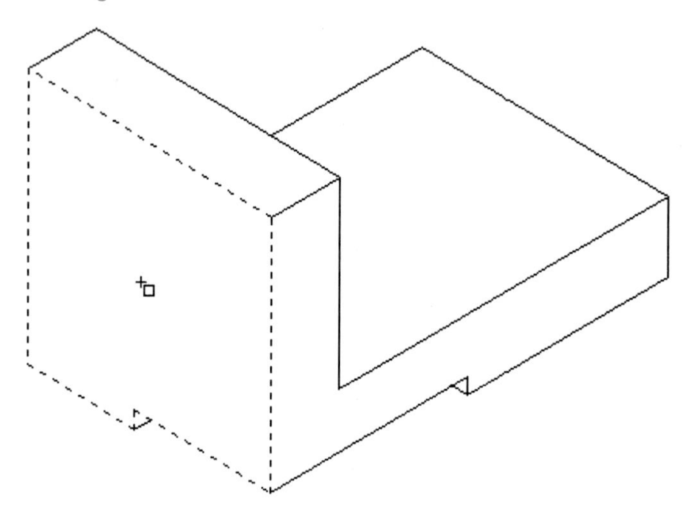

Figure 3-29 *Defining the sketch plane on the left face of the model*

Enter an option [Next/Accept] <Accept>: Enter
Select edge to align X axis or [Flip/Rotate/Origin] <Accept>: *Align the direction of the X axis using the bottom horizontal edge of the left face.*

17. Enter **9** at the Command prompt and then draw the rough sketch for the next cut feature. Profile it and then apply the required dimensions as shown in Figure 3-30.

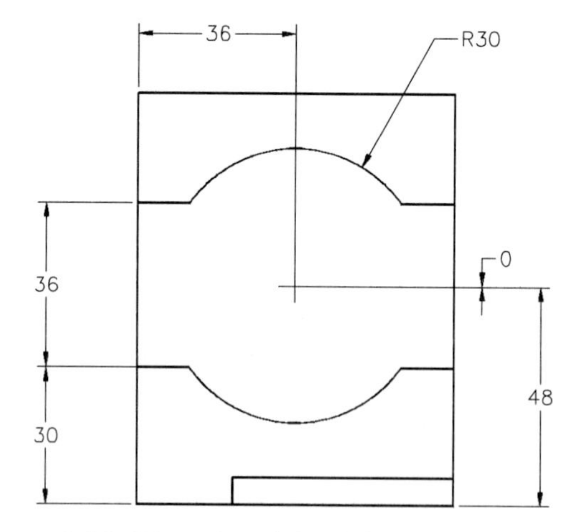

Figure 3-30 *Dimensioned sketch for the next cut feature*

18. Choose the **Sketched Features - Extrude** button from the **Part Modeling** toolbar to display the **Extrusion** dialog box. In this dialog box select the **Cut** option from the **Operation** drop-down list.

19. Set the value of the **Distance** spinner to **6**. Select the **Blind** option from the **Type** drop-down list in the **Termination** area. Choose **OK**.

20. Similarly, using the above mentioned method create the hole feature by drawing a circle and extruding it using the **Cut** operation and **Through** termination, see Figure 3-31.

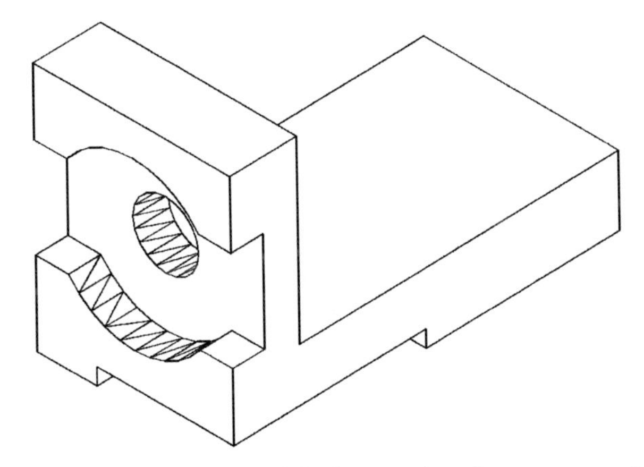

Figure 3-31 *Designer model after creating the next cut and hole*

21. Choose the **New Sketch Plane** button from the **Part Modeling** toolbar. The prompt sequence is as follows:

    Select work plane, planar face or [worldXy/worldYz/worldZx/Ucs]: *Select the top face by moving the cursor close to the top face of the model, see Figure 3-32.*

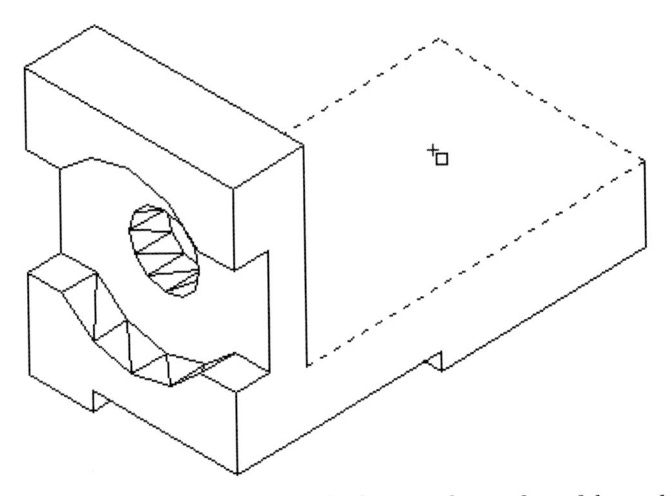

Figure 3-32 *Defining the sketch plane on the top face of the model*

Enter an option [Next/Accept] <Accept>: Enter
Select edge to align X axis or [Flip/Rotate/Origin] <Accept>: *Select the horizontal edge to ensure that the X axis direction is from left to right.*

22. Enter **9** at the Command prompt. Draw the sketch for the next cut and then profile and dimension it, see Figure 3-33.

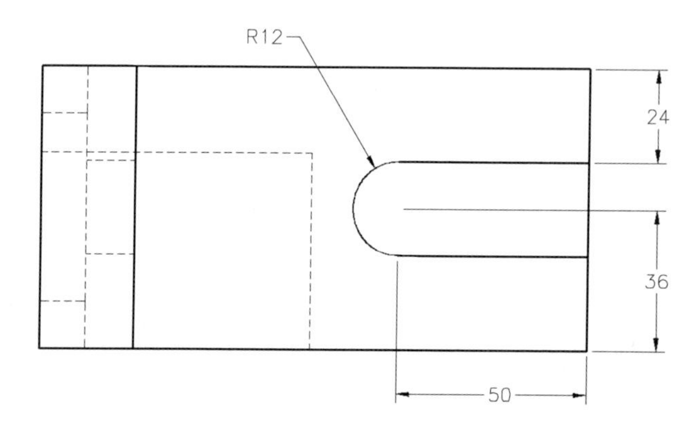

Figure 3-33 *Creating the sketch for the next cut*

23. Extrude it using the **Cut** operation and **Through** termination.

24. Enter **8** at the Command prompt. The final designer model should look similar to the one shown in Figure 3-34.

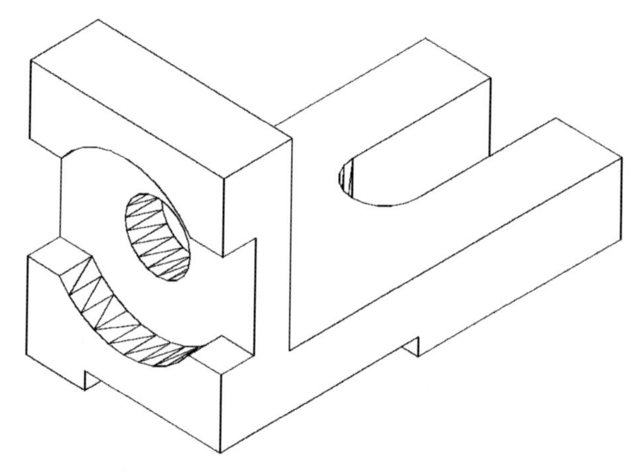

Figure 3-34 *The final designer model for Tutorial 1*

25. Save this drawing with the name given below:

    \MDT Tut\Ch-3**Tut1.dwg**

Tutorial 2

In this tutorial you will create the object shown in Figure 3-35a. The dimensions to be used are given in Figures 3-35b and 3-35c. Assume the missing dimensions.

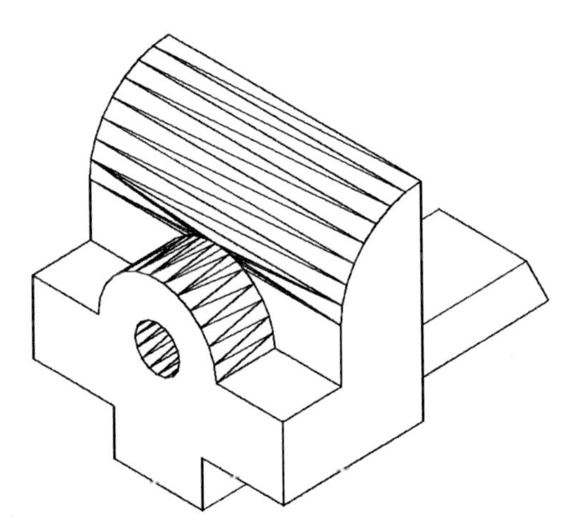

Figure 3-35a *Designer Model for Tutorial 2*

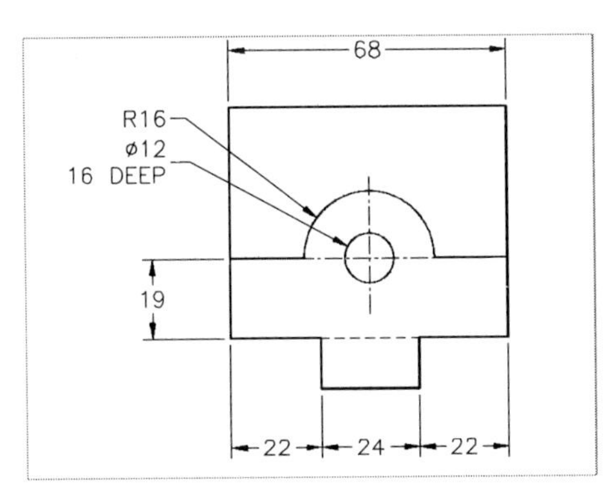

Figure 3-35b *Side view of the model*

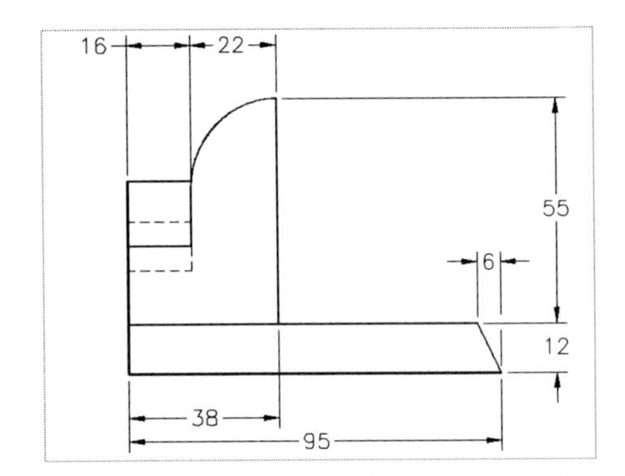

Figure 3-35c *Front view of the model*

1. Rotate the UCS through an angle of 90 degrees, increase the limits, and then create the rough sketch for the base part.

2. Profile this sketch and add the required dimension to the sketch. The final desired sketch for the base part should look similar to the one shown in Figure 3-36.

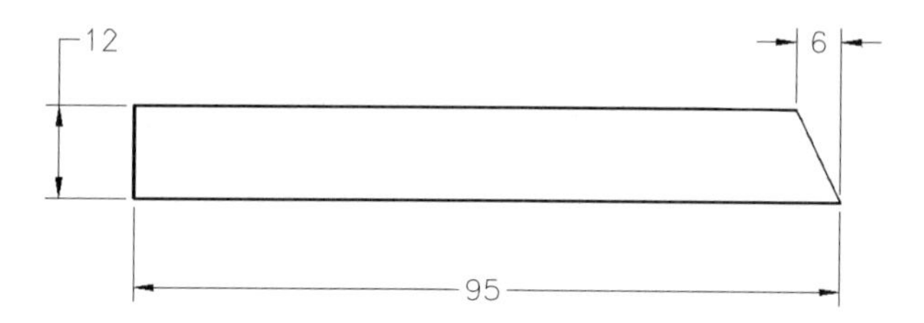

Figure 3-36 *Dimensioned sketch for the base feature*

3. Enter **8** at the Command prompt.

4. Choose the **Sketched Features - Extrude** button from the **Part Modeling** toolbar. Set the value of the **Distance** spinner to **24**.

5. Select the **Blind** option from the **Type** drop-down list in the **Termination** area. Choose **OK**. The base part should be similar to the one shown in Figure 3-37.

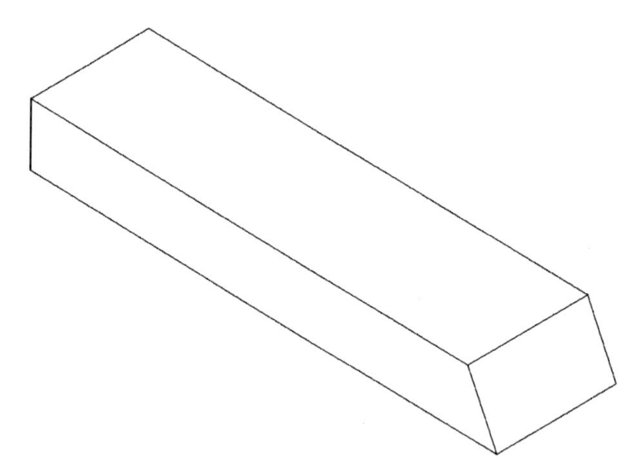

Figure 3-37 *Figure showing the base feature of the model*

6. Choose the **Work Features - Work Plane** button from the **Part Modeling** toolbar to invoke the **Work Plane** dialog box, see Figure 3-38.

7. In the **Work Plane** dialog box select the **Planar Parallel** option under the **1st Modifier** area and the **Offset** option under the **2nd Modifier** area.

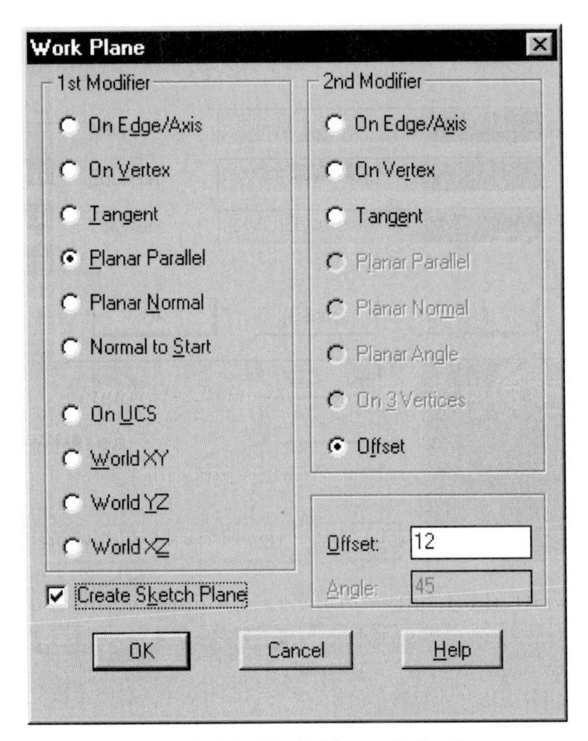

Figure 3-38 *Work Plane dialog box*

8. Enter **12** in the **Offset** edit box. Choose **OK**. The prompt sequence is as follows:

   Select work plane, planar face or [worldXy/worldYz/worldZx/Ucs]: *Select the front face of the model, see Figure 3-39.*

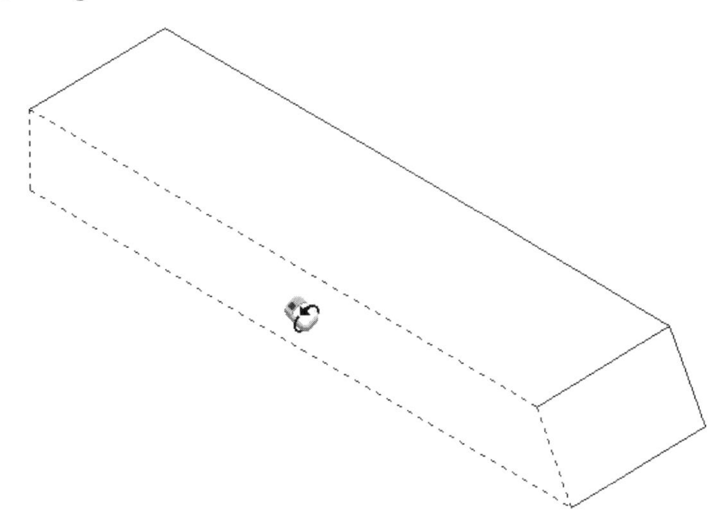

Figure 3-39 *Selecting the front face of the designer model*

Enter an option [Next/Accept] <Accept>: Enter
Enter an option [Flip/Accept] <Accept>: *Enter **F** if the arrows are pointing away from the model and press ENTER if the arrows are pointing towards the model.*
Select edge to align X axis or [Flip/Rotate/Origin] <Accept>: *Select an edge to make sure that the X axis direction is from left to right, see Figure 3-40.*

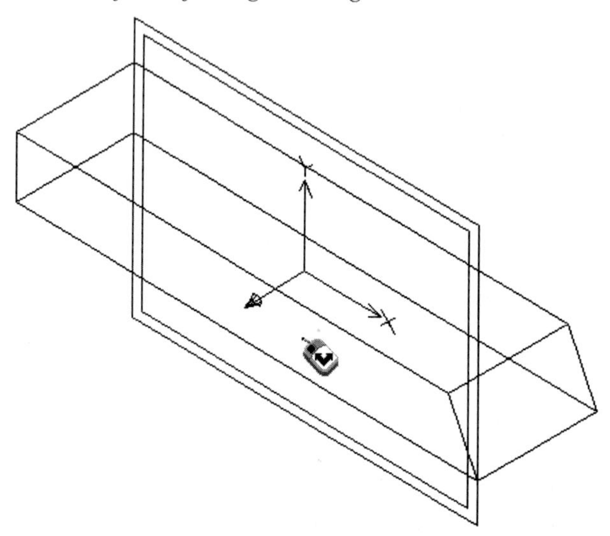

Figure 3-40 *Figure showing the X axis direction from left to right*

9. Enter **9** at the Command prompt.

10. Create the rough sketch for the next feature as shown in Figure 3-41.

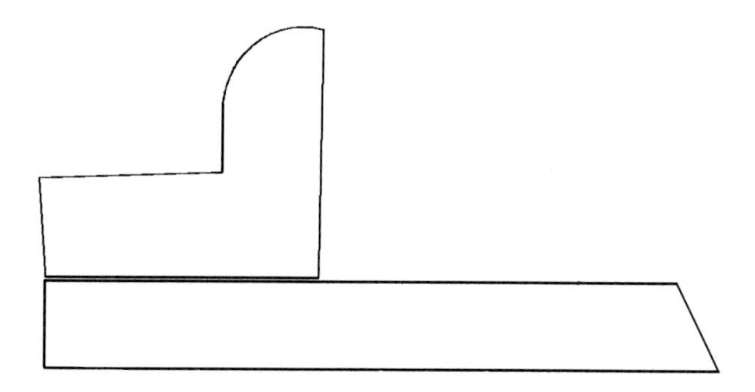

Figure 3-41 *Figure showing rough sketch for the next feature*

11. Profile this rough sketch, apply the required constraints and then add the dimensions to the sketch, Figure 3-42.

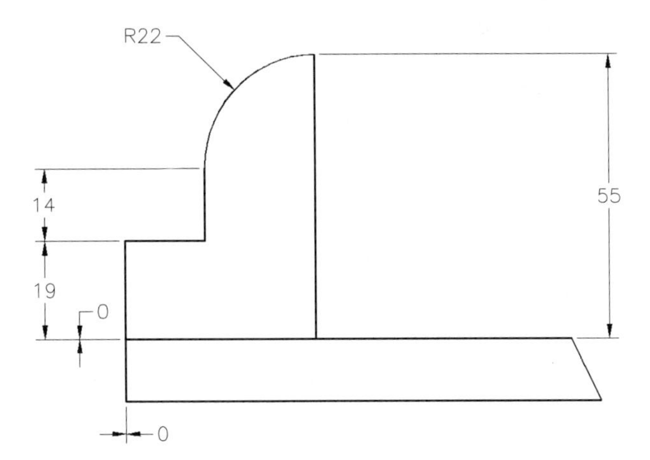

Figure 3-42 *Sketch for the next feature after adding the dimensions*

12. Enter **88** at the Command prompt.

13. Choose the **Sketched Features - Extrude** button from the **Part Modeling** toolbar to display the **Extrusion** dialog box. In the **Operation** drop-down list select the **Join** option.

14. Set the value of the **Distance** spinner to **68**. Choose **OK**.

15. Select the **MidPlane** option from the **Type** drop-down list in the **Termination** area.

16. Enter **FF** at the Command prompt. The model should be similar to the one shown in Figure 3-43.

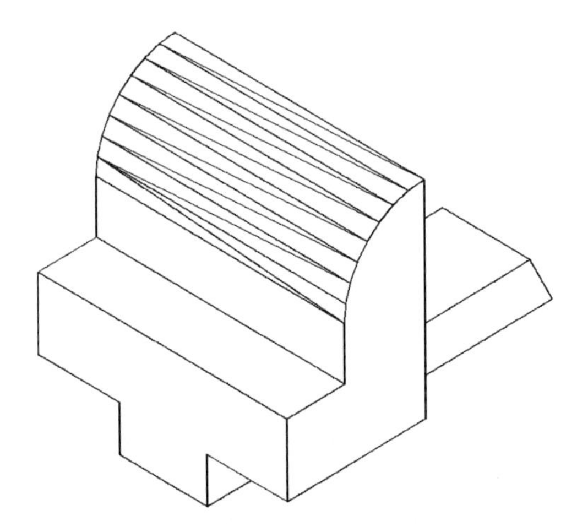

Figure 3-43 *Designer model after creating the second feature*

Tip: *If you delete a work plane after creating a feature using that work plane then the feature will also be automatically deleted. Therefore, if you do not want the work planes to be displayed, make them invisible using the desktop browser.*

17. To make the work plane invisible, right-click on the work plane in the desktop browser to display the shortcut menu, see Figure 3-44. Choose **Visible** to make the work plane invisible.

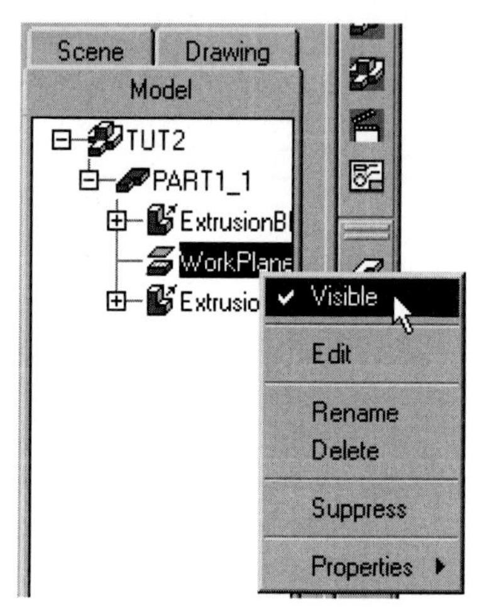

Figure 3-44 *Making the work plane invisible using the desktop browser*

18. To create the next circular feature you have to define a sketch plane on the left face. Choose the **New Sketch Plane** button from the **Part Modeling** toolbar. The prompt sequence is as follows:

Select work plane, planar face or [worldXy/worldYz/worldZx/Ucs]: *Select the left face as shown in Figure 3-45.*

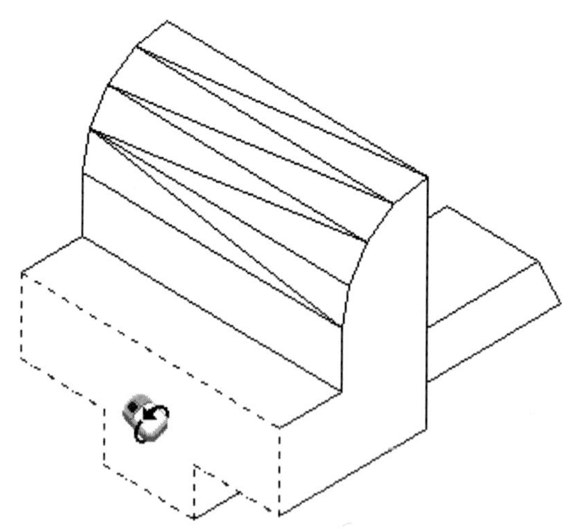

Figure 3-45 *Selecting the face to define the sketch plane*

Enter an option [Next/Accept] <Accept>: [Enter]
Select edge to align X axis or [Flip/Rotate/Origin] <Accept>: *Select the edge to make sure the X axis direction is from left to right, see Figure 3-46.*

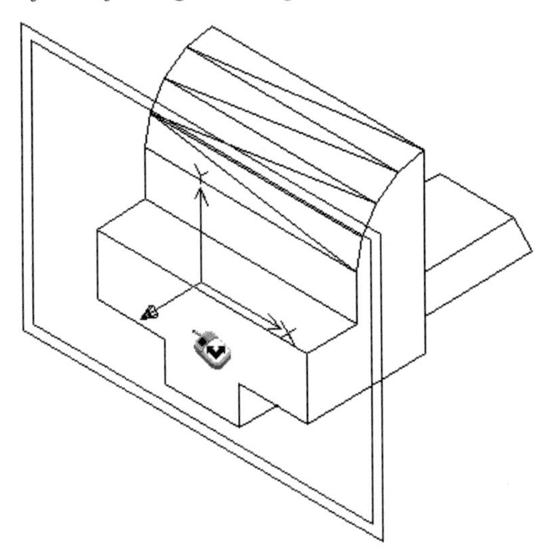

Figure 3-46 *Defining the X axis direction*

19. Enter **9** at the Command prompt. Create the rough sketch for the circular feature.

20. Profile and dimension the rough sketch. After adding the dimensioning it should be similar to the one shown in Figure 3-47.

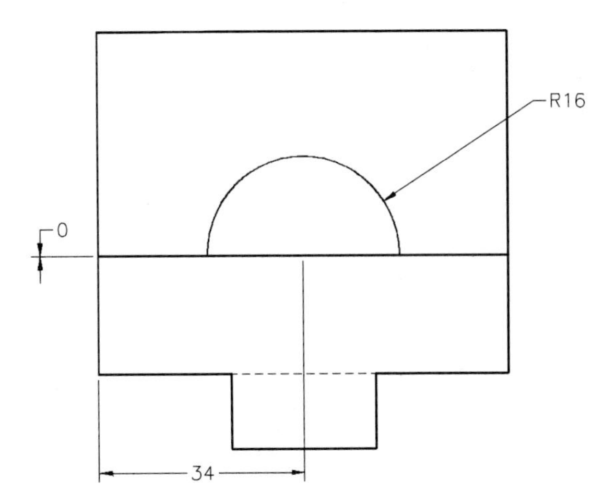

Figure 3-47 *Sketch after dimensioning*

21. Enter **88** at the Command prompt.

22. Choose the **Sketched Features - Extrude** button from the **Part Modeling** toolbar. Select **Join** from the **Operation** drop-down list.

23. Set the value of the **Distance** spinner to **18**.

24. Choose the **Flip** button to make sure the arrow is pointing towards the model.

25. Select **Blind** option from the **Type** drop-down list in the **Termination** area. Choose **OK**.

26. Enter **9** at the Command prompt.

27. Set the value of the **Distance** spinner to **16**. Choose **OK**.

Tip: *The distance value is made 18 instead of 16 as required because the upper part of the model is curved. Setting the value of the* ***Distance*** *spinner to 16 will not make the new feature merge with the model.*

28. Similarly, create the next cut feature that is a hole by drawing a circle and extruding it using the **Cut** operation and **Blind** termination. The distance should be **16**.

29. The final designer model should be similar to the one shown in Figure 3-48.

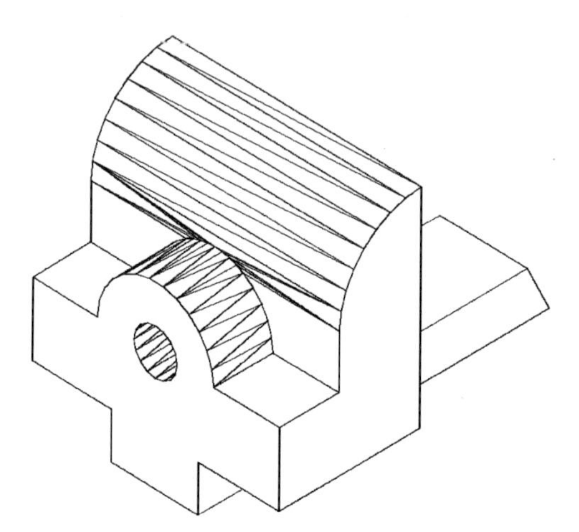

Figure 3-48 *Final designer model for Tutorial 2*

30. Save this with the name given below:

\MDT Tut\Ch-3**Tut2.dwg**

Tutorial 3

In this Tutorial you will create the model shown in Figure 3-49a. The dimensions to be used are given in Figures 3-49b, 3-49c and 3-49d. Assume the missing dimensions.

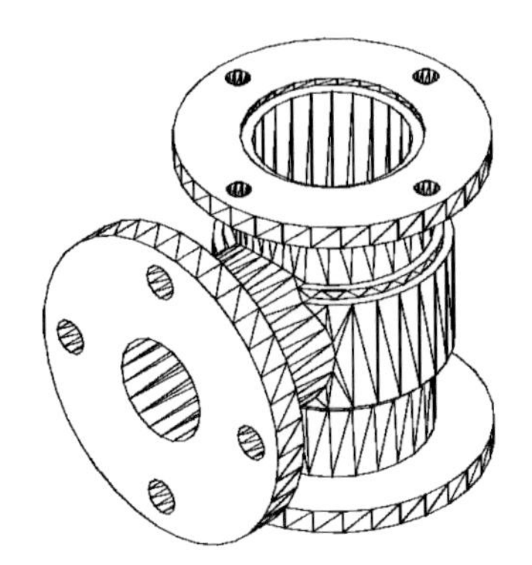

Figure 3-49a Model for Tutorial 3

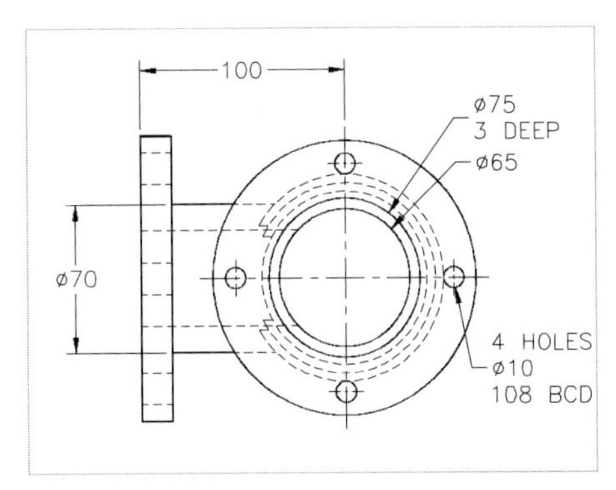

Figure 3-49b Top view of the model

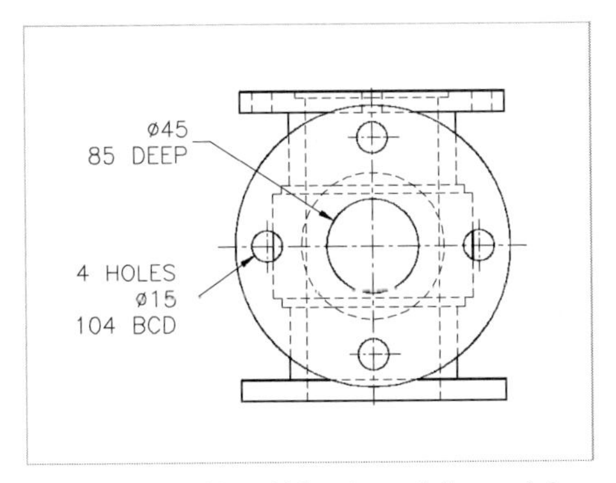

Figure 3-49c Side view of the model

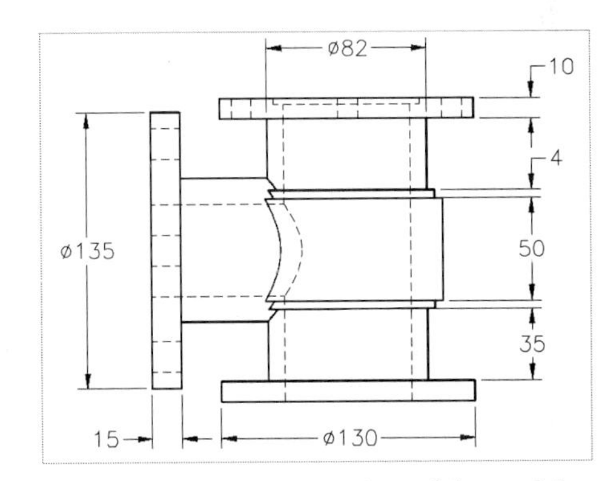

Figure 3-49d Front view of the model

1. In this tutorial the base part will be created by revolving the base sketch. Therefore, create the sketch for the base part after rotating the UCS and increasing the limits.

2. Profile this sketch and apply the required constraints to the sketch.

3. Add the dimensions to the sketch. The final desired sketch for the base part should be similar to the one shown in Figure 3-50.

4. Choose the **Revolve** button from the **Sketched Features - Extrude** flyout in the **Part Modeling** toolbar. You will be prompted to select the revolution axis. Select the right vertical line as the revolution axis. The **Revolution** dialog box will be displayed.

5. Set the value of the **Angle** spinner to **360**.

6. Select the **By Angle** option from the **Type** drop-down list. Choose **OK**.

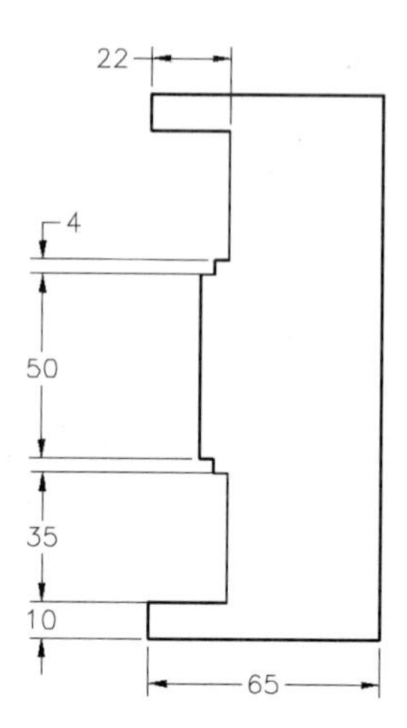

Figure 3-50 *Desired sketch for the base part*

7. The base part should be similar to the one shown in Figure 3-51.

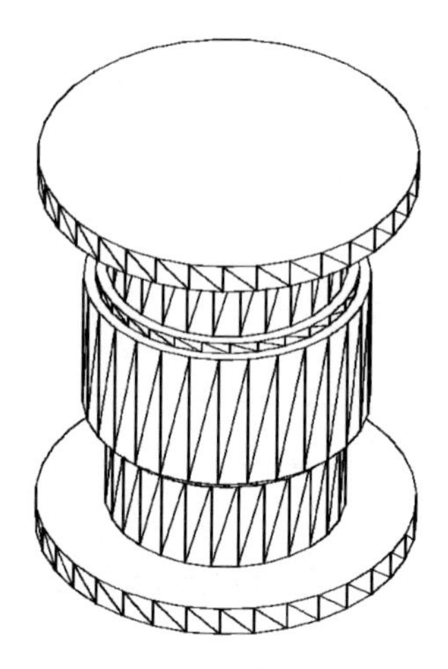

Figure 3-51 *Figure showing the base part of the designer model*

8. To create the second feature you have to first define a sketch plane or a work plane in the middle of the base part. In this case you can define a sketch plane on the UCS defined at the center of the base part. Choose **New UCS > Z Axis Vector** from the **Assist** menu. The prompt sequence is as follows:

   Enter an option [New/Move/orthoGraphic/Prev/Restore/Save/Del/Apply/?/World] <World>: _zaxis
   Specify new origin point <0,0,0>: *Select point **P1**, see Figure 3-52. (Center of the upper cylindrical part.)*
   Specify point on positive portion of Z-axis <default point>: *Select point **P2**. (Quadrant point.)*

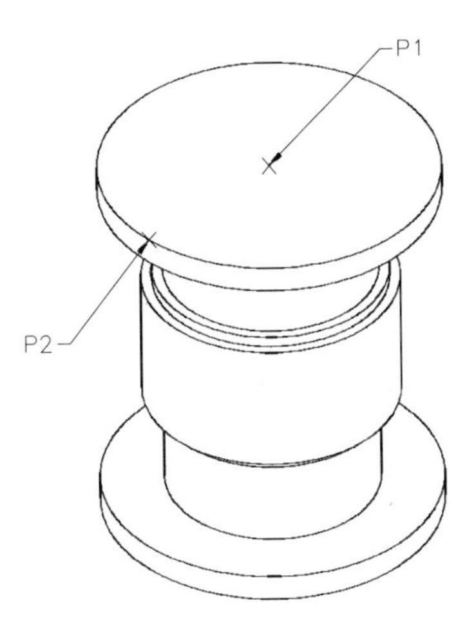

Figure 3-52 *Aligning the UCS on the top face of the model*

9. Choose the **New Sketch Plane** button from the **Part Modeling** toolbar. The prompt sequence is as follows:

Select work plane, planar face or [worldXy/worldYz/worldZx/Ucs]: **U**
Select edge to align X axis or [Flip/Rotate/Origin] <Accept>: *Make sure that the X axis direction is from left to right, see Figure 3-53.*

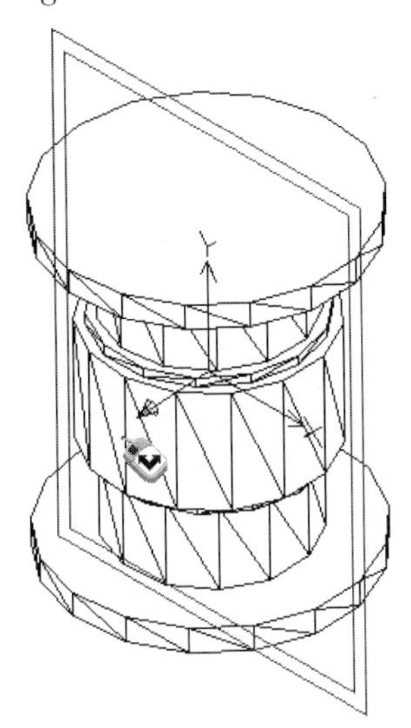

Figure 3-53 *Figure showing the direction of the X axis*

10. Enter **9** at the Command prompt.

11. Create the sketch for the second feature as shown in Figure 3-54. Profile and dimension the sketch using the required commands.

12. Choose the **Sketched Features - Extrude** button from the **Part Modeling** toolbar to display the **Extrusion** dialog box. Select the **Join** option from the **Operation** drop-down list.

13. Set the value of the **Distance** spinner to **85**.

14. Select the **Blind** option from the **Type** drop-down list.

15. Choose the **Flip** button to make sure the arrow is pointing out of the model.

16. Choose **OK**. The designer model should be similar to the one shown in Figure 3-55.

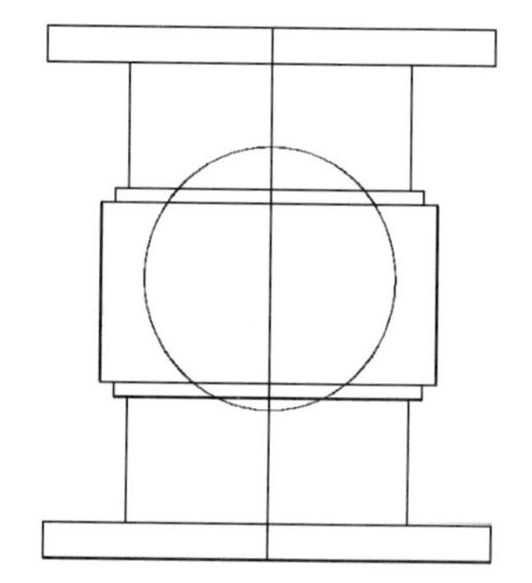

Figure 3-54 *Rough sketch for the second features*

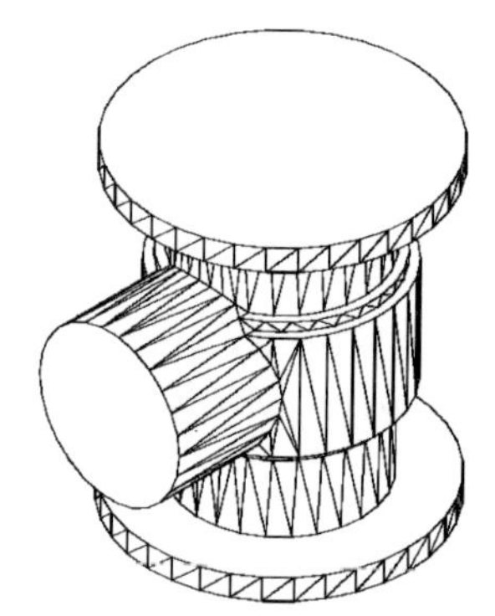

Figure 3-55 *Designer model after creating the second feature*

17. Choose the **New Sketch Plane** button from the **Part Modeling** toolbar to define sketch plane on the front face of second feature. The prompt sequence is as follows:

    Select work plane, planar face or [worldXy/worldYz/worldZx/Ucs]: *Select the front face of the second feature.*
    Select edge to align X axis or [Flip/Rotate/Origin] <Accept>: *Align the X axis direction from left to right, see Figure 3-56.*

18. Enter **9** at the Command prompt.

19. Create the rough sketch for the next feature. Profile it and add dimensions to this sketch.

20. Choose the **Sketched Features - Extrude** button to display the **Extrusion** dialog box. Select the **Join** option from the **Operation** drop-down list.

21. Set the value of the **Distance** spinner to **15**. Choose the **Flip** button to make sure the arrow is pointing out of the model.

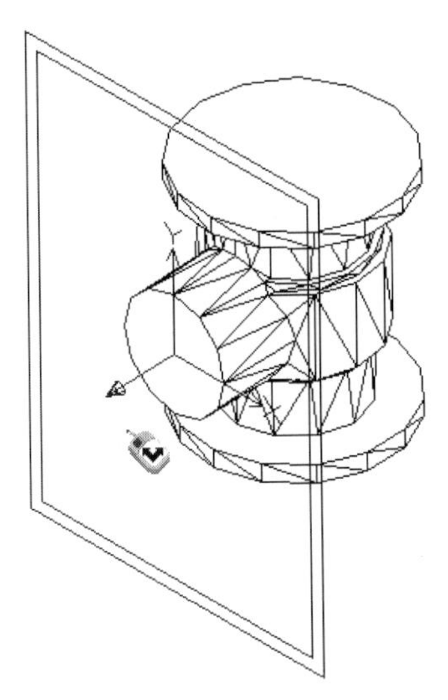

Figure 3-56 Defining the sketch plane on the front face

22. Select the **Blind** option from the **Termination** drop-down list. Choose **OK**. The model should be similar to the one shown in Figure 3-57.

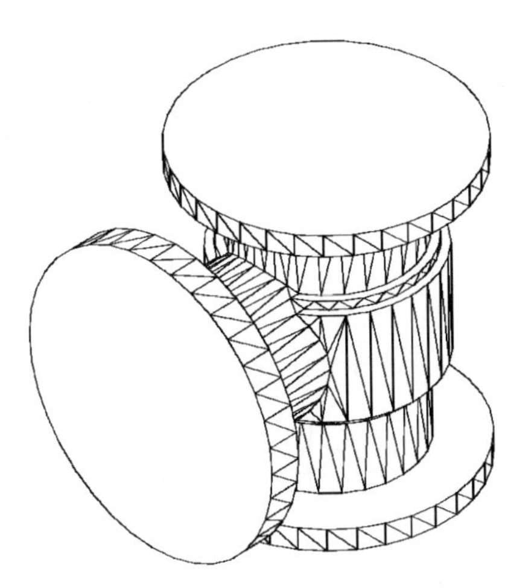

Figure 3-57 Designer model after creating the next feature

23. Choose the **New Sketch Plane** button from the **Part Modeling** toolbar and define a sketch plane on the front face of the new feature.

24. Enter **9** at the Command prompt and in this view create the rough sketch for the central hole. Profile and dimension this sketch.

25. Enter **8** at the Command prompt.

26. Choose the **Sketched Feature - Extrude** button to display the **Extrusion** dialog box. Select the **Cut** option from the **Operation** drop-down list.

27. Set the value of the **Distance** spinner to **85**.

28. Select the **Blind** option from the **Type** drop-down list. Choose **OK**.

29. Enter **9** at the Command prompt.

30. Now you have to create the smaller holes on the same feature. This feature contains four holes that have to be created one by one. Create the sketch for any one of the holes.

31. Profile and dimension this sketch.

32. Enter **8** at the Command prompt.

33. Choose the **Sketched Features - Extrude** button to display the **Extrusion** dialog box. Select the **Cut** option from the **Operation** drop-down list.

34. Set the value of the **Distance** spinner to **15**.

35. Select the **Blind** option from the **Termination** drop-down list. Choose **OK**.

36. Similarly, create the remaining three holes. The model should look similar to the one shown in Figure 3-58.

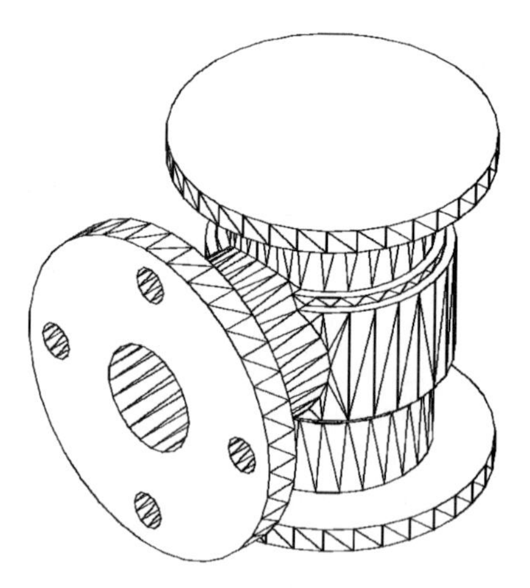

Figure 3-58 *Designer model after creating the holes on the front face*

Tip: *The holes and the array of holes can be easily created using the* ***AMHOLE*** *and* ***AMPATTERN*** *commands, respectively. Both these commands will be discussed in later chapters.*

37. Choose the **New Sketch Plane** button from the **Part Modeling** toolbar and define the sketch plane on the top face of the model.

38. Using the above mentioned steps create all the holes on the top face and the bottom face of the model. The final designer model should be similar to the one shown in Figure 3-59.

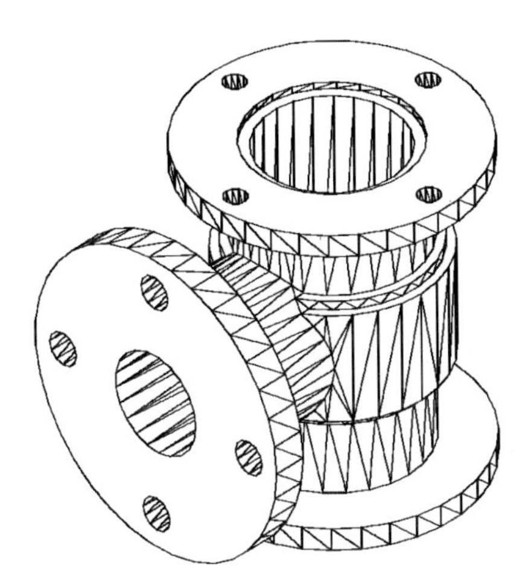

Figure 3-59 *Final designer model for Tutorial 3*

39. Save this drawing with the name given below:

   \MDT Tut\Ch-3**Tut3.dwg**

Review Questions

Answer the following questions.

1. What is the need of defining the sketch planes on different faces of the designer model?

2. What are the various options provided in the **AMSKPLN** command?

3. What are the various work features provided in Mechanical Desktop?

4. Give any three uses of the work point and the work axis.

5. Give any three differences between the sketch plane and the work plane.

6. What are the various options apart from the **Base** option in the **Operation** drop-down list of the **Extrusion** and the **Revolution** dialog boxes?

7. If you delete the work plane the feature created using that work plane will also be deleted. (T/F)

8. The work axis can be created even if the cylindrical feature does not exist. (T/F)

Exercises

Exercise 1

Create the designer model shown in Figure 3-60a. The dimensions to be used are given in Figures 3-60b, 3-60c and 3-60d. Assume the missing dimensions. Save the drawing with the name given below:

\MDT Tut\Ch-3**Exr1.dwg**

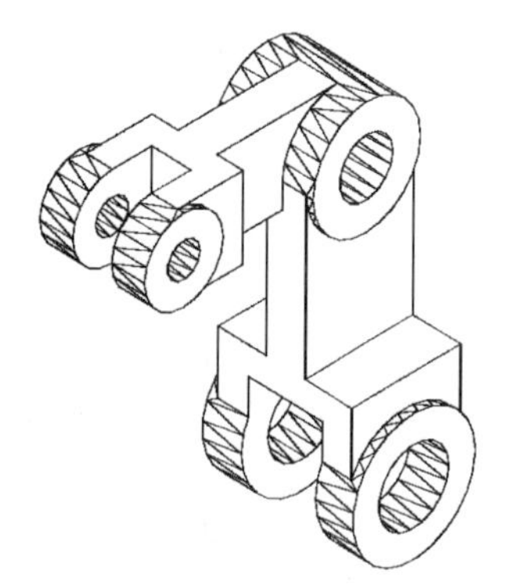

Figure 3-60a Model for Exercise 1

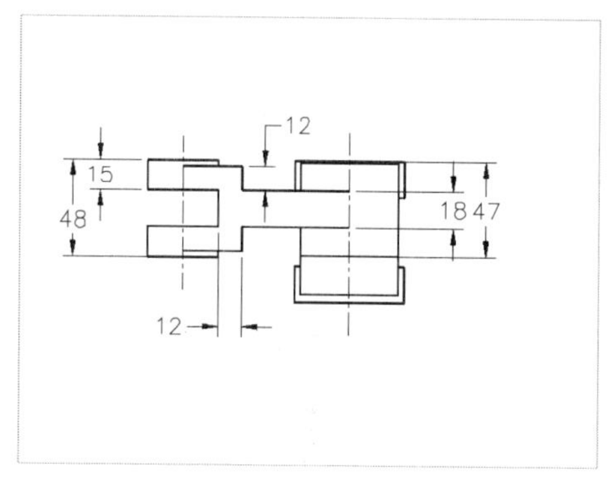

Figure 3-60b Top view of the model

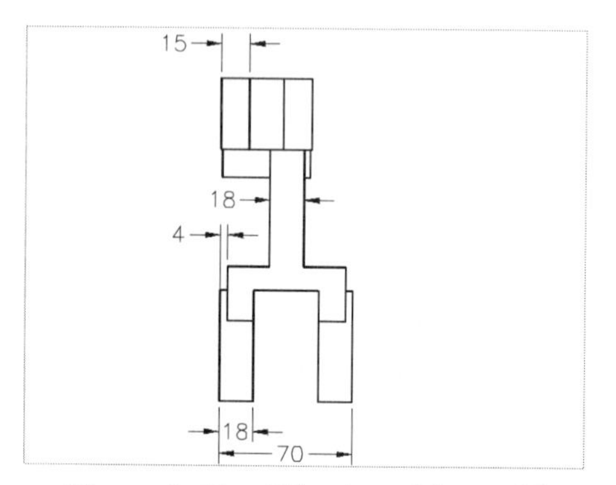

Figure 3-60c Side view of the model

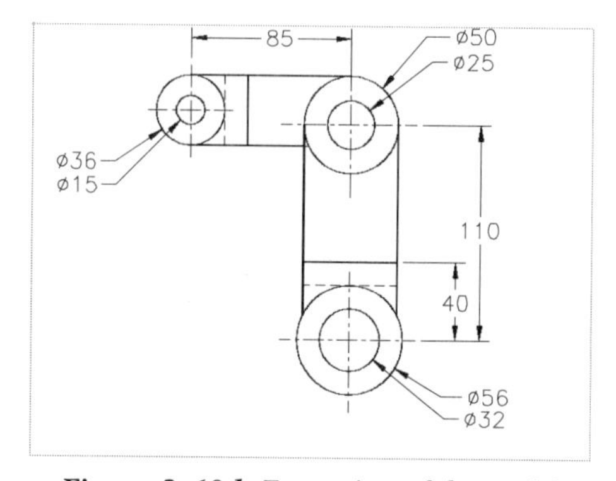

Figure 3-60d Front view of the model

Exercise 2

Draw the designer model shown in Figures 3-61a and 3-61b. The dimensions to be used are shown in Figures 3-61c and 3-61d. Assume the missing dimensions. Save the drawing with the name given below:

\MDT Tut\Ch-3**Exr2.dwg**

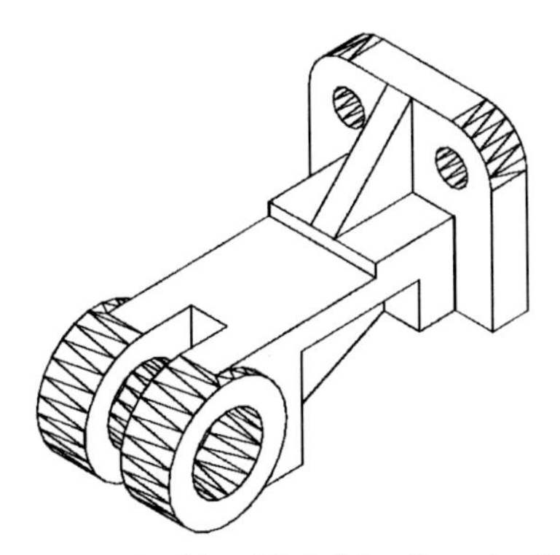

Figure 3-61a *Model for Exercise 2*

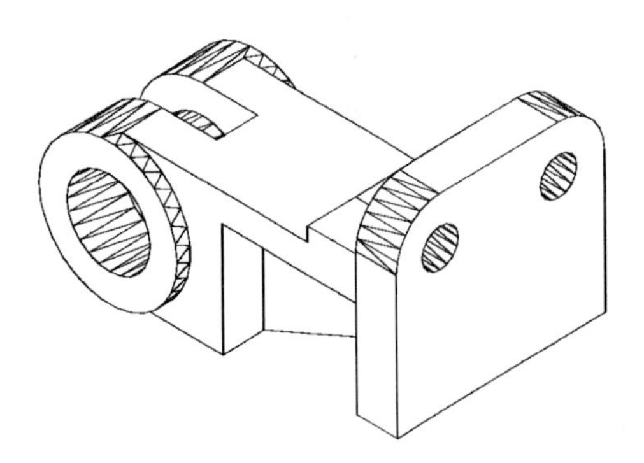

Figure 3-61b *Model for Exercise 2*

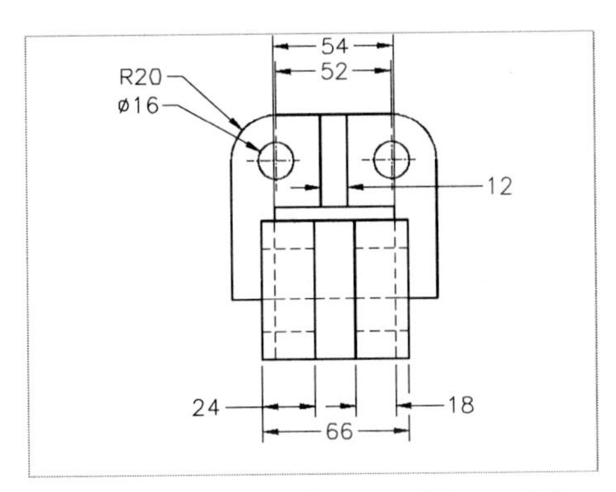

Figure 3-61c *Side view of the model*

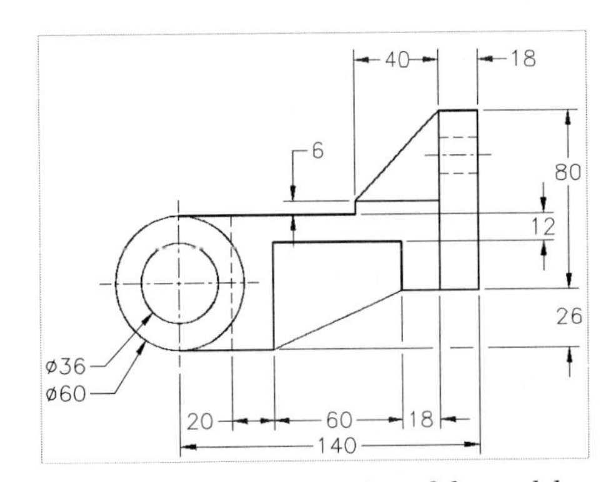

Figure 3-61d *Front view of the model*

Exercise 3

Draw the designer model shown in Figure 3-62a. The dimensions to be used are shown in Figures 3-62b, 3-62c and 3-62d. Assume the missing dimensions. Save the drawing with the name given below:

\MDT Tut\Ch-3**Exr3.dwg**

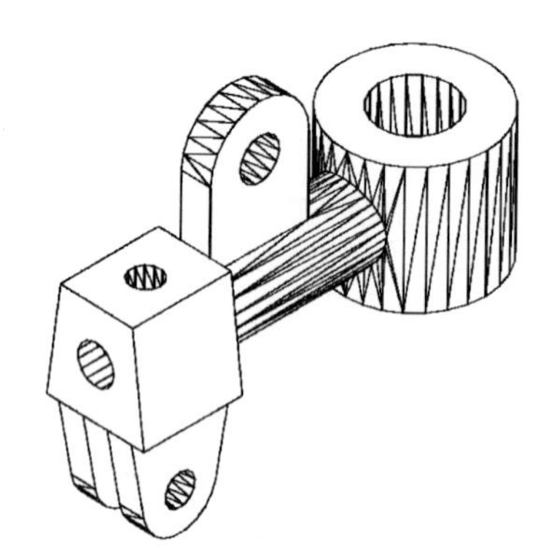

Figure 3-62a *Model for Exercise 3*

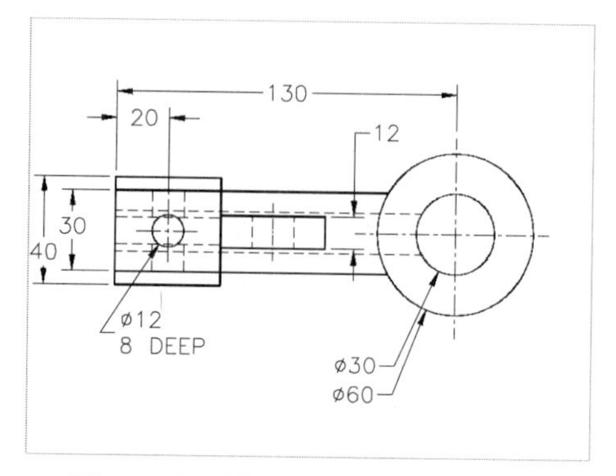

Figure 3-62b *Top view of the model*

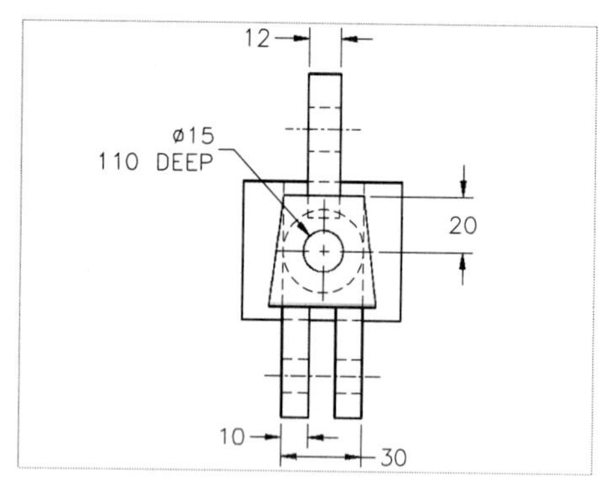

Figure 3-62c *Side view of the model*

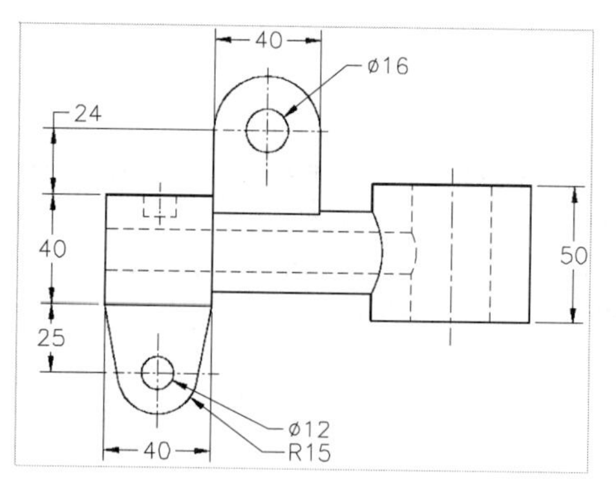

Figure 3-62d *Front view of the model*

Chapter 4

Advanced Dimensioning Techniques, Design Variables, and Visibility Options

Learning Objectives

After completing this chapter, you will be able to:

- *Dimension the sketches using the advanced dimensioning techniques like the power dimensioning and the auto dimensioning techniques.*
- *Create the local and global design variables.*
- *View the dimensions assigned to the sketches as Parameters, Numerics, or Equations.*
- *Control the visibility of different features associated with the designer models.*

Commands Covered

- ***AMPOWERDIM***
- ***AMAUTODIM***
- ***AMVARS***
- ***AMDIMDSP***
- ***AMVISIBLE***

ADVANCED DIMENSIONING TECHNIQUES

The power dimensioning and auto dimensioning are the advanced dimensioning techniques used to dimension the profiled or the unprofiled sketches in the **Model** or the **Drawing** mode. When assigned to the profiled sketches, these dimensions will act as the parametric dimensions and will drive the selected entity to the dimension value you have entered. When assigned to the unprofiled sketch, these dimensions will act as simple dimensions and will just display the dimension value you enter. In this case the actual dimension of the selected object will not change.

The auto dimensioning technique is generally used for dimensioning in the **Drawing** mode. However, you can also use this technique to dimension the profiled sketches in the **Model** mode so that the dimensions behave as parametric. While dimensioning the profiled sketch in the **Model** mode, this technique takes the help of the power dimensioning technique. Therefore, the main advanced dimensioning technique is the power dimensioning technique. The main uses of the power dimensioning technique are as follows:

1. To create linear or angular dimensions.
2. To create radius or diameter dimensions.
3. To create baseline dimensions taking the reference of a predefined linear or angular dimension.
4. To create continued dimension taking the reference of a predefined linear or angular dimension.
5. To assign tolerances to dimensions.
6. To add fits to holes or shafts.

The power dimensioning and the auto dimensioning techniques, in the context of the profiled sketches in the **Model** mode, are discussed below. Both of these techniques in the context of the **Drawing** mode will be discussed in the later chapters.

AMPOWERDIM COMMAND*

Toolbar:	Part Modeling > Power Dimensioning
Menu:	Part > Dimensioning > Power Dimensioning
Context Menu:	Dimensioning > Power Dimensioning
Command:	AMPOWERDIM

This is a very important dimensioning technique and is generally used for dimensioning the complex sketches. The options provided under this command are:

Select

The **Select** option is used to select the objects to dimension. The sub-options provided in this option varies depending upon the object selected to dimension.

For example, when you select an arc or a circle to dimension, you will be prompted to define the dimension in the terms of radius or diameter. You can select the type of radius or diameter dimensioning using the **Options** suboption. When you invoke this suboption, the **Radius &**

Diameter Options dialog box is displayed, see Figures 4-1 and 4-2. You can choose any style for the arcs or the circles using the **Radius** or **Diameter** tab of this dialog box. After choosing the dimensioning style choose **OK**. You will be prompted to place the dimension.

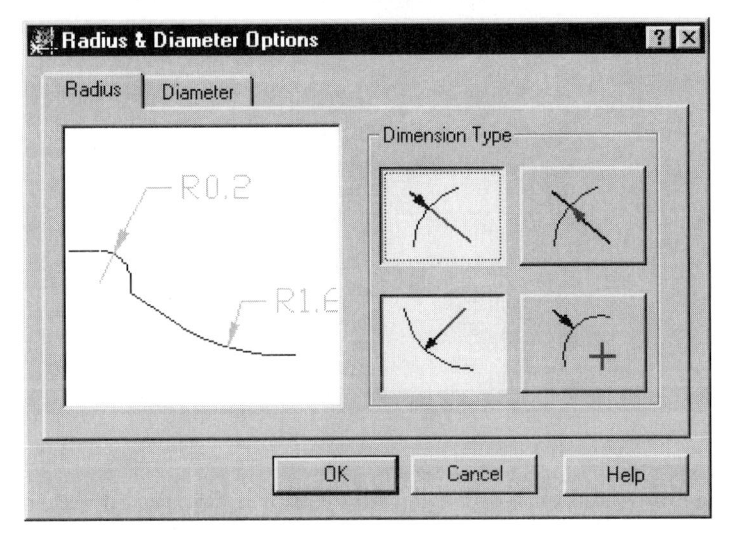

Figure 4-1 *The Radius tab of Radius & Diameter Options dialog box*

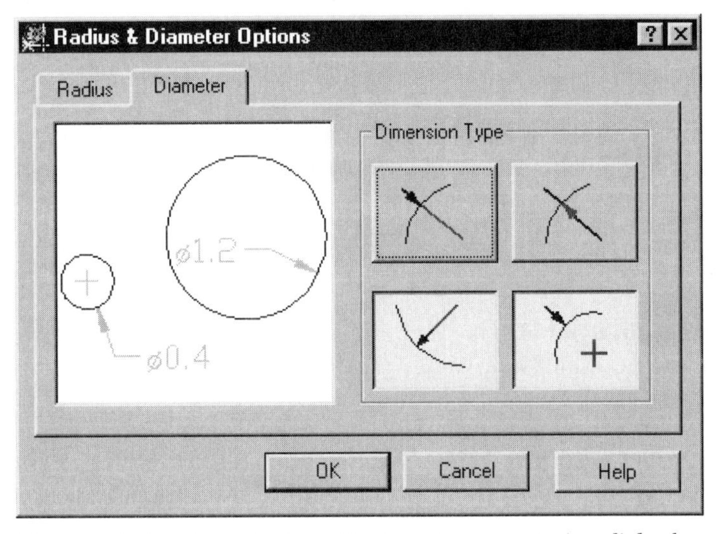

Figure 4-2 *The Diameter tab of Radius & Diameter Options dialog box*

When you place the dimension, the **Power Dimensioning** dialog box will be displayed, see Figure 4-3.

Power Dimensioning Dialog Box Options

This dialog box consists of three tabs. The options under all these tabs are discussed next.

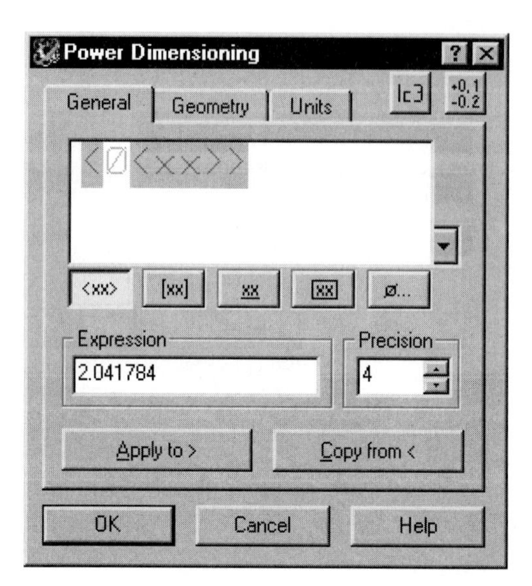

Figure 4-3 *General tab of the Power Dimensioning dialog box*

General Tab

The options provided under the **General tab** of the **Power Dimensioning** dialog box are:

Text Box. This box displays the text as it will appear upon placing the dimension. The **Text Box** displays angular brackets that actually are the default text. You can add some text to the default text or delete some text from the default text using this box. On the right of this text box you will find the button similar to the drop-down list button. When you choose this button, a predefined dimension text drop-down list is displayed. Using this drop-down list you can add some predefined text to the default dimension.

Dimension Symbol. This button is available below the **Text Box** and is used to display the dimension values on the screen. If this button is not chosen, the dimension values will not be displayed on the screen.

Alternate Unit Symbol. This button is available below the **Text Box** and is chosen to display the dimension in alternate units in addition to the default units.

Underline Dimension Text. This button is chosen to underline the dimension text.

Box Dimension Text. If this button is chosen, the dimension text will be displayed inside a box.

Special Characters. This button is chosen to display the **Special Characters** dialog box, see Figure 4-4. You can select special characters from this dialog box that can be added to the dimension text.

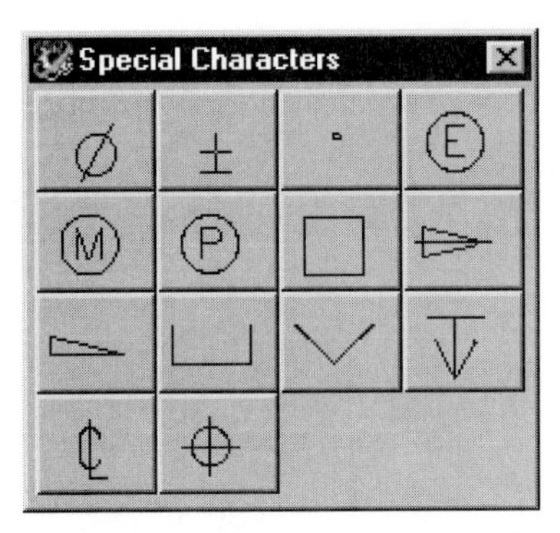

Figure 4-4 *Special Characters dialog box*

Expression. This edit box is used to specify the dimension value of the selected entity. Whatever value is entered in this edit box, the selected entity will be driven to that value.

Precision. This spinner is used to set the precision for the decimal values of the dimension text.

Apply to. This button is chosen when you want to apply the current properties to another selected dimension.

Copy from. This button is chosen to copy the properties from a predefined dimension. When you choose this button, you will be prompted to select an existing dimension from which you want to copy the properties. As soon as you select the existing dimension, the **Select Parameters to Copy** dialog box will be displayed, Figure 4-5. You can select the parameters to be copied from this dialog box.

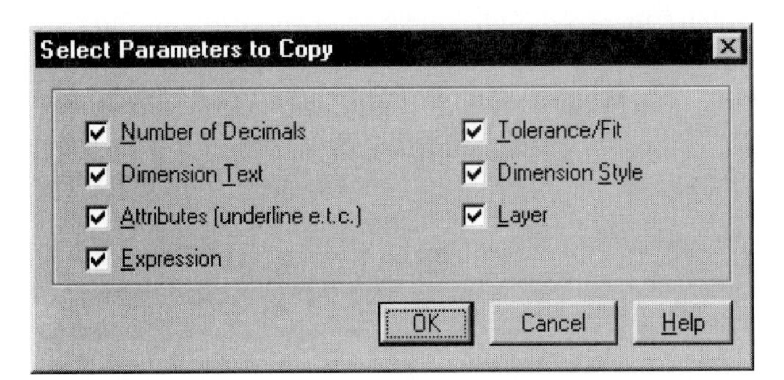

Figure 4-5 *Select Parameters to Copy dialog box*

Geometry Tab

The options under this tab are related to the dimension style, see Figure 4-6. These options change depending upon the entity selected to dimension.

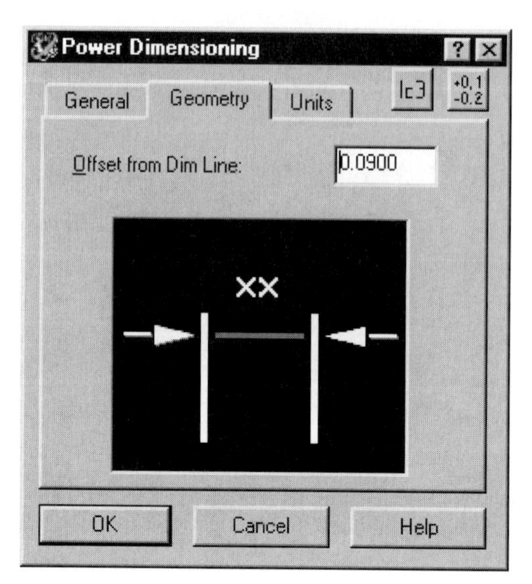

Figure 4-6 Geometry tab of the Power Dimensioning dialog box

Offset from Dim Line. This edit box is used to set the distance between the text and the dimension line.

Selection Window. This window provides all the options related to the dimension style. When you take the cursor in this window, the cursor, that was an arrow until now, is converted into a hand. You can use this hand cursor to select the parameter that you want to modify. The parameters that can be modified using this window are **First Dimension Line**, **First Arrowhead**, **First Extension Line**, **Dimension Text**, **Second Extension Line**, **Second Arrowhead** and **Second Dimension Line**. You can also force a dimension line in between using this window. To modify any of these parameters, click the required option in the window. When you click on the dimension text, the **General** tab of the **Power Dimensioning** dialog box will be displayed. When you click on any of the arrowheads, a dialog box will be displayed from which you can select the type of arrowhead required.

If the entity selected to dimension is an arc or a circle, the **Center Mark Size** edit box will be displayed in addition to the **Offset from Dim Line** edit box. This edit box is used to set the size of the center mark for the radius.

Units Tab

The options under this tab are related to the units of the dimension text, see Figure 4-7.

Units. This drop-down list is used to select the units for the primary dimension.

Linear Scale. This edit box is used to set the scale factor to be multiplied with the primary dimension value.

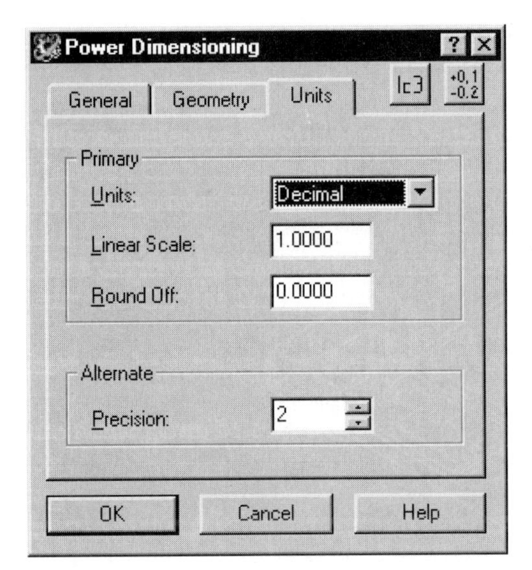

Figure 4-7 *Geometry tab of the Power Dimensioning dialog box*

Round Off. This edit box is used to set the value for rounding off the dimension text.

Precision. This spinner will be available only when the **Alternate Unit Symbol** button is chosen in the **General** tab. This spinner is used to set the precision for the dimension value of the alternate units.

Apart from these three tabs, this dialog box also provides two buttons that are available on the upper right corner of the dialog box. They are **Add Fit** button and **Add Tolerance** button.

Add Fit Button

When you choose this button, the **Power Dimensioning** dialog box is expanded, providing you the options related to the fit symbols for the mating holes and shafts. You can enter the fit symbol in the **Symbol** edit box or select it from the **Fits** dialog box that is displayed upon choosing the swatch provided on the right side of the **Symbol** edit box. When you choose the **Mating** button in this dialog box, the dialog box expands and you can view both the mating hole and shaft in same dialog box, see Figure 4-8.

The precision for the dimension values of the primary and alternate units can be set using the **Primary** and **Alternate** spinners in the **Precision** area. The **Alternate** spinner will be available only when the **Alternate Unit Symbol** button is chosen in the **General** tab of the **Power Dimensioning** dialog box.

The type of fit can be selected from the **Select Fit Type** dialog box (Figure 4-9) displayed by choosing the **Choose Fit Type** button. This button is available below the **Precision** area.

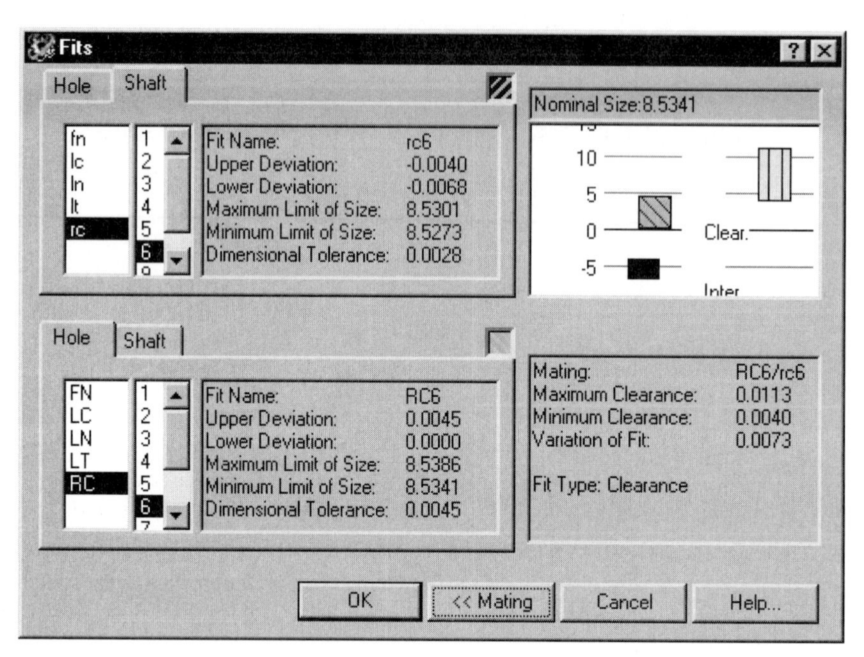

Figure 4-8 Fits dialog box

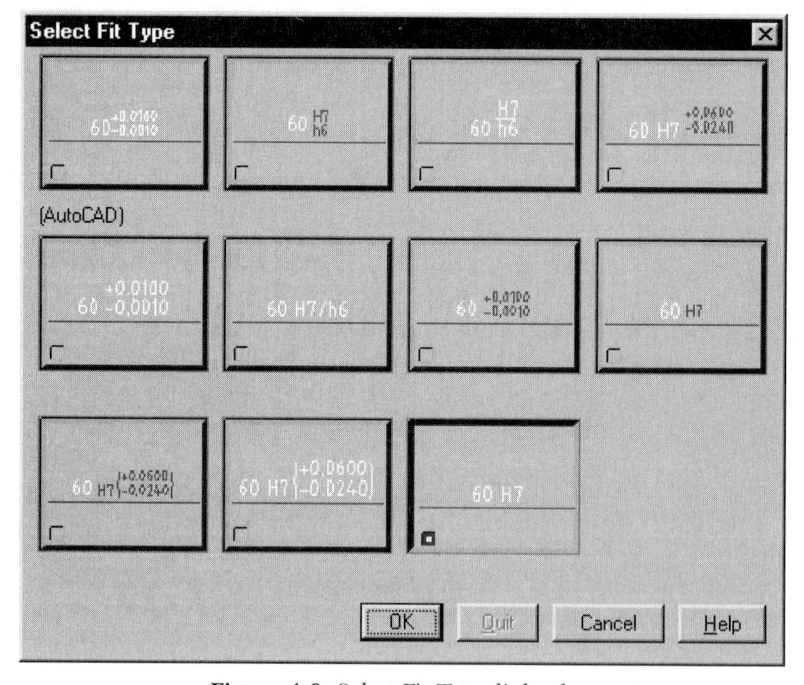

Figure 4-9 Select Fit Type dialog box

Add Tolerance Button

When you choose this button, the **Power Dimensioning** dialog box is expanded providing you the options related to the tolerance for the dimensions. The upper and lower deviation values can be set in the **Upper** and **Lower** edit boxes respectively in the **Deviation** area. You can also set the precision for the tolerances of the primary and alternate units using the **Primary** and **Alternate** edit box in the **Precision** area. The type of tolerance can be chosen from the **Select Tolerance Type** dialog box (Figure 4-10) displayed upon choosing the **Choose Tolerance Type** button provided under the **Precision** area.

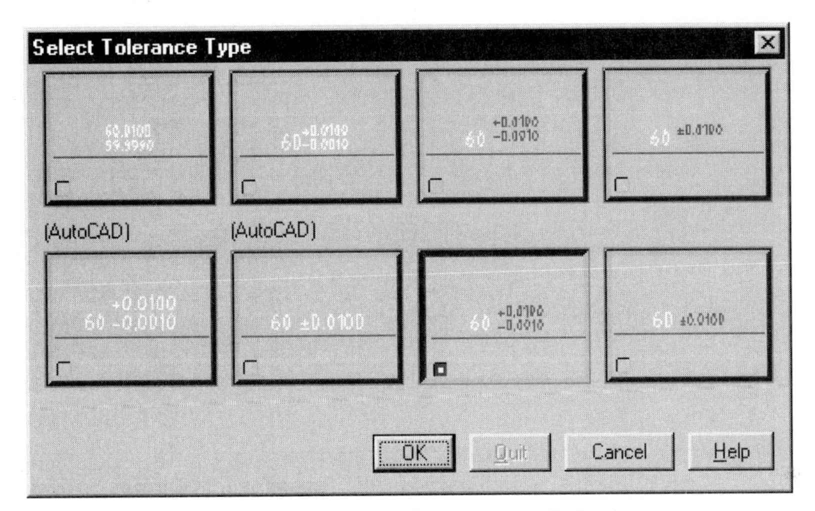

Figure 4-10 *Select Tolerance Type dialog box*

Angular

The **Angular** option is used to assign the angular dimensions to the selected set of objects. You have to select two objects to assign this dimensions. The **Power Dimensioning** dialog box is displayed when you place the dimension.

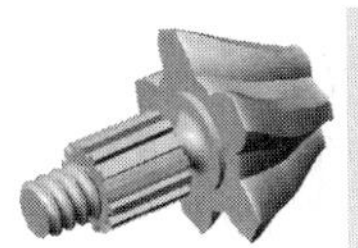

Tip: *You cannot dimension a circular object on the profiled sketch using the* ***Angular*** *option. However, you can dimension a circular entity on the unprofiled sketch using the* ***Angular*** *option of the* ***AMPOWERDIM*** *command.*

Baseline

The **Baseline** option is used to assign the baseline dimensions. Before assigning the baseline dimension you have to first assign the linear dimension to atleast one entity that will be used as the base dimension. Taking the reference of this existing linear dimension allow you to assign the baseline dimension, see Figure 4-11.

Chain

The **Chain** option is used to assign the continued dimension. You have to first assign a linear dimension to the sketch. You can now take the reference of this existing dimension and place the continued dimension using the **Chain** option, see Figure 4-12.

Baseline
Using the **Baseline** suboption you can shift to the **Baseline** option.

Single
Using the **Single** suboption you can assign a single dimension.

New
The **New** suboption is used to assign a new dimension.

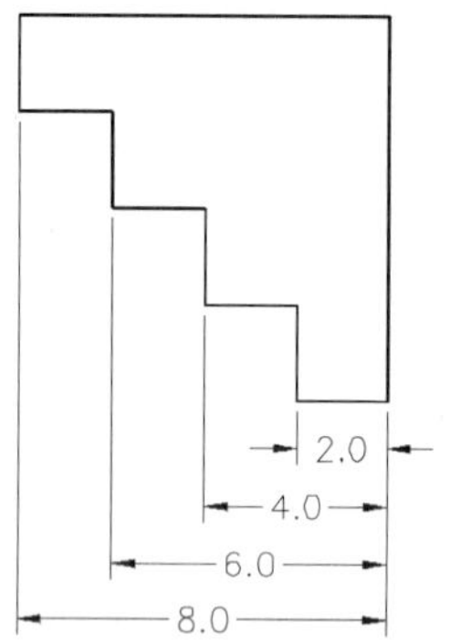

Figure 4-11 Using the baseline dimensions for dimensioning the sketch

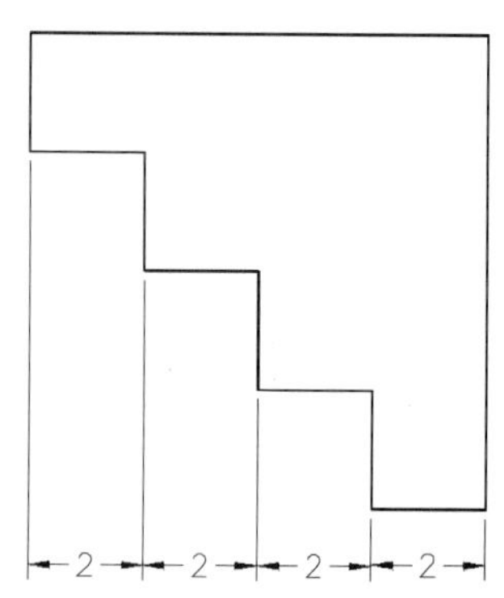

Figure 4-12 Using the chain dimensions for dimensioning the sketch

Update

The **Update** option is used to update the dimensions.

The **Power Dimensioning** dialog box will be displayed only when you place the dimensions of the first entity in the current sequence. When you place the dimension of the next entity in same sequence, you will be prompted to specify the dimension value. You will also be provided with the following options:

Associate to

The **Associate to** option is used to associate the selected dimension to an existing dimension. The value of the existing dimension will be displayed in terms of the variables. You can set the same value for the selected entity.

Equation assistant

The **Equation assistant** option is used to display the **Equation Assistant** dialog box, see Figure 4-13. This dialog box uses the design variables for defining the dimension values.

Options

This option is used to display the **Power Dimensioning** dialog box for specifying the dimension of the selected entity.

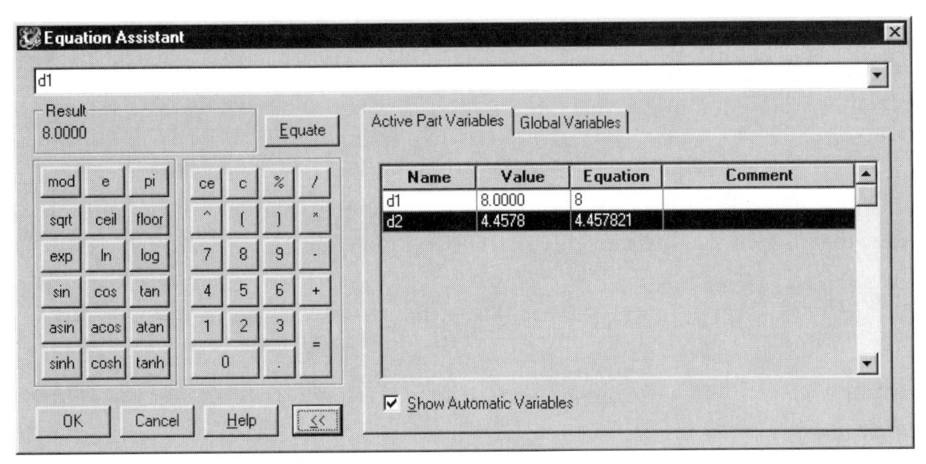

Figure 4-13 Equation Assistant dialog box

AMAUTODIM COMMAND

Menu:	Annotate > Automatic Dimensioning
Toolbar:	Drawing Layout > Power Dimensioning > Automatic Dimension
Context menu:	Dimensioning > Automatic Dimensioning
Command:	AMAUTODIM

This is the second of the automatic dimensioning techniques. This command is mainly used for dimensioning the drawing views in the **Drawing** tab. However, you can also use this command in the **Model** tab for dimensioning the profiled or the unprofiled sketch. When you invoke this command, the **Automatic Dimensioning** dialog box (Figure 4-14) is displayed.

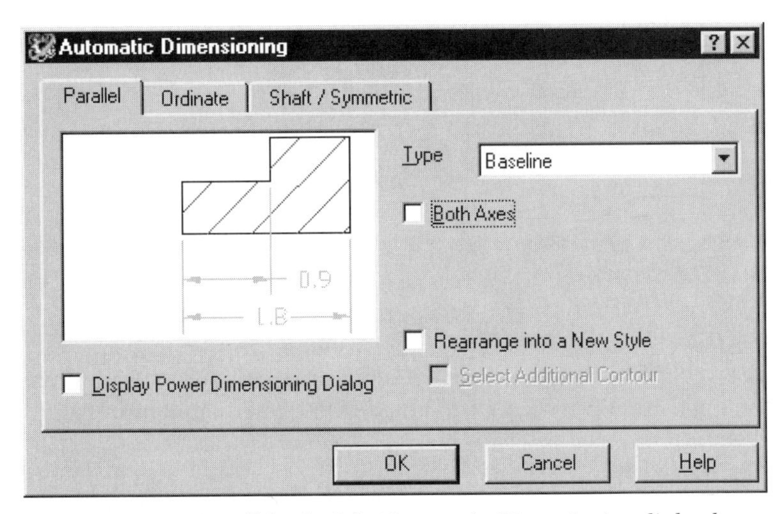

Figure 4-14 Parallel tab of the Automatic Dimensioning dialog box

Automatic Dimensioning Dialog Box Options (Parallel Tab)

The options under this tab are used to assign the linear dimensions to the profiled sketch. Various options provided under this tab are discussed below:

Type

This drop-down list is used to select the type of dimensioning. The options provided under this drop-down list are:

Baseline

The **Baseline** option is used to assign the dimensions in the baseline format starting from a single base as shown in Figure 4-15.

Chain

The **Chain** option is used to assign the dimensions in the continued dimensioning format as shown in Figure 4-16.

Both Axes

If this check box is selected, the dimensions will be assigned in two directions in a single attempt, see Figure 4-15. Once you have finished dimensioning in one axis, you will be prompted to place the dimension in the other axis. If this check box is cleared, the dimensions will be assigned only in one direction, see Figure 4-16. If you want to dimension the sketch in the other direction also then you will have to invoke this command again.

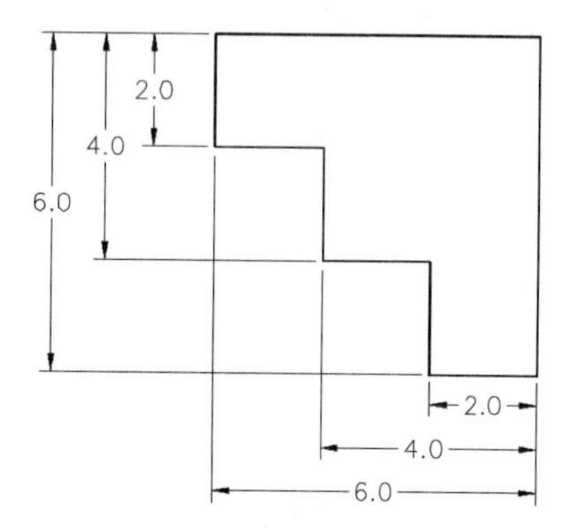

Figure 4-15 Profiled sketch dimensioned using the Baseline option in both the axes

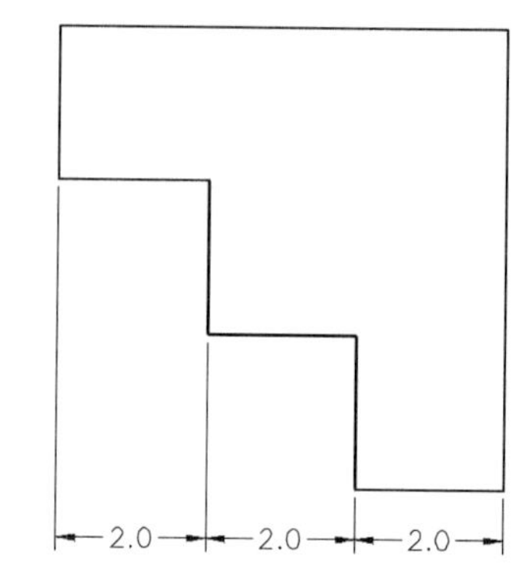

Figure 4-16 Profiled sketch dimensioned using the Chain option in one axis

Rearrange into a New Style

This check box is selected when you want to rearrange the existing automatic dimensions to a new dimension style created using the **DIMSTYLE** command.

Select Additional Contour

This check box is selected when you want to select additional edges of the sketch to dimension while the existing automatic dimensions are being rearranged to a new style. This check box is not available until the **Rearrange into a New Style** check box is selected.

Display Power Dimensioning Dialog

If this check box is selected, the **Power Dimensioning** dialog box will be displayed after you have placed the dimension. **The dimensions assigned using the AMAUTODIM command will behave as the parametric dimensions only if this check box is selected.**

Ordinate Tab

The options under the **Ordinate** tab of the **Automatic Dimensioning** dialog box are used to place the dimensions with respect to the various vertices on the outline of the selected profiled sketch, see Figure 4-17. The first point is always considered as the start point and its horizontal and vertical coordinates are taken as zero. The coordinates of the rest of the points are placed with respect of this first point. The options provided under this tab are:

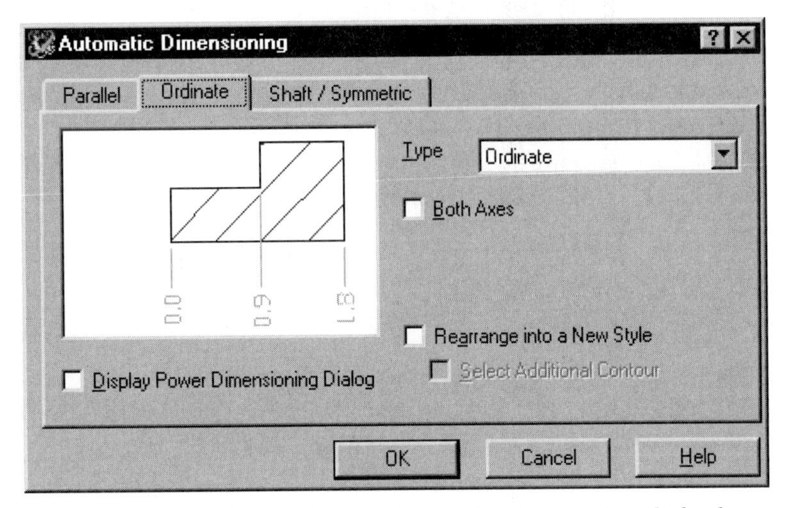

Figure 4-17 *Ordinate tab of the Automatic Dimensioning dialog box*

Type

The options provided under this drop-down list defines the type of ordinate dimensioning. They are:

Ordinate

If the **Ordinate** option is selected, all the dimensions will be placed in the same line irrespective of their start points as shown in Figure 4-18.

Equal Leader Length

If this option is selected, the length of the leader will be equal for all the dimensions. Figure 4-19 shows a sketch dimensioned using the **Equal Leader Length** option.

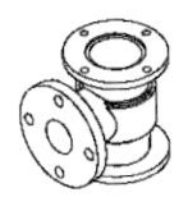

Note

The rest of the options under the ***Ordinate*** *tab are similar to those discussed under the* ***Parallel*** *tab.*

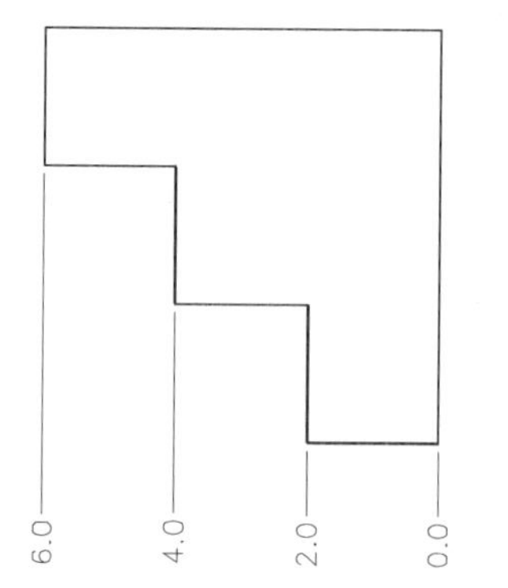

Figure 4-18 The sketch dimensioned using the ordinate dimensions

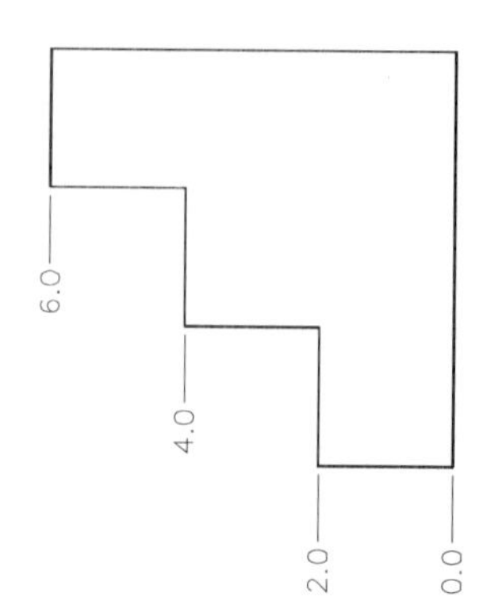

Figure 4-19 Ordinate dimensions placed using the Equal Leader Length option

Shaft/Symmetric Tab

The options under this tab are used to dimension the different views of the shafts, see Figure 4-20. The options provided under this tab are:

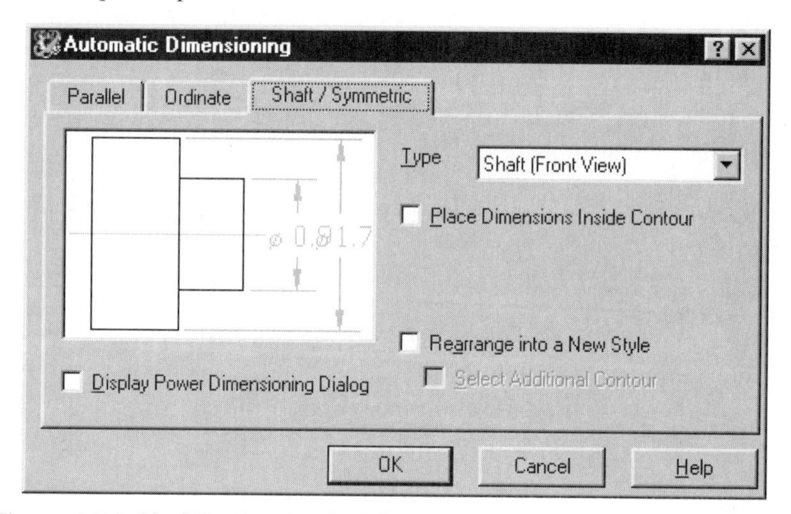

Figure 4-20 *Shaft/Symmetric tab of the Automatic Dimensioning dialog box*

Type

The options that are provided under the **Type** drop-down list are used to dimension different views of the shaft.

Shaft (Front View)

The **Shaft (Front View)** option is used to dimension the front view of the shaft. When you select the contour to be dimensioned and place the dimension, you will be prompted to select the axis of rotation. Assigning the dimensions in this manner is similar to assigning the diameter dimensions using the **AMPARDIM** command.

Shaft (Side View)

The **Shaft (Side View)** option is used to dimension the side view of the shaft. This option is mainly used to dimension the side views of the shaft in the **Drawing** tab.

Symmetric

The **Symmetric** option is used to dimension the section views of the shafts in the **Drawing** tab. Once you have selected the section view to dimension, you will be prompted to select the center line.

Place Dimensions Inside Contour

This check box is available only when you select **Shaft (Front View)** from the **Type** drop-down list. If you select this check box, the dimensions will be placed inside the sketch selected for dimensioning.

The rest of the options under this tab are similar to those under the **Parallel** or the **Ordinate** tab.

DESIGN VARIABLES

In Mechanical Desktop the dimensions can be assigned to the objects in the form of the design variables that can be numeric values, parameters (these are the unique titles associated with the dimensions), or equations (consisting of some other variables). The design variables cannot only be applied to the sketches, but even in the dialog box edit boxes you can enter the value in these forms. There are two types of the design variables; the **Active Part** design variables and the **Global** design variables. The **Active Part** design variables can only be applied to the profiled sketches or to an existing active part. The **Global** design variable can be used for any part (may or may not be active) in this drawing or even to the parts in another drawing. You can create these design variables using the **AMVARS** command discussed below.

Creating The Design Variables (AMVARS Command)

Toolbar:	2D Constraints > Design Variables
Menu:	Part > Design Variables
Command:	AMVARS

You can also invoke this command by using the shortcut menu displayed upon right-clicking on the part icon in the desktop browser. Choose **Design Variables** to invoke this command. When you invoke this command, the **Design Variables** dialog box is displayed as shown in Figure 4-21.

Design Variables Dialog Box Options (Active Part Tab)

The options provided under this tab are used for creating the active part design variables.

Text Box

The text box provided under this tab consists of seven columns. They are:

T. Indicates that the variable is a table driven.

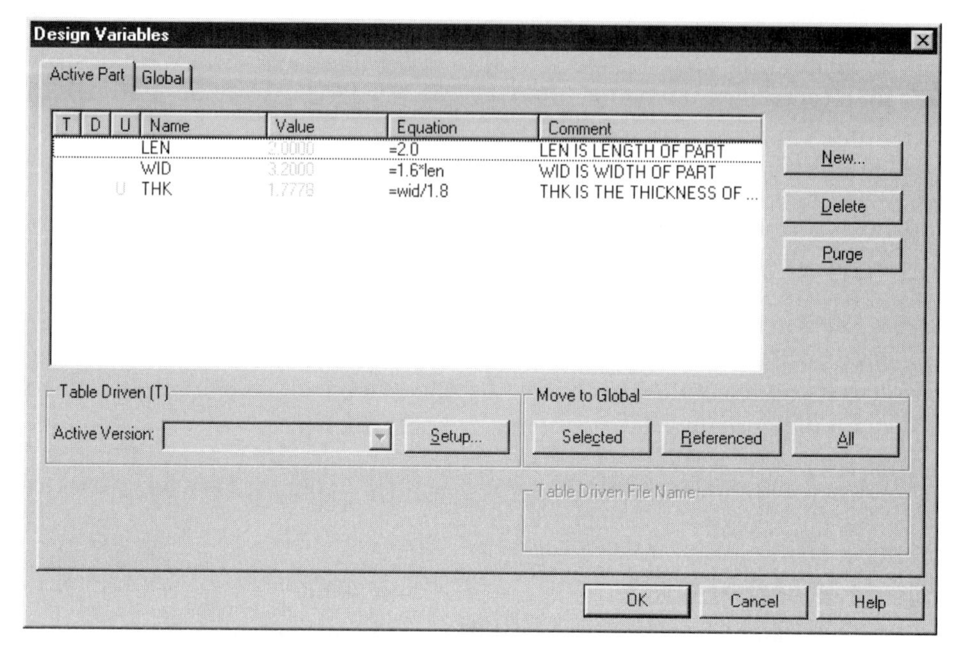

***Figure 4-21** Active Part tab of the Design Variables dialog box*

D. Indicates that there already exists a Global variable with the same name.

U. Symbolizes that the variable has not been referenced in the active part.

Name. Displays the name of the variable.

Value. Displays the numeric value of the variable.

Equation. Displays the equation used for creating the variable.

Comment. Displays the brief description regarding the variable.

New

The **New** button is used to define a new design variable. When you choose this button, the **New Part Variable** dialog box is displayed as shown in Figure 4-22.

Tip: *The **Comment** edit box of the **New Part Variable** dialog box is optional. You may or may not enter the comments regarding the design variable in this edit box.*

Delete

The **Delete** button is used to delete the design variables.

Purge

The **Purge** button is used to delete all unused design variables.

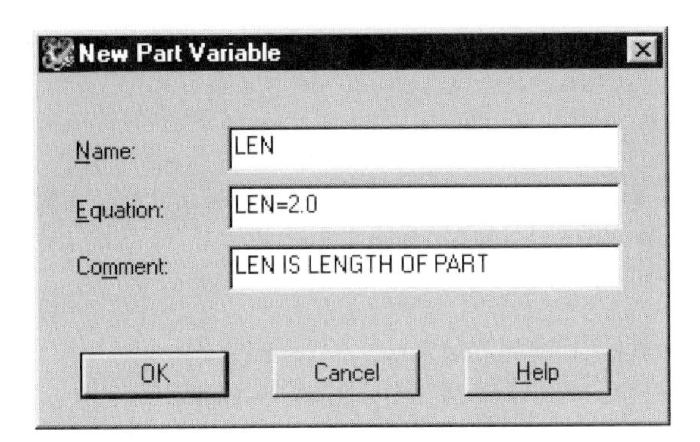

Figure 4-22 The New Part Variable dialog box

Table Driven [T] Area

This area provides the following options:

Active Version. This drop-down list displays the active version of the part.

Setup. When the **Setup** button is chosen, the **Table Driven Setup** dialog box is displayed, see Figure 4-23. The options provided in this dialog box are used to create the MS Excel spread sheet.

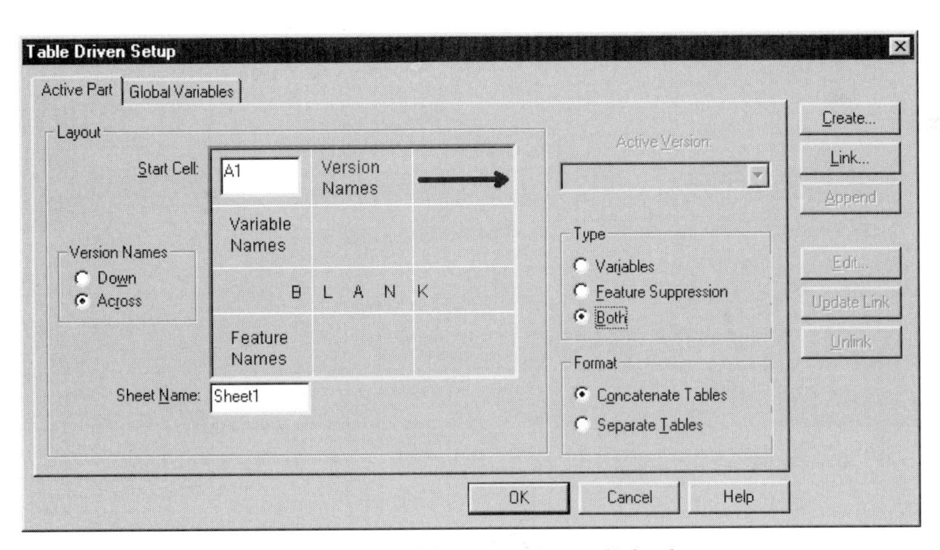

Figure 4-23 Table Driven Setup dialog box

Move to Global Area

The options provided under this area are:

Selected. The **Selected** button is chosen to convert the selected active design variable to the global variable. Once the active part variable is converted into the global design variable, it is no longer displayed in the **Text Box** of the **Active Part** tab.

Referenced. The **Referenced** button is chosen to convert all the referenced active part design variables into global design variables. They are not displayed in the **Text Box** once they are converted into global design variables.

All. The **All** button converts all the active part design variables into the global design variables. They are not displayed in the **Text Box** once they are converted into global design variables.

Table Driven File Name Area. This area displays the file name of the Microsoft Excel spread sheet along with the path in which it is saved.

Global Tab

The options provided under this tab are similar to those provided under the **Active Part** but are used to create the global design variables, see Figure 4-24.

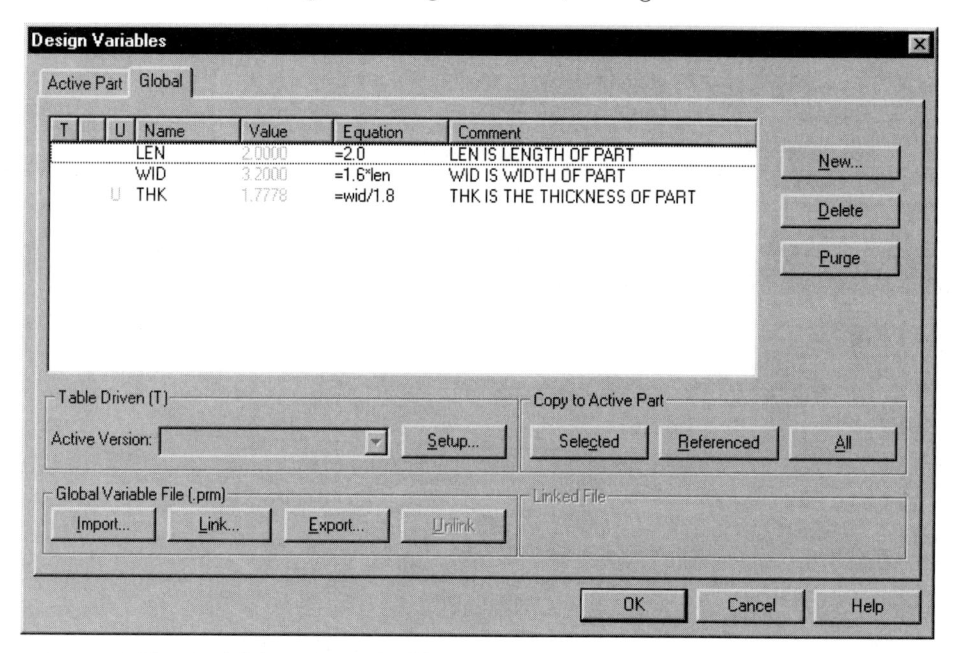

Figure 4-24 Design Variables dialog box displaying the Global tab

Global Variable File [.prm] Area

The options provided under this area are:

Import. The **Import** button is chosen to import an external parameter (.PRM) file containing the information regarding the global design variables.

Link. The **Link** button is chosen to link the external parameter file containing the information regarding the global design variables to the current drawing.

Export. The **Export** button is chosen to export the global design variables to the external parameter file (.PRM).

Unlink. This button when chosen unlinks the selected external parameter file from the current drawing.

Copy to Active Part Area

The options provided under this area are:

Selected. The **Selected** button is chosen to add the selected global design variable to the list of active part design variables. The design variable will now be displayed under both the active part and the global design variables list.

Referenced. This button is chosen to copy the referenced global design variables to the list of the active part design variables.

All. This button is chosen to add all the global design variables to the list of the active part design variables. They will now be displayed under the list of the active part as well as the global design variables.

The rest of the options under this tab are similar to those discussed under the **Active Part** tab.

DISPLAYING THE DIMENSIONS AS VARIABLES

Once you have created the design variables, you can dimension the profiled sketches using these design variables. The dimensions applied to the sketch can now be displayed in the form of the equations, parameters, or the numeric values using various options of the **AMDIMDSP** command.

Displaying Dimensions As Parameters (AMDIMDSP Command)

Toolbar: 2D Constraints > Design Variables > Display As Variables
Menu: Part > Dimensioning > Dimensions As Parameters
Context Menu: Dimensioning > Dimensions As Parameters
Command: AMDIMDSP

This command can be invoked by selecting the **Parameters** option of the **AMDIMDSP** command. When you choose this button, the dimensions applied to the sketch will be displayed as parameters as shown in Figure 4-25.

Displaying Dimensions As Numbers (AMDIMDSP Command)

Toolbar: 2D Constraints > Design Variables > Display As Numbers
Menu: Part > Dimensioning > Dimensions As Numbers
Context Menu: Dimensioning > Dimensions As Numbers
Command: AMDIMDSP

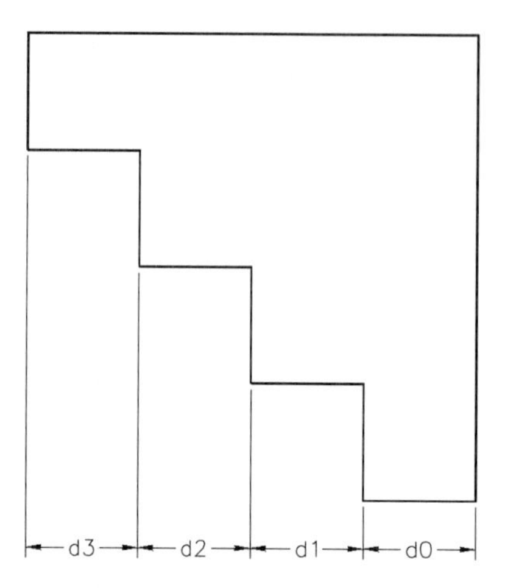

***Figure 4-25** Displaying the dimensions as Parameters*

Using this option you can display the dimensions assigned to the sketch in the form of numeric values as shown in Figure 4-26. This is the default option. You can also display the dimensions assigned to the sketch in the form of numerics by using the **Numeric** option of the **AMDIMDSP** command.

Dimensions as Equations (AMDIMDSP Command)

Toolbar:	2D Constraints > Design Variables > Display As Equations
Menu:	Part > Dimensioning > Dimensions As Equations
Context Menu:	Dimensioning > Dimensions As Equations
Command:	AMDIMDSP

When you choose this button, the dimensions assigned to the sketch are displayed in the form of the equations, see Figure 4-27. You can also display the dimensions as equations using the **Equations** option of the **AMDIMDSP** command.

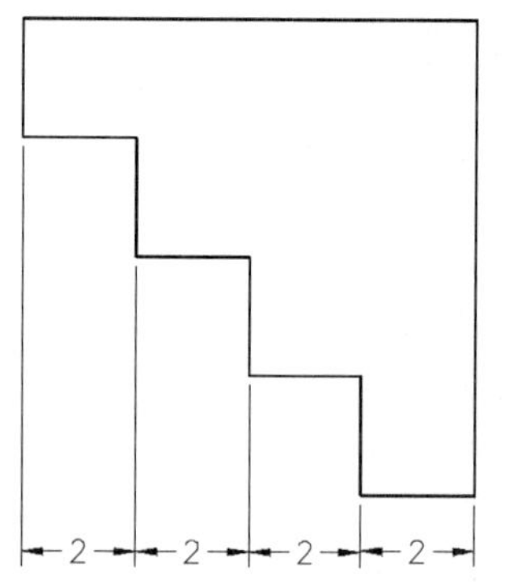

***Figure 4-26** Displaying the dimensions as Numerics*

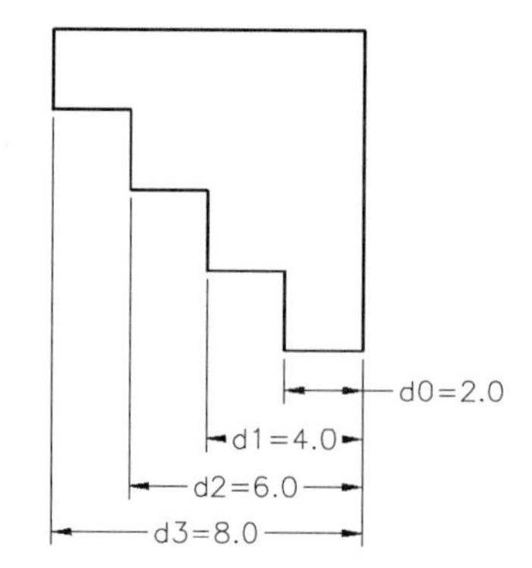

***Figure 4-27** Displaying the dimensions as Equations*

VISIBILITY OPTIONS

The visibility of the designer model or other features like work planes, work points, work axes, and so on can be controlled in Mechanical Desktop using a single dialog box; the **Desktop Visibility** dialog box. This dialog box can be displayed using the **AMVISIBLE** command discussed below.

AMVISIBLE Command

Toolbar:	Part Modeling > Part Visibility
Menu:	Part > Part Visibility
Command:	AMVISIBLE

This command can also be invoked by choosing the **Visibility** button provided on the toolbar attached to the desktop browser. When you invoke this command, the **Desktop Visibility** dialog box is displayed as shown in Figure 4-28.

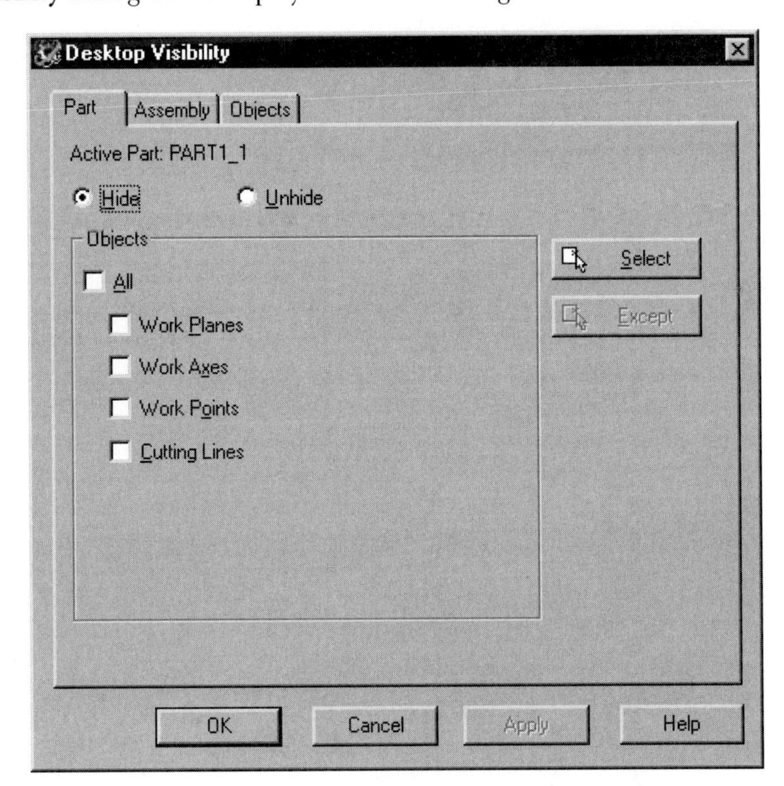

Figure 4-28 *Desktop Visibility dialog box displaying the Part tab*

Desktop Visibility Dialog Box Options (Part Tab)

The options provided under the **Part** tab of the **Desktop Visibility** dialog box are:

Hide

The **Hide** radio button is selected to hide the selected objects.

Unhide
The **Unhide** radio button is selected to unhide the selected objects.

Objects Area
This area provides you with the following check boxes:

All. This check box when selected hides or unhides all the work features and the cutting lines defined while creating the designer model.

Note
Creating the cutting lines will be described in later chapters.

Work Planes. This check box is used to control the visibility of the work planes associated with the active designer model.

Work Axes. This check box is used to control the visibility of the work axes associated with the active designer model.

Work Points. This check box is used to control the visibility of the work points associated with the active designer model.

Cutting Lines. This check box is used to control the visibility of the cutting lines in the active designer model.

Select
The **Select** button is used to select the object whose visibility you want to control. When you choose this button, the **Desktop Visibility** dialog box is temporarily closed so that you can select the object from the screen. Once the required object is selected, press ENTER. The dialog box will be displayed again.

Except
The **Except** button is used to remove the objects from a selection set created using the **All** check box. This button is available only if the **All** check box is selected. When you choose this button, the **Desktop Visibility** dialog box is closed temporarily so that you can select the objects to remove from the selection set created using the **All** check box.

Objects Tab
The options provided under the **Object** tab of the **Desktop Visibility** dialog box are:

Objects Area
The check boxes provided under this area is used to control the visibility of the respective objects.

Properties Area
The visibility of the objects can also be controlled based on their properties like color, layer, and linetype using the options provided under this area.

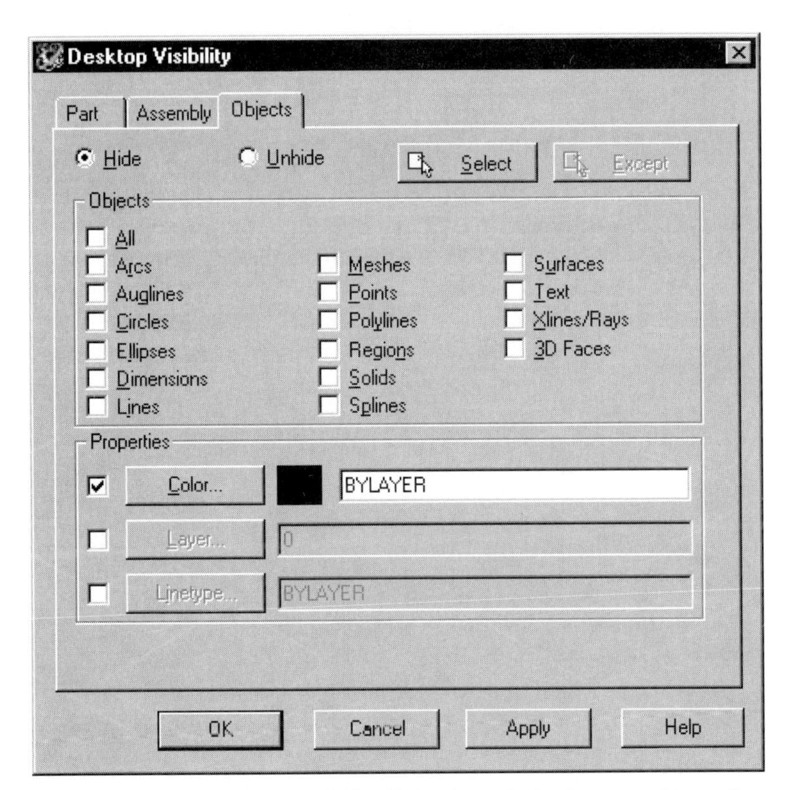

Figure 4-29 *Desktop Visibility dialog box displaying the Object tab*

Color. This option is activated by selecting the check box provided on the left of the **Color** button. You can control the visibility of the objects of a particular color. You can select that particular color from the **Select Color** dialog box or by entering the name of the color in the **Color** edit box. The **Select Color** dialog box is displayed by choosing the **Color** button.

Layer. This option is used to control the visibility of the objects in a particular layer. The layer can be selected from the **Select Layer** dialog box displayed by choosing the **Layer** button. You can also select the layer by entering the name of the layer in the **Layer** edit box.

Linetype. This option is used to control the visibility of the objects consisting of a particular linetype. The linetype can be selected from the **Select Linetype** dialog box displayed by choosing the **Linetype** button. You can also select the linetype by entering its name in the edit box.

The rest of the options under the **Objects** tab are similar to those discussed under the **Part** tab.

Note

The options that are provided under the ***Assembly*** *tab of the* ***Desktop Visibility*** *dialog box will be discussed in later chapters.*

TUTORIALS

Tutorial 1

In this tutorial you will create the designer model shown in Figure 4-30a. Dimension the sketches using the advanced dimensioning techniques. The dimensions for the model are shown in Figures 4-30b, 4-30c, and 4-30d. Assume the missing dimensions.

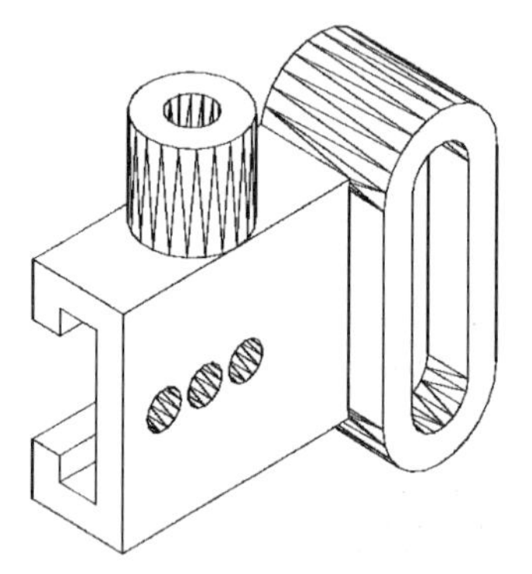

Figure 4-30a *Designer model for Tutorial 1*

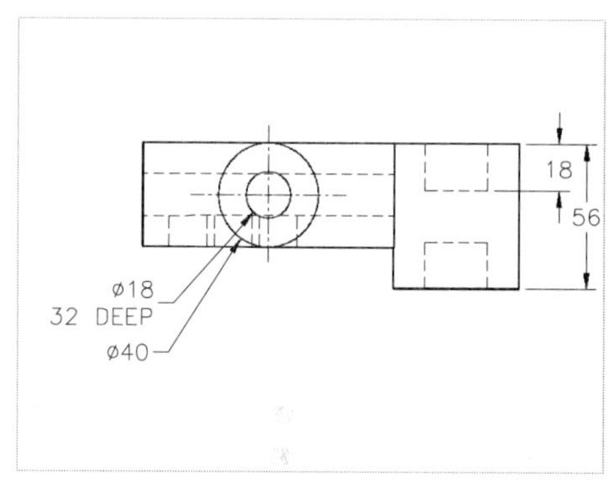

Figure 4-30b *Top view of the model*

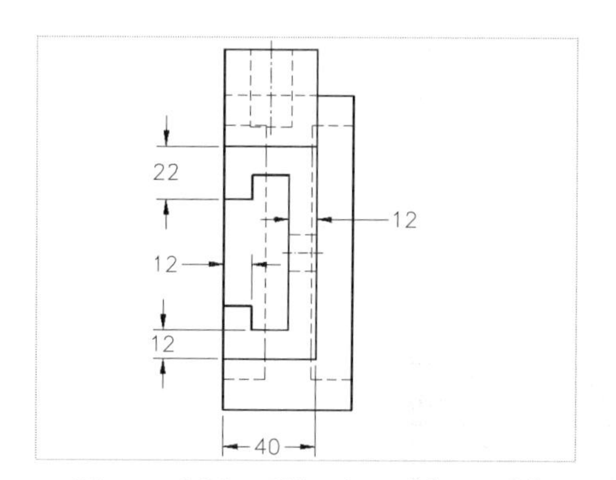

Figure 4-30c *Side view of the model*

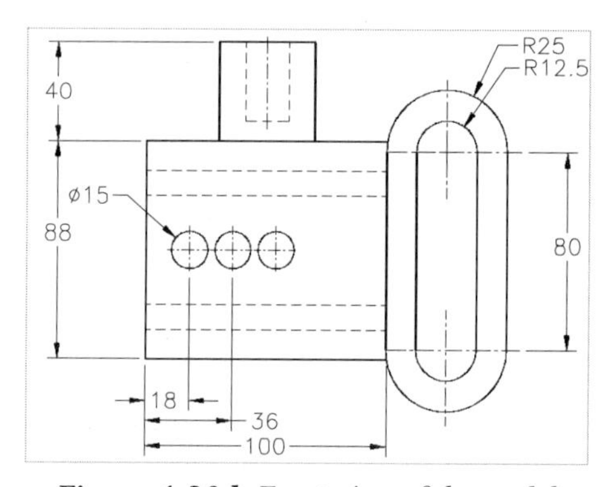

Figure 4-30d *Front view of the model*

1. Create the rough sketch for the base feature and as shown in Figure 4-31. Profile the base sketch using the **AMPROFILE** command.

2. Dimension the sketch using the power dimensioning technique. Choose the **Power Dimensioning** button from the **Part Modeling** toolbar. The prompt sequence is as follows:

(Single) Specify first extension line origin or
[Angular/Options/Baseline/Chain/Update] <Select>: Enter

Figure 4-31 *Sketch for the base feature*

Select arc, line, circle or dimension: *Select the upper arc.*
Specify dimension line location or [Diameter/Options]: *Place the dimension*

As soon as you place the dimension, the **Power Dimensioning** dialog box will bc displayed as shown in Figure 4-32.

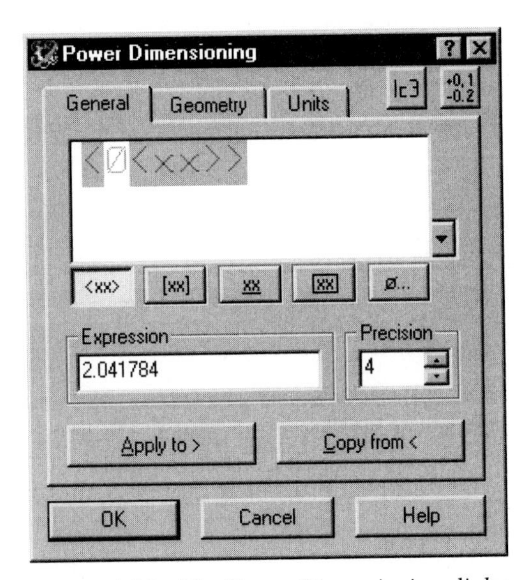

Figure 4-32 *The Power Dimensioning dialog box*

3. Enter **25** in the expression edit box. Choose **OK**. The prompt sequence is as follows:

Solved under constrained sketch requiring 1 dimension(s) or constraint(s).

(Single) Specify first extension line origin or
[Angular/Options/Baseline/Chain/Update] <Select>: Enter
Select arc, line, circle or dimension: *Select the left vertical line.*
Specify dimension line location or [Options/Pickobj]: *Place the dimension.*
Enter dimension value [Associate to/Equation assistant/Options] <default value>: **80**

Solved fully constrained sketch.
(Single) Specify first extension line origin or
[Angular/Options/Baseline/Chain/Update] <Select>: Enter
Select arc, line, circle or dimension: Enter

The desired sketch for the base part should be similar to the one shown in Figure 4-33.

4. Extrude the sketch and create the base feature as shown in Figure 4-34.

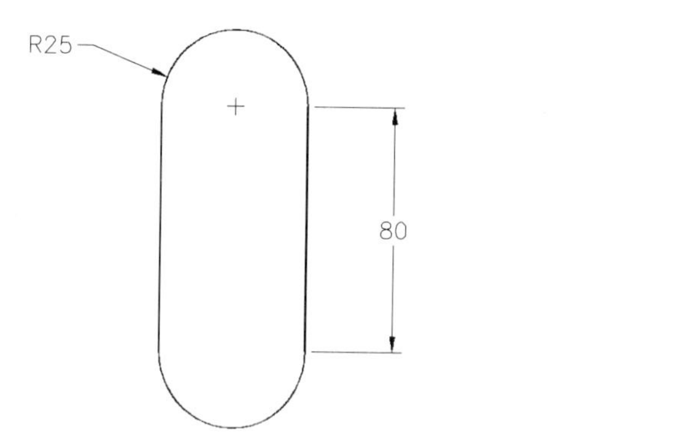

Figure 4-33 *Desired sketch for the base part*

Figure 4-34 *Base feature created upon extruding the base sketch*

5. Define the sketch plane on the front face of the base part to create the sketch for the next feature.

6. Create the sketch for the next feature on the front face and profile the sketch. Apply the required constraints.

7. Choose the **Power Dimensioning** button from the **Part Modeling** toolbar. The prompt sequence is as follows:

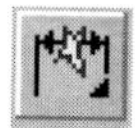

(Single) Specify first extension line origin or
[Angular/Options/Baseline/Chain/Update] <Select>: Enter
Select arc, line, circle or dimension: *Select one of the arcs.*
Specify dimension line location or [Diameter/Options]: **D**
Specify dimension line location or [Radius/Options]: *Place the dimension.*

The **Power Dimensioning** dialog box will be displayed. Enter **25** in the **Expression** edit box. Choose **OK**. The prompt sequence is as follows:

Solved under constrained sketch requiring 1 dimension(s) or constraint(s).
(Single) Specify first extension line origin or
[Angular/Options/Baseline/Chain/Update] <Select>: Enter
Select arc, line, circle or dimension: *Select one of the vertical lines.*
Specify dimension line location or [Options/Pickobj]: *Place the dimension.*
Enter dimension value [Associate to/Equation assistant/Options] <default value>: **80**

Solved fully constrained sketch.
(Single) Specify first extension line origin or
[Angular/Options/Baseline/Chain/Update] <Select>: *Press ESC.*

8. The sketch for the next feature after dimensioning should look similar to the one shown in Figure 4-35.

9. Extrude this sketch using the **Cut** operation to a distance of **18** units.

10. Using the above mentioned steps create this feature on the back face of the model. The designer model after creating the features on the front and the back face should look similar to the one shown in Figure 4-36.

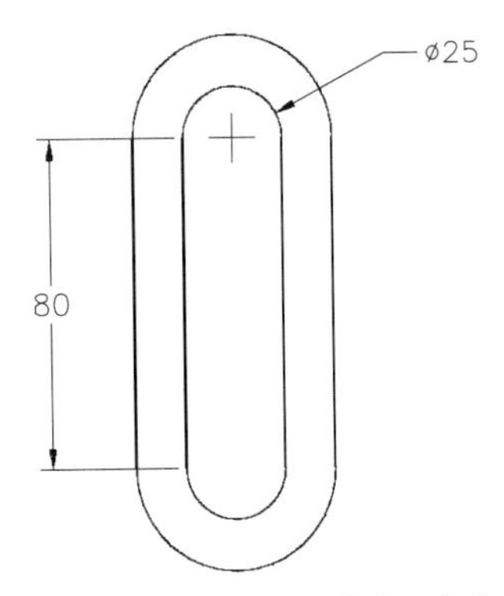

Figure 4-35 *Full dimensioned sketch for the next feature*

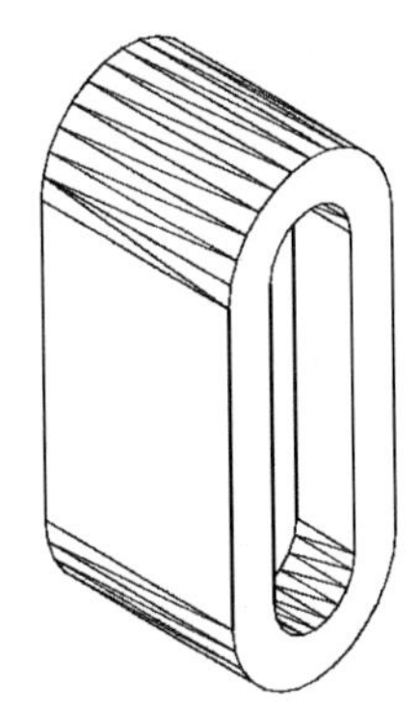

Figure 4-36 *Designer model after creating the next features on both sides*

11. Define a work plane parallel to the side face of the model and at a distance of 100 units from the side face using the **Planar Parallel-Offset** option, see Figure 4-37.

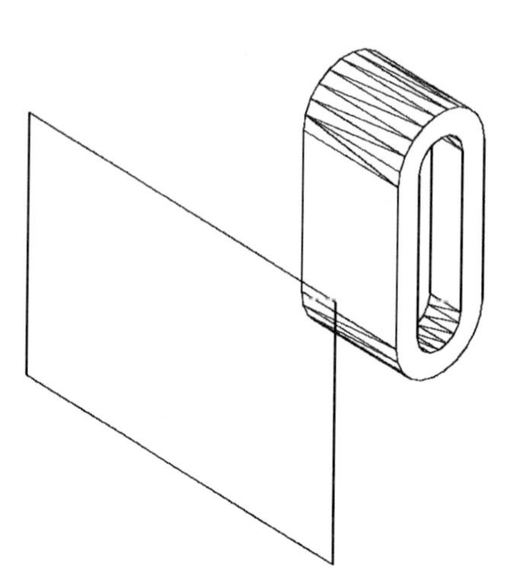

Figure 4-37 *Work plane defined at a distance of 100 units from the side face*

12. Create the rough sketch for the next feature. Profile the sketch. The lines of the sketch have been numbered for the ease of selecting, see Figure 4-38.

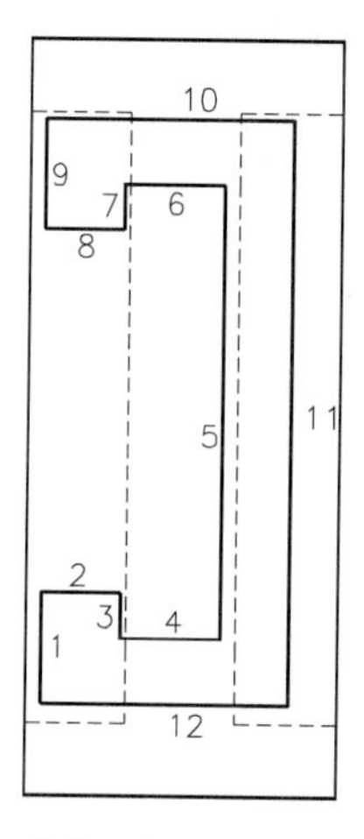

Figure 4-38 *Profiled sketch for the next feature displaying the numbers assigned to the lines of the sketch*

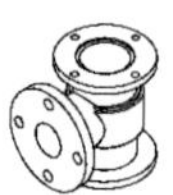

Note

The lines 1 and 9, 2 and 8, 3 and 7, 4 and 6, and 12 and 10 are assigned the equal length constraint.

13. Choose the **Power Dimensioning** button from the **Part Modeling** toolbar. The prompt sequence is as follows:

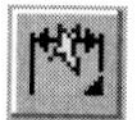

(Single) Specify first extension line origin or
[Angular/Options/Baseline/Chain/Update] <Select>: *Select the lower endpoint of line 1.*
Specify second extension line origin: *Select the center of the lower circular base feature.*
Specify dimension line location or [Options/Pickobj]: **O**

When you enter **O** at the Command prompt, the **Select Dimension Orientation** dialog box will be displayed as shown in Figure 4-39. Choose the **Horizontal** button and then choose **OK**.

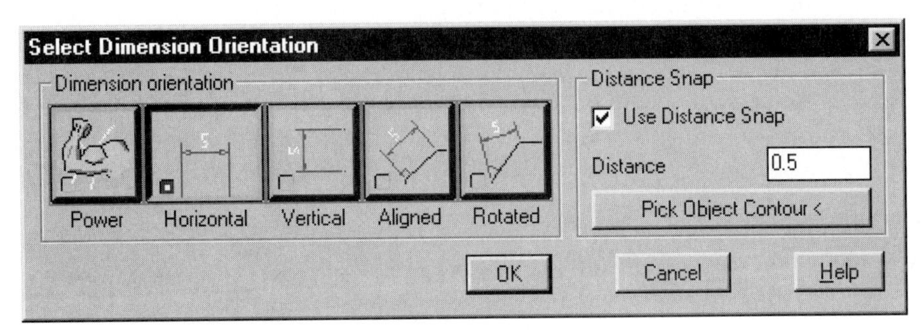

Figure 4-39 *Select Dimension Orientation dialog box*

The prompt sequence is as follows:

Specify dimension line location or [Options/Pickobj]: *Place the dimension.*

Enter **0** in the **Expression** edit box and then choose **OK**.

Solved under constrained sketch requiring 7 dimension(s) or constraint(s).
(Single) Specify first extension line origin or
[Angular/Options/Baseline/Chain/Update] <Select>: Enter
Select arc, line, circle or dimension: *Select line 1.*
Specify dimension line location or [Options/Pickobj]: *Place the dimension.*
Enter dimension value [Associate to/Equation assistant/Options] <default value>: **22**

Solved under constrained sketch requiring 6 dimension(s) or constraint(s).
(Single) Specify first extension line origin or
[Angular/Options/Baseline/Chain/Update] <Select>: Enter
Select arc, line, circle or dimension: *Select line 2.*
Specify dimension line location or [Options/Pickobj]: *Place the dimension.*
Enter dimension value [Associate to/Equation assistant/Options] <default value>: **12**

Solved under constrained sketch requiring 5 dimension(s) or constraint(s).
(Single) Specify first extension line origin or
[Angular/Options/Baseline/Chain/Update] <Select>: Enter
Select arc, line, circle or dimension: *Select line 3.*
Specify dimension line location or [Options/Pickobj]: *Place the dimension.*
Enter dimension value [Associate to/Equation assistant/Options] <default value>: **10**

Solved under constrained sketch requiring 4 dimension(s) or constraint(s).
(Single) Specify first extension line origin or
[Angular/Options/Baseline/Chain/Update] <Select>: Enter
Select arc, line, circle or dimension: *Select line 10.*
Specify dimension line location or [Options/Pickobj]: *Place the dimension.*
Enter dimension value [Associate to/Equation assistant/Options] <default value>: 16

Solved under constrained sketch requiring 3 dimension(s) or constraint(s).
(Single) Specify first extension line origin or
[Angular/Options/Baseline/Chain/Update] <Select>: Enter
Select arc, line, circle or dimension: *Select line 12.*
Specify dimension line location or [Options/Pickobj]: *Place the dimension.*
Enter dimension value [Associate to/Equation assistant/Options] <default value>: **40**

Solved under constrained sketch requiring 2 dimension(s) or constraint(s).
(Single) Specify first extension line origin or
[Angular/Options/Baseline/Chain/Update] <Select>: Enter
Select arc, line, circle or dimension: *Select line 11.*
Specify dimension line location or [Options/Pickobj]: *Place the dimension.*
Enter dimension value [Associate to/Equation assistant/Options] <default value>: **88**

Solved under constrained sketch requiring 1 dimension(s) or constraint(s).
(Single) Specify first extension line origin or
[Angular/Options/Baseline/Chain/Update] <Select>: *Select the lower endpoint of line 9.*
Specify second extension line origin: *Select the center point of the upper circular feature.*
Specify dimension line location or [Options/Pickobj]: *Place the dimension.*
Enter dimension value [Associate to/Equation assistant/Options] <default value>: **18**

Solved fully constrained sketch.
(Single) Specify first extension line origin or
[Angular/Options/Baseline/Chain/Update] <Select>: *Press ESC.*

14. Extrude this sketch using the **Join** operation and **Next** termination, see Figure 4-40.

15. Define a sketch plane on the back face of the new feature. Create three circles for creating the holes. Profile the sketches of the three holes.

16. Choose the **Power Dimensioning** button from the **Part Modeling** toolbar. The prompt sequence is as follows:

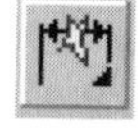

(Single) Specify first extension line origin or
[Angular/Options/Baseline/Chain/Update] <Select>: Enter
Select arc, line, circle or dimension: *Select the first circle.*
Specify dimension line location or [Radius/Options]: *Place the dimension.*

Enter **15** in the **Expression** edit box. Choose **OK**.

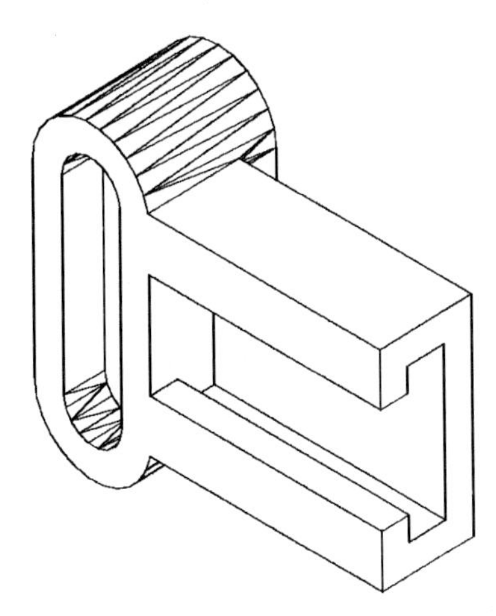

Figure 4-40 *Designer model after creating the next feature*

Solved under constrained sketch requiring 8 dimension(s) or constraint(s).
(Single) Specify first extension line origin or
[Angular/Options/Baseline/Chain/Update] <Select>: Enter
Select arc, line, circle or dimension: *Select the second circle.*
Specify dimension line location or [Radius/Options]: *Place the dimension.*
Enter dimension value [Associate to/Equation assistant/Options] <default value>: **15**

Solved under constrained sketch requiring 7 dimension(s) or constraint(s).
(Single) Specify first extension line origin or
[Angular/Options/Baseline/Chain/Update] <Select>: Enter
Select arc, line, circle or dimension: *Select the third circle.*
Specify dimension line location or [Radius/Options]: *Place the dimension.*
Enter dimension value [Associate to/Equation assistant/Options] <default value>: **15**

Solved under constrained sketch requiring 6 dimension(s) or constraint(s).
(Single) Specify first extension line origin or
[Angular/Options/Baseline/Chain/Update] <Select>: *Select the center of the first circle.*
Specify second extension line origin: *Select the upper endpoint left vertical edge of the model.*
Specify dimension line location or [Options/Pickobj]: *Place the dimension.*
Enter dimension value [Associate to/Equation assistant/Options] <default value>: **O**

Choose the **Horizontal** button from the **Select Dimension Orientation** dialog box. Choose **OK**.

Enter dimension value [Associate to/Equation assistant/Options] <default value>: **18**

Solved under constrained sketch requiring 5 dimension(s) or constraint(s).
(Single) Specify first extension line origin or
[Angular/Options/Baseline/Chain/Update] <Select>: *Select the center of the first circle.*

Specify second extension line origin: *Select the center of second circle.*
Specify dimension line location or [Options/Pickobj]: *Place the dimension to display the horizontal dimension.*
Enter dimension value [Associate to/Equation assistant/Options] <default value>: **18**

Solved under constrained sketch requiring 4 dimension(s) or constraint(s).
(Single) Specify first extension line origin or
[Angular/Options/Baseline/Chain/Update] <Select>: *Select the center of the second circle.*
Specify second extension line origin: *Select the center of the third circle.*
Specify dimension line location or [Options/Pickobj]: *Place the dimension to display the horizontal dimension.*
Enter dimension value [Associate to/Equation assistant/Options] <default value>: **18**

Solved under constrained sketch requiring 3 dimension(s) or constraint(s).
(Single) Specify first extension line origin or
[Angular/Options/Baseline/Chain/Update] <Select>: *Select the center of first circle.*
Specify second extension line origin: *Select the upper endpoint of left vertical edge.*
Specify dimension line location or [Options/Pickobj]: **O**

Choose the **Vertical** button from the **Select Dimension Orientation** dialog box. Choose **OK**.

Enter dimension value [Associate to/Equation assistant/Options] <default value>: **44**

Solved under constrained sketch requiring 2 dimension(s) or constraint(s).
(Single) Specify first extension line origin or
[Angular/Options/Baseline/Chain/Update] <Select>: *Select the center of first circle.*
Specify second extension line origin: *Select the center of second circle.*
Specify dimension line location or [Options/Pickobj]: *Place the dimension.*

Select **Move Away** button from the **Dimension Overdrawn** dialog box.

Enter dimension value [Associate to/Equation assistant/Options] <default value>: **0**

Solved under constrained sketch requiring 1 dimension(s) or constraint(s).
(Single) Specify first extension line origin or
[Angular/Options/Baseline/Chain/Update] <Select>: *Select the center of second circle.*
Specify second extension line origin: *Select the center of third circle.*
Specify dimension line location or [Options/Pickobj]: *Place the dimension.*
Enter dimension value [Associate to/Equation assistant/Options] <default value>: **0**

Solved fully constrained sketch.
(Single) Specify first extension line origin or
[Angular/Options/Baseline/Chain/Update] <Select>: *Press ESC.*

17. Extrude the fully dimensioned sketches through the existing feature using the **Cut** operation, see Figure 4-41.

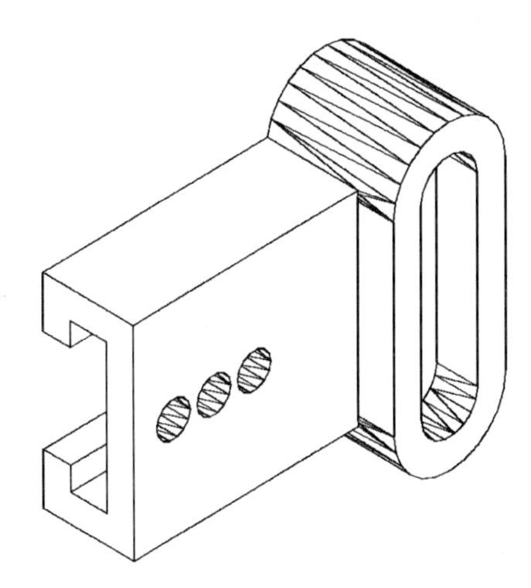

Figure 4-41 *Designer model after creating the circular cut features*

18. Define a sketch plane on top face of the second join feature for creating the cylindrical feature.

19. Draw the sketch for the cylindrical feature. Profile this sketch.

20. Choose the **Power Dimensioning** button from the **Part Modeling** toolbar. The prompt sequence is as follows:

(Single) Specify first extension line origin or
[Angular/Options/Baseline/Chain/Update] <Select>: Enter
Select arc, line, circle or dimension: *Select the circle.*
Specify dimension line location or [Radius/Options]: *Place the dimension.*

Enter **40** in the **Expression** edit box of the **Power Dimensioning** dialog box. Choose **OK.**

Solved under constrained sketch requiring 2 dimension(s) or constraint(s).
(Single) Specify first extension line origin or
[Angular/Options/Baseline/Chain/Update] <Select>: *Select the center of circle.*
Specify second extension line origin: *Select the upper endpoint of the right vertical edge.*
Specify dimension line location or [Options/Pickobj]: *Place the dimension so that it displays the vertical dimension.*
Enter dimension value [Associate to/Equation assistant/Options] <default value>: **20**

Solved under constrained sketch requiring 1 dimension(s) or constraint(s).
(Single) Specify first extension line origin or
[Angular/Options/Baseline/Chain/Update] <Select>: *Select the center of circle.*
Specify second extension line origin: *Select the lower endpoint of the right vertical edge.*
Specify dimension line location or [Options/Pickobj]: *Place the dimension so that it displays the horizontal dimension.*
Enter dimension value [Associate to/Equation assistant/Options] <default value>: **50**

Solved fully constrained sketch.
(Single) Specify first extension line origin or
[Angular/Options/Baseline/Chain/Update] <Select>: *Press ESC.*

21. Extrude this sketch using the **Join** operation up to a distance of **40**.

22. Define a sketch plane on the top face of the cylindrical feature.

23. Create the sketch for the hole and profile this sketch. Add the concentric constraint to this sketch.

24. Choose the **Power Dimensioning** button from the **Part Modeling toolbar.** The prompt sequence is as follows:

(Single) Specify first extension line origin or
[Angular/Options/Baseline/Chain/Update] <Select>: Enter
Select arc, line, circle or dimension: *Select the circle.*
Specify dimension line location or [Radius/Options]: *Place the dimension.*

Enter **18** in the **Expression** edit box of the **Power Dimensioning** dialog box. Choose **OK**.

25. Extrude this sketch using the **Cut** operation up to a distance of **32**. The final designer model should be similar to the one shown in Figure 4-42.

26. Save this drawing with the name given below:

\MDT Tut\Ch-4**Tut1.dwg**.

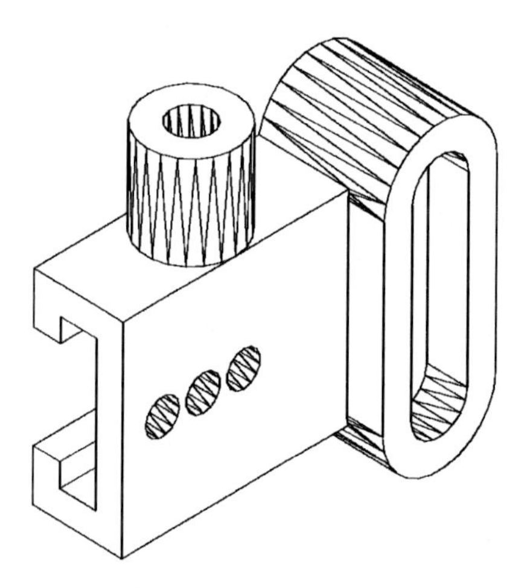

Figure 4-42 *The final designer model for Tutorial 1*

Tutorial 2

In this tutorial you will create the sketch shown in Figure 4-43. Assign the dimensions to the sketch in the form of the design variables. Create the MS Excel spread sheet. The numeric values that have to be used for the variables are given below. After assigning the dimensions, display them as equations and parameters.

LEN1 = 80, LEN2 = LEN1/5
HGT1 = 65, HGT2 = HGT1/1.85
RAD = 10

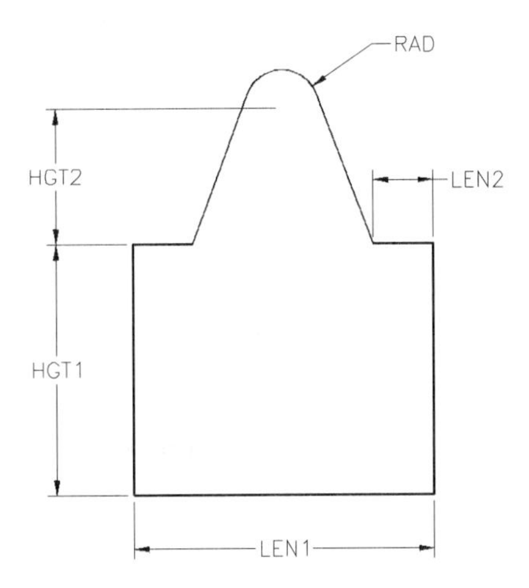

Figure 4-43 *Drawing for Tutorial 2*

1. Create the rough sketch for the object shown in Figure 4-46 and profile this sketch.

2. Choose the **Design Variables** button from the **2D Constraints** toolbar. The **Design Variables** dialog box will be displayed.

3. Choose the **New** button under the **Active Part** tab. The **New Part Variable** dialog box will be displayed as shown in Figure 4-44.

4. In the **Name** edit box enter **LEN1**. Enter **LEN1=80** in the **Equation** edit box. Enter **LEN1 IS LENGTH OF BASE** in the **Comments** edit box. Choose **OK**, see Figure 4-44.

5. Choose the **New** button from the **Design Variable** dialog box.

6. In the **Name** edit box enter **LEN2**. Enter **LEN2=15** in the **Equation** edit box. Enter **LEN1 IS SMALLER LENGTH** in the **Comments** edit box. Choose **OK**.

7. Choose the **New** button from the **Design Variable** dialog box.

8. In the **Name** edit box enter **HGT1**. Enter **HGT1=65** in the **Equation** edit box. Enter

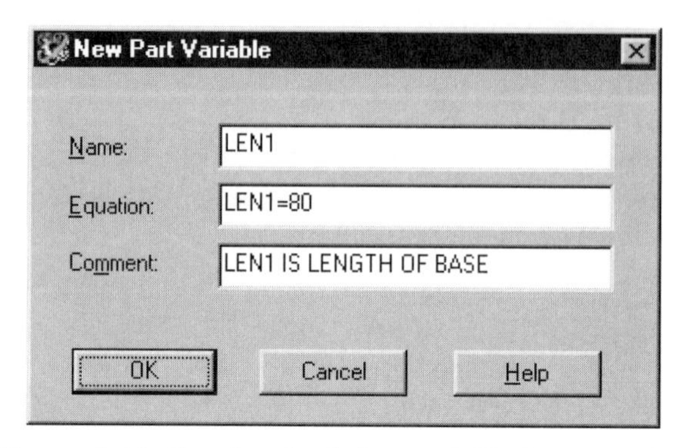

Figure 4-44 *Creating the design variable for Tutorial 2*

HGT1 IS BIGGER HEIGHT in the **Comments** edit box. Choose **OK**.

9. Choose the **New** button.

10. In the **Name** edit box enter **HGT2**. Enter **HGT2=HGT1/1.85** in the **Equation** edit box. Enter **HGT1 IS SMALLER HEIGHT** in the **Comments** edit box. Choose **OK**.

11. Choose the **New** button.

12. In the **Name** edit box enter **RAD**. Enter **RAD=10** in the **Equation** edit box. Enter **RAD IS THE RADIUS** in the **Comments** edit box. Choose **OK**.

13. All the variables created will be displayed in the **Design Variables** dialog box (Figure 4-45).

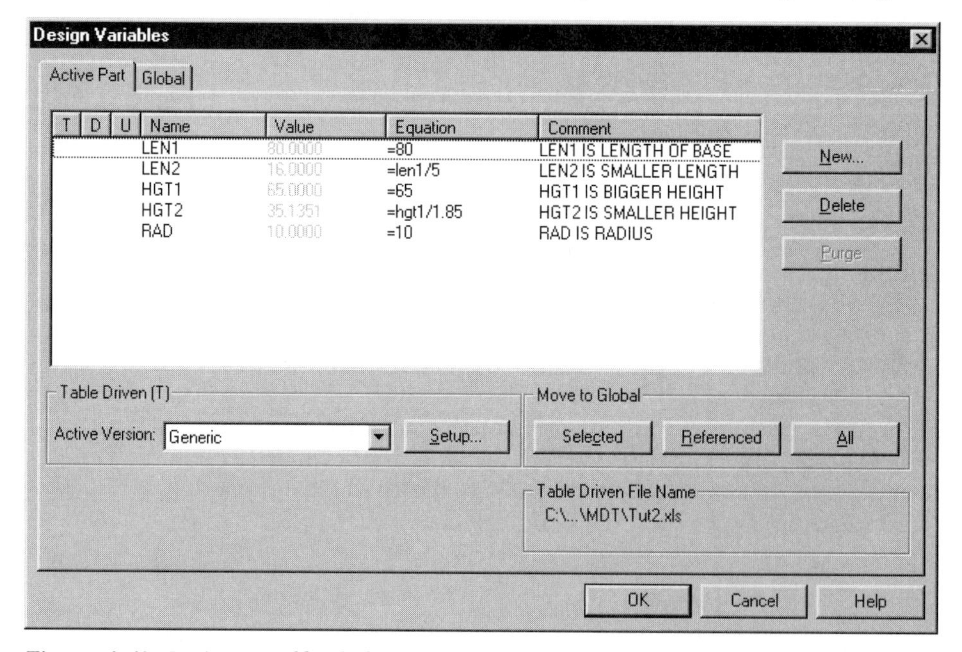

Figure 4-45 *Design Variables dialog box displaying the design variables created for Tutorial 2*

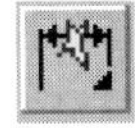

14. Choose the **Power Dimensioning** button from the **Part Modeling** toolbar. The prompt sequence is as follows:

    (Single) Specify first extension line origin or
    [Angular/Options/Baseline/Chain/Update] <Select>: Enter
    Select arc, line, circle or dimension: *Select the lower horizontal line.*
    Specify dimension line location or [Options/Pickobj]: *Place the dimension.*

Enter **LEN1** in the **Expression** edit box. Choose **OK**.

    Solved under constrained sketch requiring 4 dimension(s) or constraint(s).
    (Single) Specify first extension line origin or
    [Angular/Options/Baseline/Chain/Update] <Select>: Enter
    Select arc, line, circle or dimension: *Select the left vertical line.*
    Specify dimension line location or [Options/Pickobj]: *Place the dimension.*
    Enter dimension value [Associate to/Equation assistant/Options] <default value>: **HGT1**

    Solved under constrained sketch requiring 3 dimension(s) or constraint(s).
    (Single) Specify first extension line origin or
    [Angular/Options/Baseline/Chain/Update] <Select>: Enter
    Select arc, line, circle or dimension: *Select one of the upper horizontal lines.*
    Specify dimension line location or [Options/Pickobj]: *Place the dimension.*
    Enter dimension value [Associate to/Equation assistant/Options] <default value>: **LEN2**

    Solved under constrained sketch requiring 2 dimension(s) or constraint(s).
    (Single) Specify first extension line origin or
    [Angular/Options/Baseline/Chain/Update] <Select>: Enter
    Select arc, line, circle or dimension: *Select the arc.*
    Specify dimension line location or [Diameter/Options]: *Place the dimension.*
    Enter dimension value [Associate to/Equation assistant/Options] <default value>: **RAD**

    Solved under constrained sketch requiring 1 dimension(s) or constraint(s).
    (Single) Specify first extension line origin or
    [Angular/Options/Baseline/Chain/Update] <Select>: *Select the lower endpoint of the left vertical edge.*
    Specify second extension line origin: *Select the center of the arc.*
    Specify dimension line location or [Options/Pickobj]: *Place the dimension so that it displays the vertical dimension.*
    Enter dimension value [Associate to/Equation assistant/Options] <default value>: **HGT2**

    Solved fully constrained sketch.
    (Single) Specify first extension line origin or
    [Angular/Options/Baseline/Chain/Update] <Select>: *Press ESC.*

Tip: *While dimensioning the arcs or circles in terms of the design variables using the **AMPARDIM** command, enter the dimension for the radius and diameter as **R=RAD** and **D=DIA**, respectively. Here **R** is the radius and **D** is the diameter of the circular object and **RAD** and **DIA** are the design variables you have created.*

15. Choose the **Display As Variables** button from the **Design Variables** flyout in the **2D Constraints** toolbar. The dimensions assigned to the sketch will be displayed as parameters, see Figure 4-46.

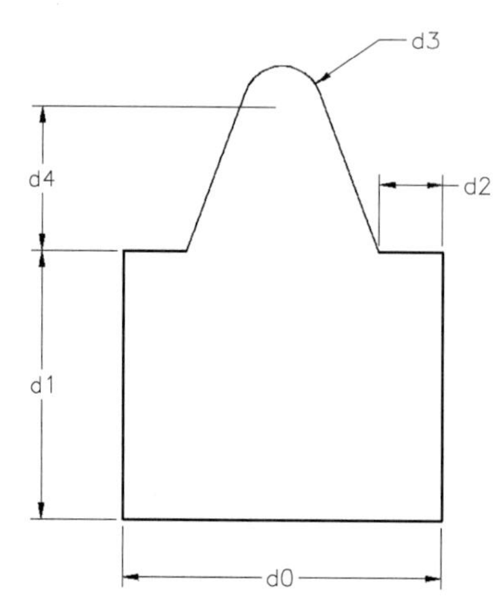

Figure 4-46 *Displaying the dimensions of the sketch in terms of the variables*

16. Choose the **Display As Equations** button from the **Design Variables** flyout in the **2D Constraints** toolbar. The dimensions assigned to the sketch will be displayed as equations, see Figure 4-47.

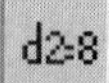

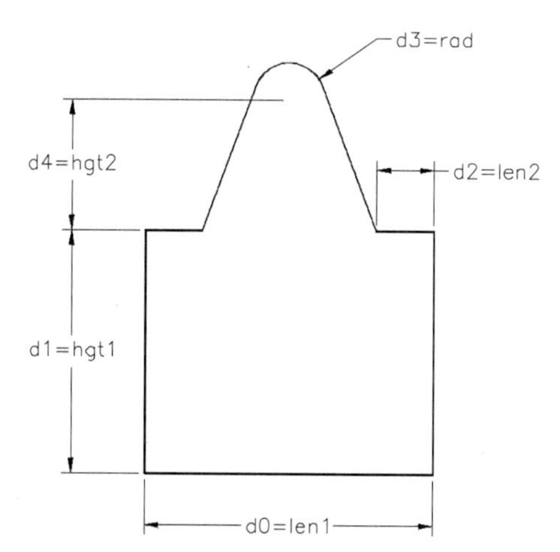

Figure 4-47 *Displaying the dimensions of the sketch as equations*

17. To create the spreadsheet, choose the **Design Variable** button from the **2D Constraints** toolbar to display the **Design Variables** dialog box.

18. Choose the **Setup** button under the **Table Driven [T]** area to display the **Table Driven Setup** dialog box.

19. Select the **Across** radio button under the **Version Names** area.

20. Select the **Variables** radio button under the **Type** area.

21. Choose the **Create** button to display the **Create Table** dialog box. Enter the name of the table in the **File Name** edit box. Choose **OK**.

22. The MS Excel spread sheet will be displayed on the screen as shown in Figure 4-48.

23. Save this drawing with the name given below:

MDT Tut\Ch-4**Tut2.dwg**

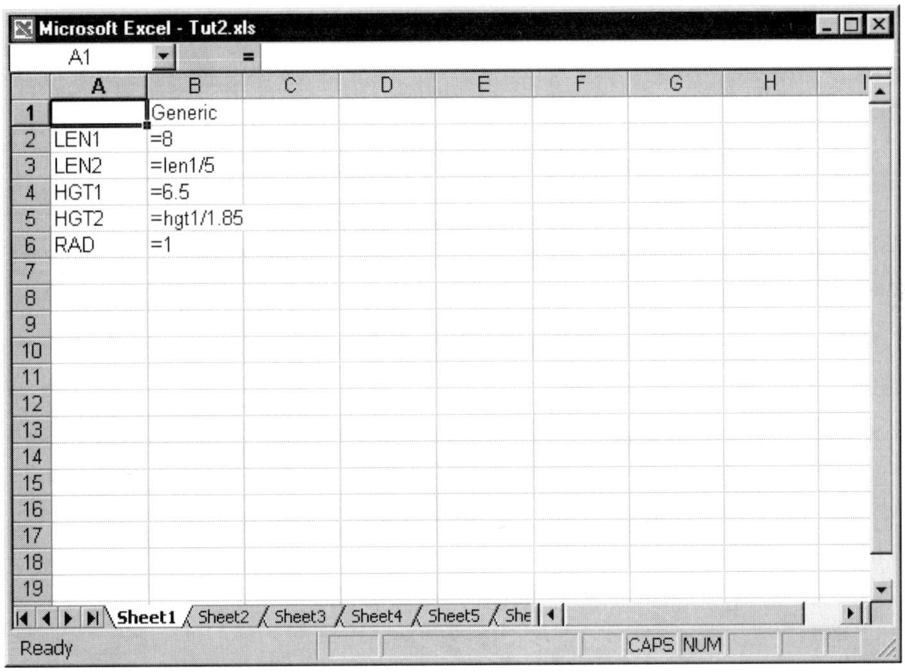

Microsoft Excel - Tut2.xls

	A	B
1		Generic
2	LEN1	=8
3	LEN2	=len1/5
4	HGT1	=6.5
5	HGT2	=hgt1/1.85
6	RAD	=1

Figure 4-48 MS Excel spreadsheet displaying the design variables

Tutorial 3

In this tutorial you will create the designer model shown in Figure 4-49a. The dimensions for the model are given in Figures 4-49b, 4-49c and 4-49d. These dimensions are in terms of design variables.

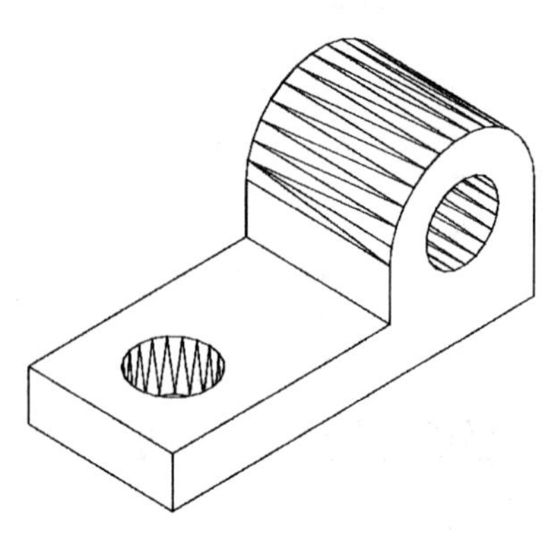

Figure 4-49a *Model for Tutorial 3*

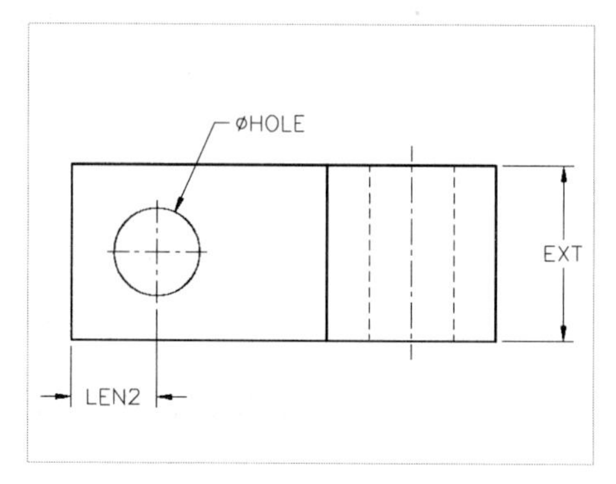

Figure 4-49b *Top view of the model*

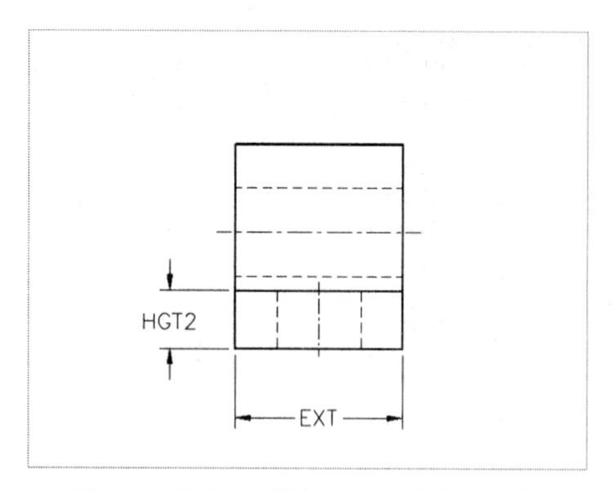

Figure 4-49c *Side view of the model*

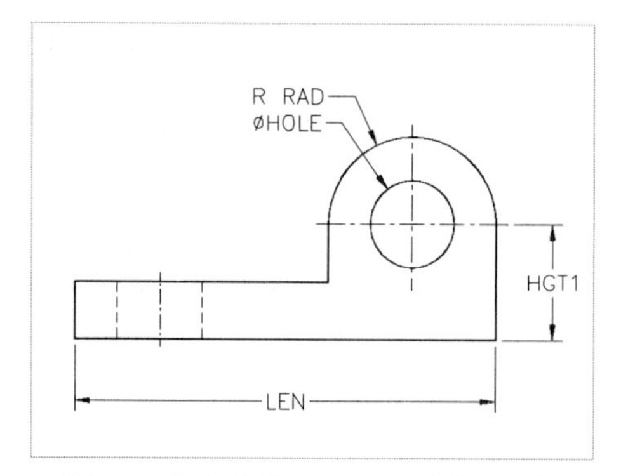

Figure 4-49d *Front view of the model*

The numeric values for the design variables are:

LEN1=150
LEN2=LEN1/5
HGT1=40
HGT2=HGT1/2
EXT=60
HOLE=30
RAD=30

1. Create the rough sketch for the base feature. Profile this rough sketch, see Figure 4-50.

2. Choose the **Design Variables** button from the **2D Constraints** toolbar to display the **Design Variables** dialog box.

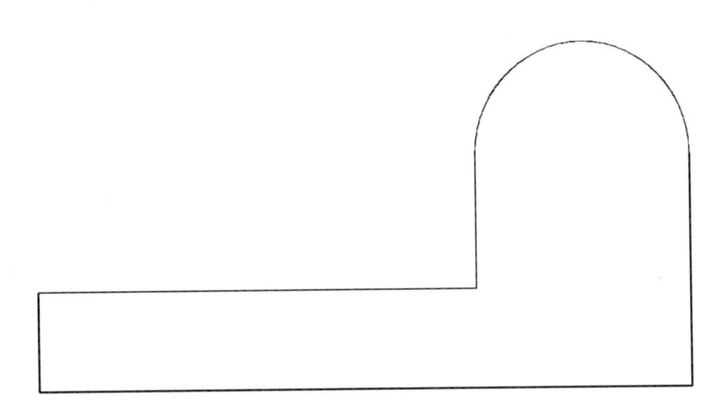

Figure 4-50 Profiled sketch for the base part

3. Choose the **New** button under the **Active Part** tab and create all the active part design variables given in the tutorial, see Figure 4-51.

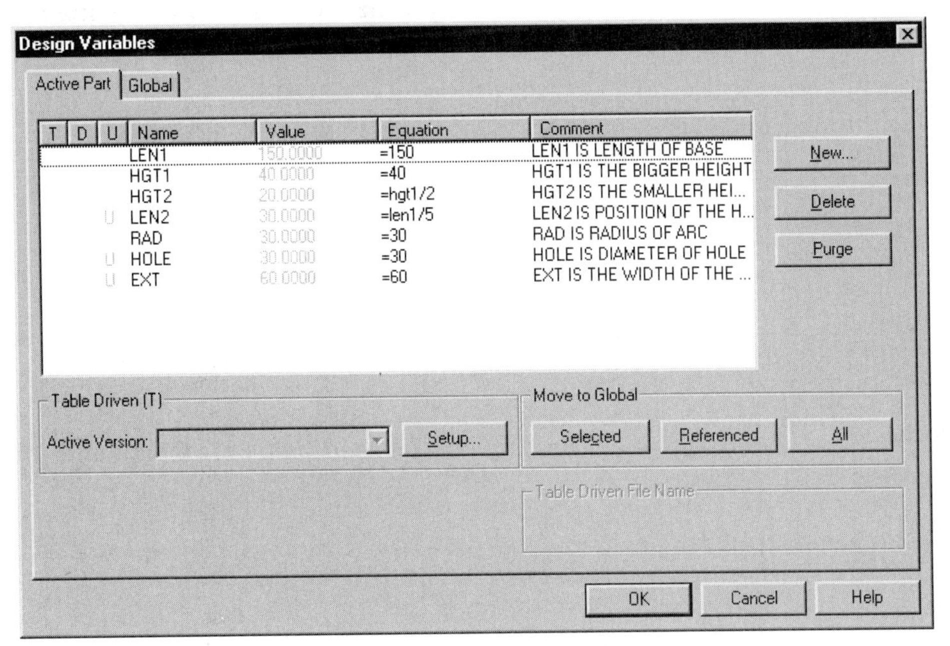

Figure 4-51 Design Variables dialog box displaying the design variables created for Tutorial 3

4. Choose the **Power Dimensioning** button from the **Part Modeling** toolbar. The prompt sequence is as follows:

(Single) Specify first extension line origin or
[Angular/Options/Baseline/Chain/Update] <Select>: Enter

Select arc, line, circle or dimension: *Select the bottom horizontal line.*
Specify dimension line location or [Options/Pickobj]: *Place the dimension.*

Enter **LEN1** in the **Expression** edit box of the **Power Dimensioning** dialog box. Choose **OK**.

Solved under constrained sketch requiring 3 dimension(s) or constraint(s).
(Single) Specify first extension line origin or
[Angular/Options/Baseline/Chain/Update] <Select>: Enter
Select arc, line, circle or dimension: *Select the right vertical line.*
Specify dimension line location or [Options/Pickobj]: *Place the dimension.*
Enter dimension value [Associate to/Equation assistant/Options] <default value>: **HGT1**

Solved under constrained sketch requiring 2 dimension(s) or constraint(s).
(Single) Specify first extension line origin or
[Angular/Options/Baseline/Chain/Update] <Select>: Enter
Select arc, line, circle or dimension: *Select the left vertical line.*
Specify dimension line location or [Options/Pickobj]: *Place the dimension.*
Enter dimension value [Associate to/Equation assistant/Options] <default value>: **HGT2**

Solved under constrained sketch requiring 1 dimension(s) or constraint(s).
(Single) Specify first extension line origin or
[Angular/Options/Baseline/Chain/Update] <Select>: Enter
Select arc, line, circle or dimension: *Select the arc.*
Specify dimension line location or [Diameter/Options]: *Place the dimension.*
Enter dimension value [Associate to/Equation assistant/Options] <default value>: **RAD**

Solved fully constrained sketch.
(Single) Specify first extension line origin or
[Angular/Options/Baseline/Chain/Update] <Select>: *Press ESC.*

5. Choose the **Display As Equations** button from the **Design Variables** flyout in the **2D Constraints** toolbar, see Figure 4-52.

6. Extrude the base sketch using the **AMEXTRUDE** command. Enter **EXT** in the **Distance** edit box.

7. Define a sketch plane on the front face of the model and create the sketch for the hole. Profile the sketch and apply the concentric constraint.

8. Choose the **Power Dimensioning** button from the **Part Modeling** toolbar. The prompt sequence is as follows:

(Single) Specify first extension line origin or
[Angular/Options/Baseline/Chain/Update] <Select>: Enter
Select arc, line, circle or dimension: *Select the circle.*
Specify dimension line location or [Radius/Options]: *Place the dimension.*

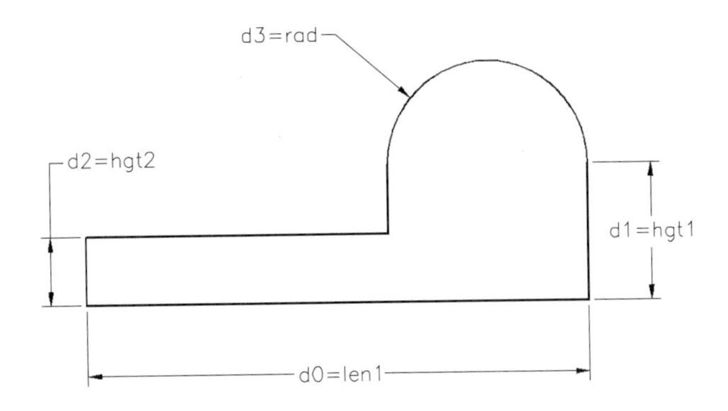

Figure 4-52 *Dimensions being displayed as equations*

Enter **HOLE** in the **Expression** edit box. Choose **OK**.

Solved fully constrained sketch.
(Single) Specify first extension line origin or
[Angular/Options/Baseline/Chain/Update] <Select>: *Press ESC.*

9. Extrude this sketch using the **Cut** operation.

10. Define the sketch plane for creating the next hole. Create the sketch for the hole and profile it.

11. Choose the **Power Dimensioning** button from the **Part Modeling** toolbar. The prompt sequence is as follows:

(Single) Specify first extension line origin or
[Angular/Options/Baseline/Chain/Update] <Select>: Enter
Select arc, line, circle or dimension: *Select the circle.*
Specify dimension line location or [Radius/Options]: *Place the dimension.*

Enter **HOLE** in the **Expression** edit box. Choose **OK**.

Solved under constrained sketch requiring 2 dimension(s) or constraint(s).
(Single) Specify first extension line origin or
[Angular/Options/Baseline/Chain/Update] <Select>: *Select the center of circle.*
Specify second extension line origin: *Select the lower endpoint of the left vertical line.*
Specify dimension line location or [Options/Pickobj]: *Place the dimension so that it displays the horizontal dimension.*
Enter dimension value [Associate to/Equation assistant/Options] <default value>: **LEN2**

Solved under constrained sketch requiring 1 dimension(s) or constraint(s).
(Single) Specify first extension line origin or
[Angular/Options/Baseline/Chain/Update] <Select>: *Select the center of circle.*
Specify second extension line origin: *Select the upper endpoint of the left vertical line.*
Specify dimension line location or [Options/Pickobj]: *Place the dimension.*
Enter dimension value [Associate to/Equation assistant/Options] <default value>: **EXT/2**

Solved fully constrained sketch.

12. Extrude this sketch using the **Cut** operation. The final designer model should be similar to the one shown in Figure 4-53.

13. Save this drawing with the name given below:

/MDT Tut/Ch-4/**Tut3.dwg**.

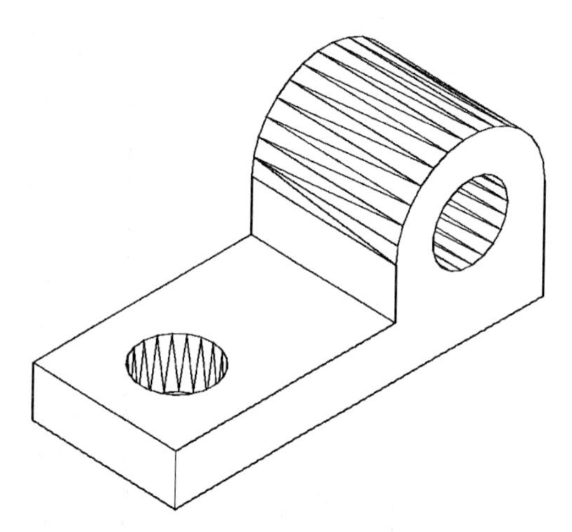

Figure 4-53 *Final designer model for Tutorial 3*

Review Questions

Answer the following questions.

1. What are the various advanced dimensioning techniques?

2. State any four uses of the power dimensioning technique.

3. You can dimension both the profiled and the unprofiled sketches using the power dimensioning and auto dimensioning techniques. (T/F)

4. The advanced dimensioning techniques do not behave as the parametric dimensions for the unprofiled sketches. (T/F)

5. What are the different types of design variables?

6. The values in the dialog box edit boxes can be entered in the form of the design variables. (T/F)

7. Which command is used to display the dimensions as variables?

8. Which dialog box is used to control the visibility of the designer model and the features associated with it?

Exercise

Exercise 1

Create the object shown in Figure 4-54a. The dimensions to be used are given in Figures 4-54b, 4-54c and 4-54d. Dimension the sketch using the advanced dimensioning techniques. Save the file with the name \MDT Tut\Ch-4**Exr1.dwg**.

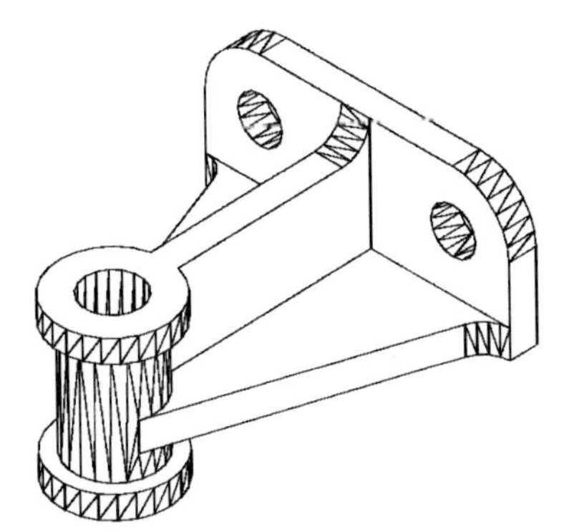

Figure 4-54a *Designer model for Exercise 1*

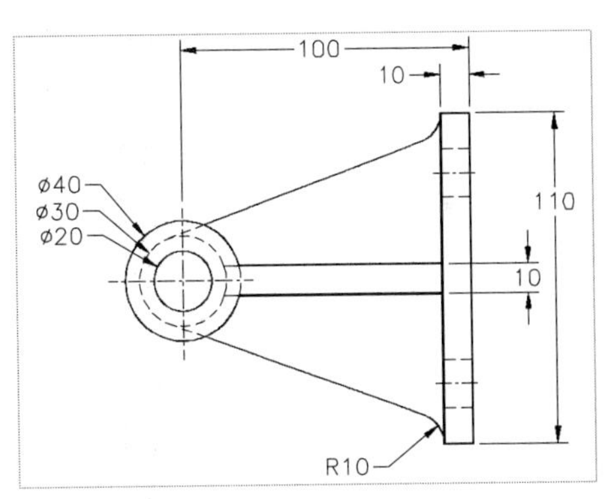

Figure 4-54b *Top view of the model*

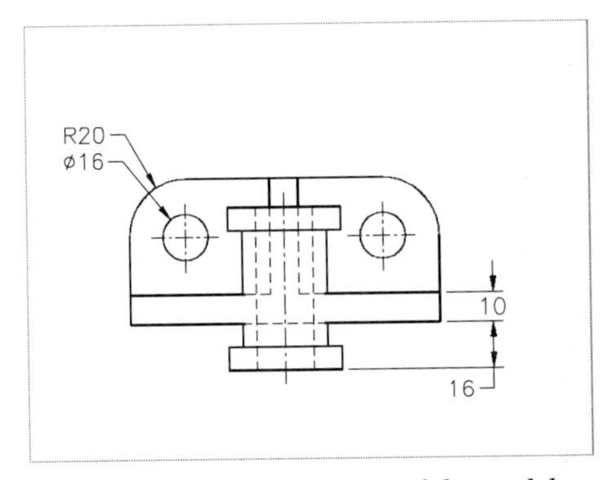

Figure 4-54c *Side view of the model*

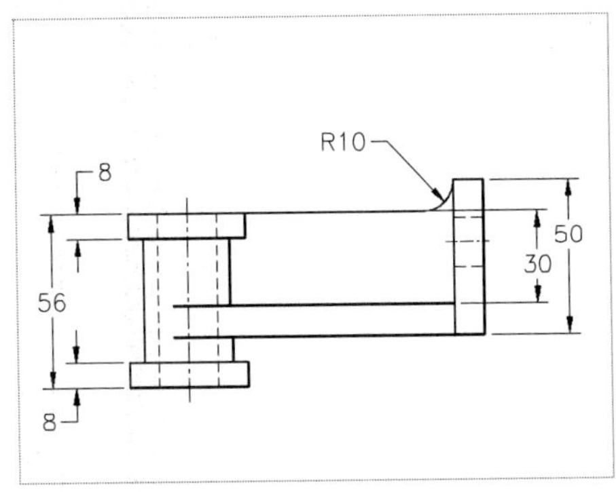

Figure 4-54d *Front view of the model*

Chapter 5

Placed Features I

Learning Objectives

After completing this chapter, you will be able to:

- *Place different type of holes using the* ***AMHOLE*** *command.*
- *Fillet the edges of the designer model using the* ***AMFILLET*** *command.*
- *Chamfer the edges of the existing model using the* ***AMCHAMFER*** *command.*
- *Add drafts to the faces of existing designer model using the* ***AMFACEDRAFT*** *command.*

Commands Covered

- *AMHOLE*
- *AMFILLET*
- *AMCHAMFER*
- *AMFACEDRAFT*

PLACED FEATURES

The placed features are the advanced features added for modifying and finishing the existing designer models or base features. All of these placed features are parametric in nature and can be edited whenever required. You do not have to define a different sketch plane or a work plane for placing these features. The placed features are of nine types. They are:

1. Hole
2. Fillet
3. Chamfer
4. Face Draft
5. Shell
6. Surface Cut
7. Patterns
8. Combine
9. Part Split.

Note

In this chapter the first four placed features are discussed and the remaining placed features will be discussed in the later chapters.

Placing The Holes (AMHOLE Command)

Toolbar:	Part Modeling > Placed Features - Hole
Menu:	Part > Placed Features > Hole
Context Menu:	Placed Features > Hole
Command:	AMHOLE

The holes are the circular grooves or cavities created on the existing designer model for the aid of assembly. The holes created using this command are parametric in nature and can be easily edited whenever required. You can create three type of holes using the **AMHOLE** command. They are:

- Drilled
- Counterbore
- Countersink

The **Hole** dialog box will be displayed when you invoke this command, see Figure 5-1.

Hole Dialog Box Options

The options that are provided under the **Hole** dialog box are:

Copy Values

The **Copy Values** button is chosen to copy the properties associated with an existing hole to the new hole.

Operation

The options provided under the **Operation** drop-down list are used to select the type

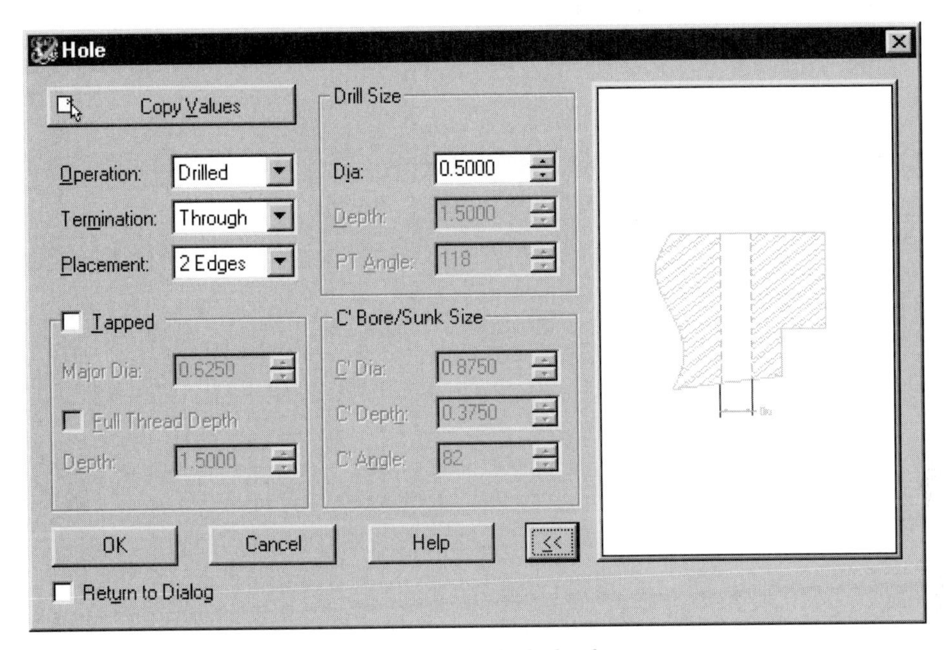

Figure 5-1 *Hole dialog box*

of the hole you want to place. The options provided under this tab are:

Drilled. The **Drilled** option is used to create a hole of a uniform diameter throughout.

C'Bore. The **C'Bore** option is used to create a counterbore hole.

C'Sink. The **C'Sink** option is used to create a countersink hole.

Termination

The options provided under the **Termination** drop-down list are used to define the termination of the hole. The options are:

Through. The **Through** option is used to create a hole cutting right through the existing model.

Blind. The **Blind** option is used to create a hole up to a certain depth as defined by you.

To-Plane. The **To-Plane** option is used to create a hole that terminates at a particular plane. The termination plane is defined by you.

Tip: *A parametric hole placed on a designer model using the* ***Through*** *option of the* ***AMHOLE*** *command will automatically adjust its depth once the thickness of the designer model is edited.*

Placement

The options provided under the **Placement** drop-down list are used to define the location of the hole. This drop-down list has following options:

2 Edges. The **2 Edges** option is used to place the hole at a specified distance from two selected edges as shown in Figures 5-2a and 5-2b.

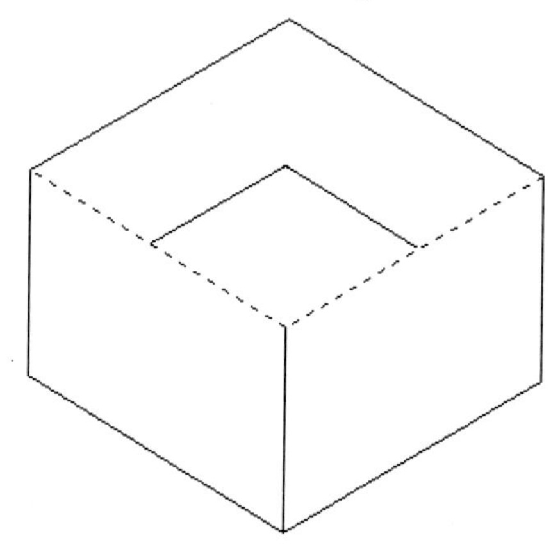

Figure 5-2a *Selecting the edges to place the holes*

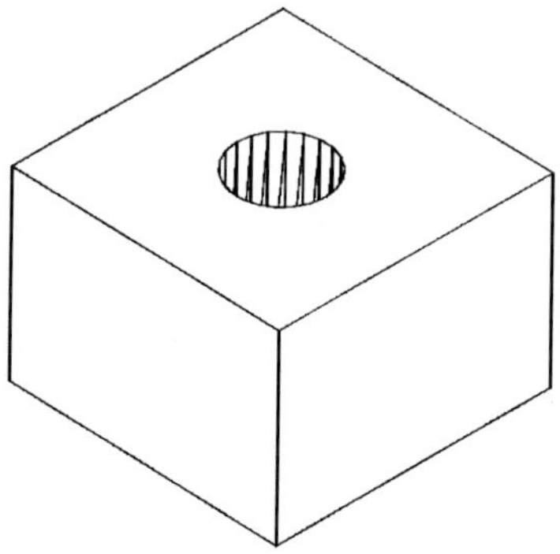

Figure 5-2b *Model after placing the holes using the 2 Edges option*

Concentric. The **Concentric** option is used to place a hole that is concentric to a selected circular face.

On Point. The **On Point** option is used to place the hole at a particular work point.

From Hole. The **From Hole** option is used to place the hole at a specified distance from an existing hole as shown in Figures 5-3a and 5-3b.

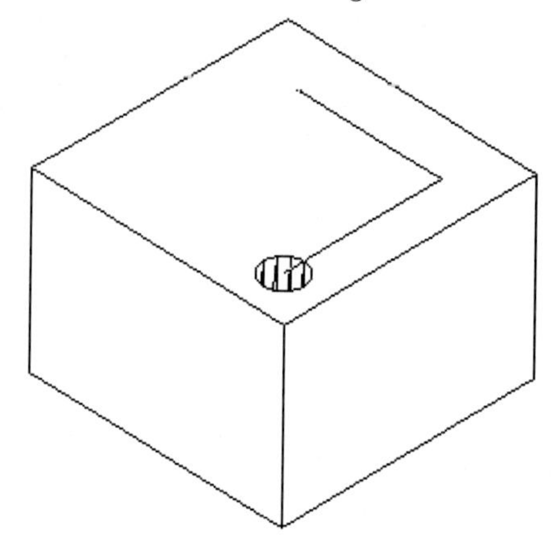

Figure 5-3a *Selecting the hole to place another hole*

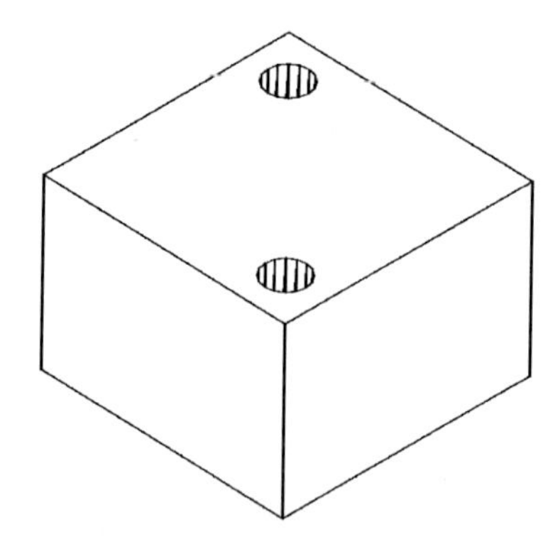

Figure 5-3b *Model after placing the holes using the From Hole option*

Drill Size Area
The options provided under this area are:

Dia. The **Dia** spinner is used to specify the diameter of the drill as shown in Figure 5-4. If the **C'Bore** or **C'Sink** option is selected from the **Operation** drop-down list then this diameter will be considered as the smaller diameter.

Depth. The **Depth** spinner is used to define the depth of the hole, see Figure 5-4. This option is available only if the **Blind** option is selected from the **Termination** drop-down list.

PT Angle. The **PT Angle** spinner is used to define the point angle at the terminating end of the hole as shown in Figure 5-5. This option is available only if the **Blind** or the **To-Plane** option is selected from the **Termination** drop-down list.

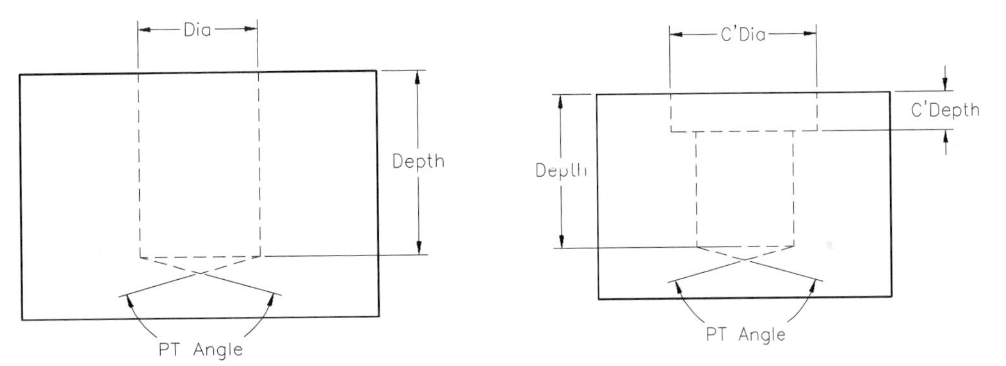

Figure 5-4 Drilled hole

Figure 5-5 Counterbore hole

Tapped Area
The options under this area are available only if the **Tapped** check box is selected. These options are used to create the drawing views of the threaded hole in the **Drawing** tab. The changes made using these options will not be displayed in the designer model in the **Model** tab.

Major Dia. The **Major Dia** spinner is used to specify the diameter of the threaded hole. This should always be greater than the drill diameter.

Full Thread Depth. If this check box is selected, the threads will run through the length of the hole.

Depth. The **Depth** spinner is used to define the depth to which the thread will run in the hole. This option is available only if the **Full Thread Depth** check box is cleared.

C'Bore/Sunk Size Area

The options provided under this area are available only if either the **C'Bore** or the **C'Sink** option is selected from the **Operation** drop-down list.

C'Dia. This spinner is used to specify the counter diameter as shown in Figure 5-3. This value has to be greater than the value of the **Dia** spinner.

C'Depth. This spinner defines the counter depth as shown in Figure 5-3. This option is available only if the **C'Bore** option is selected from the **Operation** drop-down list.

C'Angle. This spinner is used to specify the countersink angle as shown in Figures 5-6 and 5-7. This option is available only if the **C'Sink** options is selected from the **Operation** drop-down list.

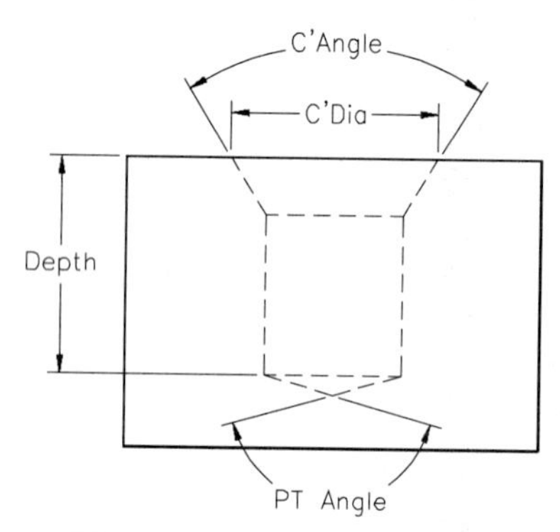

Figure 5-6 Countersink hole

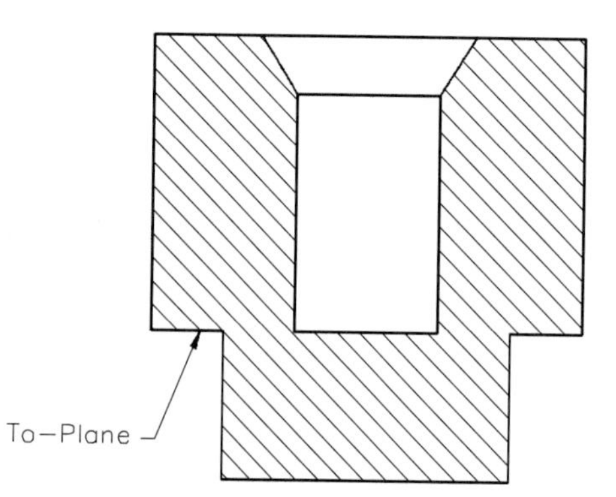

Figure 5-7 Countersink hole placed using the To-Plane termination option

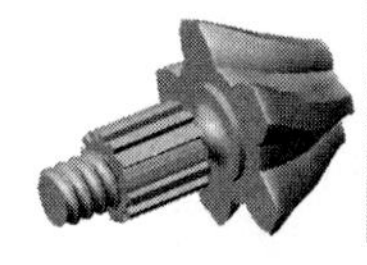

Tip: *To create a hole with a flat bottom base when the **Blind** or the **To-Plane** option is used, set the value of the **PT Angle** spinner to **180**.*

Return to Dialog

If this check box is selected, the **Hole** dialog box will be displayed again after the current hole is placed.

Filleting The Edges (AMFILLET Command)

Toolbar:	Part Modeling > Placed Features - Hole > Fillet
Menu:	Part > Placed Features > Fillet
Context Menu:	Placed Features > Fillet
Command:	AMFILLET

Generally, the sharp edges of the mechanical components tend to fail under the increased stress concentration. Therefore, the sharp edges of the models are rounded (filleted) to reduce the stress concentration. In Mechanical Desktop, the edges of the designer model can be rounded using the **AMFILLET** command. Using this command you can apply four different types of fillets. They are:

1. Constant
2. Fixed Width
3. Cubic
4. Linear

The **Fillet** dialog box is displayed when you invoke this command, see Figure 5-8.

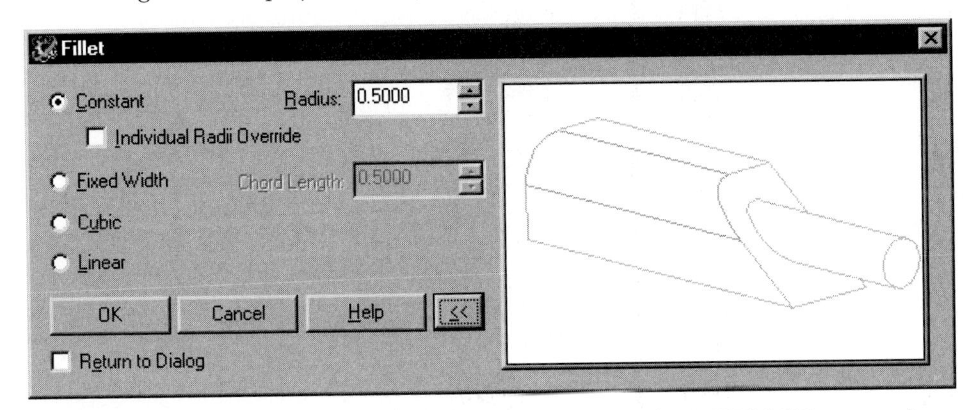

Figure 5-8 *The Fillet dialog box displayed by invoking the AMFILLET command*

Fillet Dialog Box Options

The options provided under this dialog box are discussed below:

Constant

This radio button is selected to create a fillet of constant radius. You can select as many edges to fillet and all of them will have same radius value as shown in Figures 5-9 and 5-10. However, you can also define different radius values for all the edges by selecting the **Individual Radii Override** check box.

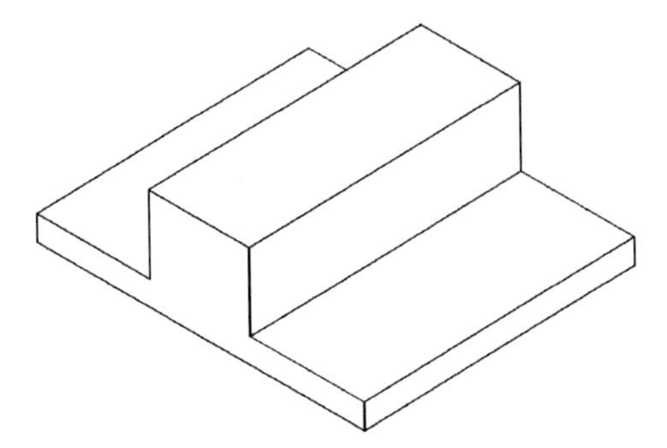

Figure 5-9 *Designer model before applying the fillet*

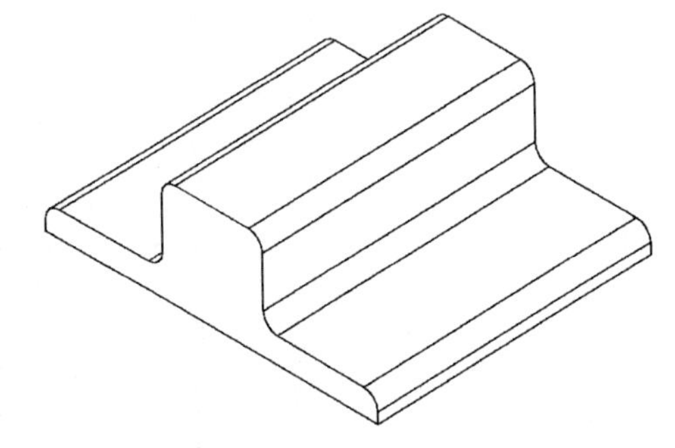

Figure 5-10 *Designer model after applying the constant fillet*

Radius

This spinner is used to set the value of the radius for the constant fillet. This spinner is available only if the **Constant** radio button is selected.

Individual Radii Override

If this check box is selected, you can specify different radius values for all the edges selected. This check box is available only when the **Constant** radio button is selected. The **Radius** spinner becomes the **Default radius** spinner when this check box is selected. Once you have selected all the edges to fillet, you will be prompted to specify the radius value for each of the edges as shown in Figures 5-11 and 5-12. The default radius value will be the one that was set in the **Default radius** spinner.

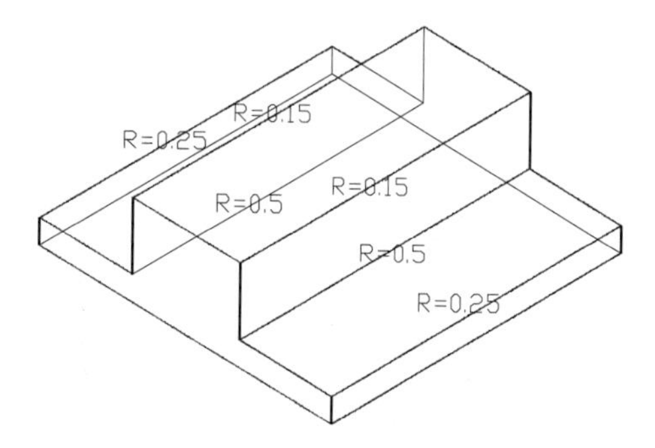

Figure 5-11 *Specifying different radius values for different edges*

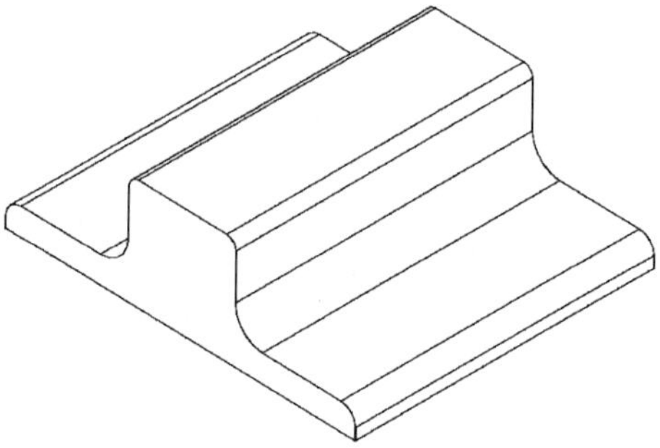

Figure 5-12 *Designer model after applying the fillet with different radius for each edge*

Fixed Width

This radio button is selected when you want to create the fillet with a fixed width. In this case the chord length is specified instead of radius. The value of the chord length is specified using the **Chord Length** spinner and this value remains constant through out the fillet. Figure 5-13 shows a model before applying the fixed width fillet and Figure 5-14 shows a model after applying the fixed width fillet.

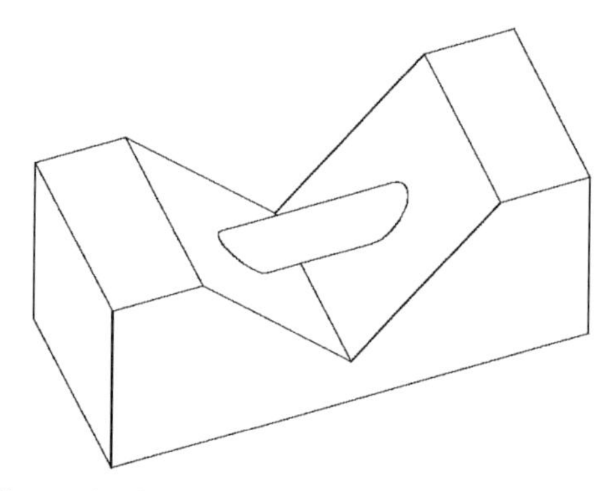

Figure 5-13 *Designer model before applying the fixed width fillet*

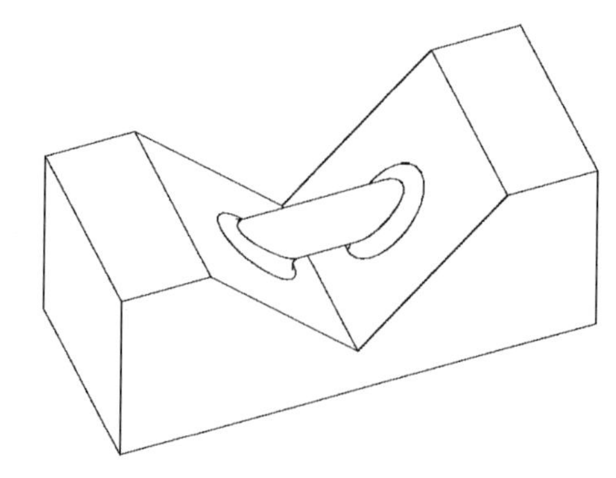

Figure 5-14 *Designer model after applying the fixed width fillet*

Chord Length

This spinner is used to specify the chord length or the width of the fillet. This spinner is available only when you select the **Fixed Width** radio button.

Cubic

This radio button is selected to create cubic fillet consisting of multiple radii along the edge selected to fillet. By default there will only be two vertices; one at the start point of the edge and other at the endpoint of the edge. You can specify as many vertices in between by defining the location of the vertex in the terms of the percentage of the edge. Once you have added the vertices, you can now specify the radius of each vertex individually, see Figures 5-15 and 5-16.

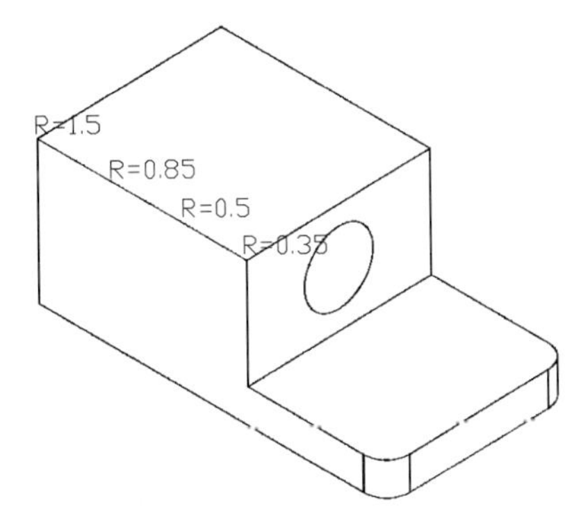

Figure 5-15 *Specifying the location of the vertices on the edge*

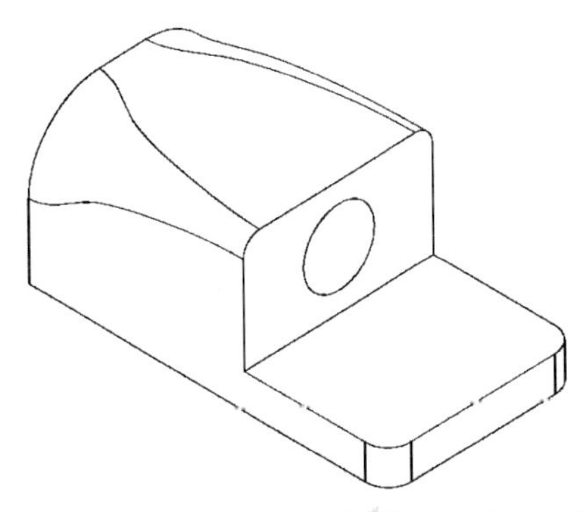

Figure 5-16 *Designer model after applying the cubic fillet on the edges*

Linear

This radio button is selected to create a linear fillet on the selected edge. When you select the edge, two vertices will appear; one at the start point of the edge and the other at the endpoint of the edge. You can now specify the radius values for both of these vertices, see Figures 5-17 and 5-18. In this case you cannot add the vertices in between the default vertices.

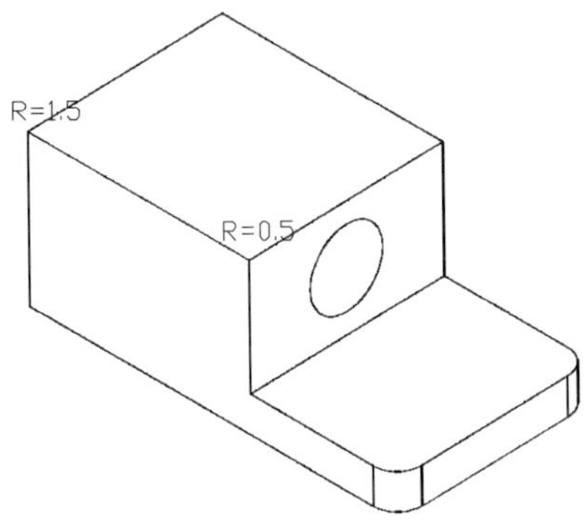

Figure 5-17 *Specifying the radius values for the linear fillet*

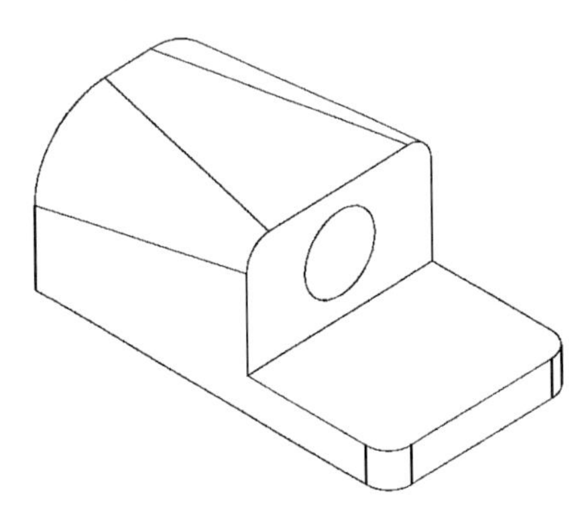

Figure 5-18 *Designer model after applying the linear fillet*

Return to Dialog

If this check box is selected, the **Fillet** dialog box will again appear once the current fillet is applied.

Chamfering The Edges (AMCHAMFER Command)*

Toolbar:	Part Modeling > Placed Features - Hole > Chamfer
Menu:	Part > Placed Features > Chamfer
Context Menu:	Placed Features > Chamfer
Command:	AMCHAMFER

This command is used to bevel the edges of the designer model. The **Chamfer Feature** dialog box is displayed when you invoke this command, see Figure 5-19. You can apply the chamfer feature to the selected edge using three methods. They are:

- Two equal distances.
- Two unequal distances.
- One distance and one angle.

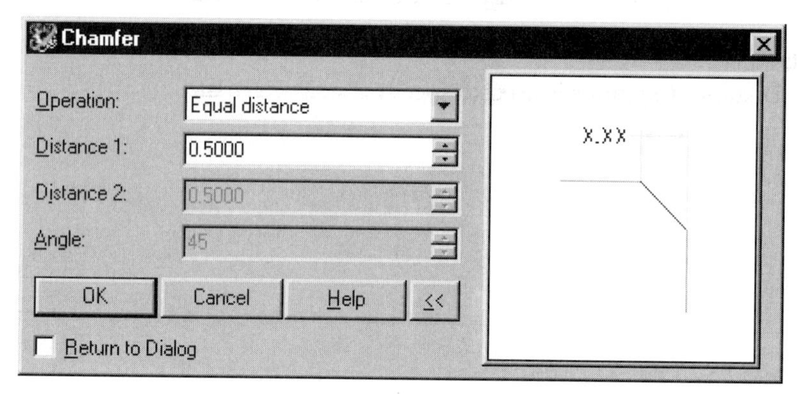

***Figure 5-19** The Chamfer dialog box displayed upon invoking the AMCHAMFER command*

Chamfer Dialog Box Options

Operation

The **Operation** drop-down list provides the methods of creating the chamfer. This drop-down list provides the following options:

Equal Distance. This option is used to create the chamfer with equal distance on both the faces from the selected edge thus creating the 45 degrees chamfer, see Figure 5-20. As the distances from both the faces are same, therefore, this option requires only one distance value. The distance value can be specified in the **Distance1** edit box.

Two Distances. This option is used to create the chamfer using two distances, see Figure 5-21. Once the edge to be chamfered is selected, you will be prompted to specify the face from which the first distance is to be calculated.

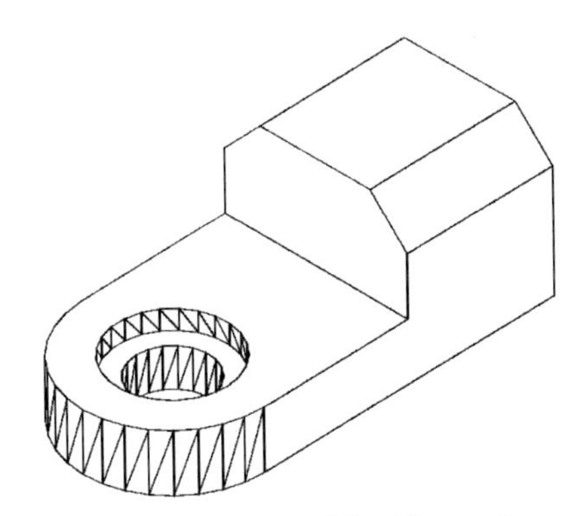

Figure 5-20 Designer model with equal distance chamfer

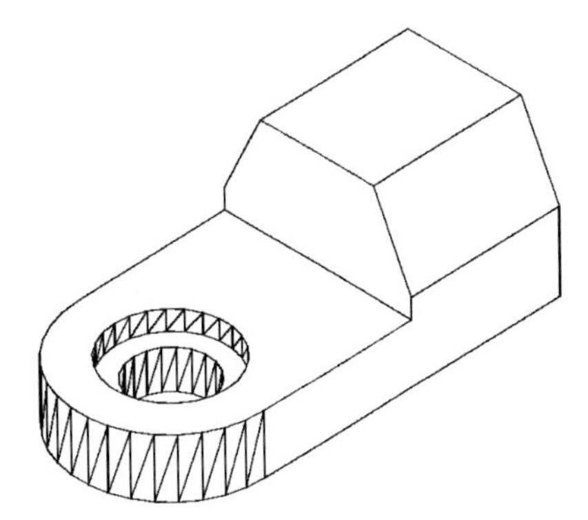

Figure 5-21 Designer model with two distance chamfer

Distance x Angle. This option is used when you want to specify the chamfer in the terms of a distance and an angle. Once you have selected the edge to chamfer, you will be asked to specify the face from which the angle is to be calculated.

Distance1

The **Distance1** spinner is used to specify the first distance for chamfer.

Distance2

The **Distance2** spinner is used to specify the second distance for chamfer. This option is available only when the **Two Distance** radio button is selected.

Angle

The **Angle** spinner is used to specify the angle value when the **Distance x Angle** radio button is selected.

Return to Dialog

This check box is selected so that the **Chamfer Feature** dialog box is again displayed after the current chamfer is created.

Applying Face Draft (AMFACEDRAFT Command)

Toolbar:	Part Modeling > Placed Features - Hole > Face Draft
Menu:	Part > Placed Features > Face Draft
Context Menu:	Placed Features > Face Draft
Command:	AMFACEDRAFT

This command is used to apply a draft angle on the selected faces of the designer model. The draft angles are applied on the faces for the easy removal of the model from the mold during manufacturing. The **Face Draft** dialog box is displayed when you invoke this command, see Figure 5-22.

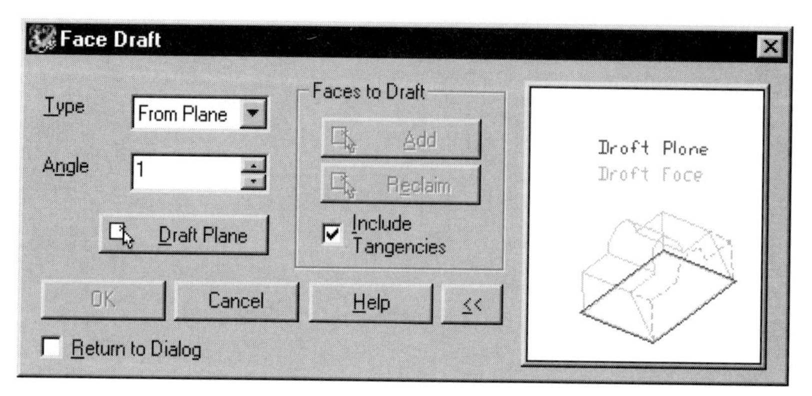

Figure 5-22 Face Draft dialog box

Face Draft Dialog Box Options

The options provided under this dialog box are:

Type

The options provided under the **Type** drop-down list are used to specify the method of creating the face draft. They are:

From Plane. The **From Plane** option is used to define the face draft from a specified plane, see Figures 5-23 and 5-24.

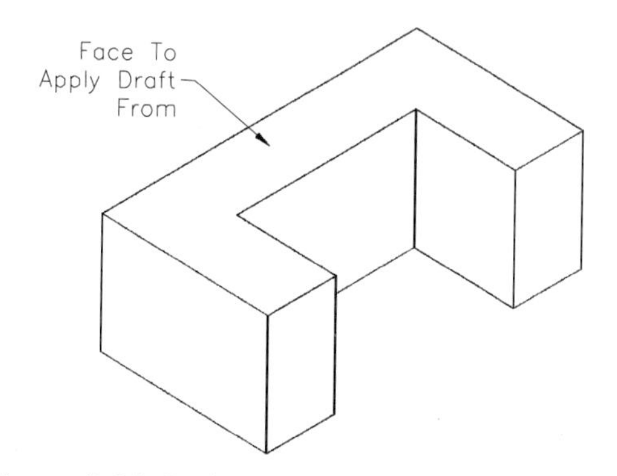

Figure 5-23 *Designer model before applying the face draft*

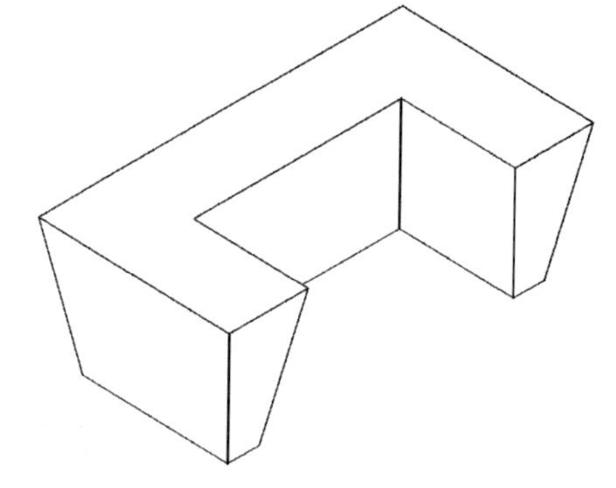

Figure 5-24 *Designer model after applying the face draft on all the side faces using the From Plane option*

From Edge. This option is used to define the face draft from a specified edge of the designer model, see Figures 5-25 and 5-26.

Shadow. This option is used to apply a face tangent to the cylindrical face, at the specified angle, in the designer model, see Figures 5-27 and 5-28.

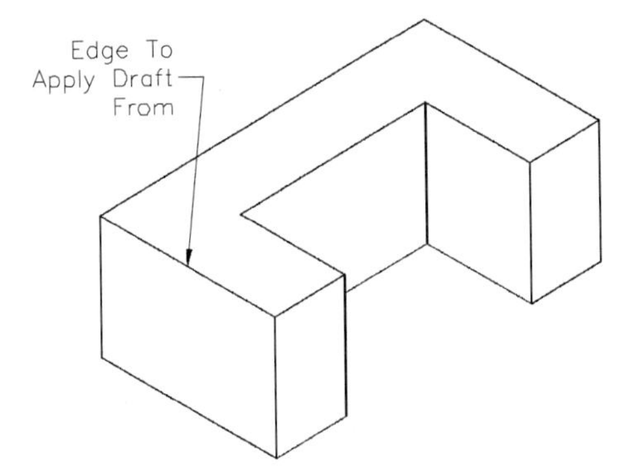

Figure 5-25 Designer model before applying the face draft

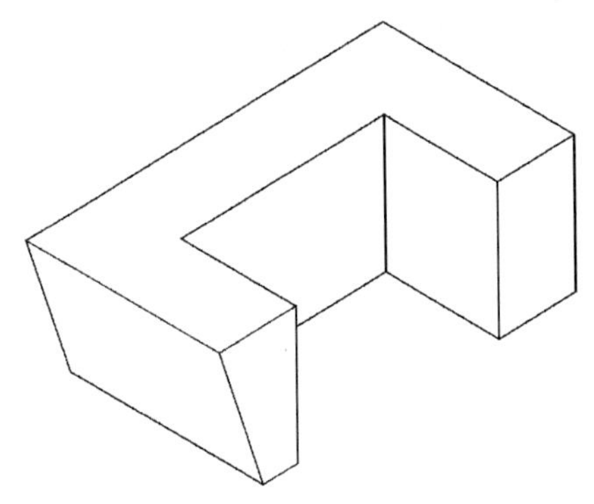

Figure 5-26 Designer model after applying the face draft on one of the side faces using the From Edge option

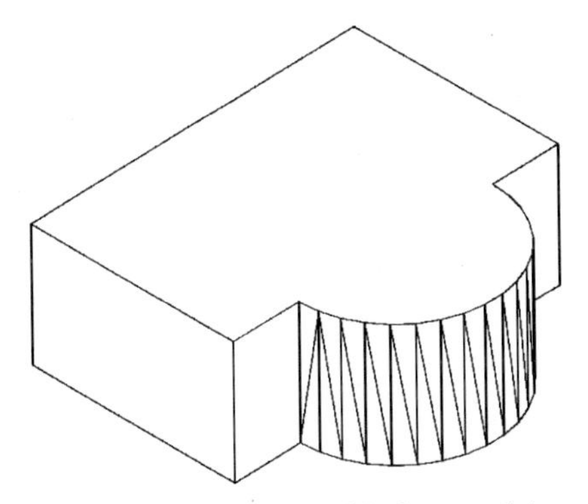

Figure 5-27 Designer model before applying the face draft

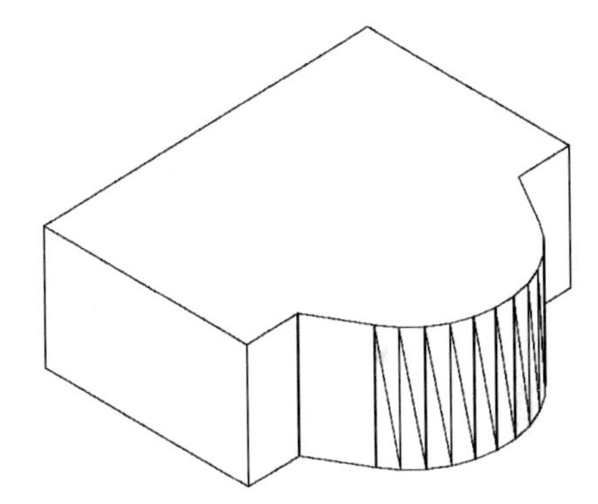

Figure 5-28 Designer model after applying the face draft using the Shadow option

Angle

This spinner is used to specify the draft angle for the face draft.

Draft Plane

This button is chosen to select the draft plane from which the draft angle is to be measured. Once you have selected the type of face draft and specified the draft angle, you have to choose this button. When you choose this button, the **Face Draft** dialog box will be temporarily closed and you will be prompted to select the draft plane. After selecting the draft plane, the draft plane direction indicator will be displayed showing the direction in which draft will be applied. Select the draft direction and then press ENTER to display the **Face Draft** dialog box again.

Faces to Draft Area

The options provided under this area are:

Add. The **Add** button is chosen to select the faces on which the draft has to be applied. This button is available only after you have selected the draft plane using the **Draft Plane** button. When you choose this button, the dialog box disappears temporarily from the screen and you will be prompted to select the faces to apply the face draft.

Reclaim. This button is chosen to remove the faces you have selected using the **Add** button. This button is not available until you have selected the faces on which the draft is to be applied using the **Add** button.

Include Tangencies. If this check box is selected, the tangent faces will also be included while applying the face drafts.

Return to Dialog

If this check box is selected, the **Face Draft** dialog box will be displayed again after the current face draft is applied.

TUTORIALS

Tutorial 1

In this tutorial you will create the designer model shown in Figure 5-29a. The dimensions for the model are given in Figures 5-29b, 5-29c, and 5-29d. Assume the missing dimensions.

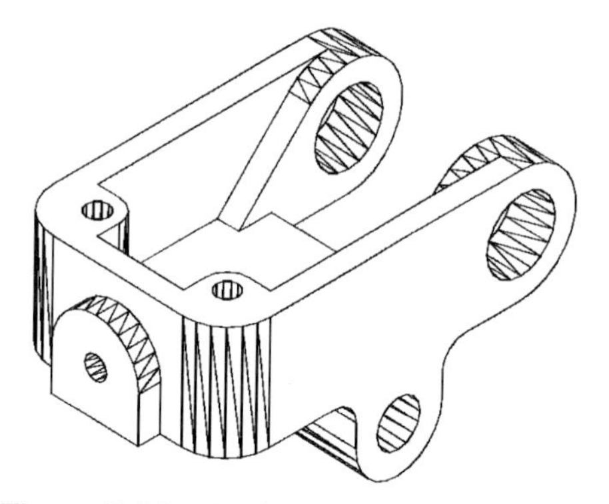

Figure 5-29a *Designer model for Tutorial 1*

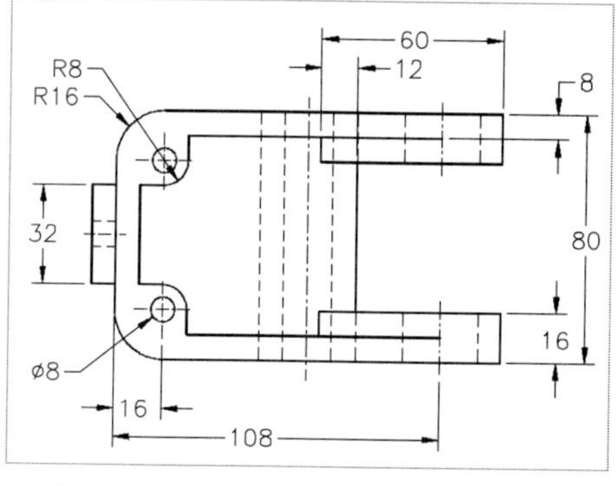

Figure 5-29b *Top view of the model*

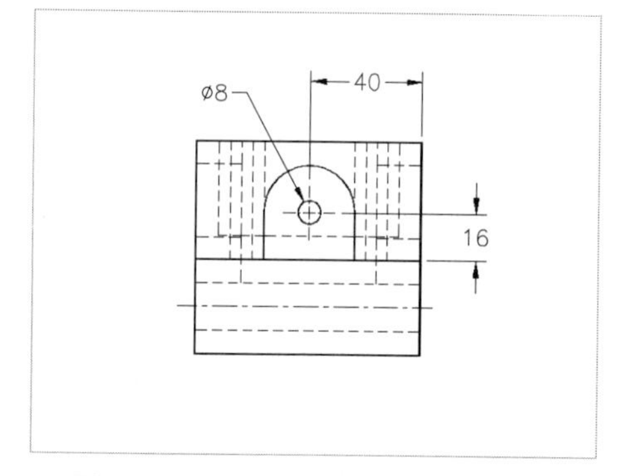

Figure 5-29c *Side view of the model*

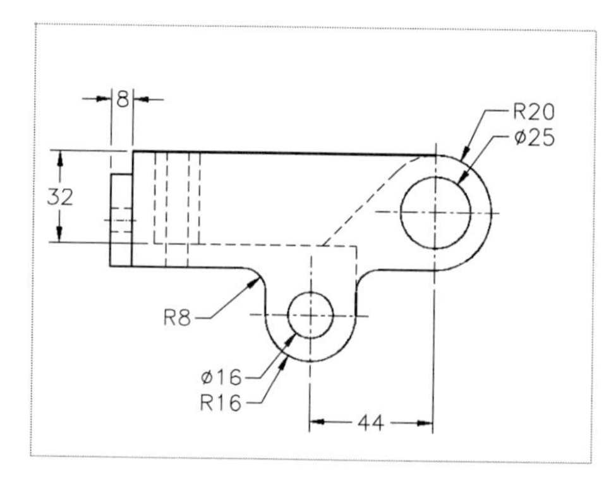

Figure 5-29d *Front view of the model*

1. Create the sketch for the base part of the designer model and dimension it using the dimensions given in the figures.

2. Extrude the base sketch to create the base feature as shown in Figure 5-30.

3. Create next features on the bottom, front and the back faces of the designer model as shown in Figure 5-31.

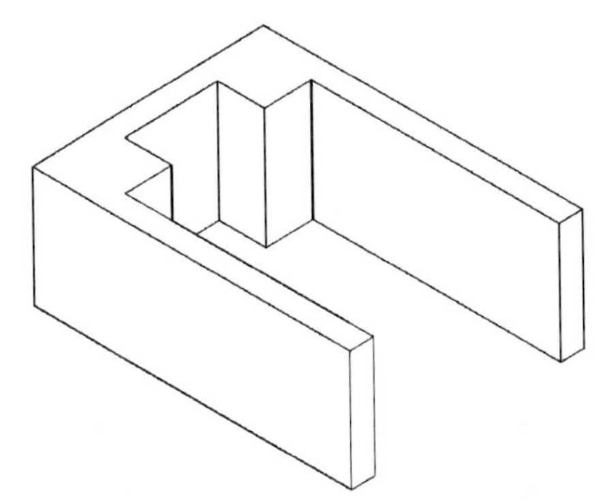

Figure 5-30 *Figure showing the base feature of the designer model*

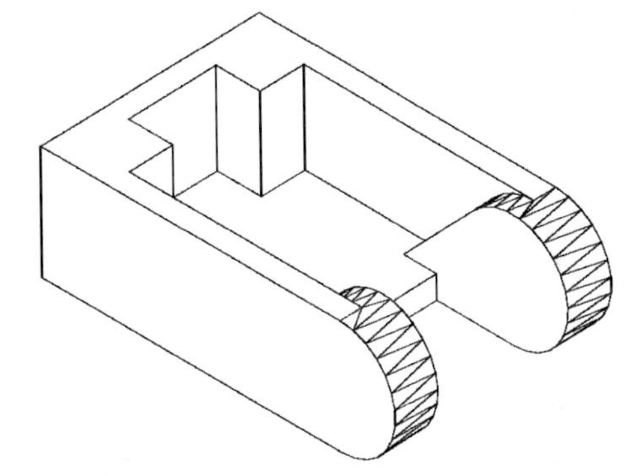

Figure 5-31 *Designer model after creating the features on the front, back and the bottom faces*

4. Create the other features on the front and the left side faces as shown in Figure 5-32.

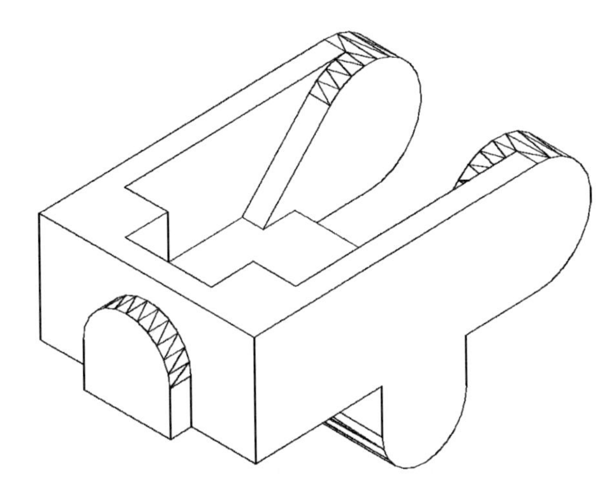

Figure 5-32 *Designer model after creating next features*

5. Choose the **Fillet** button from the **Placed Features-Hole** flyout in the **Part Modeling** toolbar to display the **Fillet** dialog box as shown in Figure 5-33.

6. Select the **Constant** radio button.

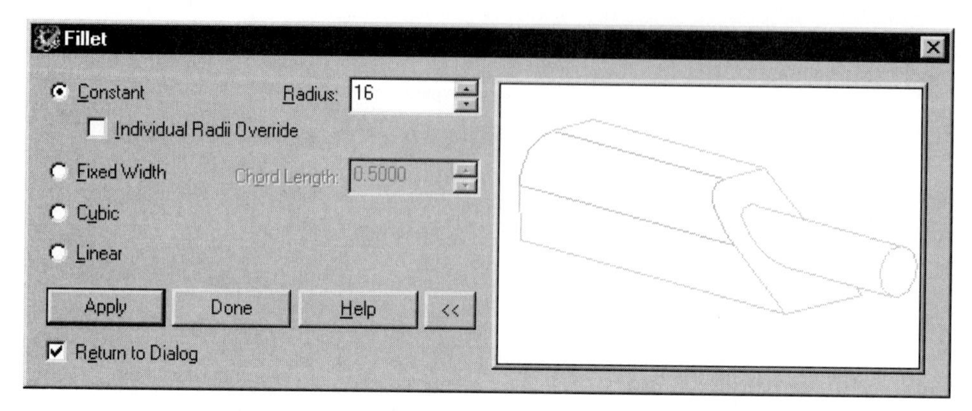

Figure 5-33 *Figure showing the Fillet dialog box*

7. Set the value of the **Radius** spinner to 16.

8. Select the **Return to Dialog** check box. Choose **Apply**. The prompt sequence is as follows:

   Select edges or faces to fillet: *Select the edge as shown in Figure 5-34.*
   Select edges or faces to fillet <continue>: *Select the next edge as shown in Figure 5-34.*

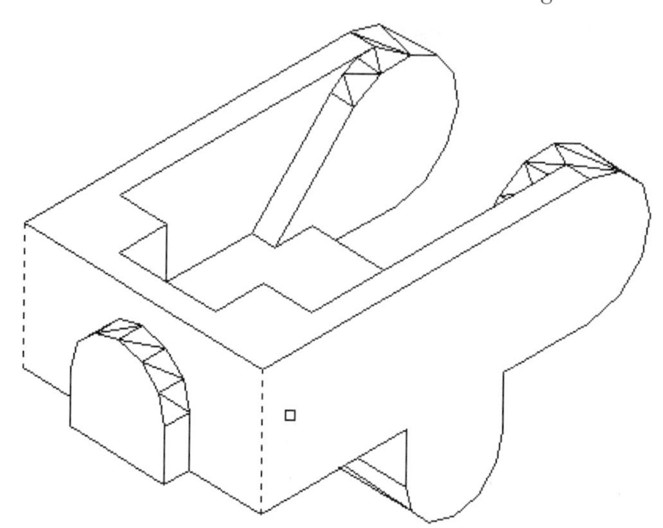

Figure 5-34 *Selecting the edges to fillet*

9. Once the fillet is created, the **Fillet** dialog box will be displayed again. Set the value of the **Radius** spinner to **8**.

10. Clear the **Return to Dialog** check box. Choose **OK**. The prompt sequence is as follows:

   Select edges or faces to fillet: *Select the edge as shown in Figure 5-35.*
   Select edges or faces to fillet <continue>: *Select the edge as shown in Figure 5-35.*
   Select edges or faces to fillet <continue>: *Select the edge as shown in Figure 5-35.*

Select edges or faces to fillet <continue>: *Select the edge as shown in Figure 5-35.*
Select edges or faces to fillet <continue>: *Select the edge as shown in Figure 5-35.*

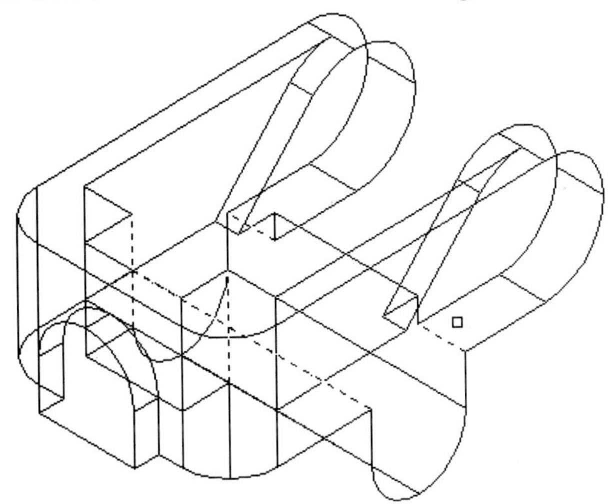

Figure 5-35 *Selecting the edges to fillet*

11. The designer model after applying all the fillets should look similar to the one shown in Figure 5-36.

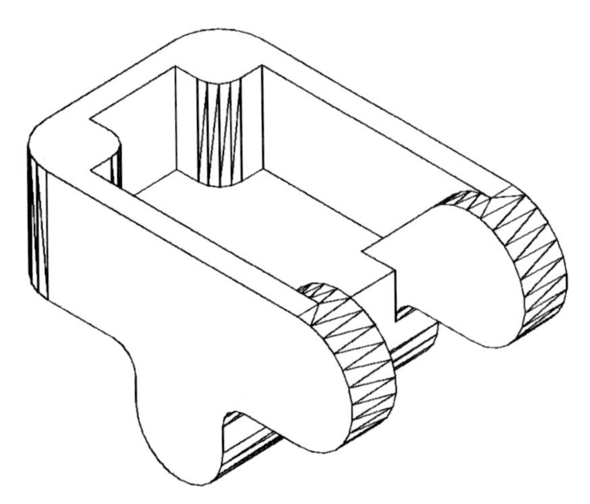

Figure 5-36 *Designer model after applying all the fillets*

12. Choose the **Placed Features-Hole** button from the **Part Modeling** toolbar to display the **Hole** dialog box as shown in Figure 5-37.

13. Select **Drilled** from the **Operation** drop-down list.

14. Select **Through** from the **Termination** drop-down list and **Concentric** from the **Placement** drop-down list.

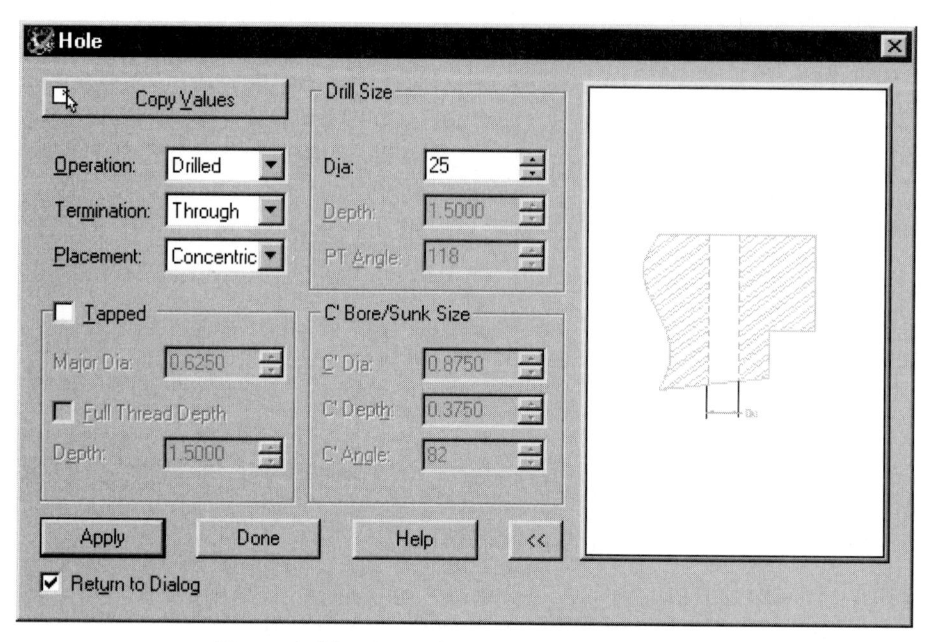

Figure 5-37 *Figure showing the Hole dialog box*

15. Set the value of the **Dia** spinner to **25**.

16. Select the **Return to Dialog** check box. Choose **Apply**. The prompt sequence is as follows:

 Select work plane, planar face or [worldXy/worldYz/worldZx/Ucs]: *Select the planar face as shown in Figure 5-38.*
 Select concentric edge: *Select the concentric edge as shown in Figure 5-39.*
 Computing ...
 Select work plane, planar face or [worldXy/worldYz/worldZx/Ucs]: Enter

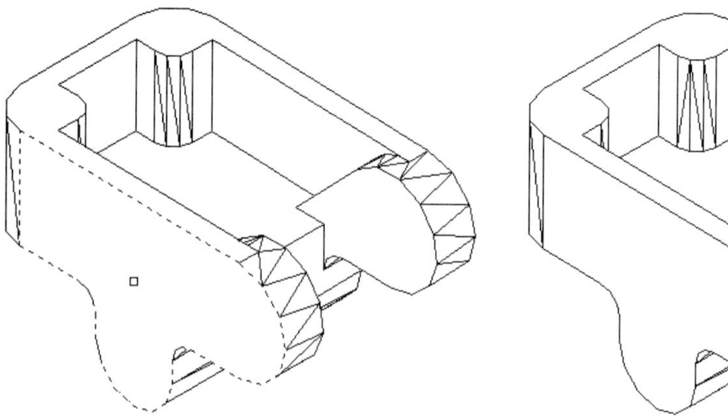

Figure 5-38 *Selecting the planar face to start the hole*

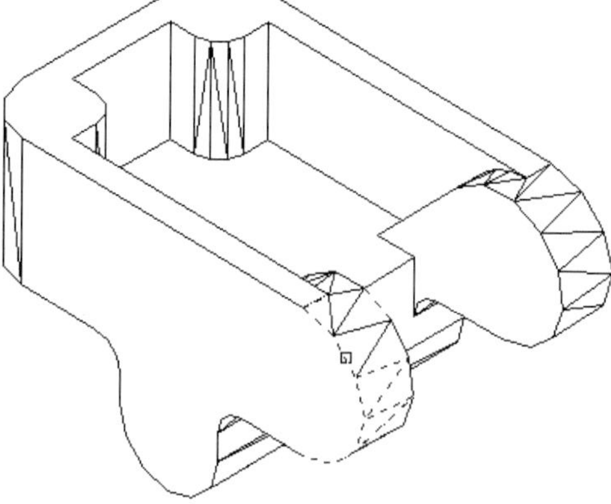

Figure 5-39 *Selecting the concentric edge to place the hole*

17. Set the value of the **Dia** spinner to **16**. Choose **Apply**. The prompt sequence is as follows:

    Select work plane or planar face [worldXy/worldYz/worldZx/Ucs]: *Select the planar face as shown in Figure 5-40.*
    Select concentric edge: *Select the concentric edge as shown in Figure 5-41.*
    Computing ...
    Select work plane or planar face [worldXy/worldYz/worldZx/Ucs]: Enter

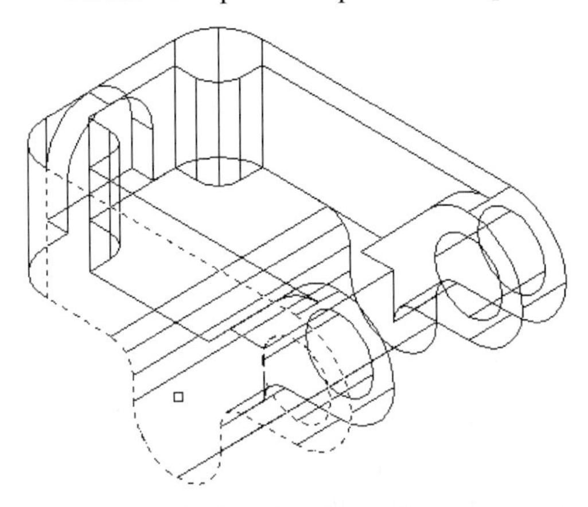

Figure 5-40 *Selecting the planar face to start the hole*

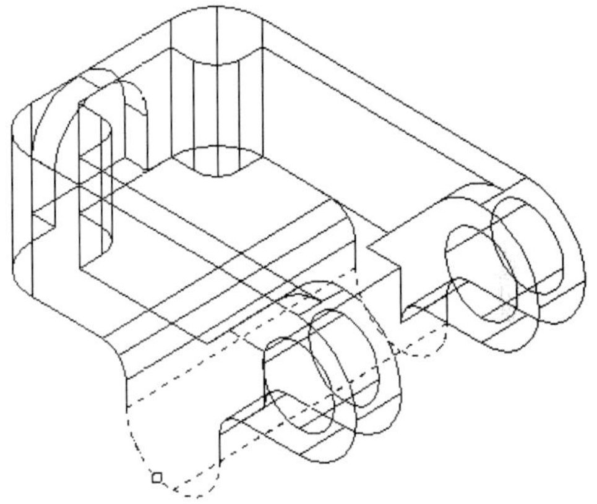

Figure 5-41 *Selecting the concentric edge to place the hole*

18. Set the value of the **Dia** spinner to **8**. Clear the **Return to Dialog** check box. Choose **OK**. The prompt sequence is as follows:

    Select work plane, planar face or [worldXy/worldYz/worldZx/Ucs]: *Select the planar face as shown in Figure 5-42.*
    Select concentric edge: *Select the concentric edge as shown in Figure 5-43.*
    Computing ...
    Select work plane, planar face or [worldXy/worldYz/worldZx/Ucs]: Enter

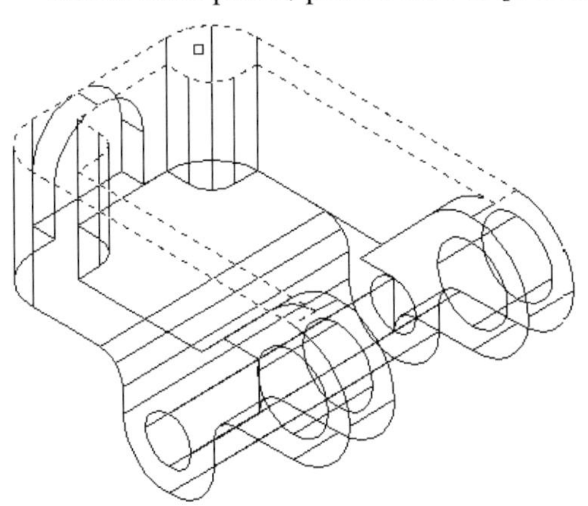

Figure 5-42 *Selecting the planar face to start the hole*

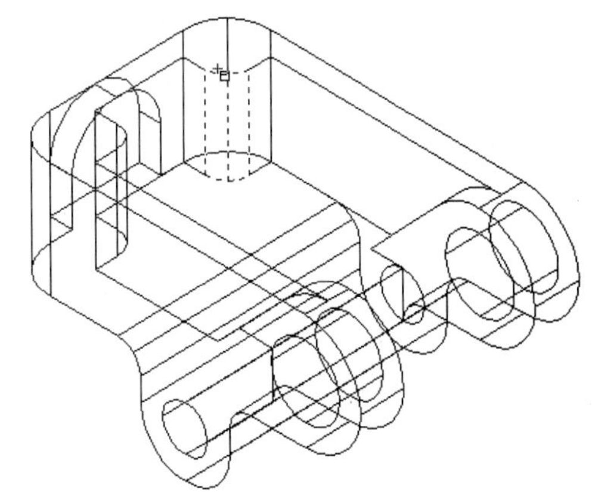

Figure 5-43 *Selecting the concentric edge to place the hole*

19. Similarly, create the hole on the other filleted edge also.

20. Enter **88** at the Command prompt.

21. Choose the **Placed Features-Hole** button.

22. Select **Drilled** from the **Operation** drop-down list. Select **Blind** from the **Termination** drop-down list.

23. Select **Concentric** from the **Placement** drop-down list. Set the value of the **Dia** spinner to **8**.

24. Set the value of the **Distance** spinner to **8** and the **PT Angle** spinner to **180**. Choose **OK**. The prompt sequence is as follows:

    Select work plane, planar face or [worldXy/worldYz/worldZx/Ucs]: *Select the planar face as shown in Figure 5-44.*
    Select concentric edge: *Select the concentric edge as shown in Figure 5-45.*
    Computing ...
    Select work plane, planar face or [worldXy/worldYz/worldZx/Ucs]: Enter

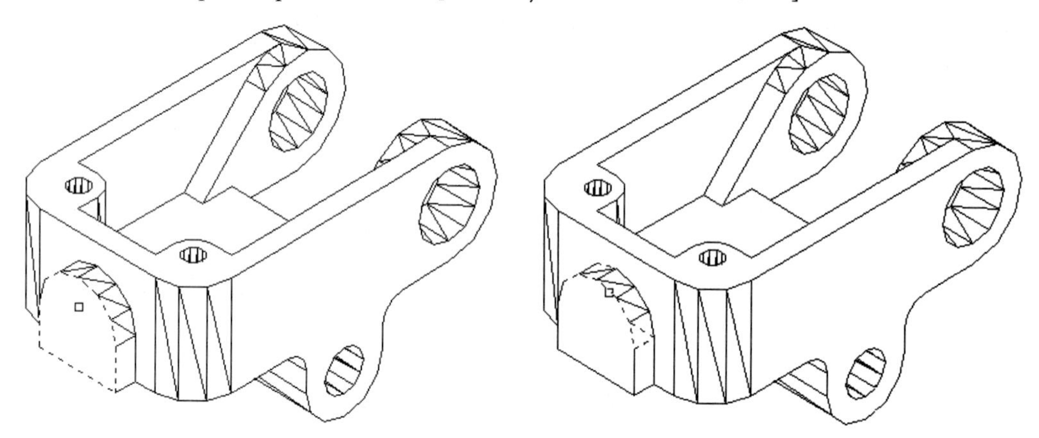

Figure 5-44 *Selecting the planar face to start the hole*

Figure 5-45 *Selecting the concentric edge to place the hole*

25. The final designer model should look similar to the one shown in Figure 5-46.

26. Save this drawing with the name given below:

    \MDT Tut\Ch-5**Tut1.dwg**

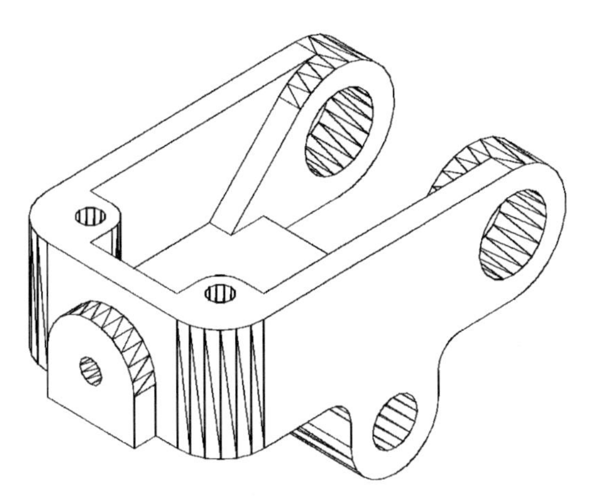

Figure 5-46 *The final designer model for Tutorial 1*

Tutorial 2

In this tutorial you will create the body of the Plummer Block as shown in Figure 5-47a. The dimensions to be used are given in Figures 5-47b, 5-47c and 5-47d.

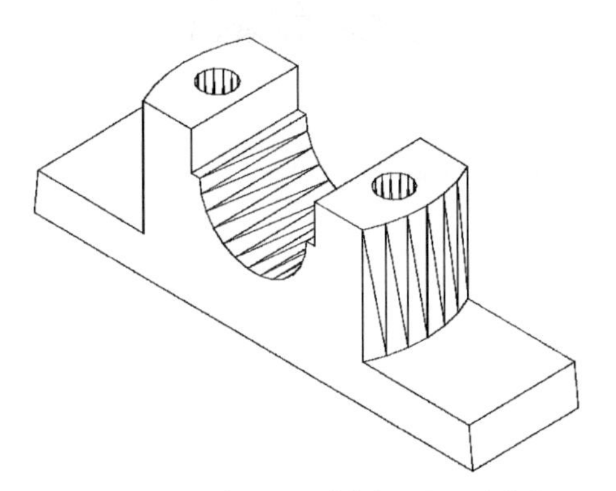

Figure 5-47a *Model for Tutorial 2*

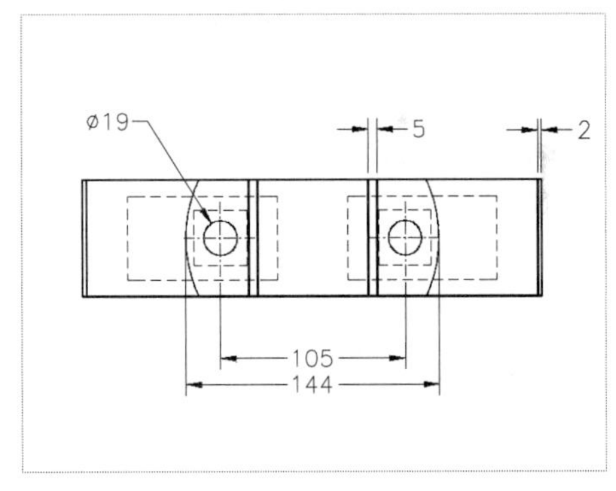

Figure 5-47b *Top view of the model*

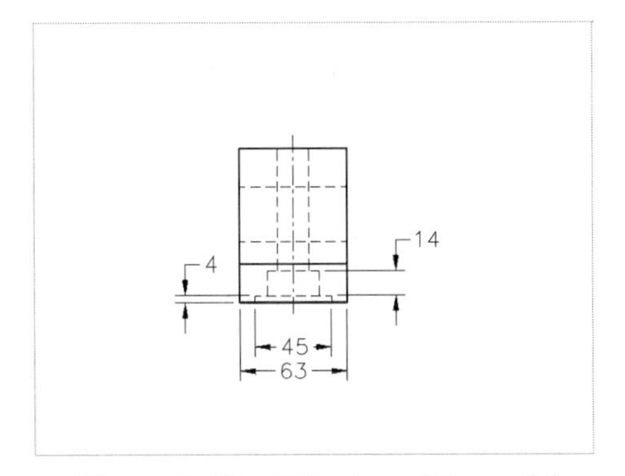

Figure 5-47c *Side view of the model*

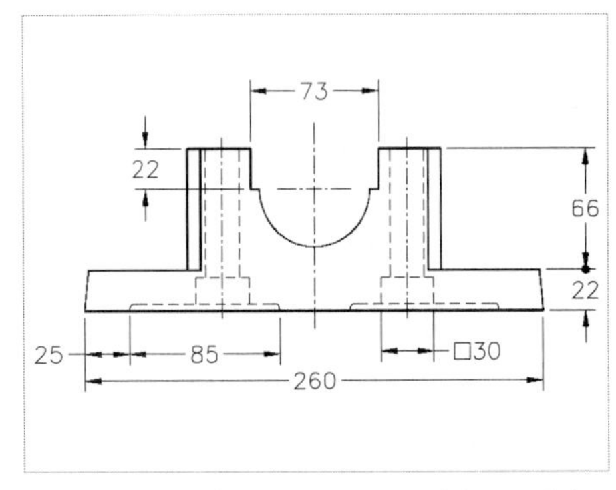

Figure 5-47d *Front view of the model*

1. Draw the sketch for the base feature and dimension it using the given dimensions.

2. Extrude it to create the base feature as shown in Figure 5-48.

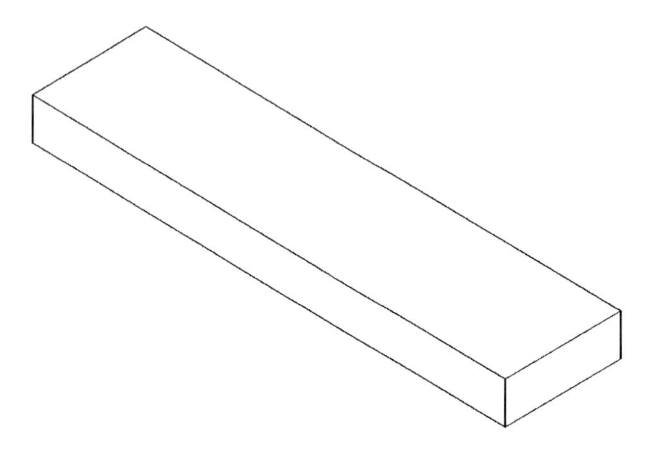

Figure 5-48 *Figure showing the base feature for Tutorial 2*

3. Choose the **Chamfer** button from the **Placed Features-Hole** flyout in the **Part Modeling** toolbar to display the **Chamfer Feature** dialog box, see Figure 5-49.

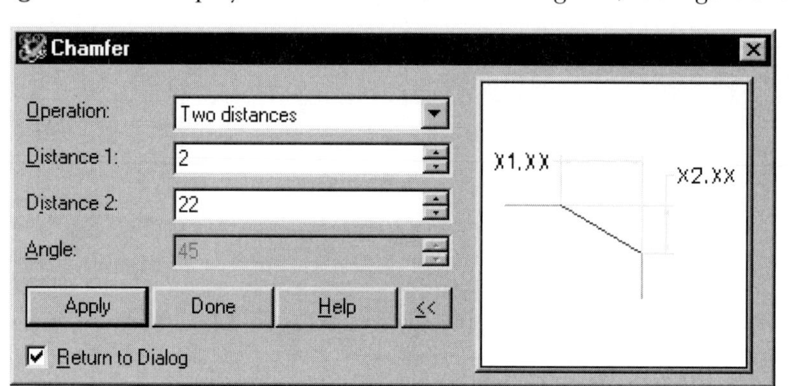

Figure 5-49 *Figure showing the Chamfer Feature dialog box*

4. Select the **Two Distance** from the **Operation** drop-down list.

5. Enter **2** in the **Distance1** edit box and **22** in the **Distance2** edit box.

6. Select the **Return to Dialog** check box. Choose **OK**. The prompt sequence is as follows:

   Select an edge or face to chamfer: *Select the edge as shown in Figure 5-50.*
   Press <ENTER> to continue:
   The specified face will be used for base distance.

Specify face for first chamfer distance (base) [Next/Accept] <Accept>: *Press **N** or **A** to make sure the top face is selected as shown in Figure 5-51.*
Computing ...

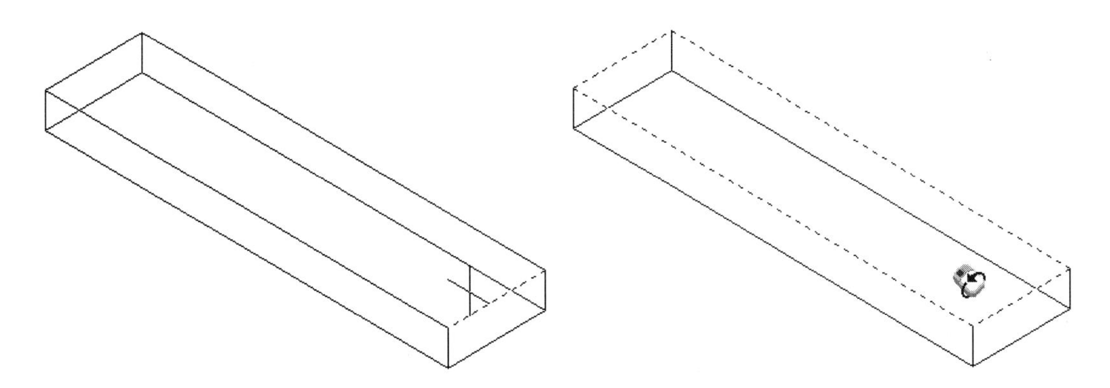

Figure 5-50 *Selecting the edge to chamfer*

Figure 5-51 *Selecting the face to apply the distance one value*

7. Similarly, create the chamfer on the other edge also. The designer model after applying the chamfer should look similar to the one shown in Figure 5-52.

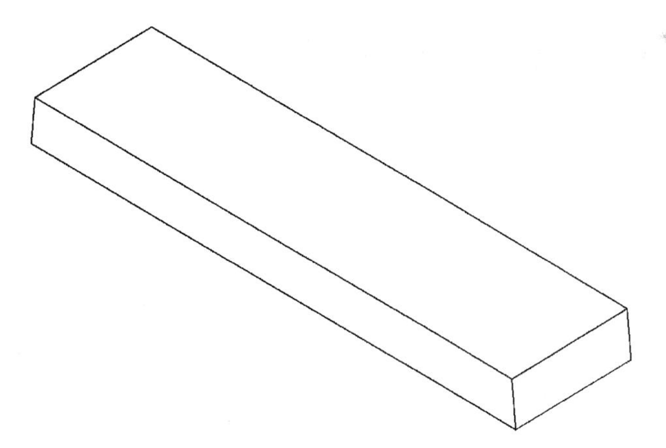

Figure 5-52 *Base feature after applying the chamfer feature*

8. Create the next feature on the top face of the base feature using the **Join** option as shown in Figure 5-53.

9. Create the cut through feature on the front face of the model as shown in Figure 5-54.

10. Create the cut feature on the bottom face of the designer model.

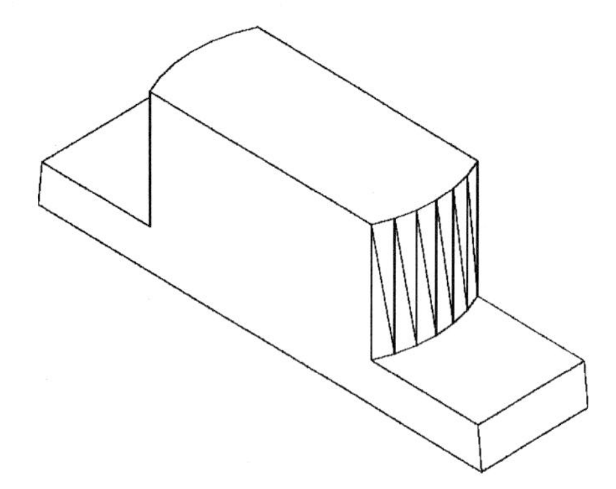

Figure 5-53 *Designer model after creating the next joined feature*

Figure 5-54 *Designer model after creating the cut through feature*

11. Choose the **Fillet** button from the **Placed Features-Hole** flyout in the **Part Modeling** toolbar to display the **Fillet** dialog box.

12. Select the **Constant** radio button. Set the value of the **Radius** spinner to **3**. Choose **OK**. The prompt sequence is as follows:

Select edges or faces to fillet: *Select the edge to fillet as shown in Figure 5-55.*
Select edges or faces to fillet <continue>: *Select the next edge to fillet.*
Select edges or faces to fillet <continue>: *Select the next edge to fillet.*
Select edges or faces to fillet <continue>: *Select the next edge to fillet.*
Select edges or faces to fillet <continue>: Enter

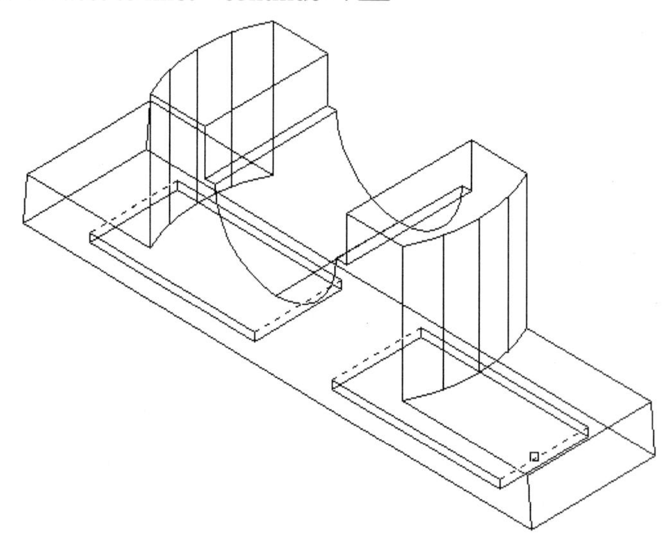

Figure 5-55 *Selecting the edges to fillet*

13. Create both the square cuts on the bottom face of the designer model.

14. Choose the **Placed Features-Hole** button from the **Part Modeling** toolbar to display the **Hole** dialog box.

15. Select **Drilled** from the **Operation** drop-down list and **Through** from the **Termination** drop-down list. Select **2 Edges** from the **Placement** drop-down list.

16. Set the value of the **Dia** spinner to **19**. Choose **OK**. The prompt sequence is as follows:

    Select first edge: *Select the edge as shown in Figure 5-56.*
    Select second edge: *Select the second edge as shown in Figure 5-56.*
    Computing ...
    Specify hole location: *Specify the placement point.*
    Enter distance from first edge (highlighted) <default value>: **15**
    Enter distance from second edge (highlighted) <default value>: **15**

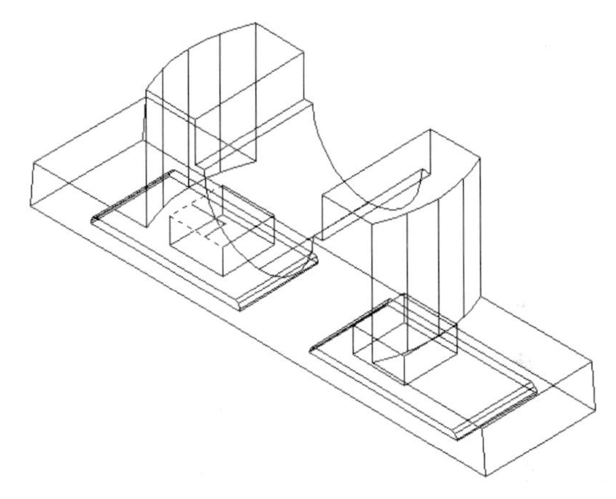

Figure 5-56 *Selecting the edges to place the hole*

17. Similarly, place the next hole. The final designer model should look similar to the one shown in Figure 5-57.

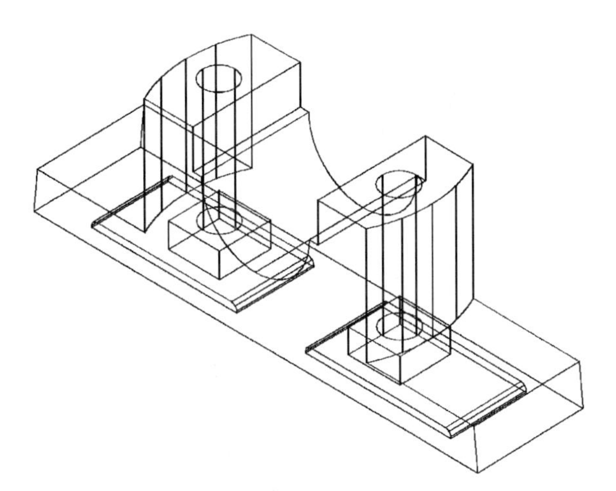

Figure 5-57 *Final designer model for Tutorial 2*

18. Save this drawing with the name

    /MDT Tut/Ch-5/**Tut2.dwg**

Tutorial 3

In this tutorial you will create the lock nut displayed in Figure 5-58. The dimensions are given in Figure 5-59. The fillet radius is 12 units.

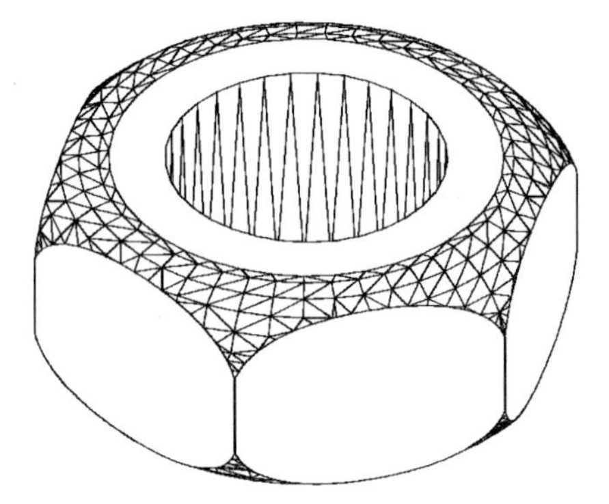

Figure 5-58 *Designer model for Tutorial 3*

Figure 5-59 *Dimensions for Tutorial 3*

1. Create the circular base feature as shown in Figure 5-60.

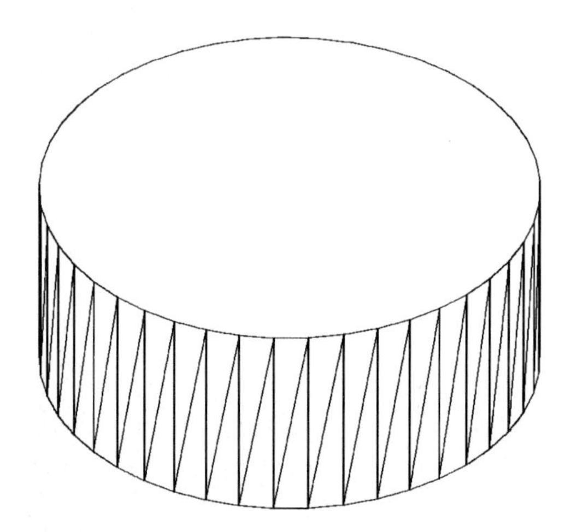

Figure 5-60 *Figure showing the base feature for Tutorial 3*

2. Choose the **Fillet** button from the **Placed Features-Hole** flyout in the **Part Modeling** toolbar to display the **Fillet** dialog box.

3. Select the **Constant** radio button and set the value of the **Radius** spinner to **3**. Choose **OK**. The prompt sequence is as follows:

   Select edges or faces to fillet: *Select the top face of the cylinder.*
   Select edges or faces to fillet <continue>: *Select the bottom face of the cylinder.*
   Select edges or faces to fillet <continue>: Enter

Computing...

4. The designer model should look similar to the one shown in Figure 5-61.

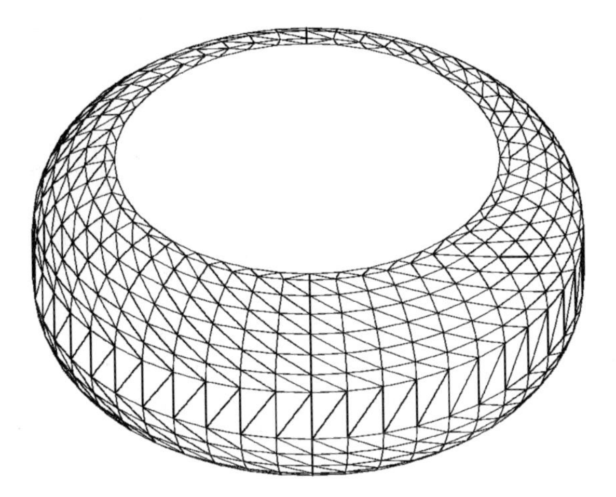

Figure 5-61 *Designer model after applying the fillet*

5. Enter **9** at the Command prompt.

6. Choose the **Polygon** button from the **2D Sketching** toolbar. The prompt sequence is as follows:

 POLYGON Enter number of sides <4>: **6**
 Specify center of polygon or [Edge]: *Select the center of model.*
 Enter an option [Inscribed in circle/Circumscribed about circle] <I>: **I**
 Specify radius of circle: **10**

7. Profile this sketch. Extrude it using the **Intersect** operation and **Through** termination type. The designer model should look similar to the one shown in Figure 5-62.

8. Choose the **Sketched Features-Hole** button from the **Part Modeling** toolbar to display the **Hole** dialog box.

9. Select **Drilled** from the **Operation** drop-down list. Select **Through** from the **Termination** drop-down list. Select **Concentric** from the **Placement** drop-down list.

10. Set the value of the **Dia** spinner to **10**. Choose **OK**. The prompt sequence is as follows:

 Select work plane, planar face or [worldXy/worldYz/worldZx/Ucs]: *Select the top face of the model.*
 Select concentric edge: *Select the circular feature on the top face.*
 Computing ...
 Select work plane, planar face or [worldXy/worldYz/worldZx/Ucs]: Enter

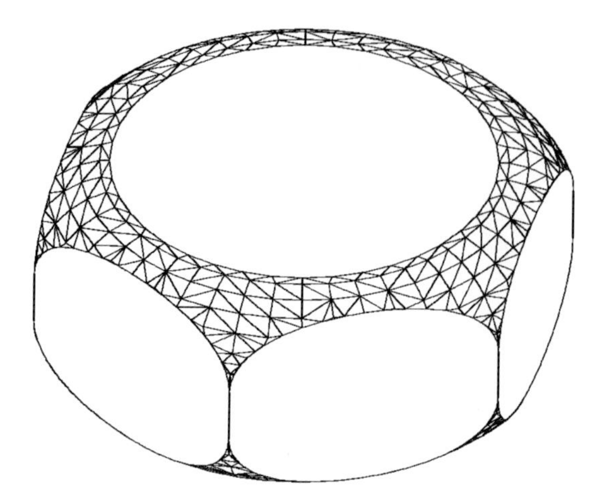

Figure 5-62 *Designer model after intersection*

11. The final designer model should look similar to the one shown in Figure 5-63. Save this drawing with the name \MDT Tut\Ch-5**Tut3.dwg**.

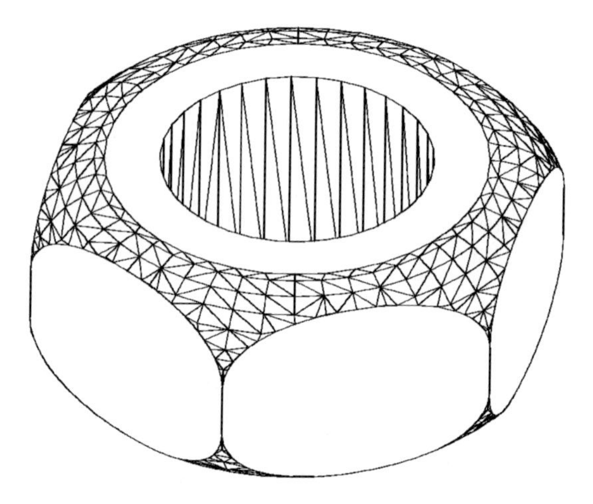

Figure 5-63 *The final designer model for Tutorial 3*

Tutorial 4

In this tutorial you will create the designer model shown in Figure 5-64a. After creating the model, apply a face draft of 2 degrees on the outer faces of the designer model. The dimensions to be used are given in Figures 5-64b, and 5-64c.

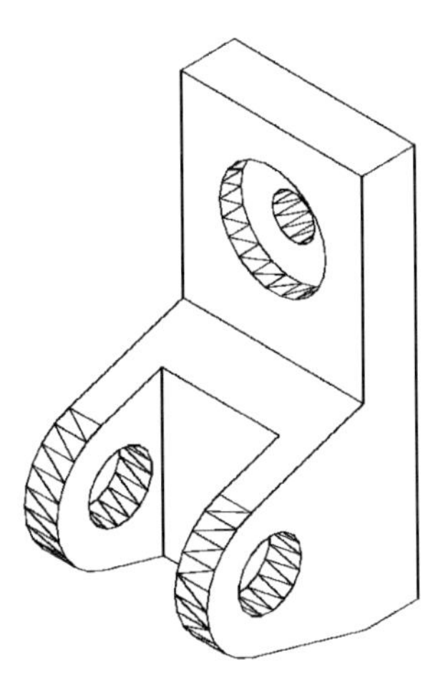

Figure 5-64a *Designer model for Tutorial 4*

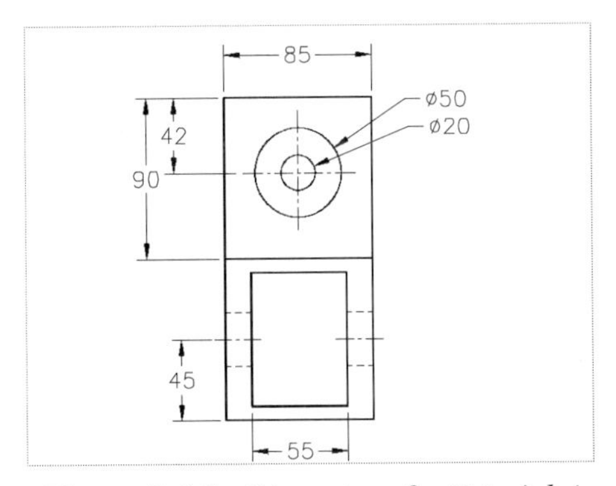

Figure 5-64b *Dimensions for Tutorial 4*

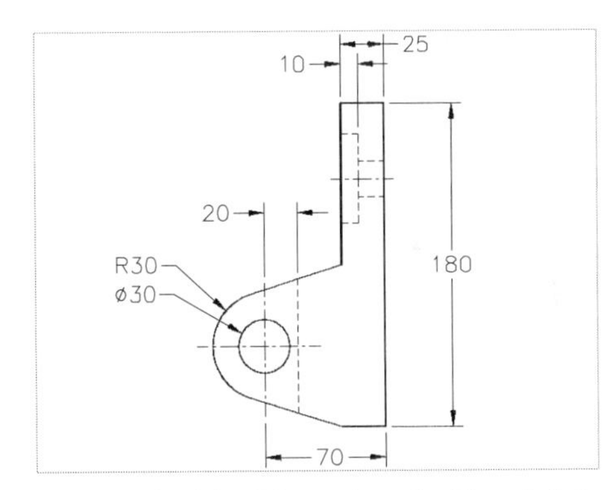

Figure 5-64c *Dimensions for Tutorial 4*

1. Draw the base feature for the designer model as shown in Figure 5-65.

2. Create the cut feature as shown in Figure 5-66.

3. Choose the **Sketched Feature-Hole** button to display the **Hole** dialog box.

4. Select **Drilled** from the **Operation** drop-down list and **Through** from the **Termination** drop-down list. Select **Concentric** from the **Placement** drop-down list.

5. Set the value of the **Dia** spinner to **30**. Select the **Return to Dialog** check box. Choose **Apply**. The prompt sequence is as follows:

   Select work plane, planar face or [worldXy/worldYz/worldZx/Ucs]: *Select the front face of the model.*
   Select concentric edge: *Select the circular edge.*

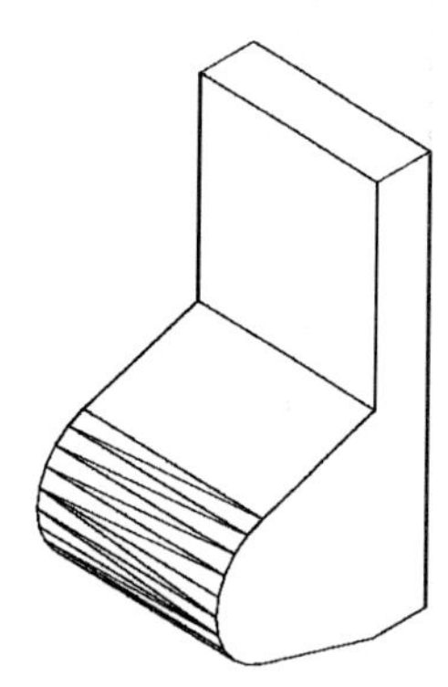

Figure 5-65 Base feature for the designer model

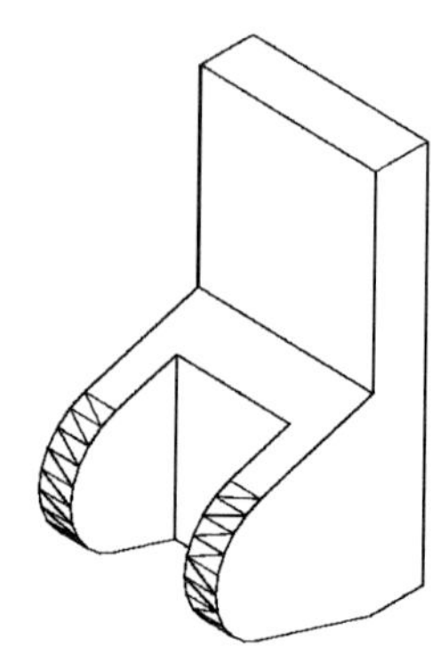

Figure 5-66 Designer model after creating the cut feature

Computing ...
Select work plane, planar face or [worldXy/worldYz/worldZx/Ucs]: Enter

6. Select **C'Bore** from the **Operation** drop-down list, **Through** from the **Termination** drop-down list, and **2 Edges** from the **Placement** drop-down list.

7. Set the value of the **Dia** spinner to **20**. Set the value of the **C'Dia** spinner to **50** and value of the **C'Depth** to **10**.

8. Clear the **Return to Dialog** check box. Choose **OK**. The prompt sequence is as follows:

Select first edge: *Select the top most edge of the model.*
Select second edge: *Select the right edge of the model.*
Computing ...
Specify hole location: *Specify the placement point.*
Enter distance from first edge (highlighted) <default value>: **42**
Enter distance from second edge (highlighted) <default value>: **45**
Computing ...
Select first edge: Enter

9. The designer model should look similar to the one shown in Figure 5-67.

10. Choose the **Face Draft** button from the **Placed Features-Hole** flyout in the **Part Modeling** toolbar to display the **Face Draft** dialog box.

11. Choose the **Draft Plane** button. The **Face Draft** dialog box will be temporarily closed. The prompt sequence is as follows:

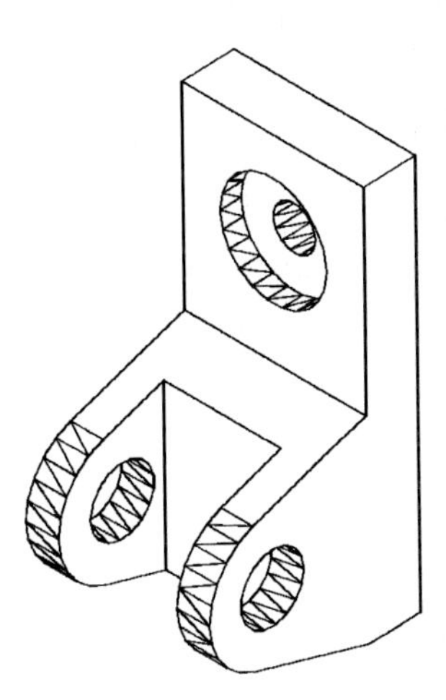

Figure 5-67 *Designer model for Tutorial 4*

Select draft plane (planar face or work plane): *Select the top face as shown in Figure 5-68.*
Enter an option [Next/Accept] <Accept>: Enter
Draft direction [Flip/Accept] <Accept>: *Enter **A** to make sure the draft direction is downwards.*

12. The **Face Draft** dialog box will be redisplayed. Choose the **Add** button. The prompt sequence is as follows:

Select faces to draft (ruled faces only): *Select the left face of the model as shown in Figure 5-69.*
Select faces to draft (ruled faces only): *Select the right face of the model as shown in Figure 5-69.*
Select faces to draft (ruled faces only): Enter

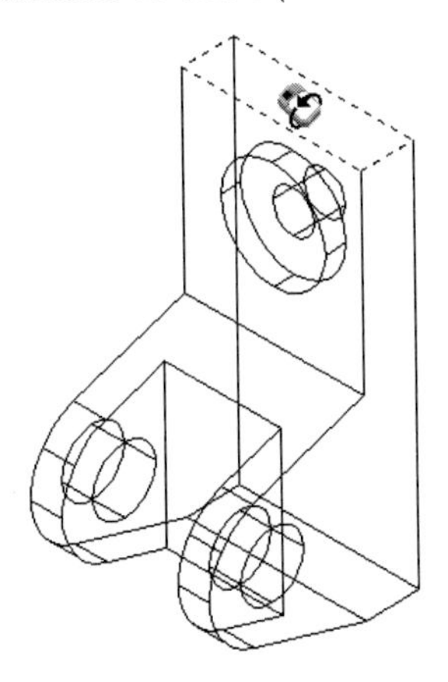

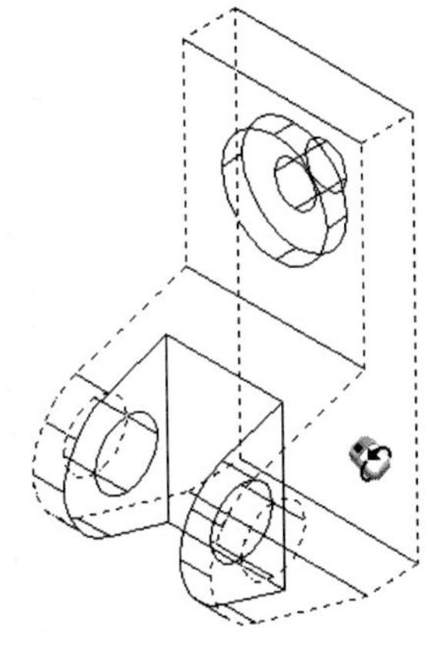

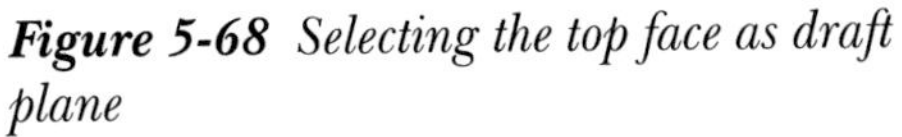

Figure 5-68 *Selecting the top face as draft plane*

Figure 5-69 *Selecting the faces to apply the draft*

13. Set the value of the angle spinner to **2**. Choose **OK**. The designer model after applying the face draft should look similar to the one shown in Figure 5-70.

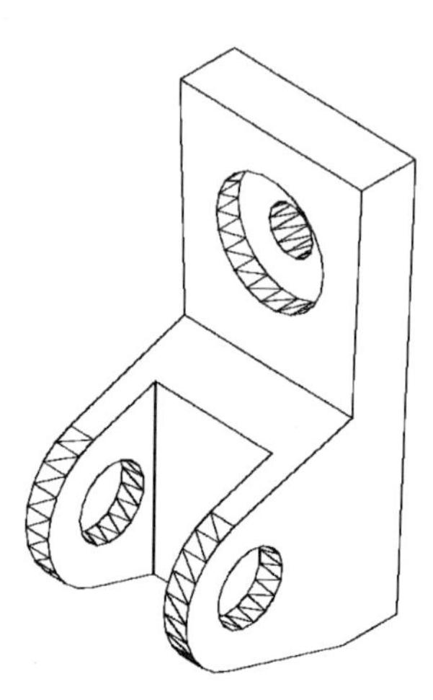

Figure 5-70 *Designer model after applying the face draft*

14. Save this file with the name given below:

    \MDT Tut\Ch-5**Tut4.dwg**

Review Questions

Answer the following questions.

1. Define placed feature. How many types of placed features are available in Mechanical Desktop?
2. Define a hole. How many types of holes can be placed in Mechanical Desktop?
3. What is the difference between counterbore and countersink holes?
4. The drill diameter can be greater then the counter diameter. (T/F)
5. What is the need of rounding or bevelling the edges?
6. How many types of fillet can be placed using the **AMFILLET** command?
7. What are the various chamfering operations available in Mechanical Desktop?
8. What is the need for applying the face draft to the faces of the designer model?

Exercises

Exercise 1

Draw the designer model shown in Figure 5-71a. The dimensions are given in Figures 5-71b, 5-71c and 5-71d. Assume the missing dimensions.

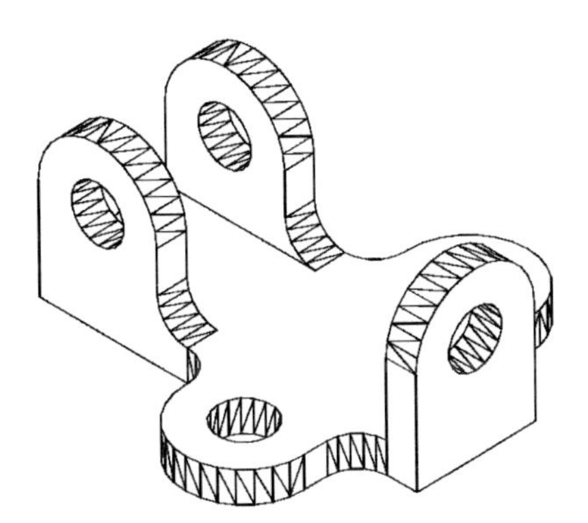

Figure 5-71a *Model for Exercise 1*

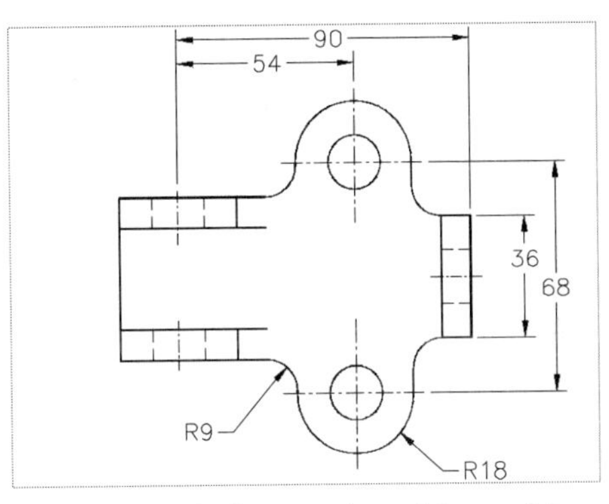

Figure 5-71b *Top view of the model*

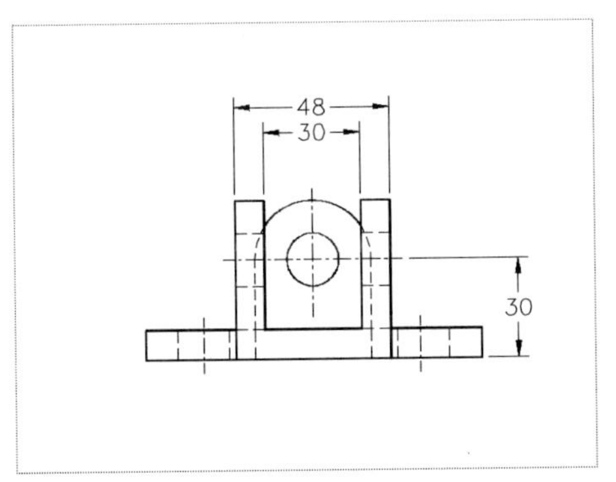

Figure 5-71c *Side view of the model*

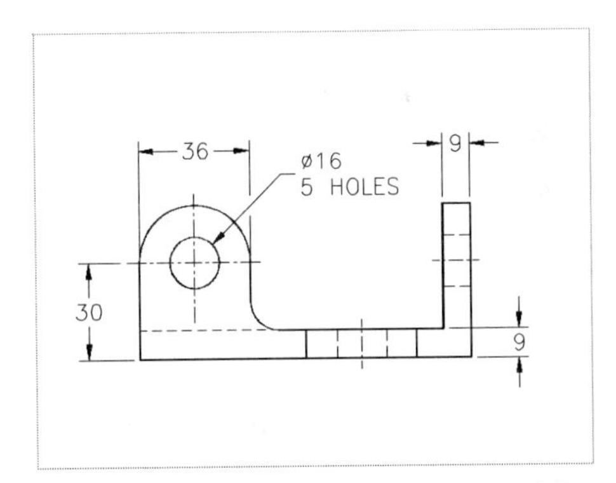

Figure 5-71d *Front view of the model*

Exercise 2

Draw the designer model shown in Figure 5-72a. The dimensions to be used are given in Figures 5-72b, 5-72c and 5-72d. Assume the missing dimensions.

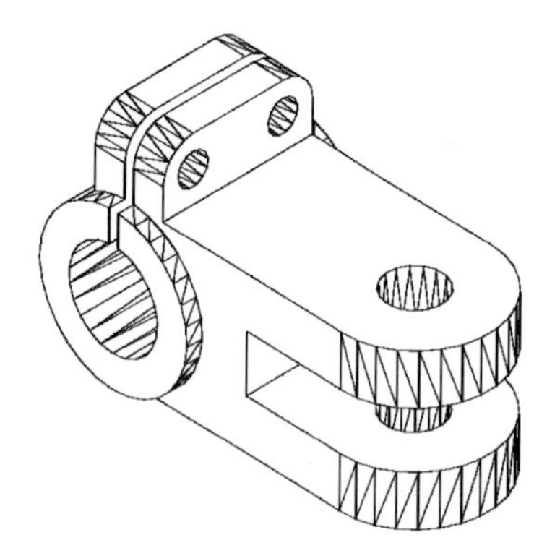

Figure 5-72a *Designer model for Exercise 2*

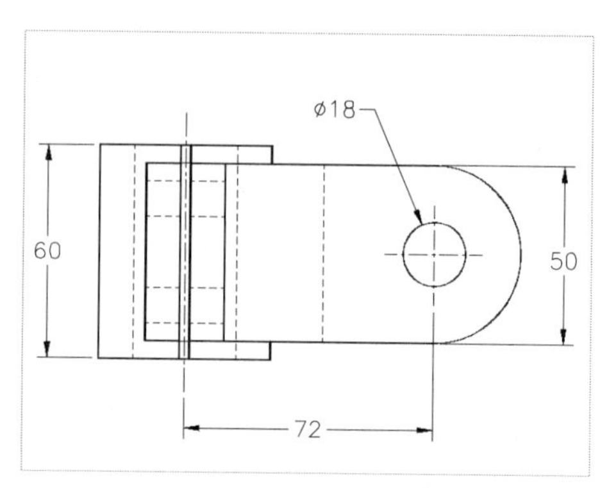

Figure 5-72b *Top view of the model*

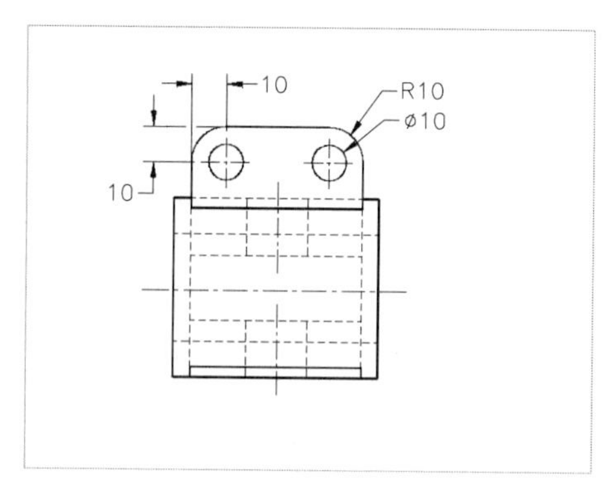

Figure 5-72c *Side view of the model*

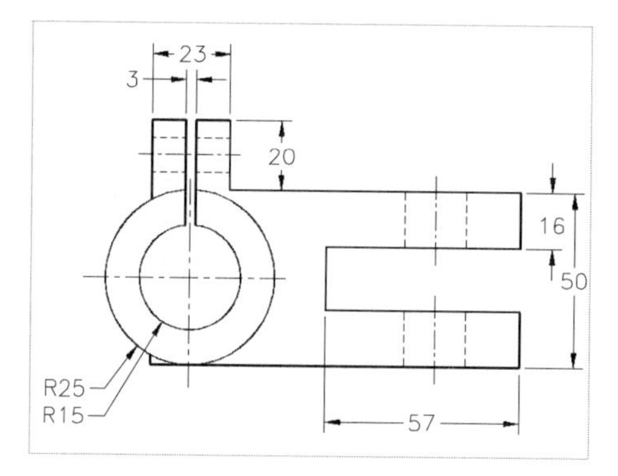

Figure 5-72d *Front view of the model*

Chapter 6

Placed Features II, Bend Features, and Rib Features

Learning Objectives

After completing this chapter, you will be able to:

- *Create a hollow designer model using the **AMSHELL** command.*
- *Create multiple copies of an existing feature using the **AMPATTERN** command.*
- *Create Bend features.*
- *Create Rib features.*
- *Create cuts using the AutoSURF surfaces.*

Commands Covered

- *AMSHELL*
- *AMPATTERN*
- *AMBEND*
- *AMRIB*
- *AMSURFCUT*

PLACED FEATURES

In this chapter some of the other placed features will be discussed.

Creating Hollow Models (AMSHELL Command)

Toolbar:	Part Modeling > Placed Features - Hole > Shell
Menu:	Part > Placed Features > Shell
Context Menu:	Placed Features > Shell
Command:	AMSHELL

This command is used to create hollow designer models by defining some wall thickness and removing the remaining inside material. You can define a uniform wall thickness or multiple wall thickness and at the same time remove the unwanted faces, thus converting it into an open model. This command is used on an existing designer model. The **Shell** dialog box will be displayed when you invoke this command, see Figure 6-1.

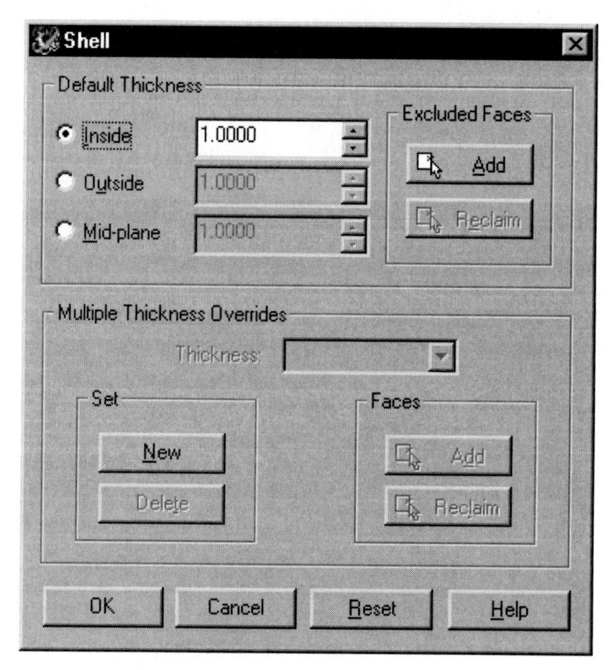

Figure 6-1 *The Shell dialog box*

Shell Dialog Box Options

The options provided under the **Shell** dialog box are:

Default Thickness Area

The options provided under this area are used to define the wall thickness for the shell feature. The wall thickness can be defined in the following three ways:

Inside. If the **Inside** radio button is selected, the wall thickness will be calculated into the model taking the outer face as the outside wall of the model. The value of the wall thickness can be specified in the **Inside** spinner. This spinner be will be available only when the **Inside** radio button is selected.

Outside. If the **Outside** radio button is selected, the wall thickness will be calculated out of the model taking the outer face as the inner wall of the shell feature. The value of the wall thickness can be specified in the **Outside** spinner which will be available only if the **Outside** radio button is selected.

Mid-plane. If the **Mid-plane** radio button is selected, the wall thickness will be calculated equally on both sides of the outer face. The wall thickness can be specified in the **Mid-plane** spinner available only if the **Mid-plane** radio button is selected.

Excluded Faces Area

Add. The **Add** button is chosen to select the faces to be excluded from the shell. The selected face will now be left open, see Figures 6-2, and 6-3.

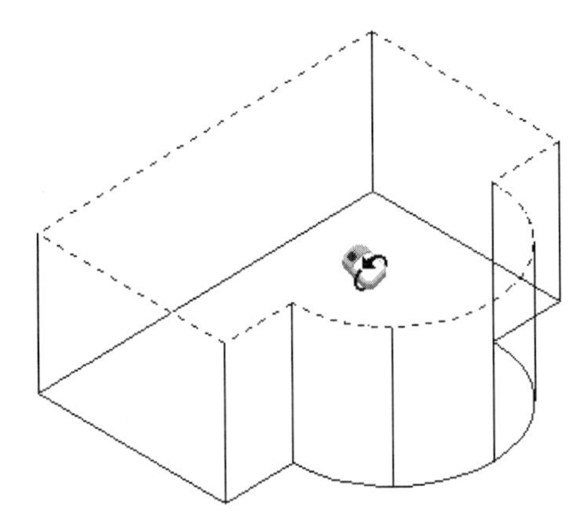

Figure 6-2 Selecting the face to exclude for shelling using the Add button

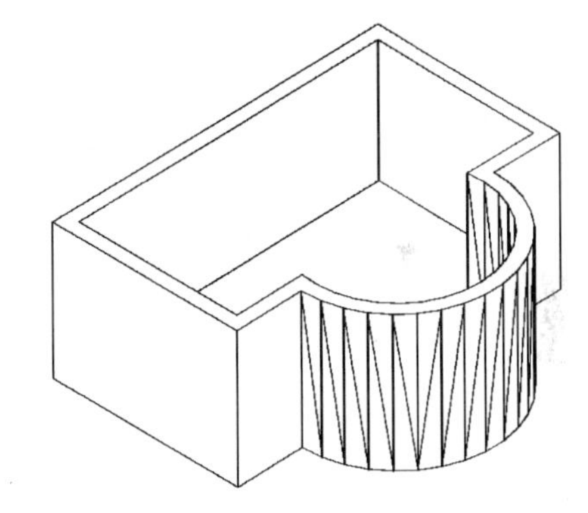

Figure 6-3 After creating the shell feature with the top face excluded

Reclaim. This button is chosen to reclaim the faces excluded using the **Add** button. This button is not available until you have selected the faces to exclude using the **Add** button.

Multiple Thickness Overrides Area

The options provided under this area are used to specify multiple wall thickness for the shell feature.

Thickness. This drop-down list is not available until the **New** button under the **Set** area is selected. The value entered in this drop-down list will be applied to the walls selected using the **Add** button under the **Faces** area.

New. The **New** button under the **Set** area is used to add multiple wall thickness.

When this button is chosen, the **Thickness** drop-down list is activated.

Delete. The **Delete** button is used to clear the multiple wall thickness option. This button is available only after some thickness value is specified in the **Thickness** drop-down list.

Add. The **Add** button under the **Faces** area is chosen to select the faces on which the new wall thickness will be applied as shown in Figures 6-4 and 6-5. This button is available only after some thickness value is specified in the **Thickness** drop-down list.

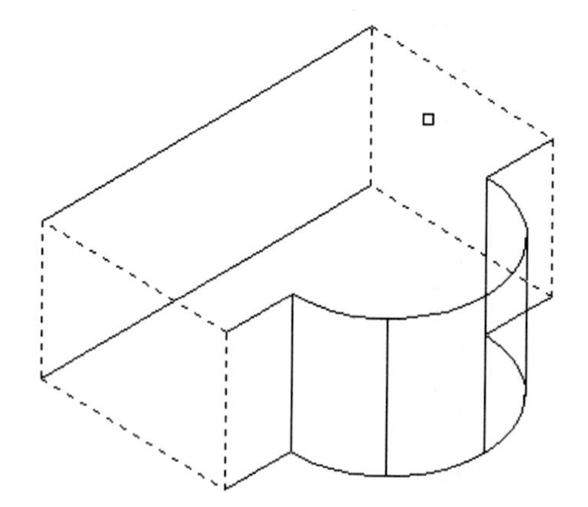

Figure 6-4 *Selecting the faces to apply the new wall thickness*

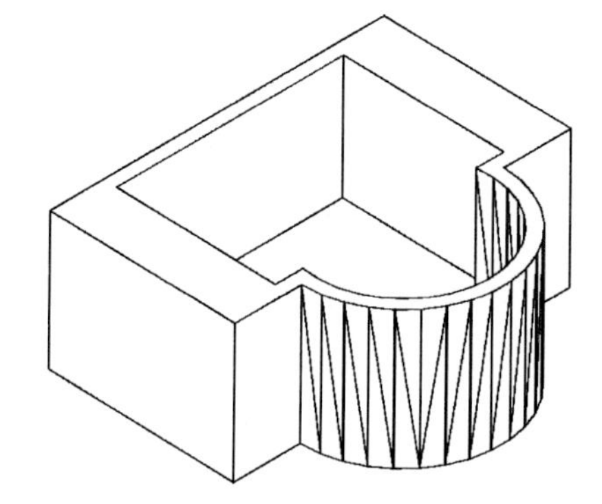

Figure 6-5 *Shell feature with the multiple wall thickness*

Reclaim. This button is chosen to reclaim the faces selected to apply the new wall thickness.

Reset

This button is chosen to reset the values of the **Shell Feature** dialog box to the initial setting.

AMPATTERN Command*

Sometimes, while creating complex designs, you may need to create multiple copies of an existing feature. This can be done easily using the **AMPATTERN** command. This is a new command introduced in Mechanical Desktop 5 and it replaces the **AMARRAY** command of previous releases. Using this command, you can create the following three patterns:

- Rectangular
- Polar
- Axial

Creating The Rectangular Patterns

Toolbar: Part Modeling > Placed Features - Hole > Rectangular Pattern
Menu: Part > Placed Features > Rectangular Pattern
Context Menu: Placed Features > Rectangular Pattern
Command: AMPATTERN

Creating patterns in the rectangular fashion is the method of copying the selected feature about the edges of an imaginary rectangle. You can also create patterns using the shortcut menu displayed upon right-clicking on the feature to be patterned in the desktop browser. Choose **Pattern > Rectangular** from the shortcut menu to invoke this option. When you invoke this command, you will be prompted to select the feature to be patterned. The options that are provided when you select the feature to be patterned are:

liSt

This option is used to list the features that are selected to pattern.

Remove

This option is used to remove the feature from the pattern. You can list the features that can be removed from the pattern, remove all the features or add the features pattern using this option.

Accept

This option is used to accept the selection of the features to pattern. Once you have accepted the selection set, the **Pattern** dialog box will be displayed, see Figure 6-6.

Pattern Dialog Box Options (Pattern Control Tab)

The options under this tab are:

Type

This drop-down list provides type of patterns that can be created. Here only the **Rectangular** option will be discussed.

Suppress Instance

This button is chosen to suppress the selected instances from the pattern. You can select the instances to be suppressed from the preview on the designer model.

Plane Orientation

This button is chosen to select the plane for the distribution of pattern. When you choose this button, the **Pattern** dialog box is temporarily closed and you will be prompted to select the face to orient the distribution of pattern. You can orient the direction of the rows and columns as per your convenience.

Column Placement Area

The options provided under this area are:

Incremental Spacing. If this button is chosen, the value of the spacing between

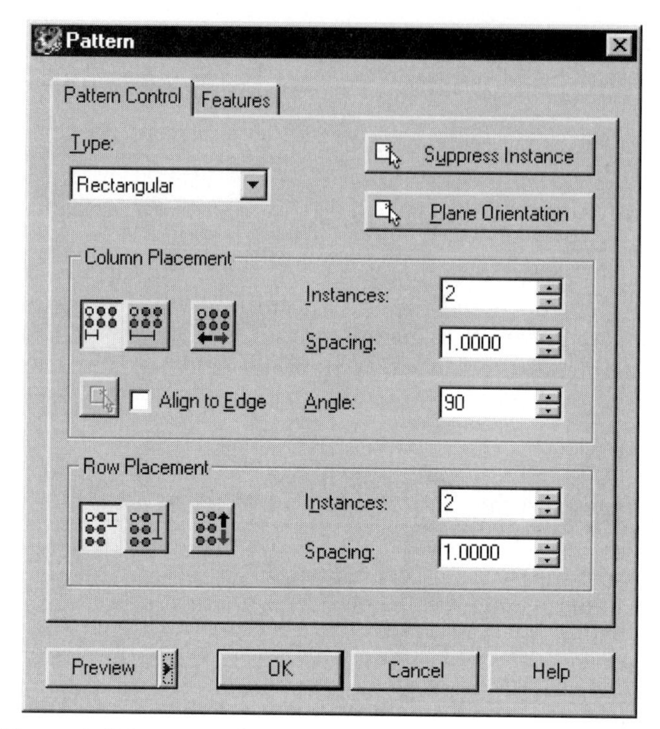

***Figure 6-6** Pattern dialog box displaying the Pattern Control tab*

the columns will be specified incrementally as shown in Figure 6-7.

Included Spacing. If this button is chosen, the value specified in the **Spacing** spinner in this area will be considered as the total spacing between all the instances of the column, see Figure 6-8.

Flip Column Direction. This button is chosen to reverse the direction of the column creation.

Instances. This spinner is used to specify the number of columns in the pattern.

Spacing. This spinner is used to specify the column spacing.

Angle. This spinner is used to specify the angle of positioning the columns.

Align to Edge. This check box is selected to align the direction of the column placement using an existing edge or a work axis. Choose the **Select edge to align column to** button to select the edge or work axis to align the column direction.

Row Placement Area

The options provided under this area are:

Incremental Spacing. If this button is chosen, the value of the spacing between the rows will be specified incrementally, see Figure 6-7.

Included Spacing. If this button is chosen, the value specified in the **Spacing** spinner in this area will be considered as the total spacing between all the instances of the column, see Figure 6-8.

Flip Column Direction. This button is chosen to reverse the direction of the column creation.

Instances. This spinner is used to specify the number of columns in the pattern.

Spacing. This spinner is used to specify the column spacing.

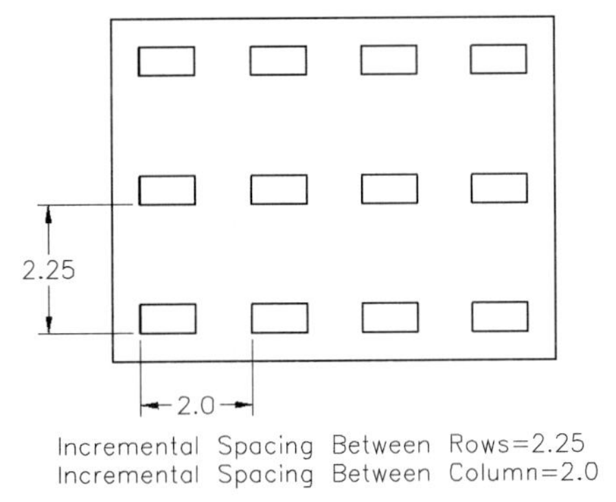

Figure 6-7 Rectangular pattern created using the incremental spacing

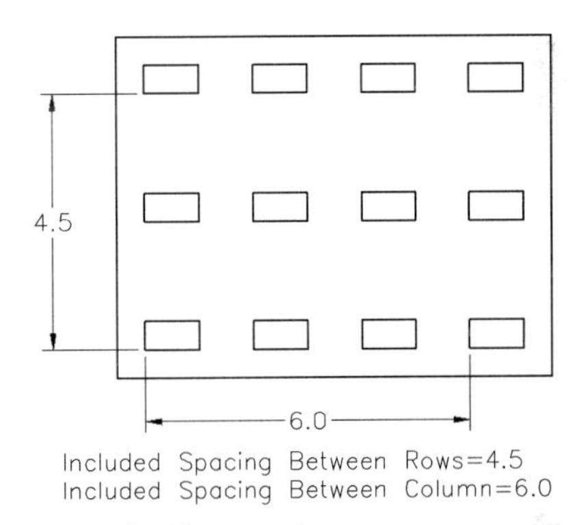

Figure 6-8 Rectangular pattern created using the included spacing

Features Tab

This tab displays all the features that are included in the current pattern. The features included in the current pattern are displayed in the list box. You can add a feature in the current pattern list or remove a feature from the current pattern list using the **Add** or **Delete** button.

Preview

This button is chosen to preview the pattern before it is actually created. Mechanical Desktop allows you to preview the pattern features in two ways that are discussed next.

True Preview

You can view a true or a simplified representation of the pattern before it is actually created using this option.

Dynamic Preview

If this option is selected, any changes in the pattern are reflected dynamically in the preview. If this option is cleared, then the changes are not dynamically reflected in the preview.

Creating The Polar Patterns

Toolbar: Part Modeling > Placed Features - Hole > Polar Pattern
Menu: Part > Placed Features > Polar Pattern
Context Menu: Placed Features > Polar Pattern
Command: AMPATTERN

The polar pattern is the method of copying the selected feature in circular fashion about a specified rotational center. When you invoke this command, you will be prompted to select the feature to pattern. However, if you invoke this command using the desktop browser, you will not be prompted to select the feature to be patterned. The options that are provided once you select the feature to pattern are:

liSt

This option is used to list the feature that are selected to pattern.

Remove

This option is used to remove the feature from the pattern. You can list the features that can be removed from the pattern, remove all the features or add the features pattern using this option.

Accept

This option is used to accept the selection of the features to pattern. Once you have accepted the selection set, you will be prompted to select the rotational center for the polar pattern. The **Pattern** dialog box will be displayed upon selecting the rotational center, see Figure 6-9.

Pattern Dialog Box Options (Pattern Control Tab)

The options under this tab are:

Type

This drop-down list provides type of patterns that can be created. Here only the **Polar** option will be discussed.

Suppress Instance

This button is chosen to suppress the selected instances from the pattern. You can select the instances to be suppressed from the preview on the designer model.

Rotational Center

This button is chosen to redefine the rotational center for the polar pattern.

Polar Placement Area

The options provided under this area are:

Maintain Orientation. This check box is selected to ensure that the new features maintain their orientation when they are copied about an existing polar or axial pattern. You will have to select a work point about which the features will maintain

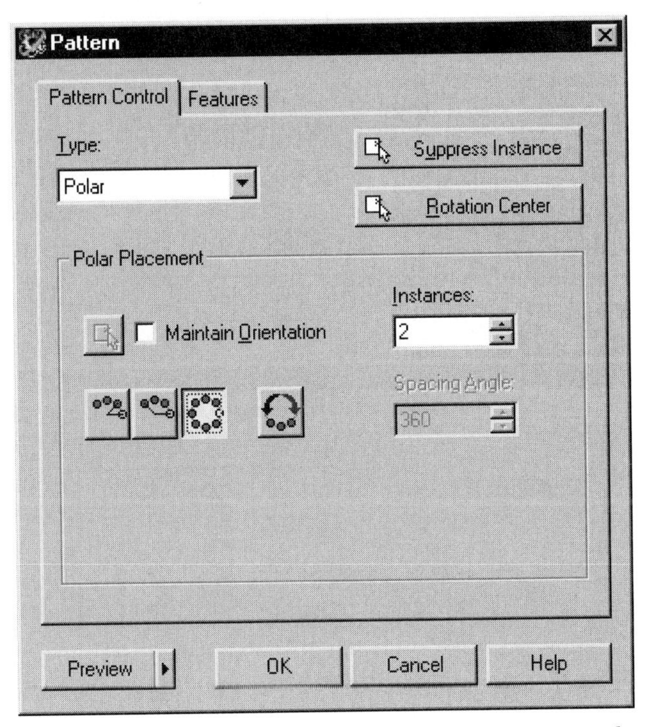

Figure 6-9 *Pattern dialog box displaying the Pattern Control tab*

their orientation. The work point can be chosen using the **Select work point** button that is available only when you select the **Maintain Orientation** check box.

Instances. This spinner is used to specify the total number of instances in the pattern.

Incremental Angle. This button is chosen when you want to create a polar pattern by specifying the incremental angle between instances, see Figure 6-10.

Included Angle. This button is chosen to create a polar pattern by specifying the total included angle in which all the instances are evenly arranged, see Figure 6-11.

Full Circle. This button is chosen to create a full circle polar pattern as shown in Figure 6-12.

Flip Rotation Direction. This button is chosen to reverse the direction in which the polar pattern is created.

Spacing Angle. This spinner is used to specify the incremental or included angle for the polar patterns.

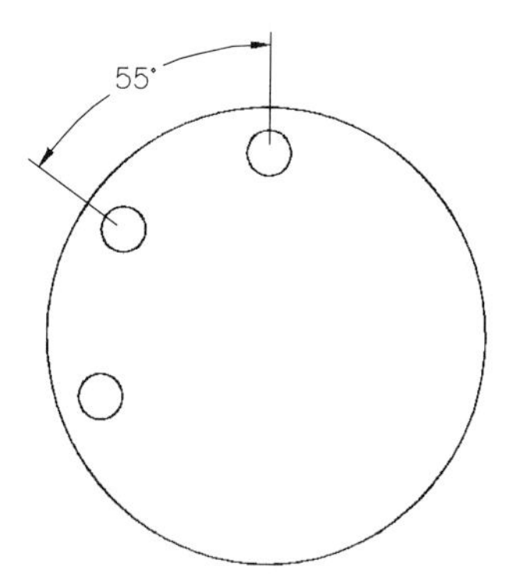

Figure 6-10 The polar pattern created through an incremental angle of 55 degrees

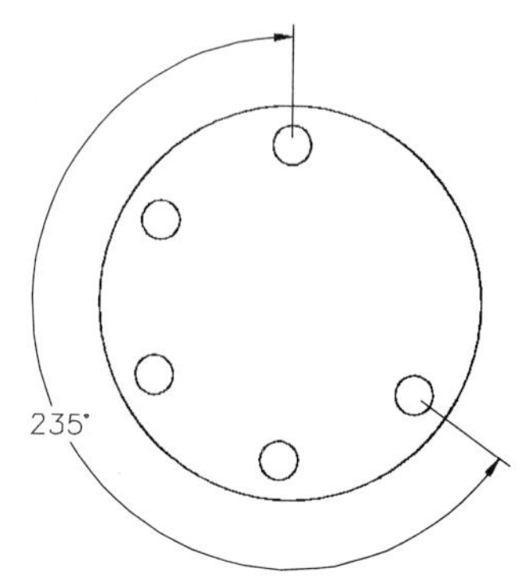

Figure 6-11 The polar pattern created through an included angle of 235 degrees

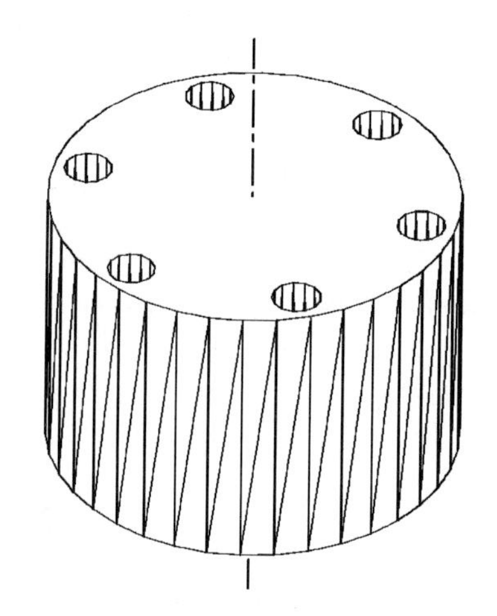

Figure 6-12 Full circle polar pattern

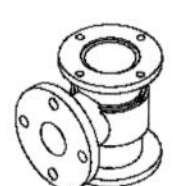

Note

*The options under the **Features** tab and the **Preview** button are the same as those discussed in the **Rectangular** pattern.*

Creating The Axial Patterns

Toolbar:	Part Modeling > Placed Features - Hole > Axial Pattern
Menu:	Part > Placed Features > Axial Pattern
Context Menu:	Placed Features > Axial Pattern
Command:	AMPATTERN

The axial pattern is the method of copying the selected feature in the helical fashion about a specified rotational center. Figure 6-13 shows a model before creating an axial pattern and Figure 6-14 shows a model after creating the axial pattern. When you invoke this command, you will be prompted to select the feature to pattern. However, if you

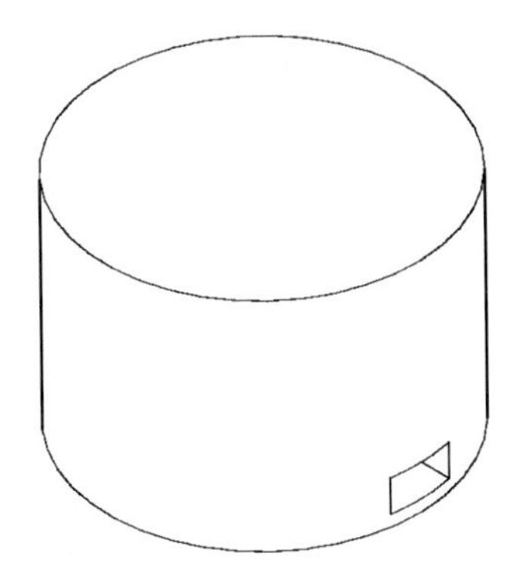

Figure 6-14 *Designer model before creating the axial pattern*

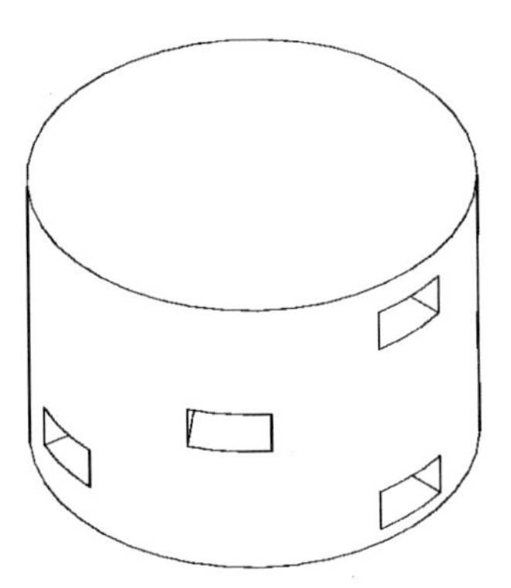

Figure 6-15 *Designer model after creating the axial pattern*

invoke this command using the desktop browser, you will not be prompted to select the feature to be patterned. The options that are provided once you select the feature to pattern are:

liSt

This option is used to list the feature that are selected to pattern.

Remove

This option is used to remove the feature from the pattern. You can list the features that can be removed from the pattern, remove all the features or add the features pattern using this option.

Accept

This option is used to accept the selection of the features to pattern. Once you have accepted the selection set, you will be prompted to select the rotational center for the axial pattern. The **Pattern** dialog box will be displayed upon selecting the rotational center, see Figure 6-15.

Pattern Dialog Box Options (Pattern Control Tab)

The options under this tab are:

Type

This drop-down list provides type of patterns that can be created. Here only the **Axial** option will be discussed.

Suppress Instance

This button is chosen to suppress the selected instances from the pattern. You can select the instances to be suppressed from the preview on the designer model.

Rotational Center

This button is chosen to redefine the rotational center for the axial pattern.

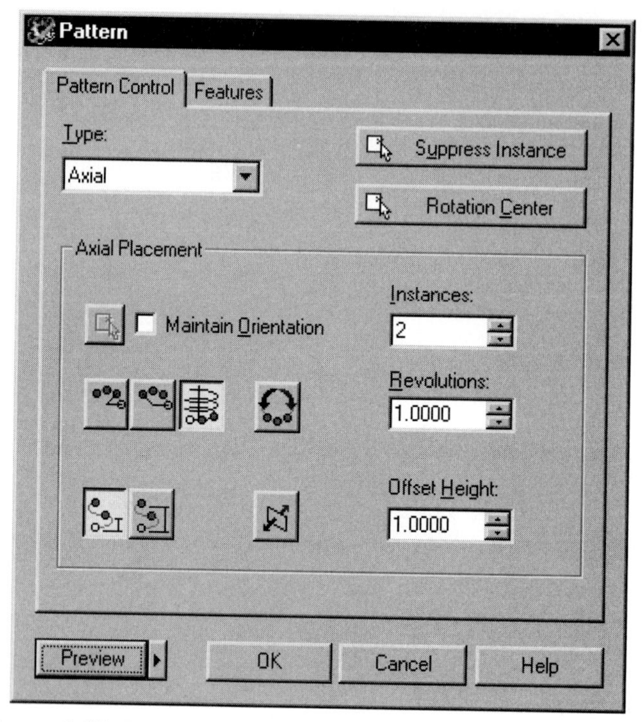

Figure 6-15 Pattern dialog box displaying the Pattern Control tab

Axial Placement Area

The options that are provided under this area are:

Maintain Orientation. This check box is selected to ensure that the new features maintain their orientation when they are copied about an existing polar or axial pattern. You will have to select a work point about which the features will maintain their orientation. The work point can be selected using the **Select work point** button that is available only when you select the **Maintain Orientation** check box.

Instances. This spinner is used to specify the total number of instances in the pattern.

Incremental Angle. This button is chosen when you want to create an axial pattern by specifying the incremental angle between instances.

Included Angle. This button is chosen to create an axial pattern by specifying the total included angle in which all the instances are evenly arranged.

Revolutions. This button is chosen to specify the total number of revolutions in the axial pattern. The number of revolutions can be specified in the **Revolutions** spinner that appears in place of the **Spacing Angle** spinner when you choose the **Revolutions** button.

Spacing Angle. This spinner is used to specify the incremental or included angle for the axial pattern.

Revolutions. This spinner appears in place of the **Spacing Angle** spinner when you choose the **Revolutions** button. You can specify the number of revolutions in the axial pattern using this spinner.

Incremental Offset. This button is chosen to specify the incremental offset between the instances of the axial pattern.

Included Offset. This button is chosen to specify the total included angle between all the instances of the axial pattern.

Flip Offset Direction. This button is chosen to reverse the direction of the pattern creation.

Offset Height. This button is chosen to specify the incremental or included offset height between the instances of the axial pattern.

Note
*The options under the **Features** tab and the **Preview** button are the same as those discussed in the **Rectangular** pattern.*

CREATING BEND FEATURES

The bend feature is one of the major enhancements of Mechanical Desktop 5 over the previous releases. Using this option you can bend an existing feature with the help of an open profile. This is done using the **AMBEND** command.

AMBEND Command*

Toolbar: Part Modeling > Sketched Features - Extrude > Bend
Menu: Part > Sketched Features > Bend
Context Menu: Sketched & Worked Features > Bend
Command: AMBEND

This command is used to bend an existing feature. To create a bend feature you must have a feature to bend and an open profile that you can use to bend the existing feature. Figure 6-16 shows a base feature to bend and an open profile used to bend the feature. Figure 6-17 shows the feature after bending.

You can also invoke this command using the shortcut menu displayed upon right-clicking on the open profile and choosing **Bend**. When you invoke this command, the **Bend** dialog box will be displayed, see Figure 6-18.

Bend Dialog Box Options

The options provided under this dialog box are:

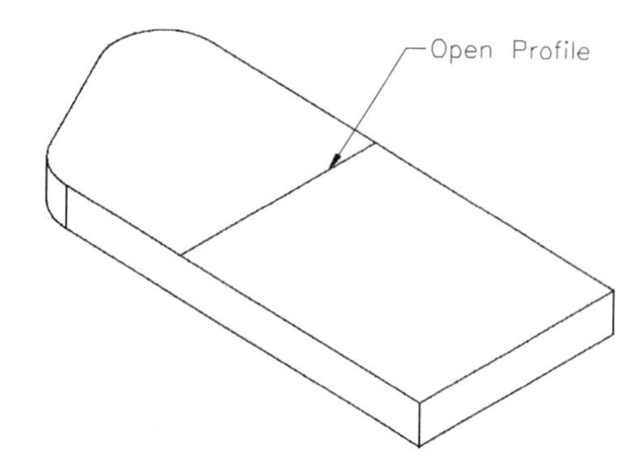

Figure 6-16 Designer model before creating the bend feature

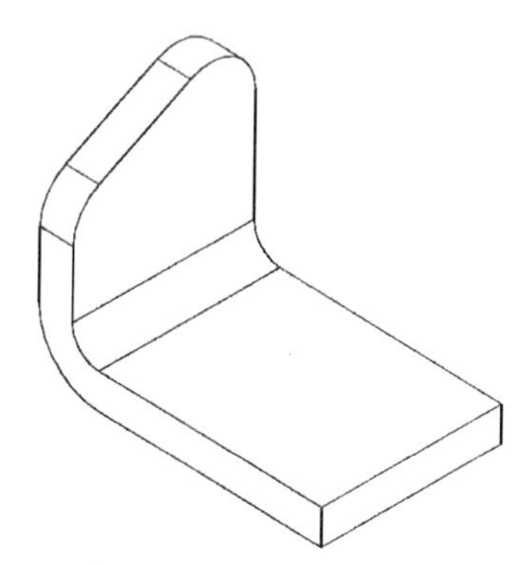

Figure 6-17 Designer model after creating the bend feature

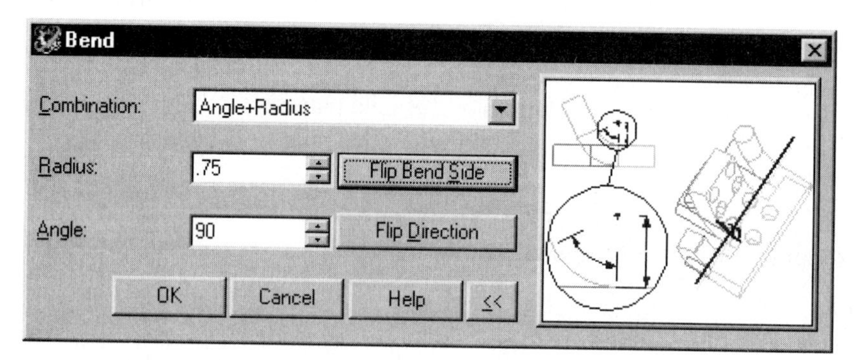

Figure 6-18 Bend dialog box

Combination

This drop-down list provides the various input options that can be used to create the bend feature. The options provided under this drop-down list are:

Angle+Radius. This option is used to create the bend by specifying the angle of the bend and the radius of the curve at the bend.

Radius+ArcLen. This option is used to create the bend by specifying the radius of the curve at bend and the length of the arc at bend.

ArcLen+Angle. This option is used to create the bend by specifying the length of the arc at the bend and the angle of the bend.

Radius

This spinner is used to specify the radius of the curve at the bend.

Angle

This spinner is used to specify the angle of the bend.

Arc Length

This spinner is used to specify the length of the arc at the bend. This spinner appears in place of the **Radius** or the **Angle** spinner when you select any option requiring the length of the arc from the **Combination** drop-down list.

Flip Bend Side

This button is chosen to reverse the side of the feature or the part that you are bending.

Flip Direction

This button is chosen to reverse the direction of the bend feature.

CREATING RIB FEATURES

The ribs are thin wall like structures used as supports to strengthen a joint. These are used to bind or tie features and provided proper support to them so that the part does not fail under increased loading. In Mechanical Desktop 5, a new command called **AMRIB** has been introduced using which you can directly convert an open profile into a rib feature.

AMRIB Command*

Toolbar:	Part Modeling > Sketched Features - Extrude > Rib
Menu:	Part > Sketched Features > Rib
Context Menu:	Sketched & Worked Features > Rib
Command:	AMRIB

This command is used to convert an open profile into a rib feature. Before you invoke this command, you must have a base feature and an open profile that has to be converted into the rib feature. You can also invoke this command using the shortcut menu displayed upon right-clicking on the open profile and choosing **Rib**. While creating the rib feature this has to be kept in mind that the rib feature finds a termination face when the profile is extruded. Figure 6-19 shows an open profile before creating the rib and Figure 6-20 shows a model after creating the rib feature.

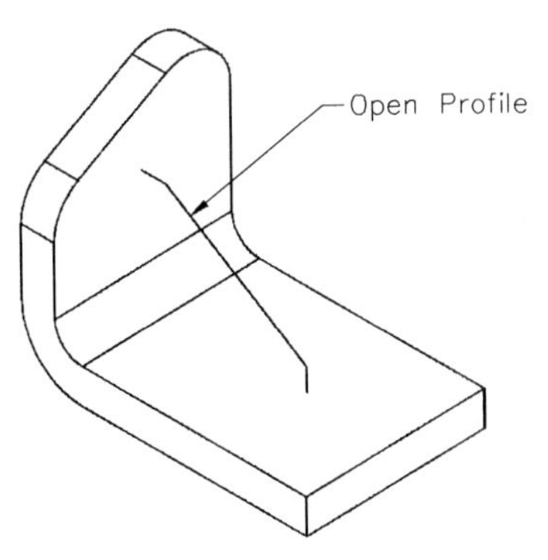

Figure 6-19 *Designer model before creating the rib feature*

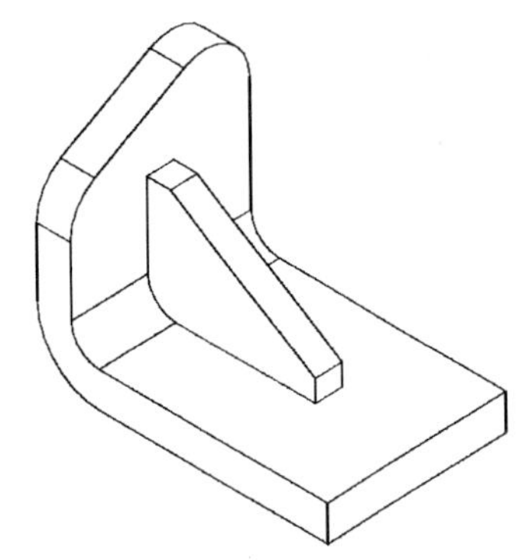

Figure 6-20 *Designer model after creating the rib feature*

When you invoke this command, the **Rib** dialog box will be displayed, see Figure 6-21.

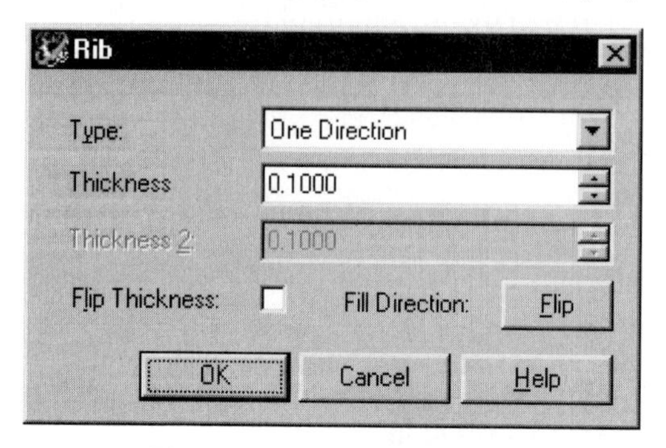

Figure 6-21 *Rib dialog box*

Rib Dialog Box Options

The options that are provided under the **Rib** dialog box are:

Type

The **Type** drop-down list provides you with the methods for specifying the direction of the rib feature. The options provided under this drop-down list are:

One Direction. This option is used to create the rib feature by specifying the thickness of the rib feature in any one direction from the open profile.

Two Directions. This option is used to create the rib feature by specifying the thickness of the feature in both the direction of the open profile. The thickness values can be specified in the **Thickness 1** and the **Thickness 2** spinners.

Midplane. This option is used to create the rib feature by specifying the thickness of the rib equally in the either direction of the plane of the rib feature. The thickness of the rib can be specified in the **Thickness** spinner.

Thickness

The **Thickness** spinner is used to specify the thickness of the rib when you select the **One Direction** or **Midplane** option from the **Type** drop-down list.

Thickness 1

The **Thickness 1** spinner is used to specify the thickness value in the first direction when you select the **Two Directions** option from the **Type** drop-down list. This spinner appears in place of the **Thickness** spinner when you select the **Two Directions** option from the **Type** drop-down list.

Thickness 2

The **Thickness 2** spinner is used to specify the thickness value in the second direction when you select the **Two Directions** from the **Type** drop-down list.

Flip Thickness

This check box is selected to reverse the direction in which the thickness is applied. This check box is available only when you select **One Direction** from the **Type** drop-down list.

Fill Direction Flip

This button is chosen to reverse the direction of creation of the rib feature.

AMSURFCUT Command

Toolbar:	Part Modeling > Placed Features - Hole > Surface Cut
Menu:	Part > Placed Features > Surface Cut
Context Menu:	Placed Features > Surface Cut
Command:	AMSURFCUT

This command is used to create a cut or a join feature on the existing designer model using the AutoSurf surfaces. This command allows you to associate the surface to a work point so that their position can be edited later. The type of surfcuts are:

Protrusion

This option is used to join the selected designer part with the portion of the AutoSurf surface that extends beyond the designer part.

Cut

This option is used to cut the active designer part using the AutoSurf surface. The portion from which the material is to be removed is specified by you.

TUTORIALS

Tutorial 1

In this tutorial you will create the Piston shown in Figure 6-22a. The dimensions to be used are given in the Figures 6-22b, and 6-22c.

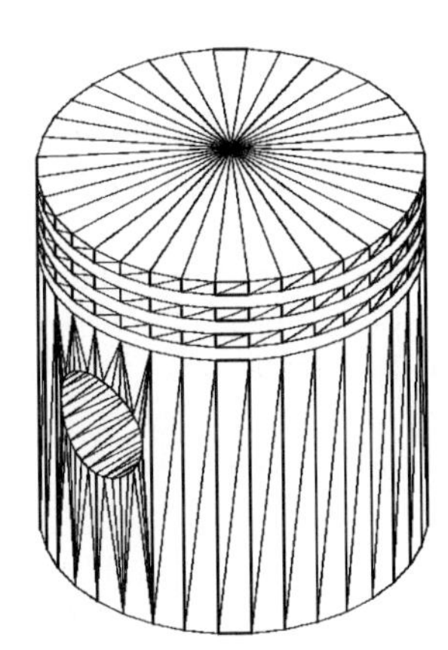

Figure 6-22a *Designer model for Tutorial 1*

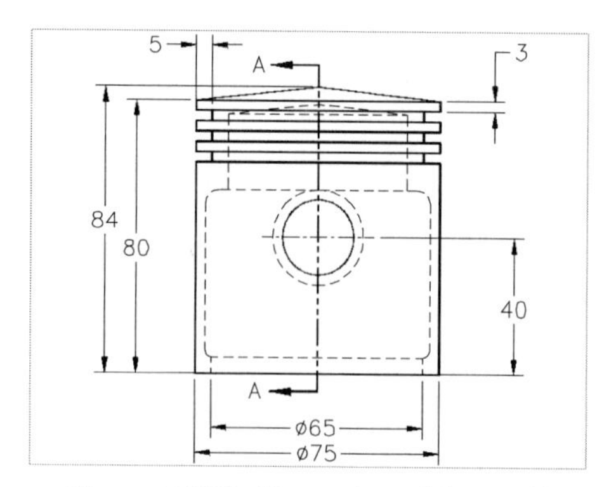

Figure 6-22b *Front view of the model*

Figure 6-22c *Sectioned side view of the model*

1. Create the sketch for the base feature and then revolve it to create the base feature for the designer model as shown in Figure 6-23.

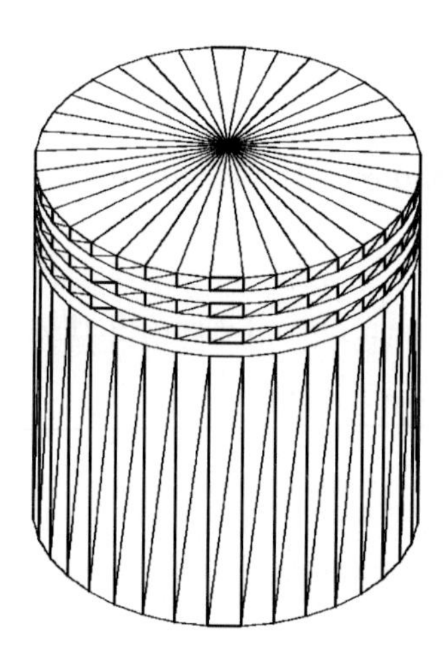

Figure 6-23 *Base feature for Tutorial 1*

2. Choose the **Shell** button from the **Placed Features - Hole** flyout in the **Part Modeling** toolbar to display the **Shell Feature** dialog box.

3. Set the value of the **Inside** spinner under the **Default Thickness** to **5**.

4. Choose the **New** button under the **Set** area.

5. Set the value of the **Thickness** drop-down list to **10**.

6. Choose the **Add** button under the **Faces** area. When you choose this button, the **Shell Feature** dialog box will be temporarily closed and you will be asked to select the face on which the 10 unit thickness should be applied. The prompt sequence is as follows:

Select faces to add: *Select the face as shown in Figure 6-24.*
Enter an option [Next/Accept] <Accept>: Enter
Select faces to add: Enter
Computing ...

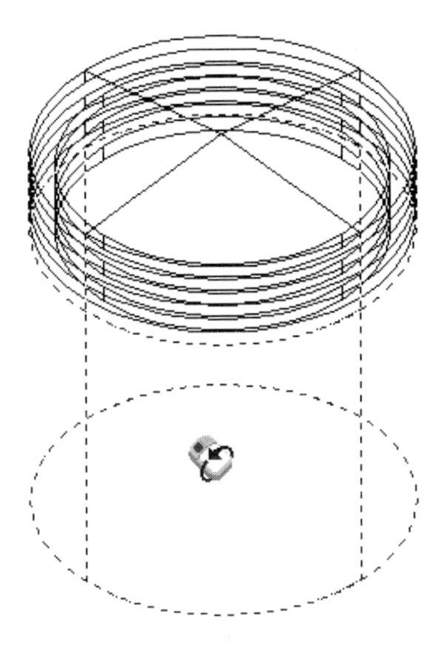

Figure 6-24 *Selecting the face*

7. The **Shell Feature** dialog box will be displayed on the screen again. Choose **OK**.

8. The designer model after creating the shell feature should be similar to the one shown in Figure 6-25.

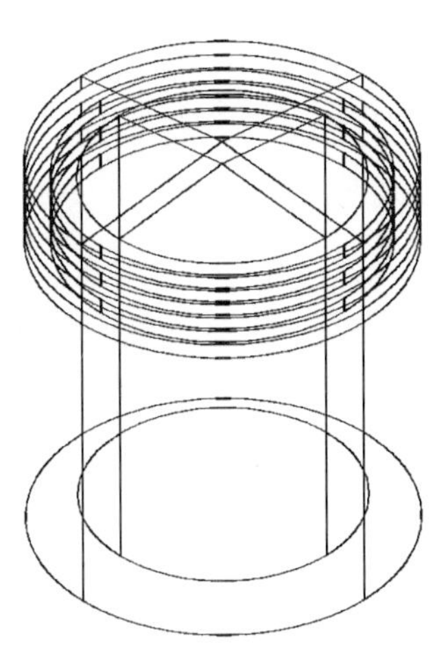

Figure 6-25 *Designer model after creating the shell feature*

9. Create a hole feature on the bottom face of the designer model of 65 units diameter.

10. Define a work plane at an offset of 5 units inside the model.

11. Create the cut feature of 49 units inside the model from the new work plane.

12. Create the required fillets in the designer model.

13. Create the join feature of 28 units diameter and at a distance of 40 units from the bottom face in the designer model as shown in Figure 6-26.

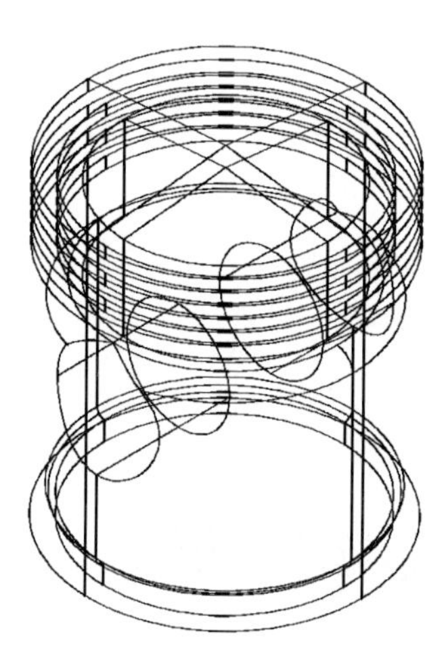

Figure 6-26 *Designer model after creating the join feature*

14. Create the hole in the new join feature.

15. The final designer model should look similar to the one shown in Figure 6-27.

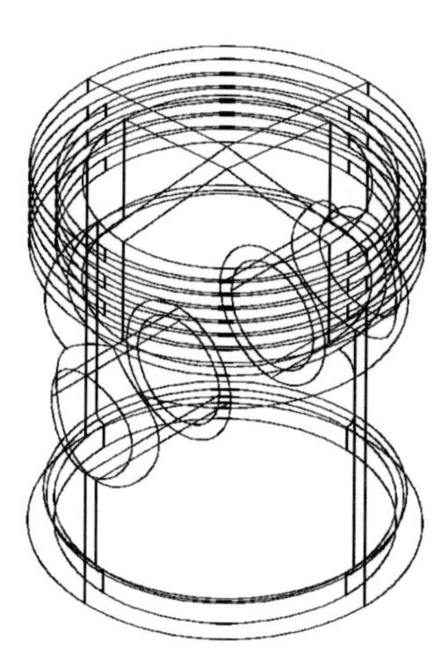

Figure 6-27 *Final designer model for Tutorial 1*

16. Save this drawing with the name given below:

    \MDT Tut\Ch-6**Tut1.dwg**

Tutorial 2

In this tutorial you will create the designer model shown in Figure 6-28a. The dimensions to be used are given in Figures 6-28b, and 6-28c.

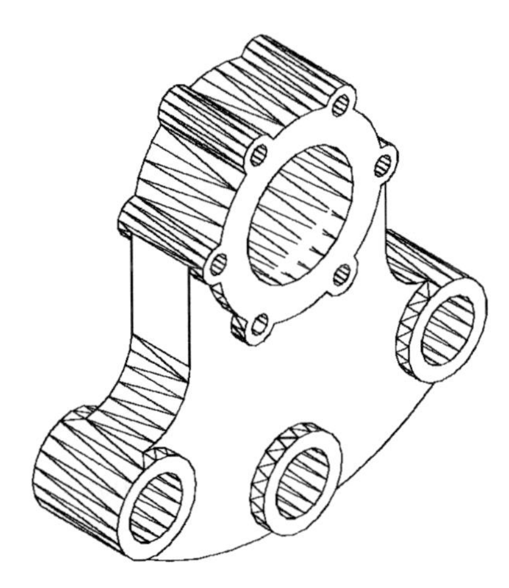

***Figure 6-28a** Designer model for Tutorial 2*

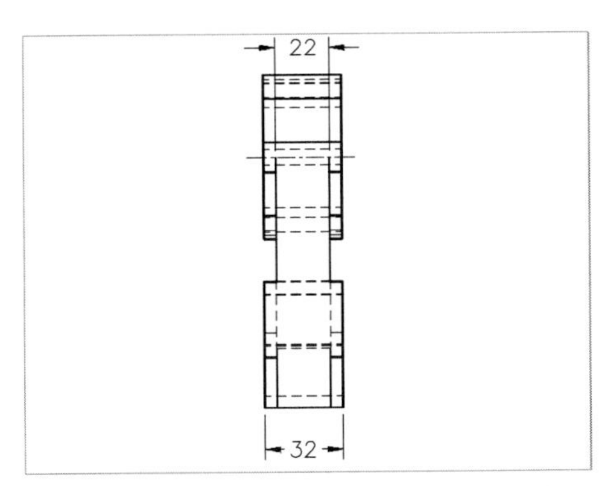

***Figure 6-28b** Dimensions for Tutorial 2*

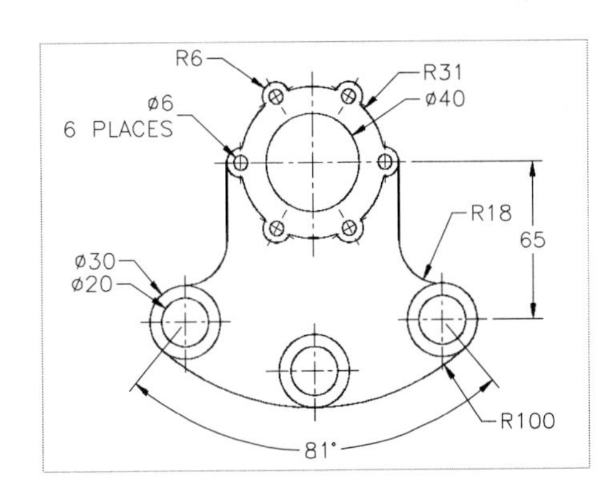

***Figure 6-28c** Dimensions for Tutorial 2*

1. Create the sketch for the base feature and then extrude it to create the base feature as shown in Figure 6-29.

2. Define a work plane at an offset distance of 11 units inside the model from the front face of the base feature.

3. Create the next cylindrical feature taking the new work plane as the mid plane for extrusion. Now create one of the smaller cylindrical feature.

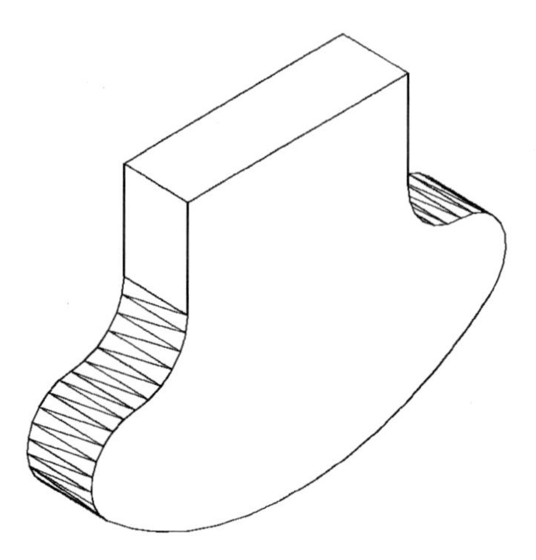

Figure 6-29 *Base feature for Tutorial 2*

4. Create the hole on the smaller cylindrical feature, see Figure 6-30.

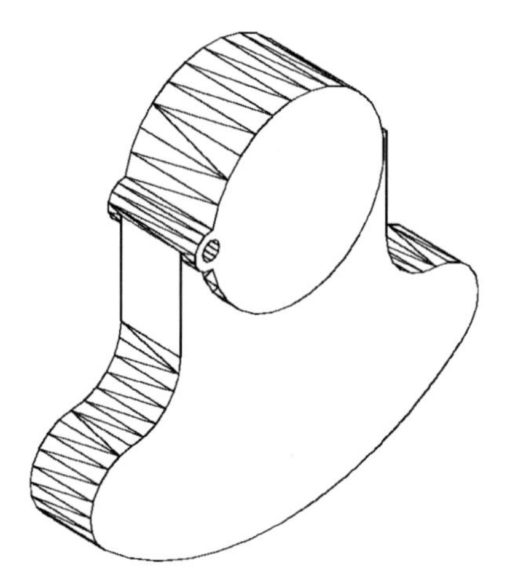

Figure 6-30 *Designer model after creating the cylindrical features*

5. Choose the **Polar Pattern** button from the **Placed Features - Hole** flyout in the **Part Modeling** toolbar. The prompt sequence is as follows:

   Select features to pattern or [liSt/Remove] <Accept>: *Select the smaller cylindrical feature.*
   Select features to pattern or [liSt/Remove] <Accept>: *Select the hole.*
   Select features to pattern or [liSt/Remove] <Accept>: Enter
   Valid selections: work point, work axis, cylindrical edge/face
   Select rotational center: *Select the bigger cylindrical feature.*

6. The **Pattern** dialog box will be displayed.

7. Set the value of the **Instances** spinner to **6**.

8. Choose the **Full Circle** button. Choose **OK**.

9. The designer model after creating the pattern should look similar to the one shown in Figure 6-31.

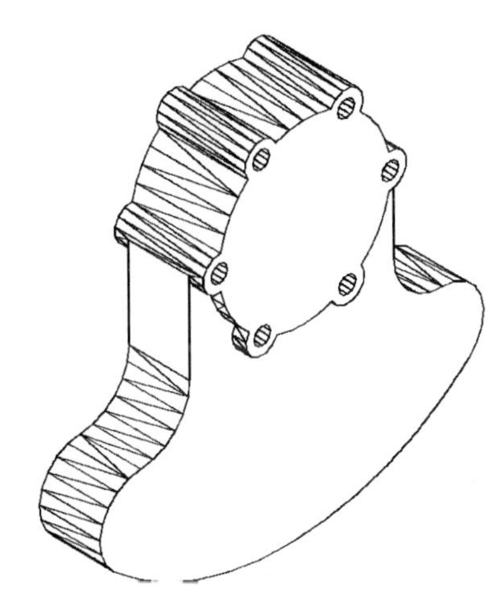

Figure 6-31 *Designer model after creating the polar pattern*

10. Create the central drilled through hole.

11. Create the next cylindrical feature and place a drilled through hole in it, see Figure 6-32.

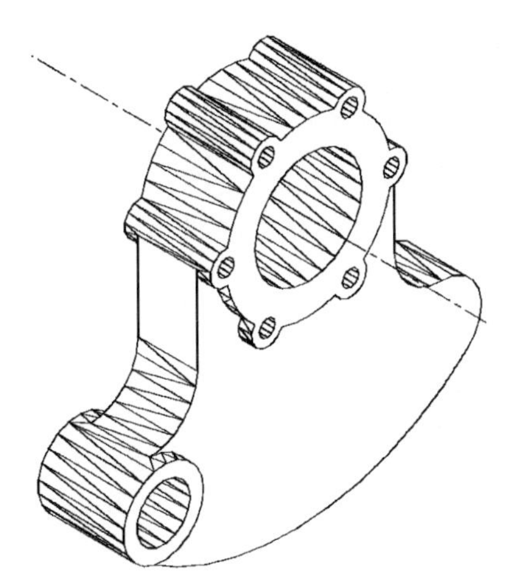

Figure 6-32 *Designer model after creating the next feature*

12. Choose the **Polar Pattern** button from the **Placed Features - Hole** flyout in the **Part Modeling** toolbar. The prompt sequence is as follows:

Select features to pattern or [liSt/Remove] <Accept>: *Select the new cylindrical feature.*
Select features to pattern or [liSt/Remove] <Accept>: *Select the hole.*
Select features to pattern or [liSt/Remove] <Accept>: Enter
Valid selections: work point, work axis, cylindrical edge/face
Select rotational center: *Select the bigger cylindrical feature earlier used as the rotational center.*

13. Set the value of the **Instances** spinner to **3**.

14. Choose the **Included Angle** button. Set the value of the **Angle** spinner to **81**.

15. Make sure that the pattern is created in the counterclockwise direction. If not, choose the **Flip Rotation Direction** button. Choose **OK**.

16. Make the work axis invisible using the desktop browser. The final designer model should be similar to the one shown in Figure 6-33.

17. Save this drawing with the name given below:

\MDT Tut\Ch-6**Tut2.dwg**

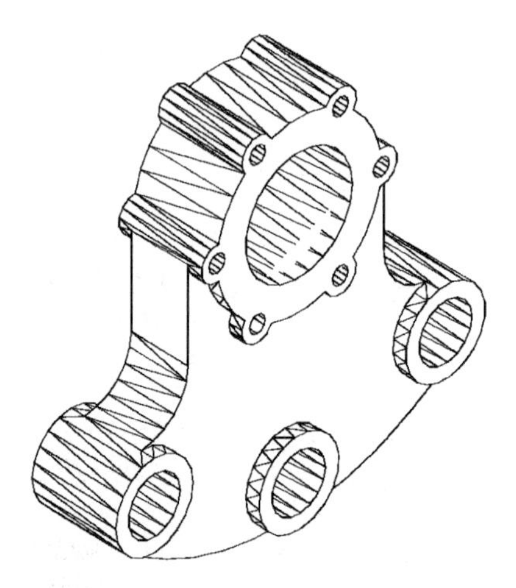

Figure 6-33 *Final designer model for Tutorial 2*

Tutorial 3

In this tutorial you will create the designer model shown in Figure 6-34a. The dimensions to be used are given in Figures 6-34b, 6-34c and 6-34d.

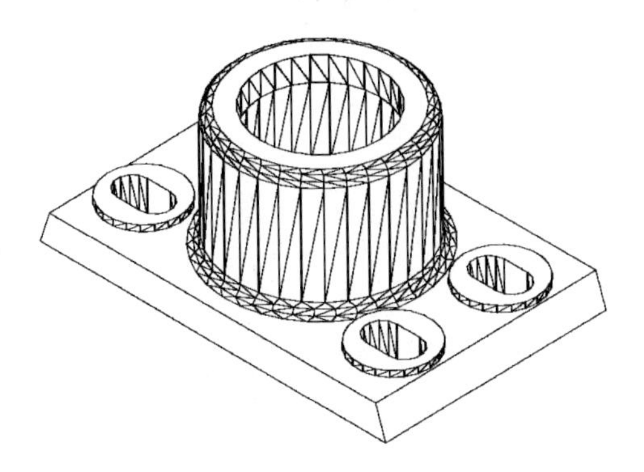

Figure 6-34a Model for Tutorial 3

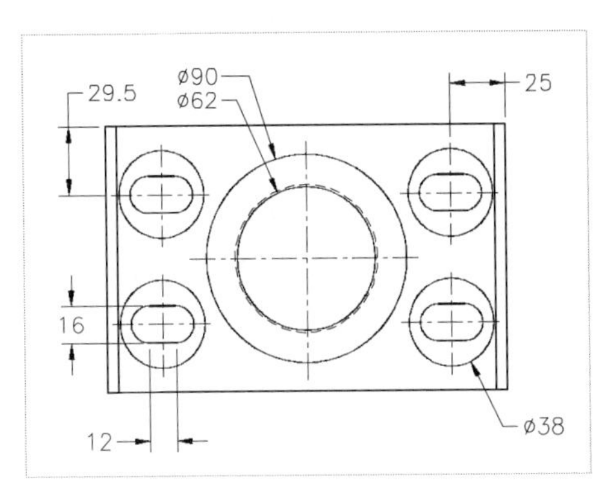

Figure 6-34b Top view of the model

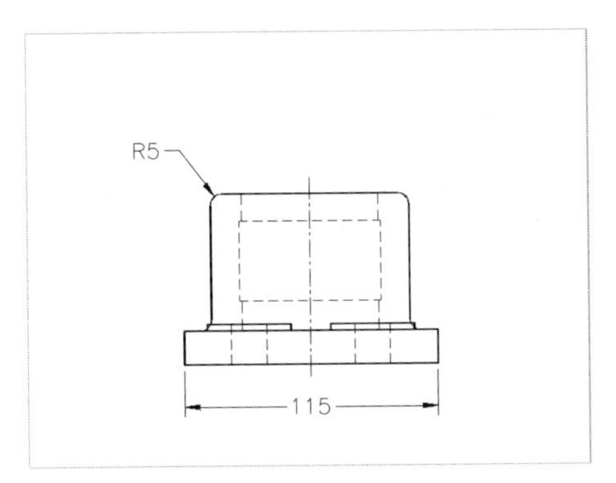

Figure 6-34c Side view of the model

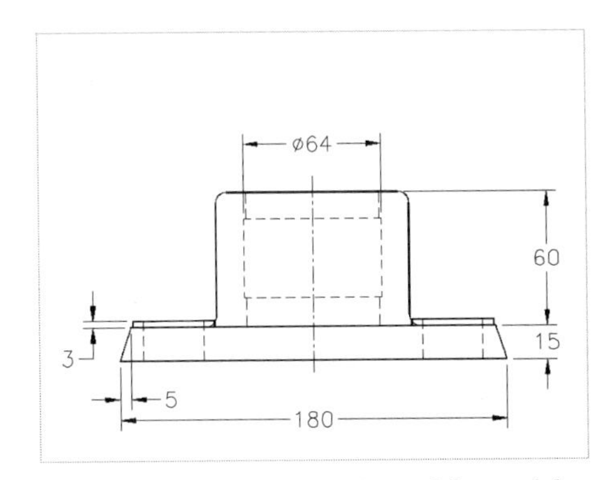

Figure 6-34d Front view of the model

1. Create the base feature and then apply the chamfer on the base feature, see Figure 6-35.
2. Create the next cylindrical feature and then apply the fillets, see Figure 6-36.

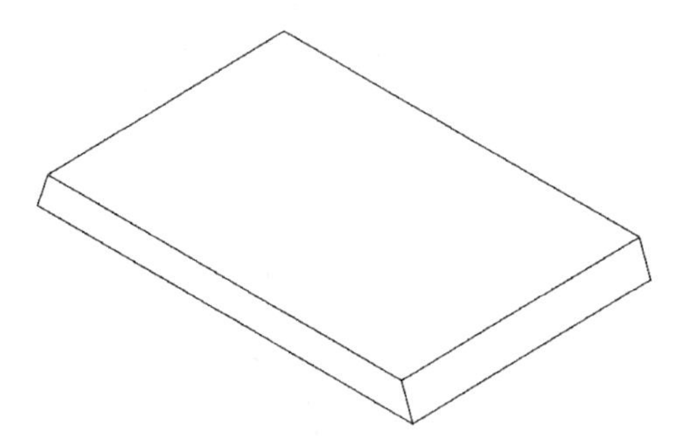

Figure 6-35 Base feature after applying the chamfer feature

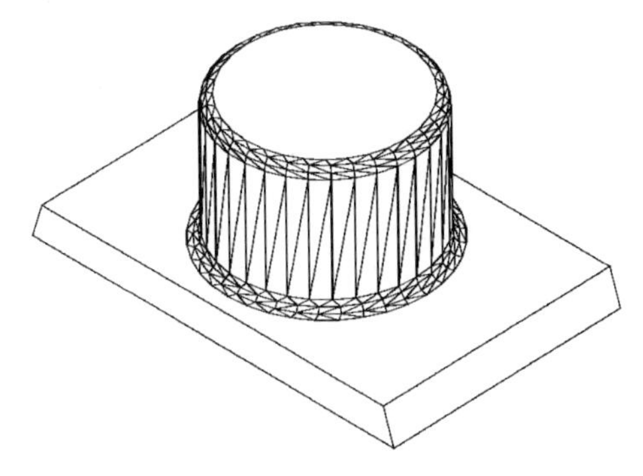

Figure 6-36 Designer model after creating the cylindrical feature and applying the fillets

3. Create the central hole and the central cut in the cylindrical feature and apply fillet.

4. Create the next feature on the top face of the base feature as shown in Figure 6-37.

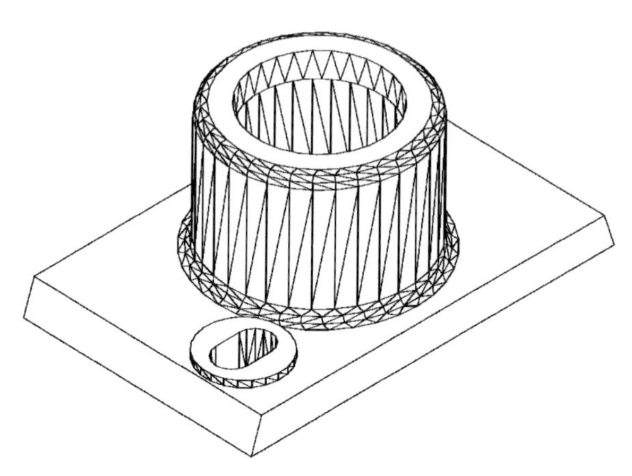

Figure 6-37 *Designer model after creating the next feature on the top face of the base feature*

5. Choose the **Rectangular Pattern** button from the **Placed Features - Hole** flyout in the **Part Modeling** toolbar. The prompt sequence is as follows:

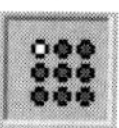

Select features to pattern: *Select the new feature on the top face.*
Select features to pattern or [liSt/Remove] <Accept>: *Select the cut in the new feature.*
Select features to pattern or [liSt/Remove] <Accept>: Enter

6. Choose the **Included Spacing** button in the **Column Placement** area. Set the value of the **Instances** spinner to **2** and the value of the **Spacing** spinner to **130**.

7. Choose the **Included Spacing** button in the **Row Placement** area. Set the value of the **Instances** spinner to **2** and the value of the **Spacing** spinner to **56**. Choose **OK**.

8. The final designer model should look similar to the one shown in Figure 6-38.

9. Save this drawing with the name given below:

\MDT Tut\Ch-6**Tut3.dwg**

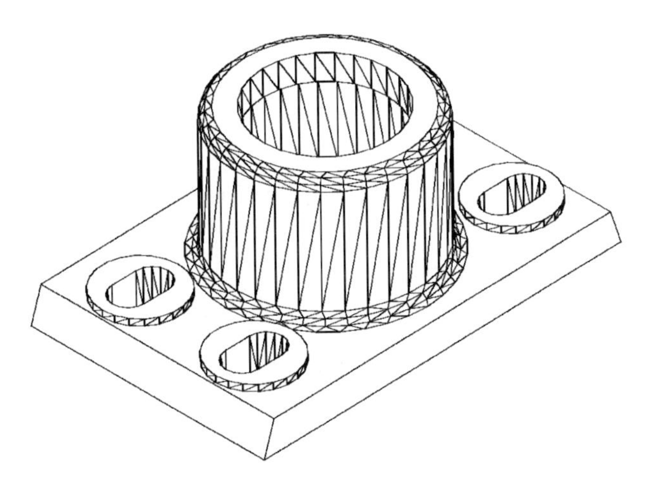

Figure 6-38 *Final designer model for Tutorial 3*

Tutorial 4

In this tutorial you will create the designer model shown in Figure 6-39a. The dimensions to be used are given in Figures 6-39b and 6-39c. Assume the missing dimensions. Create the base feature using bend. The length of the base before bending is **8** and **10** respectively. The bend is create at distance of **3** from the left corner of the smaller edge.

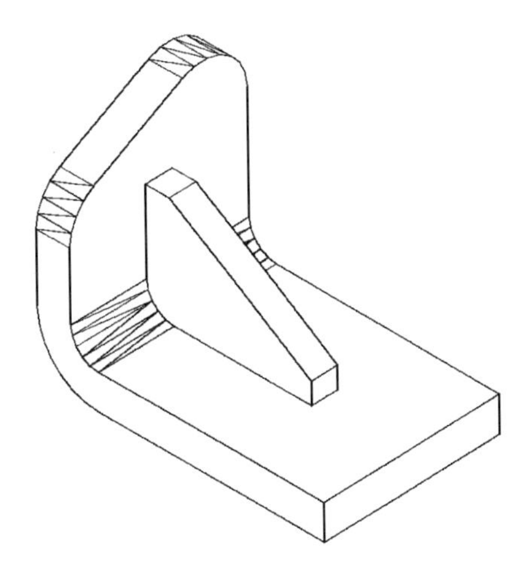

Figure 6-39a *Designer model for Tutorial 4*

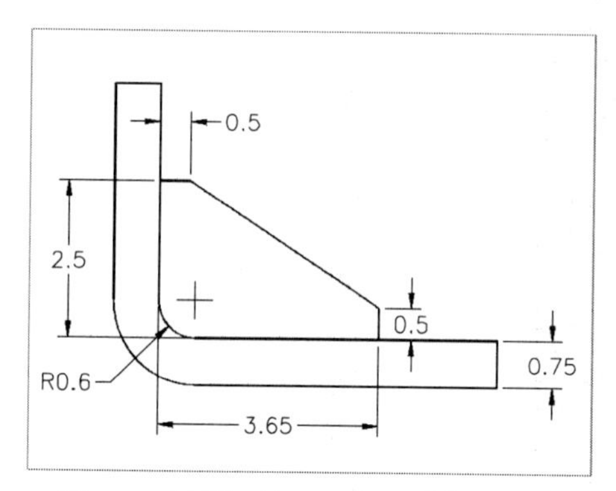

Figure 6-39b *Front view of the model*

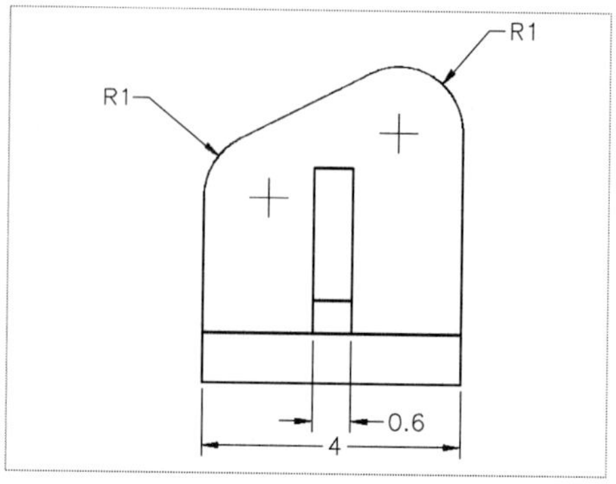

Figure 6-39c *Side view of the model*

1. Create the base feature of the designer model. The sketch for the base feature is shown in Figure 6-40 and the base feature after extrusion is shown in Figure 6-41.

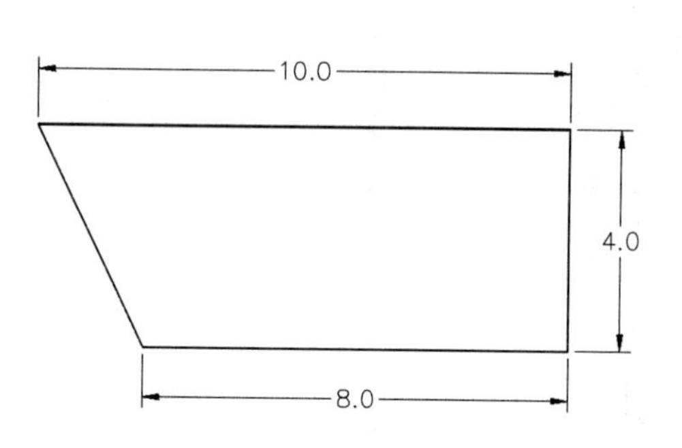

Figure 6-40 *Sketch for the base feature*

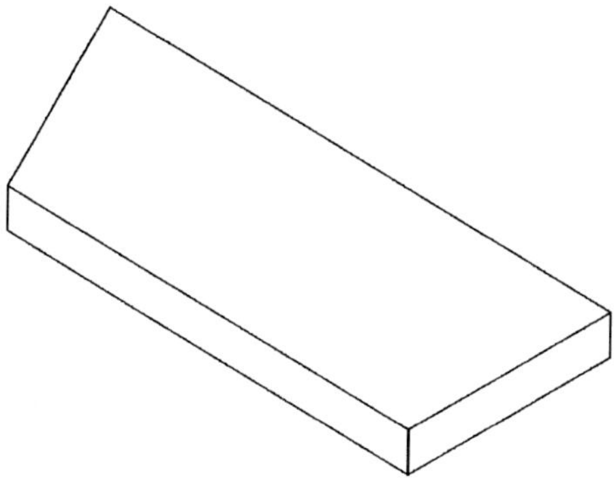

Figure 6-41 *Figure showing the base feature*

2. Define the sketch plane on the top face of the model.
3. Draw a line and then convert it into an open profile.
4. Add the dimensions to the open profile, see Figure 6-42.
5. Right-click on the open profile in the desktop browser to display the shortcut menu.
6. Choose **Bend** to display the **Bend** dialog box.
7. Select **Angle+Radius** from the **Combination** drop-down list.
8. Set the value of the **Radius** spinner to **0.6**.
9. Set the value of the **Angle** spinner to **90**. Choose **OK**. The model after bending should look similar to the one shown in Figure 6-43.

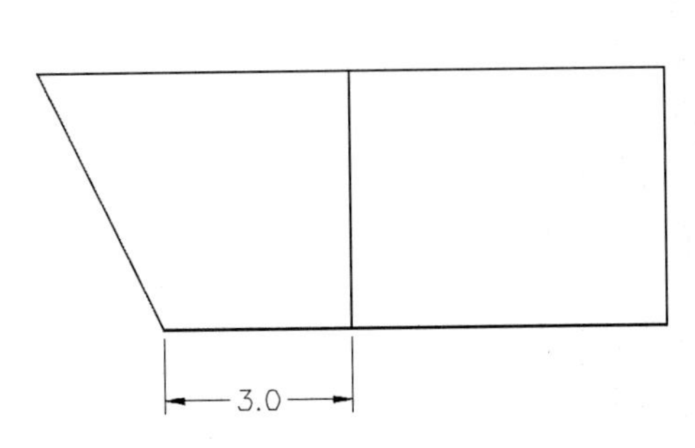

Figure 6-42 Open profile for bend feature

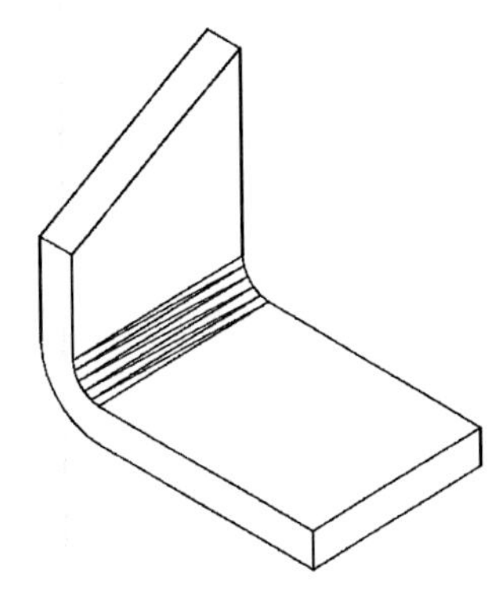

Figure 6-43 Model after creating the bend feature

10. Apply the fillet and then define a work plane parallel to the front face of the model and at an offset of 2 units inside the model.

11. Draw the sketch for the rib feature. Convert it into an open profile and then dimension this sketch, see Figure 6-44.

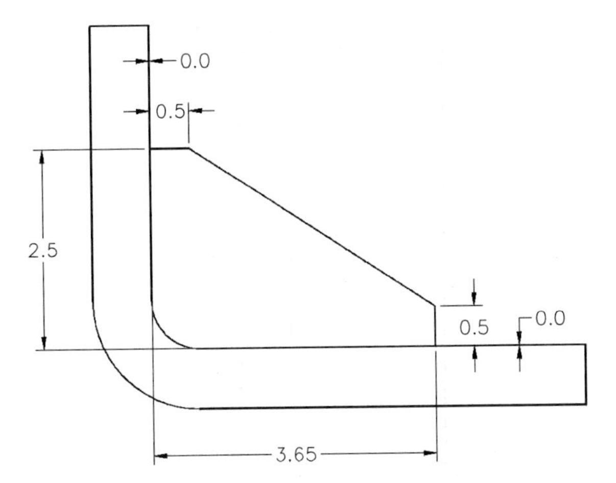

Figure 6-44 Sketch for the rib feature

12. Right-click on the open profile in the desktop browser to display the shortcut menu.

13. Choose **Rib** to display the **Rib** dialog box. Select **Midplane** from the **Type** drop-down list. Set the value of the **Thickness** spinner to **0.6**. Choose **OK**. The final designer model should look similar to the one shown in Figure 6-45.

14. Save this drawing with the name \MDT Tut\Ch-6**Tut4.dwg**.

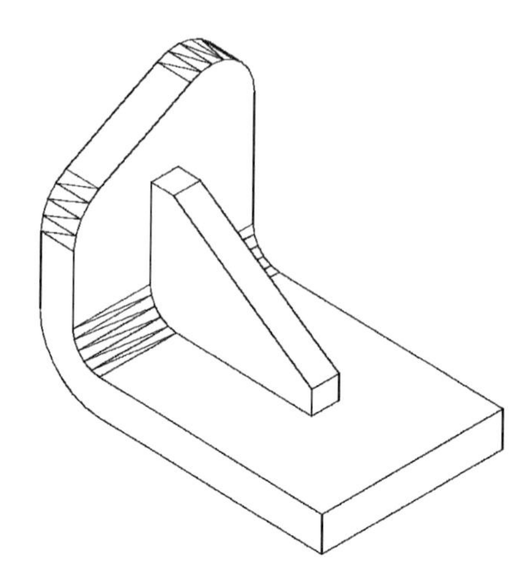

Figure 6-45 *Final designer model for Tutorial 4*

Review Questions

Answer the following questions.

1. What is the use of the **AMSHELL** command?
2. What are the various methods of defining the wall thickness for the shell?
3. The __________ button under the __________ area is used to remove the unwanted faces in the shell feature.
4. Define the term multiple thickness override for the shell features.
5. Define rectangular pattern.
6. What is the difference between the polar and axial pattern?
7. What is the difference between full circle, included angle, and incremental angle type?
8. What are the rib features? What are they used for? Which command is used to create the rib feature in Mechanical Desktop 5?

Exercise

Exercise 1

Create the Pump Body shown in Figure 6-46a. The dimensions to be used are given in Figures 6-46b, 6-46c, and 6-46d.

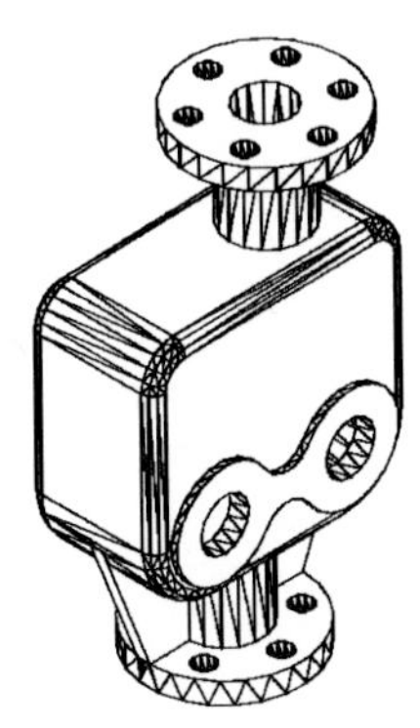

Figure 6-46a *Designer model for Exercise 1*

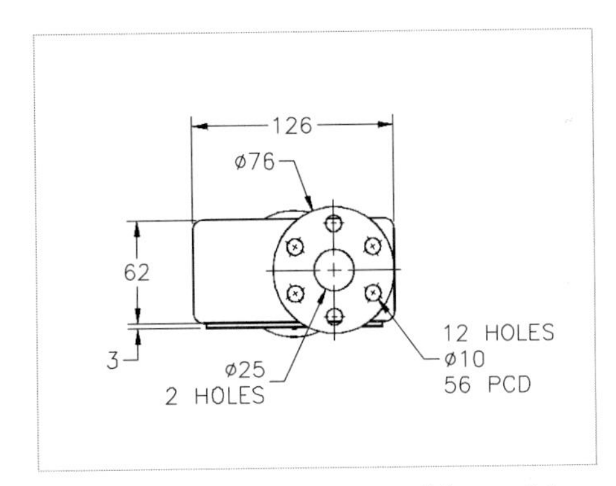

Figure 6-46b *Top view of the model*

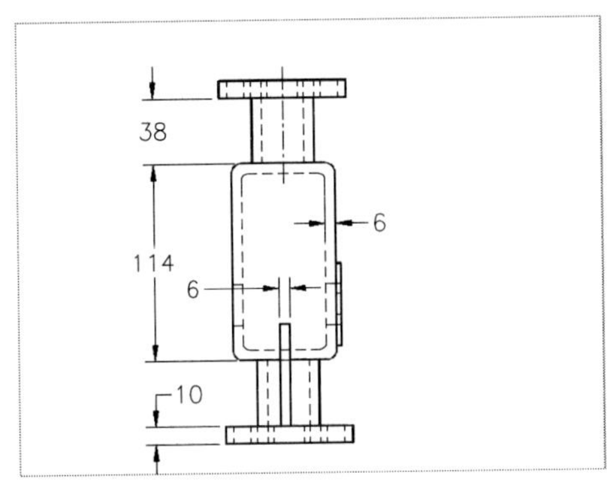

Figure 6-46c *Side view of the model*

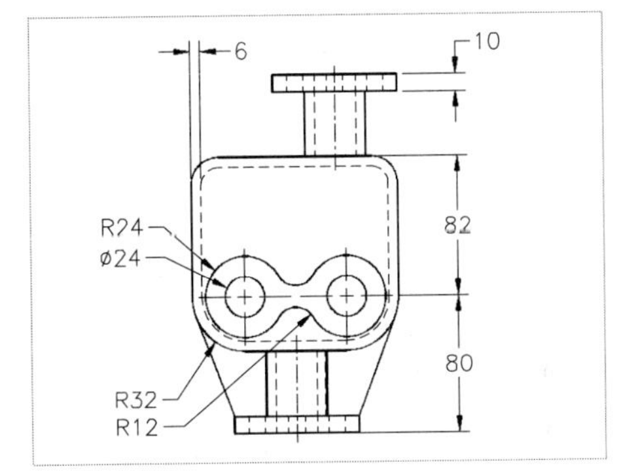

Figure 6-46d *Front view of the model*

Chapter 7

Editing, Suppressing, and Reordering the Features

Learning Objectives

After completing this chapter, you will be able to:

- *Edit features.*
- *Edit basic sketches of the designer model.*
- *Update edited sketches or features.*
- *Delete features in the designer model.*
- *Copy sketches of features in the designer model.*
- *Copy features in the designer model.*
- *Suppress and Unsuppress the features.*
- *Change the order of feature creation.*

Commands Covered

- ***AMEDITFEAT***
- ***AMUPDATE***
- ***AMDELFEAT***
- ***AMCOPYSKETCH***
- ***AMPARTEDGE***
- ***AMCOPYFEAT***
- ***AMSUPPRESSFEAT***
- ***AMUNSUPPRESSFEAT***
- ***AMREORDFEAT***

EDITING THE FEATURES

Generally, every designer model needs to be edited while creating or after completion. Not only the basic sketches but also the other features (even the placed or the worked features) may sometimes need to be edited. You can edit these features using the **AMEDITFEAT** command or by using the desktop browser.

Editing The Features Using The AMEDITFEAT Command

Toolbar:	Part Modeling > Edit Feature
Menu:	Part > Edit Feature
Context Menu:	Edit Features > Edit
Command:	AMEDITFEAT

This command is used to restructure the sketches of the features or alter various parameters assigned to the features. The options provided under this command are:

Independent array instances

This option is used to edit individual rectangular or polar array instances. However, when you use this option, the relationship that exist between the instances of the array is broken and the selected feature becomes an independent feature.

Sketch

This option is used to edit the basic sketch of the designer model or the basic sketch of the features created on the base feature. After editing the features you will have to update the part to view the effects of the editing operation.

select Feature

This option is used to select the feature to edit. Depending upon the feature to edit, the related dialog box will be displayed.

surfCut

This option is used to edit the feature created using the **AMSURFCUT** command.

Toolbody

This option is used to edit the feature created using the **AMCOMBINE** command. When you invoke this option, all the designer models used to create the combine features will be displayed so that you can make the necessary changes in the models.

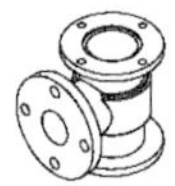

Note

*The **AMCOMBINE** command will be discussed in the later chapters.*

Editing Features Using The Desktop Browser

To edit the feature using the desktop browser, right-click on the feature you want to edit so that the shortcut menu is displayed. Choose the **Edit** option as shown in Figure 7-1 to display the related dialog box or dimensions for editing. However, only the dimensions assigned using the parametric dimensioning technique will be displayed to edit. You can also edit a

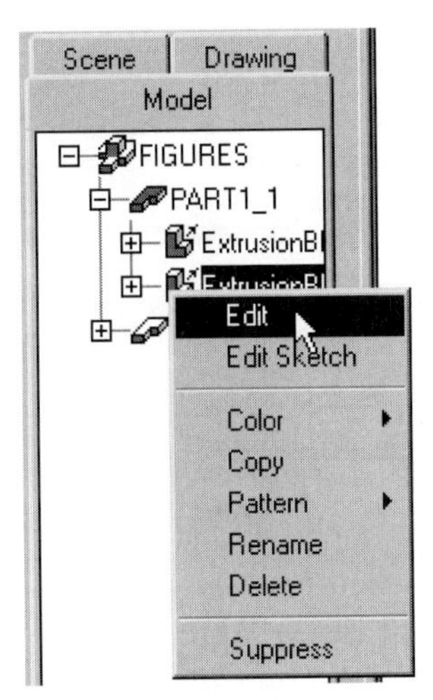

Figure 7-1 *Editing the feature using the desktop browser*

feature using the desktop browser by double-clicking on the feature to be edited in the tree view.

EDITING THE SKETCH OF THE FEATURE

Mechanical Desktop provides you the flexibility of editing the basic sketch of the sketched features. You can add additional geometry to the sketch or remove certain entities from the basic sketch. The most important advantage in this is that you do not have to extrude or revolve the sketch. Instead, you just have to update the part and it will take all the extrusion or revolution values of the initial feature.

For example, consider a designer model as shown in Figure 7-2. The basic sketch of the designer model is shown in Figure 7-3.

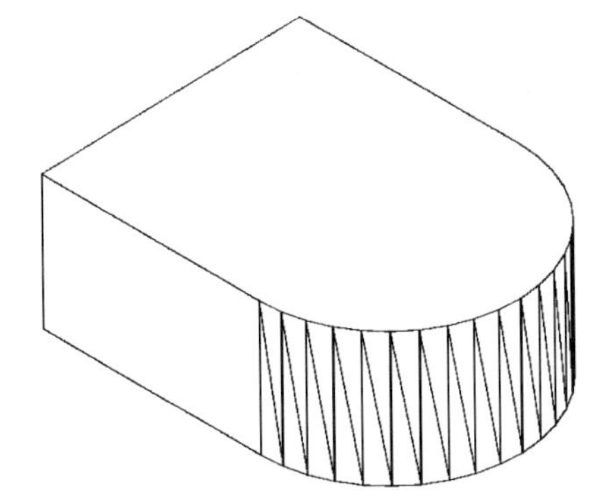

Figure 7-2 *The original designer model*

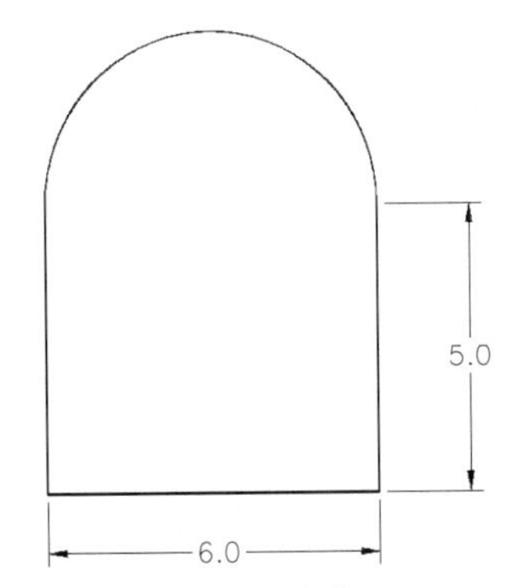

Figure 7-3 *The basic sketch that was used to create the designer model*

To edit the basic sketch of the sketched feature, right-click on that feature under the part name in the tree view in desktop browser. Choose the **Edit Sketch** option from the shortcut menu,

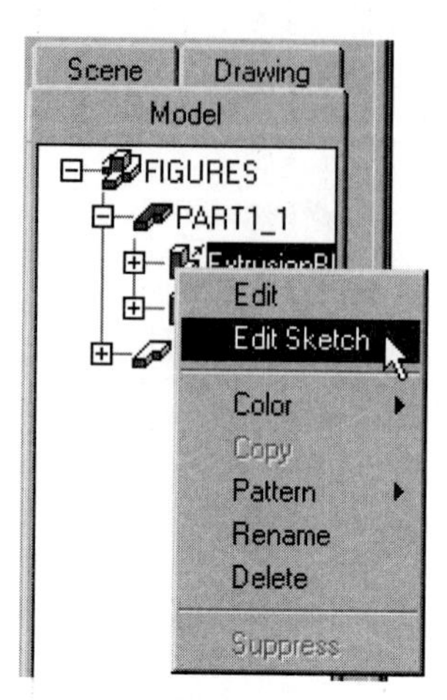

Figure 7-4 *Selecting Edit Sketch option from the shortcut menu in the desktop browser*

see Figure 7-4. When you choose the **Edit Sketch** option, the feature is reduced to the basic sketch displaying all the dimension values added using any of the parametric dimensioning techniques. You can now delete any object and add any additional geometry to the sketch. Once you have added the new geometry or deleted the unwanted geometries from the sketch, choose the **Append to Sketch** button from the **Profile a Sketch** flyout in the **Part Modeling** toolbar. The geometries that were the part of the initial sketch will be highlighted and you will be prompted to select the new geometries to profile the complete sketch. After profiling, you can add dimensions to the new geometries using any of the parametric dimensioning techniques, see Figure 7-5. Now, choose the **Update Part** button from the **Part Modeling** toolbar and select the **active Part** option. The initial extrusion or revolution values will be applied to the new sketch, see Figure 7-6.

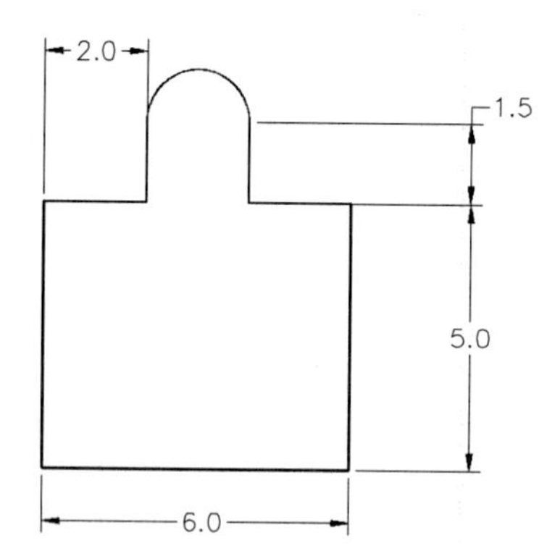

Figure 7-5 *The dimensions assigned to the sketch with new geometries*

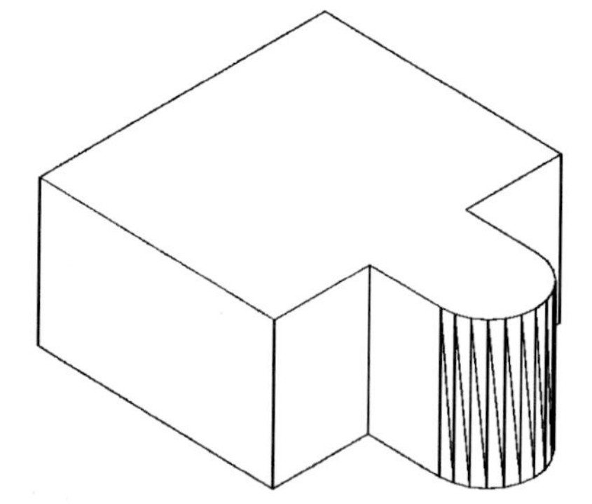

Figure 7-6 *Designer model created after updating the new sketch*

Tip: *The new geometry should be added to the sketch in such a manner that the sketch still remains a closed entity. If the sketch is an open entity, it will not be profiled and you will be prompted to select the edge to close the profile.*

UPDATING EDITED SKETCHES AND FEATURES

Once you have modified the basic sketch of the feature or the feature itself, you will have to update them using the **AMUPDATE** command to view the effect of the editing operation. The **AMUPDATE** command is discussed below.

AMUPDATE Command

Toolbar:	Part Modeling > Update Part
Menu:	Part > Update Part
Command:	AMUPDATE

This command is used to update the edited features. The options provided under this command are:

active Part

This option is used to update the sketch of the base feature or the feature itself in the active part. This is the default option.

Assembly

This option is used to update the current assembly.

aLl parts

This option is used to update all of the local parts in the current drawing.

linKs

This option is used to update the links and the variables associated with all the local parts in the current drawing.

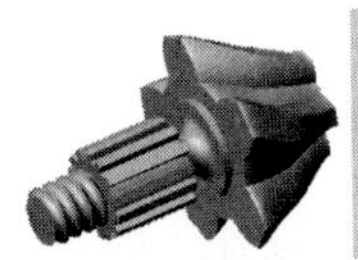

Tip: *When you edit any feature by double-clicking on it, you do not have to use the **AMUPDATE** command to view the affect of editing. The feature is automatically edited when you complete the editing.*

DELETING THE FEATURES

You can delete the unwanted features by using the desktop browser or with the help of the **AMDELFEAT** command.

Note

To delete a feature using the desktop browser, use the same method as for editing features.

AMDELFEAT Command

Toolbar:	Part Modeling > Edit Feature > Delete Feature
Menu:	Part > Delete Feature
Context Menu:	Edit Features > Delete
Command:	AMDELFEAT

This command is used to delete the unwanted features from the active designer model. When you select the feature to delete, it will be highlighted and you will be prompted to confirm the deletion process.

COPYING THE SKETCHES OF THE FEATURES

Copying the sketches of features is one of the methods of reducing the time taken in creating the designer model. The similar complicated features can be created by copying their base sketches using the **AMCOPYSKETCH** command.

AMCOPYSKETCH Command

Toolbar:	Part Modeling > Profile a Sketch > Copy Sketch
Menu:	Part > Sketch Solving > Copy Sketch
Context Menu:	Sketch Solving > Copy Sketch
Command:	AMCOPYSKETCH

You can copy the sketch when it is still a sketch or even after you have created feature from that sketch. Once you have selected the sketch to copy or the feature whose sketch is to be copied, you will be prompted to specify the center for the placement of the sketch. The new sketch will be a profiled sketch and will carry along all the constraints and dimensions assigned to it except the dimensions used to define the location of the original sketch. However, it has to be kept in mind that the new sketch will be placed in the current sketch plane irrespective of the actual position of the original sketch. The options provided under this command are:

Sketch

This option is used to copy the sketches when the initial sketch is not converted into a feature and is still a sketch.

Feature

This option is used to copy the basic sketch of a feature. Once you have selected the feature, the basic sketch of that feature will be displayed and you will be asked to specify the center for the location of the new sketch.

AMPARTEDGE Command

Toolbar:	Part Modeling > Profile a Sketch > Copy Edge
Menu:	Part > Sketch Solving > Copy Edge
Context Menu:	Sketch Solving > Copy Edge
Command:	AMPARTEDGE

This command is used to copy a single edge or all the edges comprising a face of an existing designer model. However, this has to be kept in mind that the edges or the faces thus copied are not profiled. They will be the basic AutoCAD entities like lines, circles, arcs and so on. The only advantage of using this command is that you may ignore adding dimensions as they will have the dimensions exactly the same as that of the edge or the face from which they are copied. The options provided under this command are:

select Edge

This is the default option and is used to select the edges to be copied one by one from the designer model.

Face

This option is used to select the entire face of the designer model.

COPYING THE FEATURES

You can directly copy the features of the designer model using the **AMCOPYFEAT** command.

AMCOPYFEAT Command

Toolbar:	Part Modeling > Placed Feature-Hole > Copy Feature
Menu:	Part > Copy Feature
Context Menu:	Edit Feature > Copy
Command:	AMCOPYFEAT

This command is used to copy the features of a designer model from their actual location to a new point specified on the current sketch plane. The features can be copied from any part in the drawing. Once you have selected the feature to be copied, you will be prompted to specify the center for the location of the feature on the current sketch plane of the active part. The new feature can be an independent feature or can be a dependent feature. You can also rotate the new feature as per your convenience before it is actually placed.

SUPPRESSING THE FEATURES

Generally, when you do not want some of the features to be displayed in the printout, you tend to delete them. In Mechanical Desktop, you can simply suppress them. This is similar to turning off the features. A complicated model can be made easier to work on by suppressing some of its features. Later on you can unsuppress the features, thus getting the desired model. The features can be suppressed using the desktop browser or with the help of the **AMSUPPRESSFEAT** command.

Suppressing Features Using The Desktop Browser

To suppress the features using the desktop browser, right-click on the feature you want to suppress so that the shortcut menu is displayed. In the shortcut menu, choose **Suppress**. The selected feature will be highlighted and you will be prompted to confirm the feature suppression.

AMSUPPRESSFEAT Command

Toolbar:	Part Modeling > Edit Feature > Suppress Features
Menu:	Part > Feature Suppression > Suppress Feature
Context Menu:	Edit Feature > Suppress Feature
Command:	AMSUPPRESSFEAT

When you invoke this command, you will be prompted to select the feature to suppress. As you take the cursor close to the feature, it will be highlighted for the ease of selecting.

The options provided in this command are:

By type

If this option is selected, the **Suppress By Type** dialog is displayed as shown in Figure 7-7. This dialog box is used to suppress the placed, sketched, or the work features associated with the active part or all the parts available in the drawing.

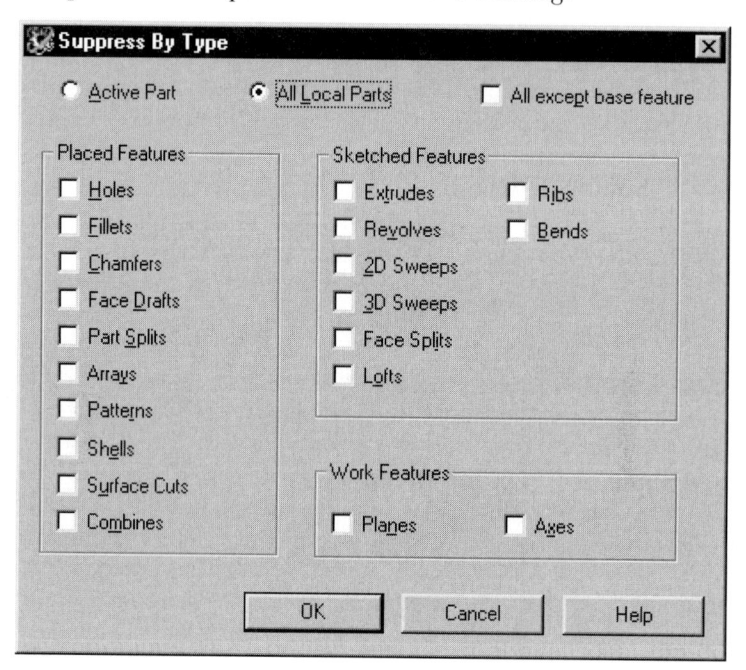

Figure 7-7 *Suppress By Type dialog box*

Suppress By Type Dialog Box Options

Active Part

If this radio button is selected, all the placed, sketched, or the work features available in the active part will be activated. You can select any of these features to suppress.

All Local Parts

If this radio button is selected, all the placed, sketched, and work features will be activated. You can specify the features to be suppressed by selecting them.

All except base feature

If this check box is selected, all the features on the designer model except the base feature will be suppressed.

Placed Features Area

This area contains all the placed features. You can select the placed features associated with the active part or all the parts in the drawing to be suppressed.

Sketched Features Area

This area contains all the sketched features. You can select the sketched features associated with the active part or all the parts in the drawing to suppress.

Work Features Area

This area provides you the option of suppressing the work planes or the work axes.

Table

When you select this option, the **Table Driven Setup** dialog box will displayed, see Figure 7-8.

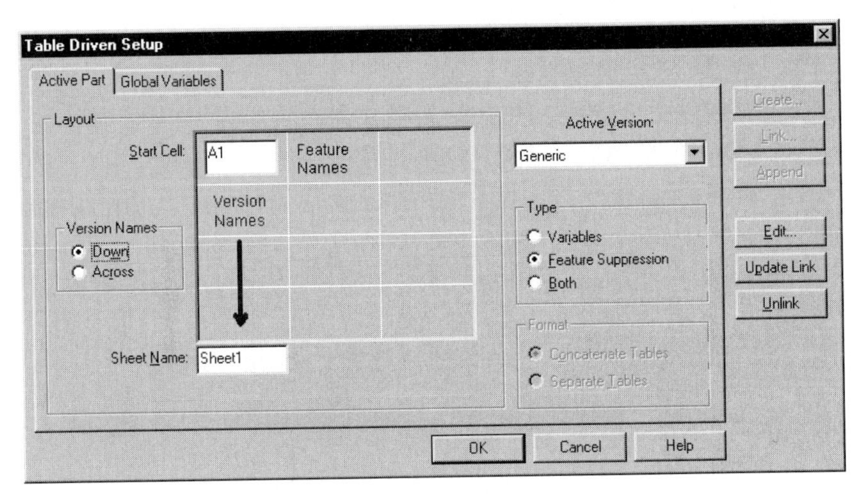

Figure 7-8 *Table Driven Setup dialog box*

This dialog box is used to create a Microsoft Excel spreadsheet for maintaining the records of the suppressed features as shown in Figure 7-9. These spreadsheets are similar to the spreadsheets created using the **AMVARS** command discussed in Chapter 4.

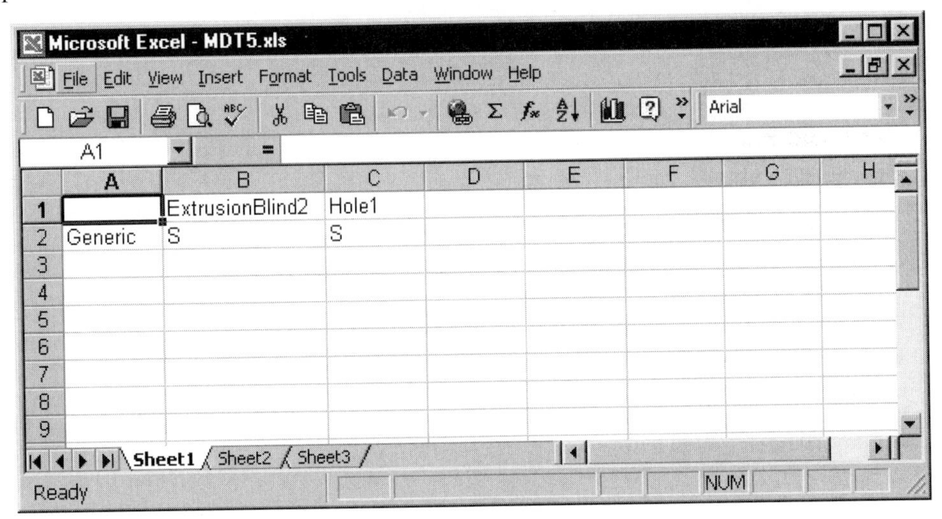

Figure 7-9 *Spreadsheet created for suppressed features*

Tip: *The* ***By type*** *or the* ***Table*** *option of the* ***AMSUPPRESSFEAT*** *command can be directly invoked by choosing the* ***Suppress By Type*** *or the* ***Table Driven suppression access*** *button from the* ***Edit Feature*** *flyout in the* ***Part Modeling*** *toolbar.*

UNSUPPRESSING THE FEATURES

You can unsuppress the features suppressed earlier by using the **AMUNSUPPRESSFEAT** command or using the desktop browser.

AMUNSUPPRESSFEAT Command

Toolbar: Part Modeling > Edit Feature > Unsuppress Features
Menu: Part > Feature Suppression > Unsuppress Feature
Context Menu: Edit Feature > Unsuppress Feature
Command: AMUNSUPPRESSFEAT

This command is used to unsuppress the features that were suppressed earlier. The options provided under this command are:

By type

This option is used to unsuppress the placed, sketched or worked features with the help of the **Unsuppress By Type** dialog box shown in Figure 7-10.

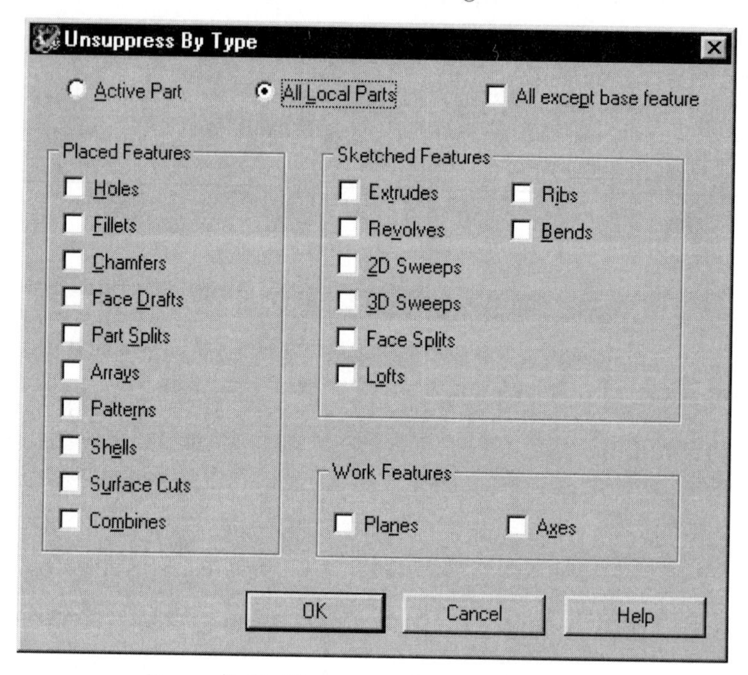

Figure 7-10 *Unsuppress By Type dialog box*

All

This option is used to unsuppress the features on the all parts, active part or selected part.

Unsuppressing Features Using The Desktop Browser

You can also unsuppress the features using the desktop browser by right-clicking on the suppressed feature to display the shortcut menu. Choose the **Unsuppress** option to unsuppress the feature.

CHANGING THE ORDER OF THE FEATURE CREATION

Sometimes, after creating the model you may need to change the order in which the features on the model were created. Mechanical Desktop provides the flexibility of reordering the feature creation. You can change the order using the **AMREORDFEAT** command or by using the desktop browser.

AMREORDFEAT Command

Toolbar:	Part Modeling > Edit Feature > Reorder Feature
Menu:	Part > Reorder Feature
Context Menu:	Edit Feature > Reorder
Command:	AMREORDFEAT

This command is used to change the order in which the features were created on the designer model. When you invoke this command you will be asked to select the feature you want to reorder. Once you have selected the feature, you will be prompted to select the destination feature after which you want to place the selected feature.

Reordering The Features Using The Desktop Browser

To reorder the features using the desktop browser, all you have to do is to drag and drop the feature you want to reorder below or above the destination feature. For example, consider a designer model shown in Figure 7-11. It consists of a rectangular pattern of 3 rows and 4 columns. Now, a shell feature is created on this model with the top and front faces excluded, as shown in Figure 7-12.

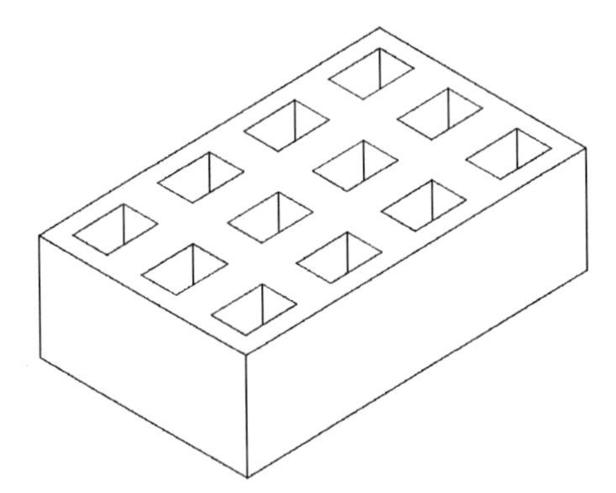

Figure 7-11 *Designer model with rectangular array*

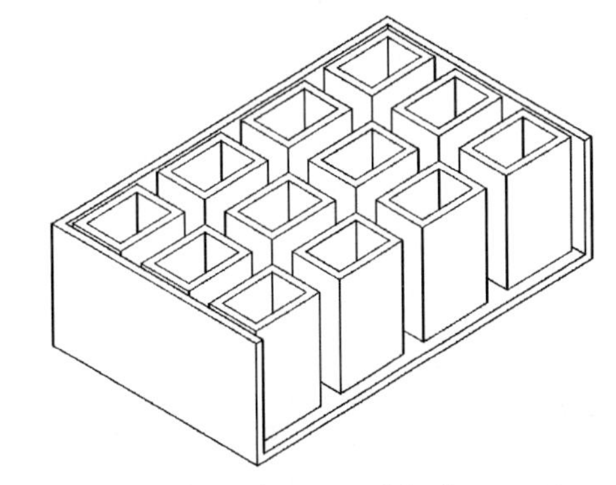

Figure 7-12 *Designer model after creating shell feature*

Figure 7-12 shows that the material equal to the wall thickness is added around the cut features.

This is the result that is not desired. Therefore, you have to change the order of the feature creation so that the shell feature is created before the parent extrusion using which the array is created. To reorder, press the shell feature in the desktop browser and drag it above the extrusion as shown in Figure 7-13.

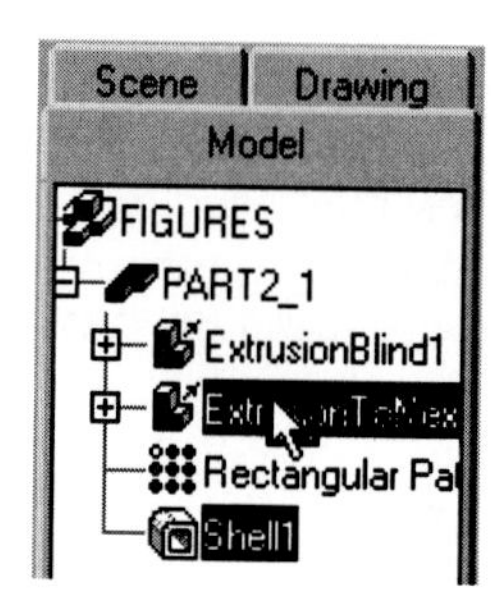

Figure 7-13 *Reordering the shell feature by dragging it and dropping it before extrusion*

When you drag and drop the shell feature above the parent extrusion, all the features will be automatically adjusted in the new order as shown in Figure 7-14.

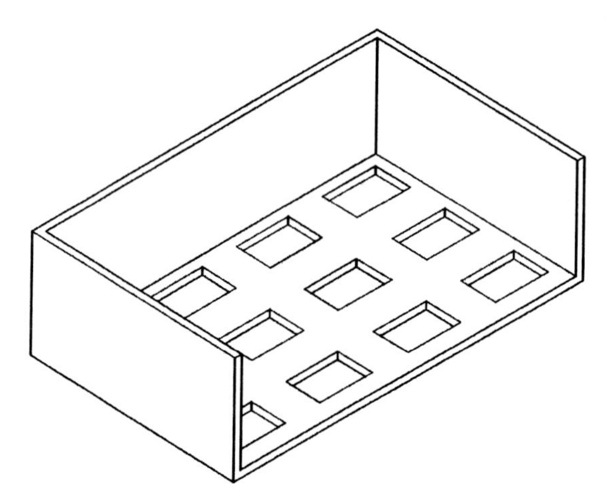

Figure 7-14 *Designer model after reordering the features*

TUTORIAL

Tutorial 1

In this tutorial you will create the designer model shown in Figure 7-15a. The dimensions to be used are given in Figures 7-15b, 7-15c, and 7-15d. Later change the drilled hole of 30 units diameter to counter bore hole with counter diameter of 30, counter depth of 8 and drill diameter to 24.

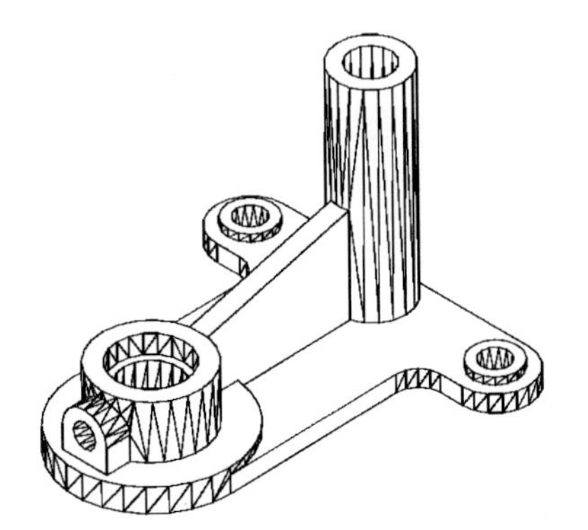

Figure 7-15a Designer model for Tutorial 1

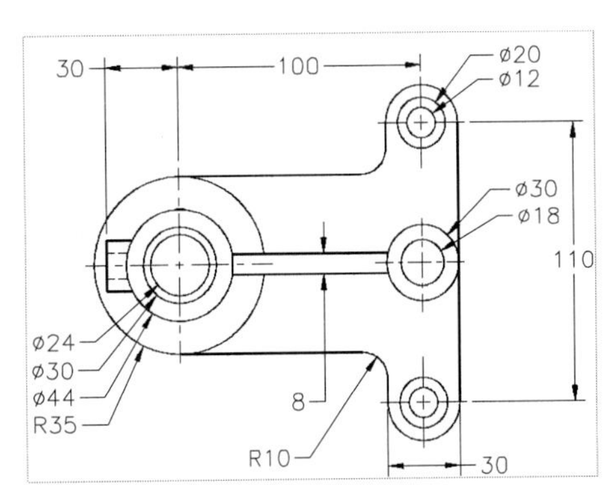

Figure 7-15b Top view of the model

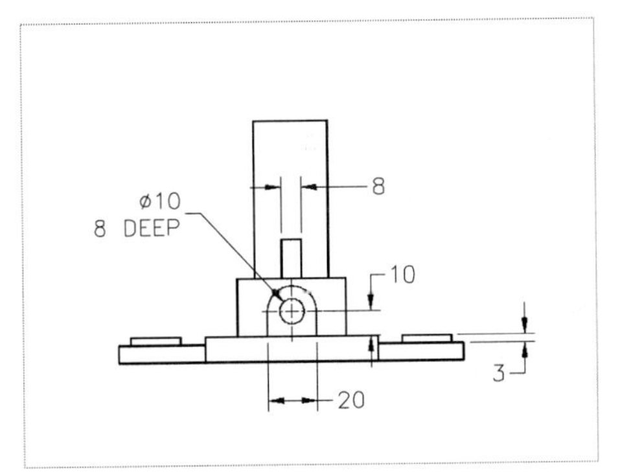

Figure 7-15c Side view of the model with the hidden lines suppressed for clarity

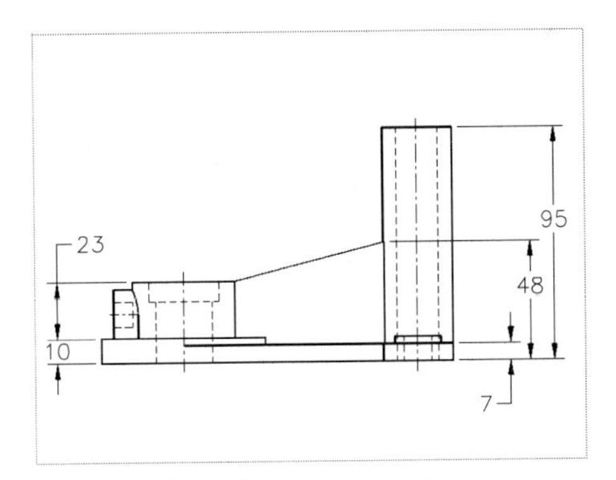

Figure 7-15d Front view of the model

1. Create the base feature as shown in Figure 7-16.

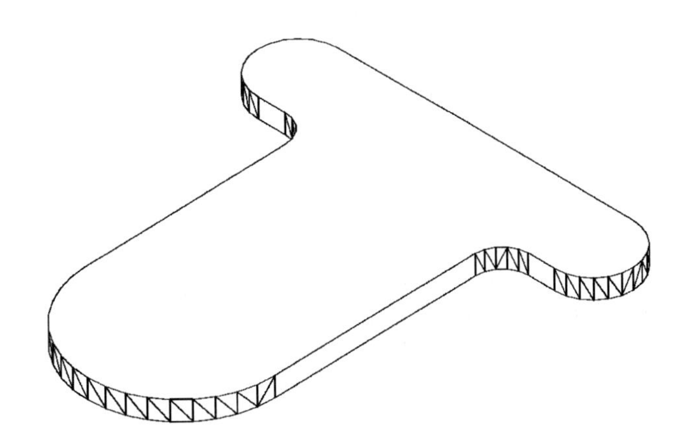

Figure 7-16 Base feature for Tutorial 1

2. Create both of the next cylindrical features and central rib.

3. Create the drilled through holes in both the cylindrical features, see Figure 7-17.

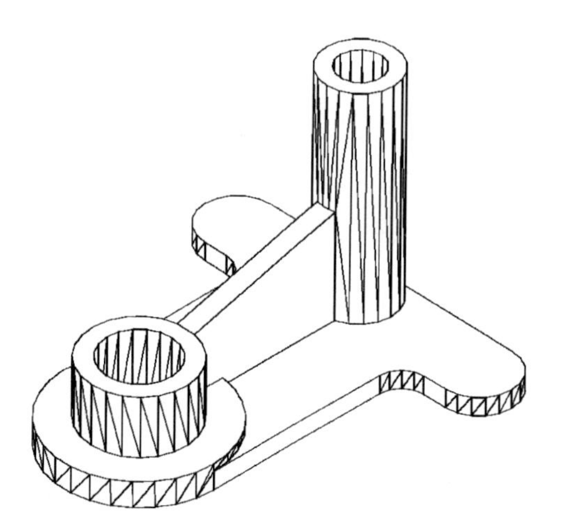

Figure 7-17 *Designer model after creating the drilled through holes*

4. Now create the next cylindrical feature on the base and then place a drilled hole in it, see Figure 7-18.

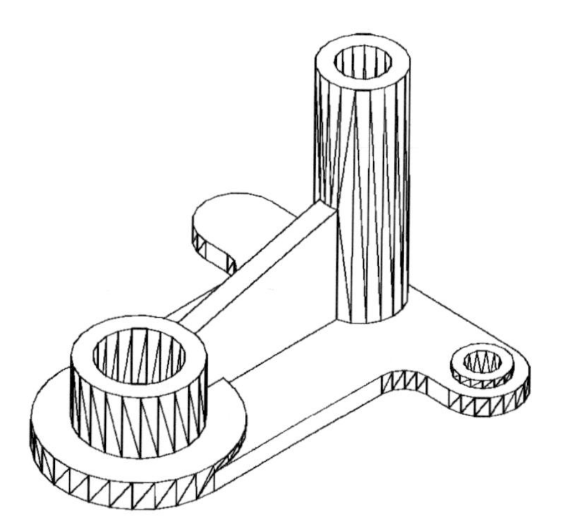

Figure 7-18 *Designer model after creating next feature on the base*

5. Choose the **Copy Feature** button from the **Placed Features - Hole** flyout in the **Part Modeling** toolbar. The prompt sequence is as follows:

Select feature to be copied (from any part): *Select the feature along with the hole, see Figure 7-19*
Enter an option [Next/Accept] <Accept>: Enter

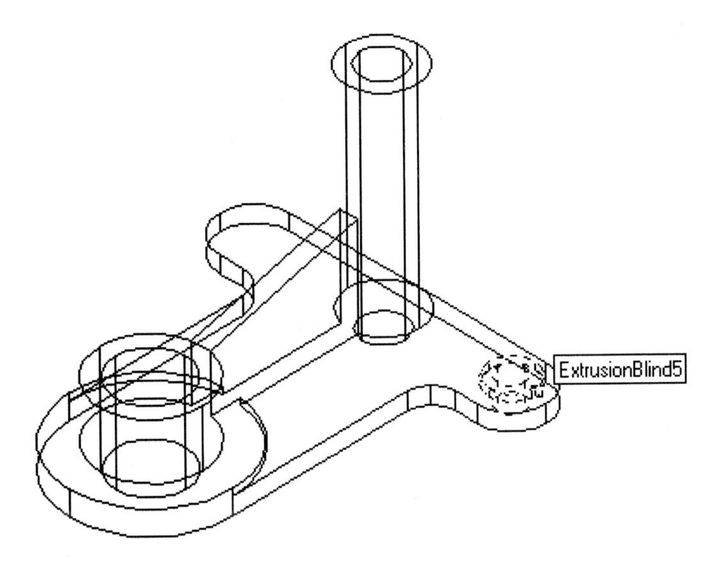

Figure 7-19 *Selecting the feature to copy*

Parameters=Independent
Specify location on the active part or [Parameters]: *Select the center on the other side of the base feature.*
Parameters=Independent
Specify location on the active part or [Parameters/Rotate/Flip]: *Make sure the feature is getting placed at the center of the cylindrical portion of the base feature, see Figure 7-20.*
Computing ...

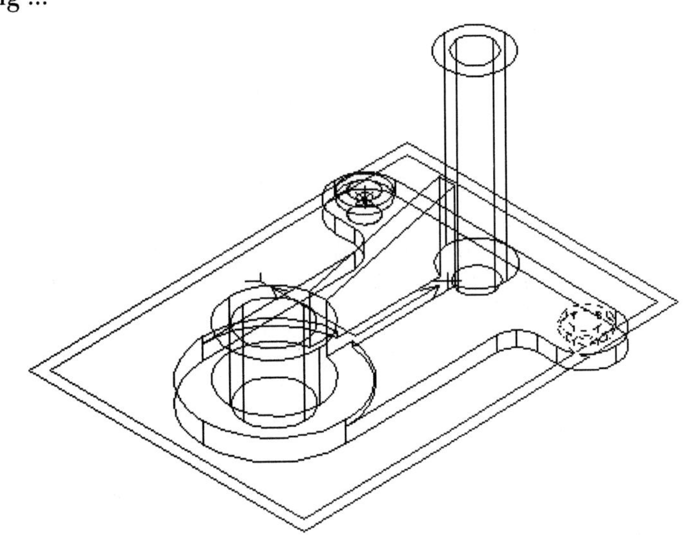

Figure 7-20 *Specifying the location of the new feature*

6. Create the next feature and place a hole of diameter 10 units and 10 deep in it as shown in Figure 7-21.

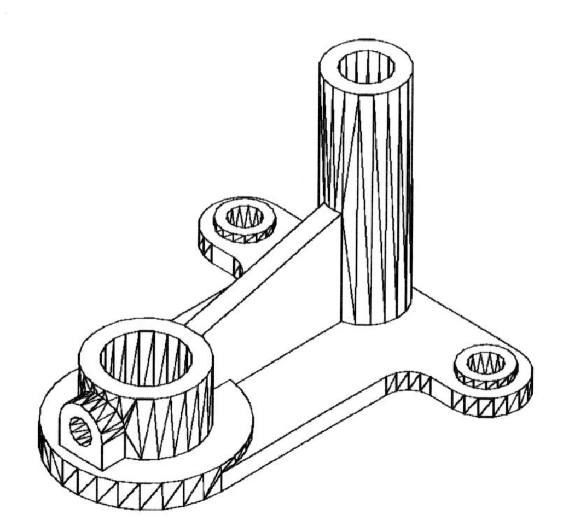

Figure 7-21 *Designer model after creating the next feature*

7. Double-click on the hole of 30 units diameter in the tree view of the desktop browser.

8. In the **Hole** dialog box displayed, select **C'Bore** from the **Operation** drop-down list.

9. Set the value of the **Dia** spinner under the **Drill Size** area to **24**.

10. Set the value of the **C'Dia** spinner to **30** and **C'Depth** spinner to **8** under the **C'Bore/Sunk Size** area. Choose **OK**. The prompt sequence is as follows:

    Select object: Enter

11. The hole will be automatically updated.

12. The final designer model should look similar to the one shown in Figure 7-22.

13. Save this drawing with the name given below:

    \MDT Tut\Ch-7**Tut1.dwg**

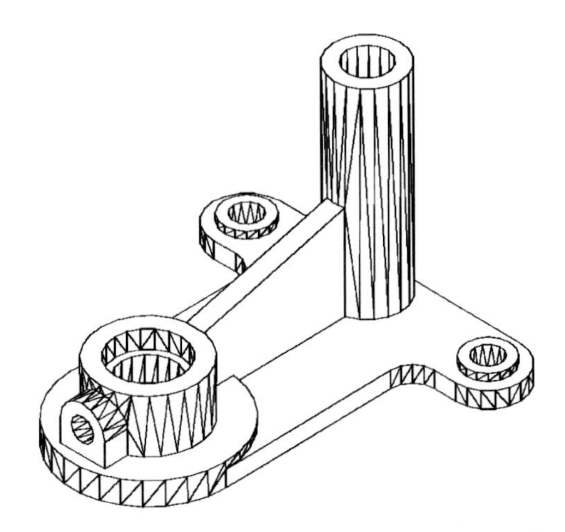

Figure 7-22 Final designer model for Tutorial 1

Tutorial 2

In this tutorial you will create the designer model shown in Figure 7-23a. The dimensions to be used are given in Figures 7-23b, 7-23c and 7-23d.

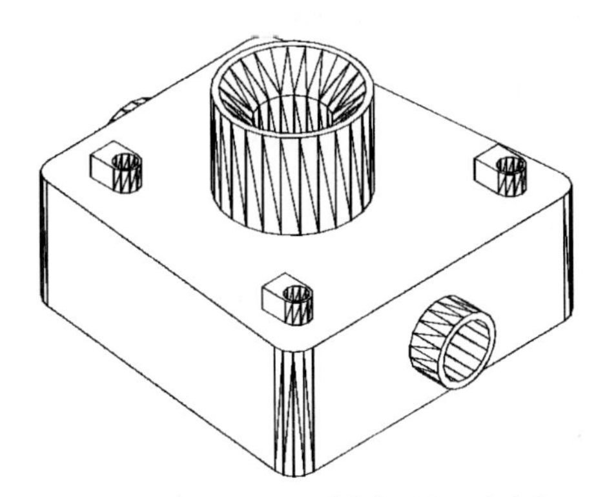

Figure 7-23a Model for Tutorial 2

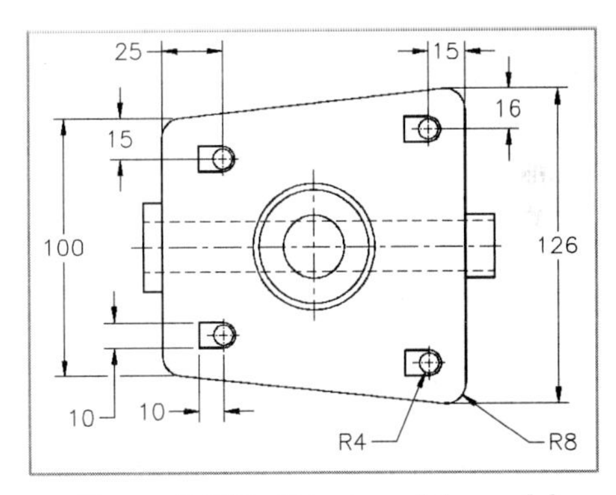

Figure 7-23b Top view of the model

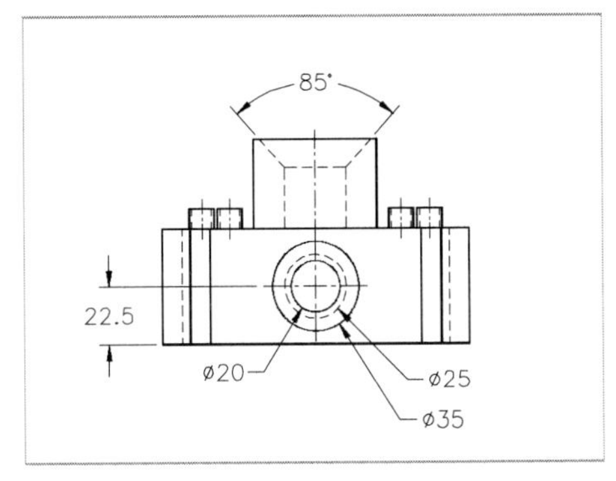

Figure 7-23c Side view of the model

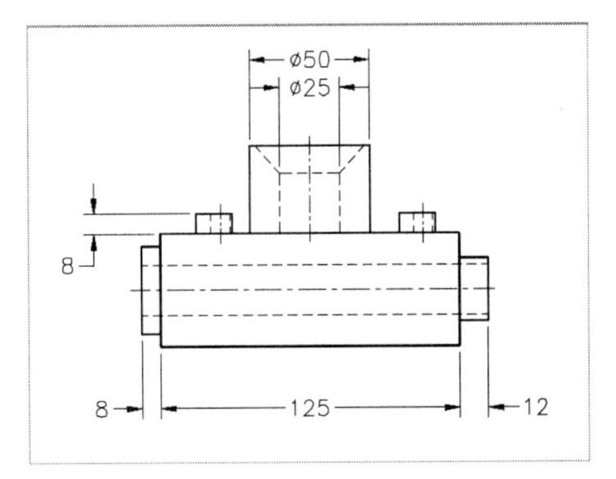

Figure 7-23d Front view of the model

1. Create the base feature for the designer mode. Add fillet on all the four vertical edges.

2. Create both the horizontal cylindrical features on the front and the back face of the designer model. Create the through hole in them, see Figure 7-24.

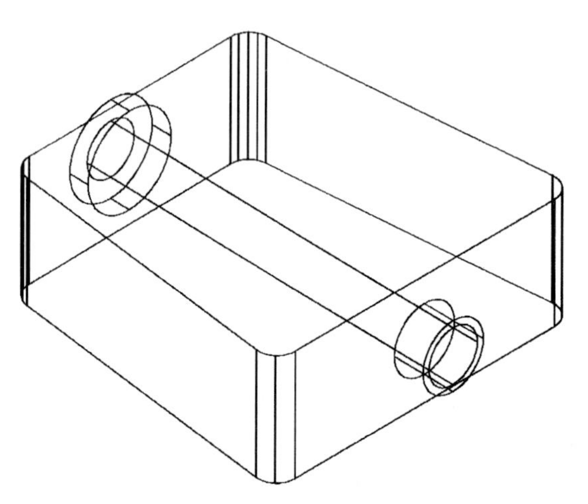

Figure 7-24 *Designer model after creating the horizontal cylindrical features*

3. Now, for adding the dimensions to the next cylindrical feature on the top face, you need to suppress the fillet. Right-click on the fillet in the desktop browser to display the shortcut menu. Choose the **Suppress** option as shown in Figure 7-25. The prompt sequence is as follows:

   The highlighted features will be suppressed.
   Continue? [Yes/No] <Yes>: Enter

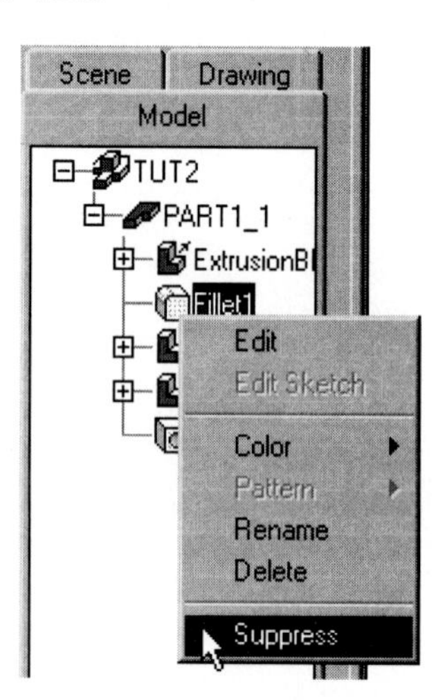

Figure 7-25 *Suppressing the fillet using the desktop browser*

4. Create the cylindrical feature on the top face and then place a countersink hole in it as shown in Figure 7-26.

5. Create the next feature on the to face and then place the hole in it as shown in Figure 7-27.

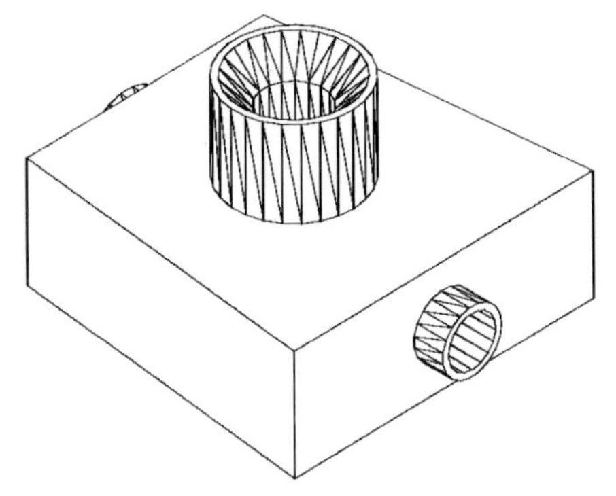

Figure 7-26 *Designer model after creating the cylindrical feature on the top face*

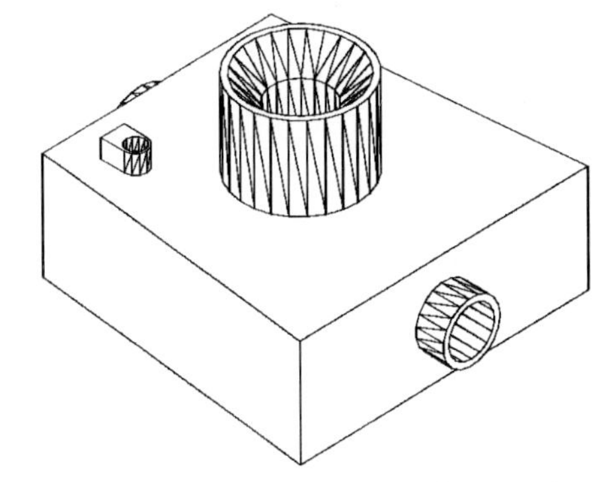

Figure 7-27 *Designer model after creating next feature on top face*

6. Choose the **Copy Sketch** button from the **Profile a Sketch** flyout in the **Part Modeling** toolbar. The prompt sequence is as follows:

Enter an option [Feature/Sketch] <Sketch>: **Feature**
Select feature: *Select the new feature.*
Sketch center: *Place the sketch.*
Sketch center: Enter
Computing ...

7. The new sketch will be a profiled sketch and will carry all the dimensions except for the alignment. Therefore, align the new sketch and then extrude it. Place the hole in it as shown in Figure 7-28.

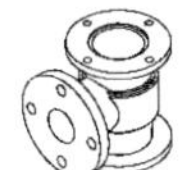

Note
Make sure the current sketch plane is on the top face of the designer model as the copied sketch will be placed only the current sketch plane.

8. Similarly, copy the sketch at remaining two places and complete the model by extruding them and placing the holes in them.

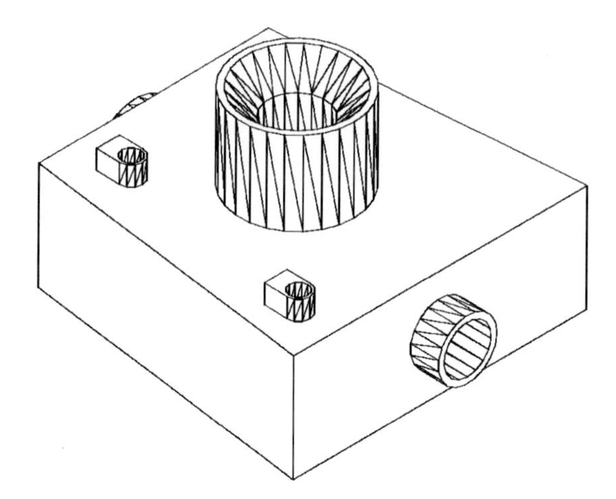

Figure 7-28 *Designer model after extruding the copied sketch and placing the hole in it*

9. Unsuppress the fillet using the desktop browser. The final designer model should be similar to the one shown in Figure 7-29.

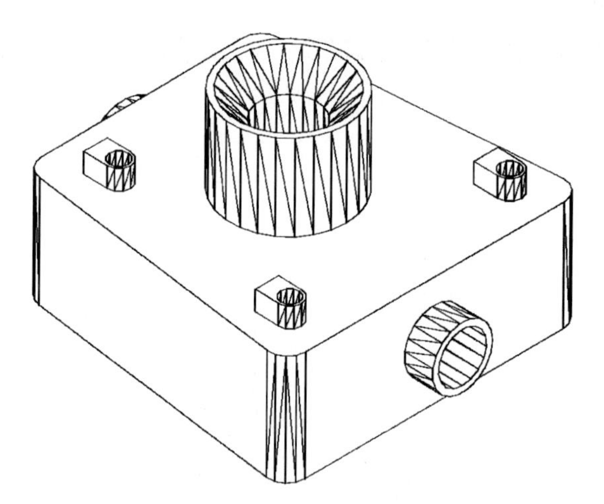

Figure 7-29 *Final designer model for Tutorial 2*

10. Save this drawing with the name given below:

   \MDT Tut\Ch-7**Tut2.dwg**

Review Questions

Answer the following questions.

1. Which command is used to alter the dimension values assigned to the features?
2. The basic sketch of the features cannot be changed. (T/F)
3. Which command is used to view the effect of the editing operation?
4. The sketches of the features can also be copied. (T/F)
5. Which command is used to copy the features?
6. What is the need of suppressing the features?
7. Which command is used to unsuppress the features?
8. The order of feature creation can be changed. (T/F)

Exercise

Exercise 1

In this tutorial you will create the body of the Tailstock shown in Figure 7-30a. The dimensions to be used are given in Figures 7-30b, and 7-30c. Assume the missing dimensions.

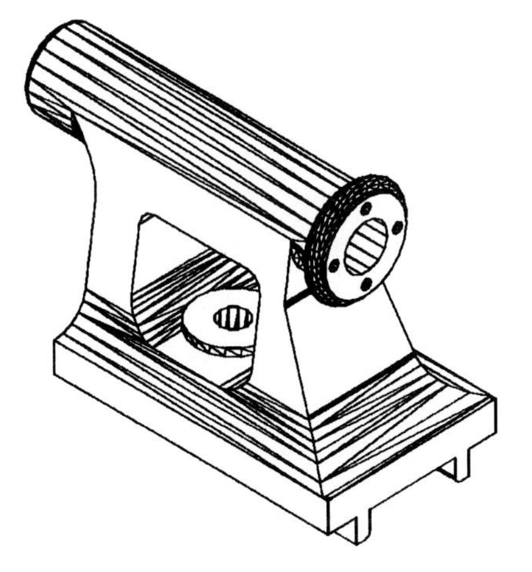

***Figure 7-30a** Designer model for Exercise 1*

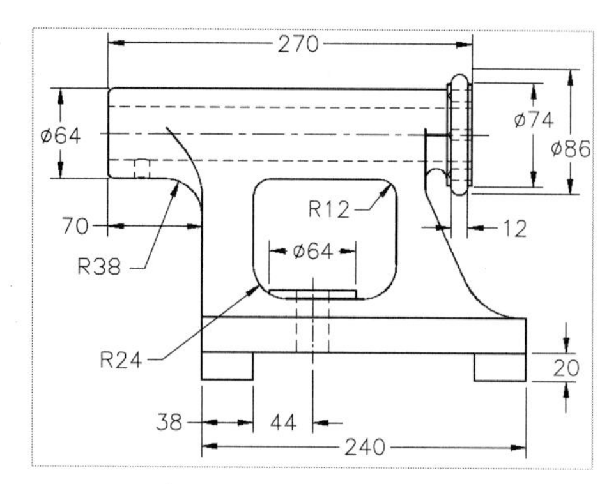

Figure 7-30b Front view of the model

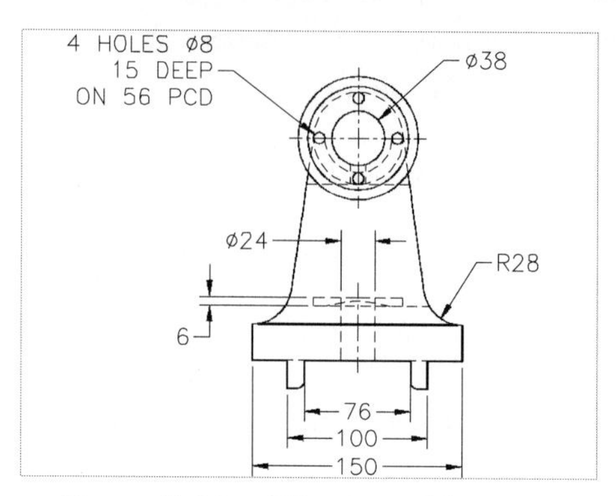

Figure 7-30c Side view of the model

Chapter 8

2D Path, 3D Path, Sweep, and Loft

Learning Objectives

After completing this chapter, you will be able to:

- *Define 2D paths using the* ***AM2DPATH*** *command.*
- *Define different types of 3D paths using the* ***AM3DPATH*** *command.*
- *Sweep 2D profiles about a path using the* ***AMSWEEP*** *command.*
- *Create complex designer model using the* ***AMLOFT*** *command.*

Commands Covered

- *AM2DPATH*
- *AM3DPATH*
- *AMSWEEP*
- *AMLOFT*

DEFINING 2D PATHS

Sometimes, you may need to profile some entities that can act as a path for sweeping the profiled sketch in a single plane. This can be done using the **AM2DPATH** command.

AM2DPATH Command

Toolbar:	Part Modeling > Profile a Sketch > 2D Path
Menu:	Part > Sketch Solving > 2D Path
Context Menu:	Sketch Solving > 2D Path
Command:	AM2DPATH

This command is used to profile an open entity in a single plane, thus converting it into a parametric two dimensional path that can be used to sweep a profiled sketch. When you select the sketch, you will be prompted to specify the start point of the path. A work point will be placed at the start point of the path. You will be provided an option of defining a work plane normal to the path for drawing the sketch to be swept.

3D PATHS

The 3D paths are created in three dimensional space and are not restricted to a single plane. In Mechanical Desktop, the 3D paths are created using the **AM3DPATH** command. Using this command you can create the following four types of 3D paths:

1. Edge Path
2. Helix Path
3. Pipe Path
4. Spline Path

Creating The Edge Path (AM3DPATH Command)

Toolbar:	Part Modeling > Profile a Sketch > 3D Path
Menu:	Part > Sketch Solving > 3D Edge Path
Context Menu:	Sketch Solving > 3D Edge Path
Command:	AM3DPATH

This option of the **AM3DPATH** command is used to create 3D path along tangentially continuous edges of the designer model (Figure 8-1). When you select the tangentially continuous edges about which you want to define the path, you will be prompted to specify the start point of the path. A work point will be placed at the start point of the path and you will be provided an option of defining a work plane normal to the start point of the path. Depending upon your requirement you can specify the X, Y, and Z directions of this work plane.

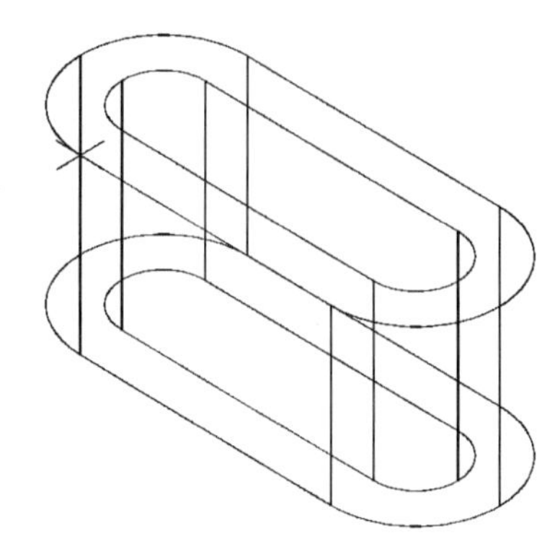

Figure 8-1 *Edge path on the top face*

Creating The 3D Helical Path (AM3DPATH Command)

Toolbar:	Part Modeling > Profile a Sketch > 3D Path
Menu:	Part > Sketch Solving > 3D Helix Path
Context Menu:	Sketch Solving > 3D Helix Path
Command:	AM3DPATH

This option of the **AM3DPATH** command is used to create helical 3D paths. These paths can be used to create internal threaded components, external threaded component, springs and so on. The helical paths can be of circular or elliptical shape. You will be prompted to select a circular edge, circular face, or a work axis to be used as helical center when you invoke this option. As soon as you select any one of the said features, the **Helix** dialog box will be displayed as shown in Figure 8-2.

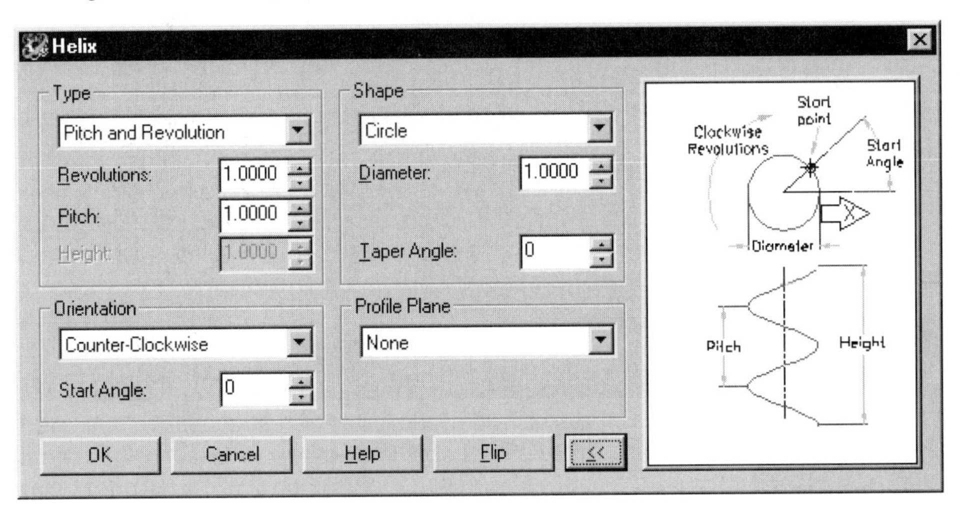

Figure 8-2 Helix dialog box for creating helical path

Helix Dialog Box Options

The options provided under the **Helix** dialog box are:

Type Area

This area is used to define the method of creating the helical path. The methods for creating the helical path can be selected from the drop-down list provided under this area. The options that are provided in this drop-down list are:

Pitch and Revolution. In this method the helical path is created by specifying the number of revolutions of the helical path and the pitch between the path.

Revolution and Height. This method is used to create the helical path by specifying the number of revolutions in the path and the overall height of the path.

Height and Pitch. This is the method of creating the helical path by specifying the overall height of the path and the pitch between the revolutions.

Spiral. This method is used to create a spiral path on a single plane.

This area also provides you with the options of defining the parameters associated with the selected method. These parameters are:

Revolution. This spinner is used to specify the number of revolutions in the helical path. This spinner is available only when the **Pitch and Revolution**, **Revolution and Height**, or the **Spiral** method is selected.

Pitch. This spinner is used to specify the pitch between the revolution of the helical path as shown in Figure 8-3. This spinner is available only when the **Pitch and Revolution**, **Height and Pitch**, or the **Spiral** method is selected.

Height. This spinner is used to specify the overall height of the helical path as shown in Figure 8-3. This spinner is available only when the **Revolution and Height** or the **Height and Pitch** method is selected.

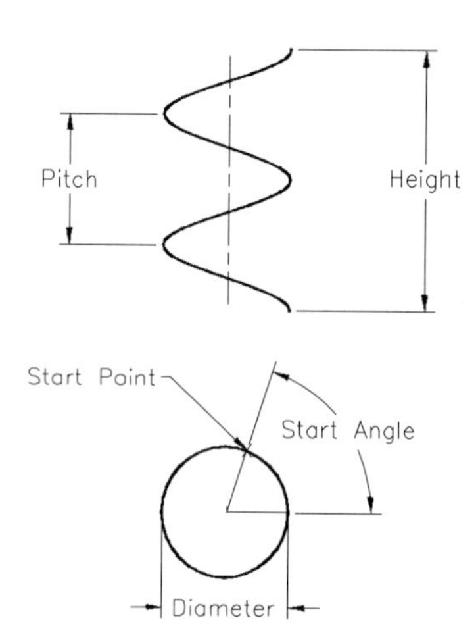

Figure 8-3 *Description of the helical path*

Orientation Area

The option provided under this area is used to define the orientation of the helical path. The path can be created in the clockwise or the counterclockwise direction by selecting it from the drop-down list provided under this area. The other option provided under this area is:

Start Angle. This spinner is used to specify the start point angle of the helical path.

Shape Area

The option provided under this area is used to specify the shape of the helical path. You can create a **circular** or an **elliptical** path (Figures 8-4 and 8-5) by selecting it from the drop-down list provided under this area. The other options provided under this area are:

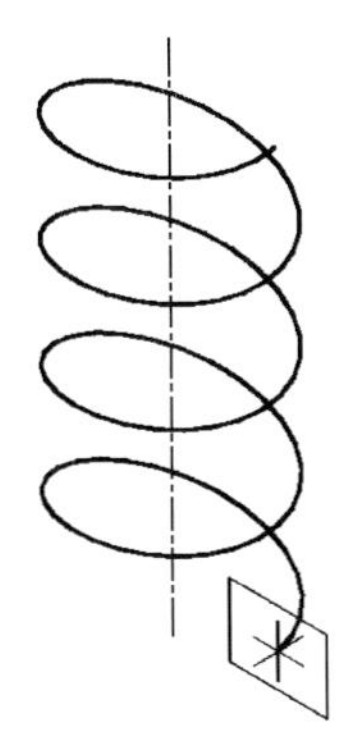

Figure 8-4 Elliptical helical path with a work plane aligned to the start point of the path and the work axis

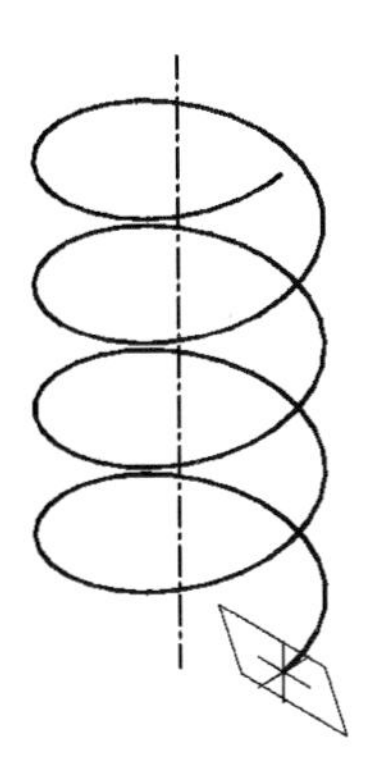

Figure 8-5 Circular helical path with a work plane normal to the start

Diameter. This spinner is used to specify the diameter of the helical path. When the Ellipse shape is selected, the **Major Axis** and **Minor Axis** spinners appear in the place of the **Diameter** spinner for specifying the major and minor axes values.

Taper Angle. This spinner is used to specify the taper angle for the helical path. This spinner is not available when the **Spiral** method is selected from the **Type** area.

Profile Plane Area

The options provided under this area are used to specify the location of the work plane created at the start point of the helical path. The drop-down list provided under this area has the following three options

None. If this option is selected, no work plane will be created.

Center Axis/Path. If this option is selected, the work plane created will be aligned with the work point placed at the start point of the path and work axis place at the center of the helix.

Normal to Path. If this option is selected, the work plane will be created normal to the start point of the helical path.

Creating The Pipe Path (AM3DPATH Command)

Toolbar:	Part Modeling > Profile a Sketch > 3D Path
Menu:	Part > Sketch Solving > 3D Pipe Path
Context Menu:	Sketch Solving > 3D Pipe Path
Command:	AM3DPATH

The 3D pipe paths are used to create pipe lines in the 3D space. You can fillet the corners of the pipe paths to create a smooth pipe path, see Figures 8-6 and 8-7. When you invoke this option, you will be prompted to select the polyline that has to be

converted into a pipe path. Then you will be prompted to specify the start point of the path.

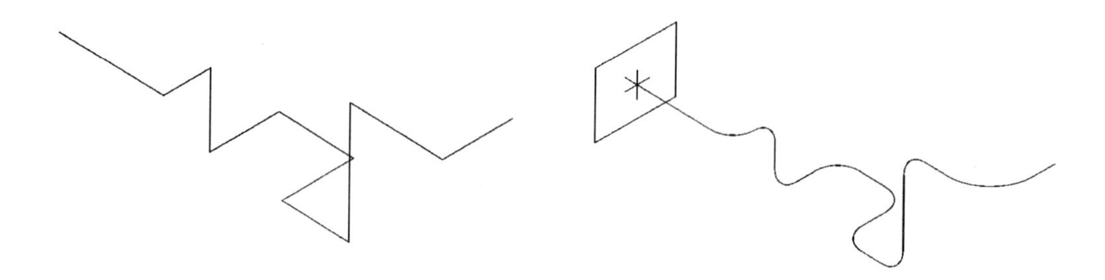

Figure 8-6 *The polyline before converting it into a pipe path*

Figure 8-7 *The pipe path created from the polyline*

As soon as you specify the start point of the path, the **3D Pipe Path** dialog box will be displayed as shown in Figure 8-8.

3D Pipe Path

No.	C	From	Delta X	Delta Y	Delta Z	Length	Angle XY	Angle Z	Radius
1			-2.4130	-3.3005	0.0000	10.5015	90.4268	0.0000	
2		1	-0.0782	10.5012	0.0000	8.7831	0.6281	0.0000	0.0000
3		2	8.7825	0.0963	0.0000	8.0396	-86.8931	0.0000	0.0000
4		3	0.4357	-8.0278	0.0000	8.9818	-0.2461	0.0000	0.0000
5		4	8.9818	-0.0386	0.0000	12.1173	94.4935	0.0000	0.0000
6		5	-0.9493	12.0800	0.0000	5.0000	0.0000	90.0000	0.0000
7		6	0.0000	0.0000	5.0000	8.0000	-90.0000	0.0000	0.0000

Create Work Plane   Closed   OK   Cancel   Help

Figure 8-8 *3D Pipe Path dialog box*

3D Pipe Path Dialog Box Options

The text area of this dialog box is divided into ten columns. The number of rows depend upon the number of segments the path consists of. You can modify the value of any of these boxes by selecting them. The box is converted into a spinner upon selecting and you can modify the value using the spinner. All ten columns in the text area are described next.

No.

This column specifies the number of vertices the path consists of. These vertices are called the fit points. This value includes the start point and the endpoint of the path. All these points, by default are constrained to the previous fit point of the path. You can constrain these points to a work point, unconstrain these fit points, delete these fit points or insert a fit point using the shortcut menu displayed by right-clicking on these columns, see Figure 8-9.

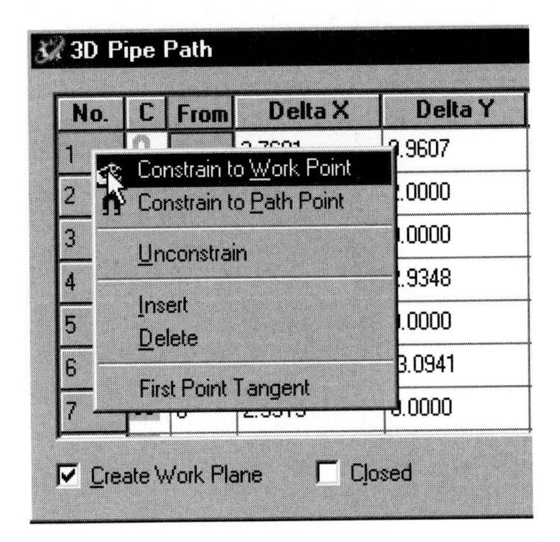

Figure 8-9 *Shortcut menu displayed upon right-clicking on the column*

C
This column displays the kind of constraint applied to the fit point. Depending upon whether the fit point is constrained to the work point or the path point, the symbol displayed in this column changes.

From
This column displays the point to which the current fit point is constrained.

Delta X
This column specifies the delta X value of the current fit point. This value is calculated from the fit point to which the current fit point is constrained. The delta X value for the start point is calculated from the origin point.

Delta Y
This column specifies the delta Y value for the current fit point. This value is calculated from the fit point to which the current fit point is constrained. The delta Y value for the start point is calculated from the origin point.

Delta Z
This column specifies the delta Z value for the current fit point. This value is calculated from the fit point to which the current fit point is constrained. The delta Z value for the start point is calculated from the origin point.

Length
This column displays the length of the segment. You can modify this value using the spinner displayed upon selecting this field.

Angle XY
This column displays the angle of the current segment in the XY plane.

Angle Z

This column displays the angle of the current segment from the XY plane in the space.

Radius

This column displays the fillet radius for the current fit point.

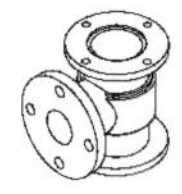

Note

All the columns in the white base can be modified using the spinner that appears on them and by selecting them using the left mouse button.

Apart from the text box, the **3D Pipe Path** dialog box contains the following two check boxes:

Create Work Plane

This check box is selected to create a work plane normal to the segment containing the start point of the path.

Closed

If this check box is selected, a segment is added from the endpoint of the path to the start point of the path, thus closing the pipe path.

Creating The Spline Path (AM3DPATH Command)

Toolbar:	Part Modeling > Profile a Sketch > 3D Path
Menu:	Part > Sketch Solving > 3D Spline Path
Context Menu:	Sketch Solving > 3D Spline Path
Command:	AM3DPATH

The spline paths are to be used for creating complex electrical coils, frames of wires, or tubes. You have to create a spline first for creating a spline path, see Figures 8-10 and 8-11.

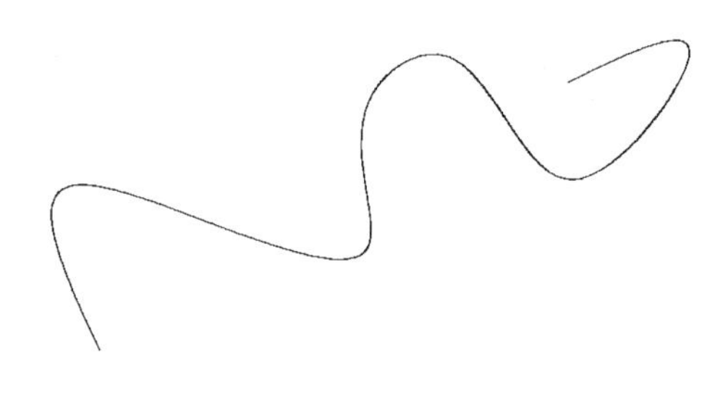

Figure 8-10 The spline before it is converted into spline path

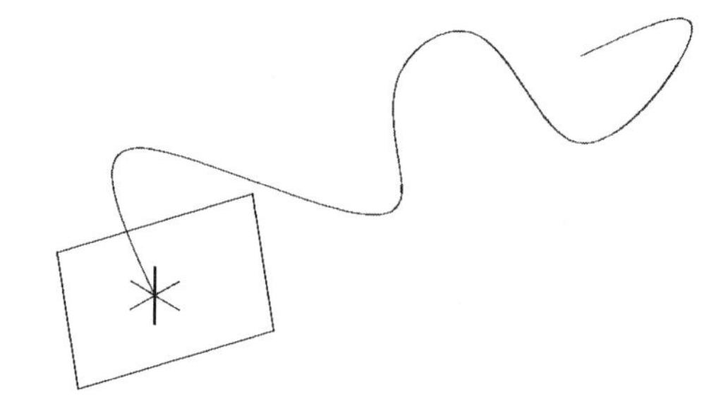

Figure 8-11 The spline path with the work plane normal to the start of the path

Once you have selected the spline and specified the start point of the path, the **3D Spline Path** dialog box will be displayed as shown in Figure 8-12.

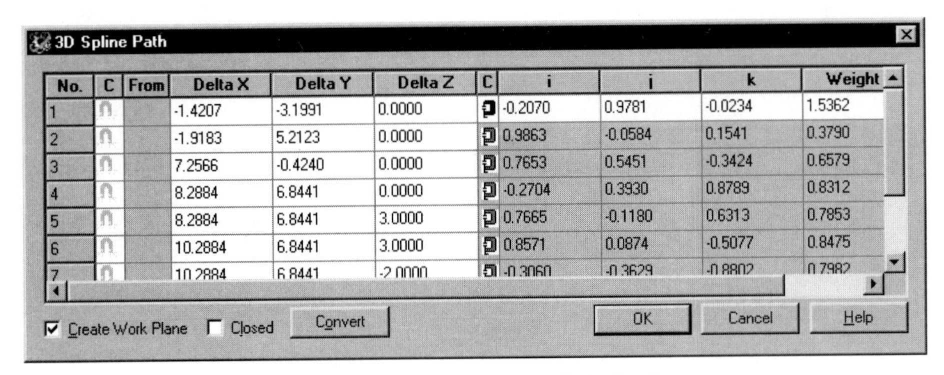

Figure 8-12 3D Spline Path dialog box

3D Spline Path Dialog Box Options

This dialog box is similar to the **3D Pipe Path** dialog box. The text box of this dialog box also contain ten columns. The first five columns are similar to those discussed in the **3D Pipe Path** dialog box. The rest of the five columns are discussed below:

C

The second **C** column specifies whether or not the tangent constraint is applied to the related fit point. Generally, the tangent constrain is applied to the start point and the endpoint of the spline.

i

This column is used to control the tangency vector in the X axis direction for the start and the endpoint of the spline path.

j

This column is used to control the tangency vector in the Y axis direction for the start and the endpoint of the spline path.

k

This column is used to control the tangency vector in the Z axis direction for the start and the endpoint of the spline path.

Weight

This column is used to specify the distance up to which the current spline segment will maintain the tangency to the previous fit point.

Convert

This button is chosen to convert a fit point spline into a control point spline or a control point spline into a fit point spline. When you convert a fit point into a control point spline, you will be prompted to confirm the conversion. When you convert a control point spline into a fit point spline, the **Fit Point Conversion** dialog box will be displayed. You can specify the **Tolerance** or the **Point Count** conversion type using this dialog box.

Note

*The **Create Work Plane** and the **Closed** options are similar to those discussed in the **3D Pipe Path** dialog box.*

SWEEPING THE PROFILES

Once you have created the required path, you need to sweep the profile about that path. The profiles can be swept about the paths using the **AMSWEEP** command.

AMSWEEP Command

Toolbar:	Part Modeling > Sketched Features-Extrude > Sweep
Menu:	Part > Sketched Features > Sweep
Context Menu:	Sketched & Worked Features > Sweep
Command:	AMSWEEP

This command is used to sweep a profiled sketch about a parametric path created using the **AM2DPATH** or the **AM3DPATH** command. It is similar to extruding a profiled sketch about a specified path in 2D plane or 3D space. The **Sweep** dialog box is displayed when you invoke this command, see Figure 8-13.

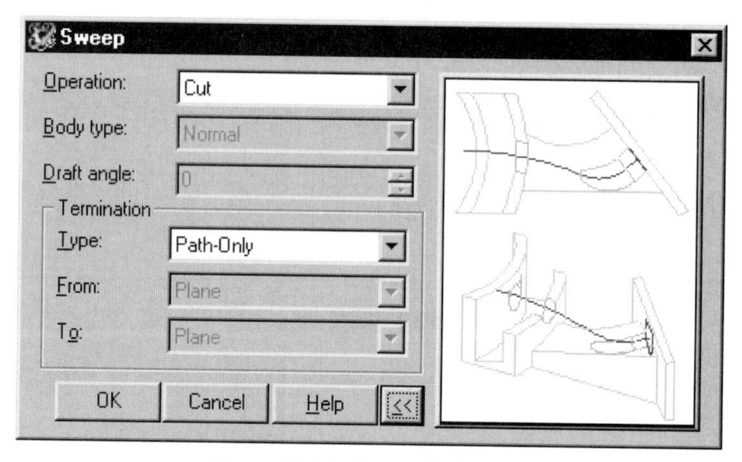

Figure 8-13 *Sweep dialog box*

Sweep Dialog Box Options

The options provided under the **Sweep** dialog box are:

Operation

The options provided under this drop-down list are used to specify the kind of operation the profile will be used for. If the sweep is the first feature, this drop-down list will not be available. If the sweep is not the first feature then this drop-down list will be available and the options provided under this list are:

Cut. This option when selected creates a swept feature by removing the material, thus creating the swept groove or cavity.

Join. This option is used to create a new feature by joining a swept feature to an existing feature.

Intersect. This option is used to create a new feature by retaining the material common to the swept feature and the existing feature.

Split. This option is used to create a new part using the area common to the swept feature and the existing feature. The name of the new part can be specified in the **Enter name of the new part <default name>** prompt.

Body type

The options provided under this drop-down list are used to define the shape of the new swept feature. This drop-down list has the following two options:

Normal. If this option is used, the profile will be swept normal to the parametric path, see Figures 8-14 and 8-15.

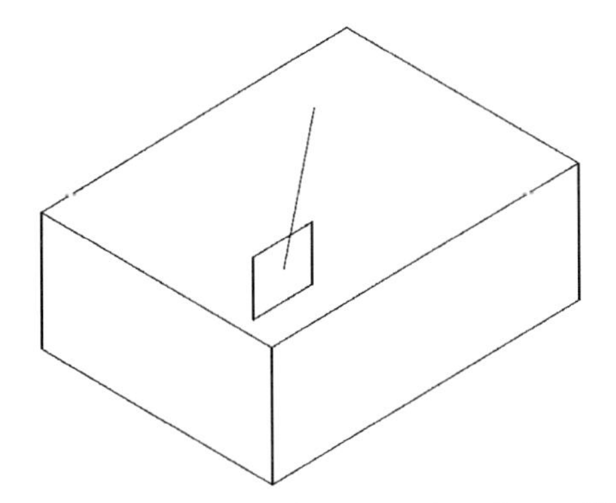

Figure 8-14 *Model before sweeping the profile normal to the path*

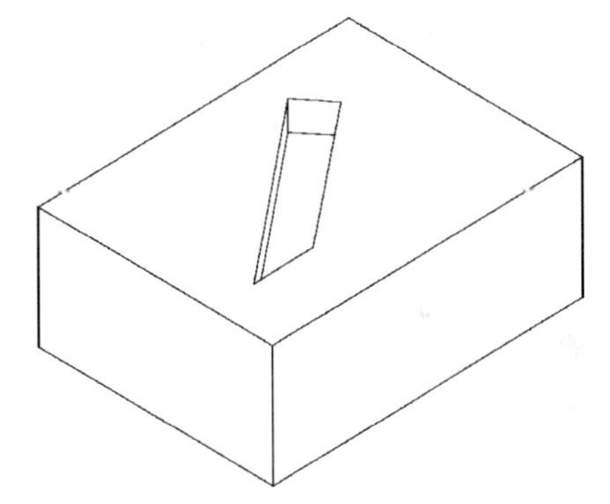

Figure 8-15 *Model after sweeping the profile normal to the path*

Parallel. If this option is selected then the profile will be swept parallel to the parametric path, see Figures 8-16 and 8-17.

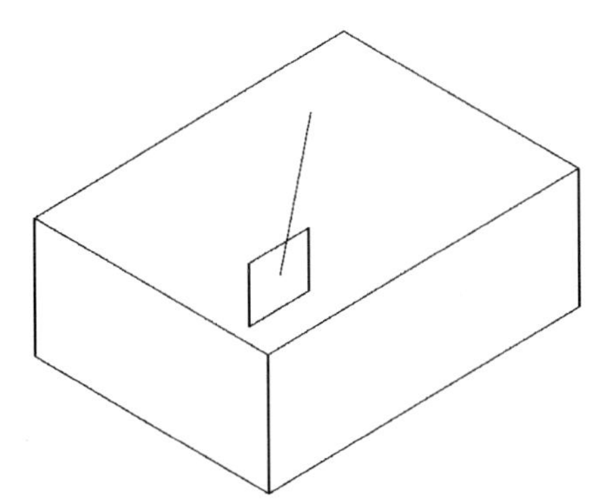

Figure 8-16 *Model before sweeping the profile normal to the path*

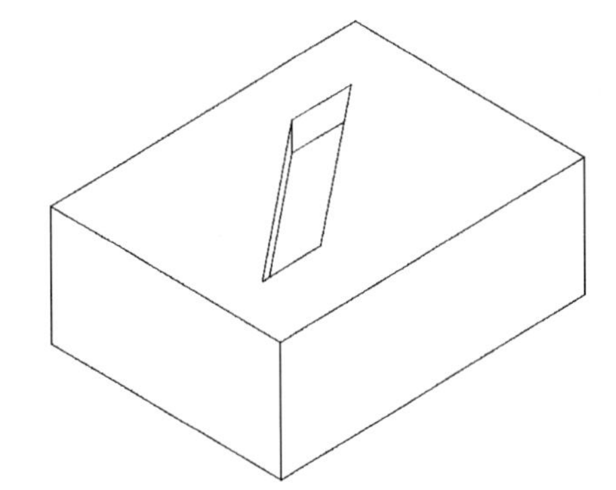

Figure 8-17 *Model after sweeping the profile parallel to the path*

Draft Angle

This spinner is used to specify the draft angle for the sweep feature.

Termination Area

Type

The options under this drop-down list are used to specify the type of termination. The options under this drop-down list are:

Path-Only. If this termination is selected then the sweep will be terminated at the end of the parametric path.

Next. This option is used to specify the termination of the sweep using the next face that has full intersection with the profile.

Plane. This option is used to specify the termination point of the sweep feature using a work plane or a planar face.

Face. This option is used to terminate the sweep feature using a specified face.

Extended Face. This option is used to specify the termination of the sweep using a face that may or may not actually intersect the selected profile. This option is generally used when the profile is drawn somewhere above or below the face to be used for termination.

From-To. This option is used to create a sweep feature that is restricted between two faces, planes or extended faces. When you invoke this option, the **From** and **To** drop-down lists are activated. You can select face, plane or extended face for restricting the sweep using these drop-down lists.

CREATING THE LOFTED FEATURES (AMLOFT COMMAND)

Toolbar:	Part Modeling > Sketched Feature-Extrude > Loft
Menu:	Part > Sketched Features > Loft
Context Menu:	Sketched & Worked Features > Loft
Command:	AMLOFT

Lofting is defined as the method of blending two or more dissimilar geometries thus forming a complex designer solid. This is done using the **AMLOFT** command. This command is used to create the complex designer model consisting of a combination of a number of different geometries blended together. The geometries that can be blended together include a closed profiled sketch, a planar face of the active designer model or a work point. When you invoke this command, you will be prompted to select the profiles or the planar faces to be used for lofting. Once you have selected them, the **Loft** dialog box will be displayed, see Figure 8-18.

Loft Dialog Box Options

The options provided in the **Loft** dialog box are:

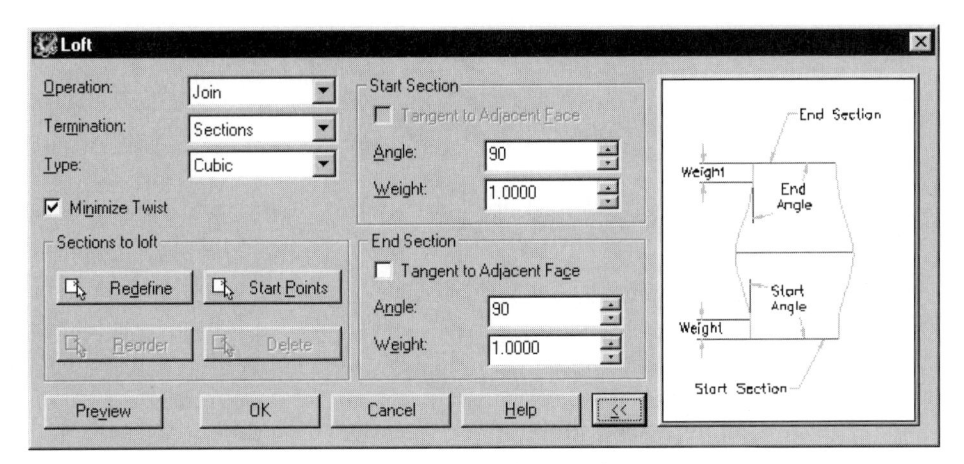

Figure 8-18 *Loft dialog box*

Operation

This drop-down list will provide only the **Base** option if the loft is the first feature. If the base feature already exists, this drop-down list will provide the following options:

Join

This option creates a lofted feature by adding the material to the existing feature.

Cut

This drop-down list creates a feature by removing the material from the existing feature thus creating a lofted cavity or a slot.

Intersect

This option creates a lofted feature by retaining the material common to the existing feature and the loft.

Split

This option is used to create a new part using the material common to the existing feature and the loft. The name of the new part can be entered in the **Enter name of the new part <default name>** prompt.

Termination

The options provided under this drop-down list are used to specify the termination of the loft feature. It has the following options:

Sections

This option is used to restrict the loft feature between the profiles or the planar faces initially selected. The loft feature will start from the first profile or planar face selected and will end at the last profile or planar face selected.

To Face/Plane

This option is used to specify the termination of the lofted feature using a planar face.

From To
This option is used to create a lofted feature restricted between specified starting face and an ending face.

Type
The options provided under this drop-down list are used to define the type of lofting. It has the following options:

Linear
This option creates a linear blending between the selected profiles or planar faces. This option can be used for only two profiles or planar faces. Figures 8-19 and 8-20 shows the linear lofting between a pentagon and a circle. When this type of loft is selected, the options related to the start and end section will not be available.

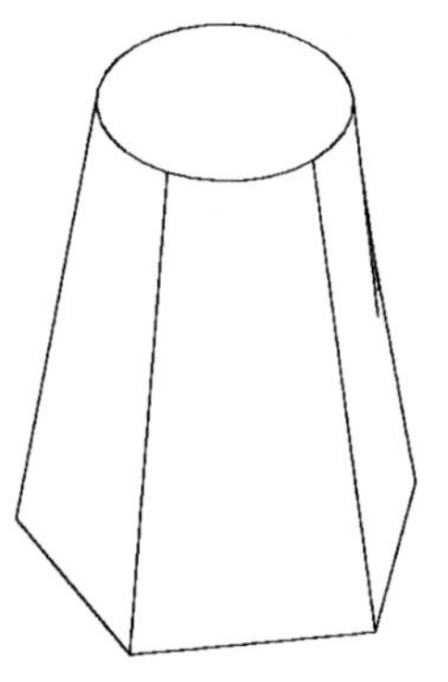

Figure 8-19 *Isometric view of linear loft*

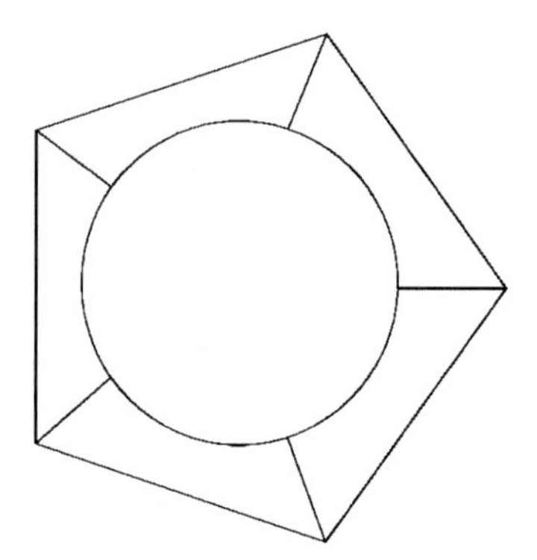

Figure 8-20 *Plan view of linear loft*

Cubic
This type of loft is used to create a smooth blend between the profiles or the planar faces selected. Figure 8-21 shows three different geometries selected to create a cubic loft and Figure 8-22 shows the loft created using the selected geometries.

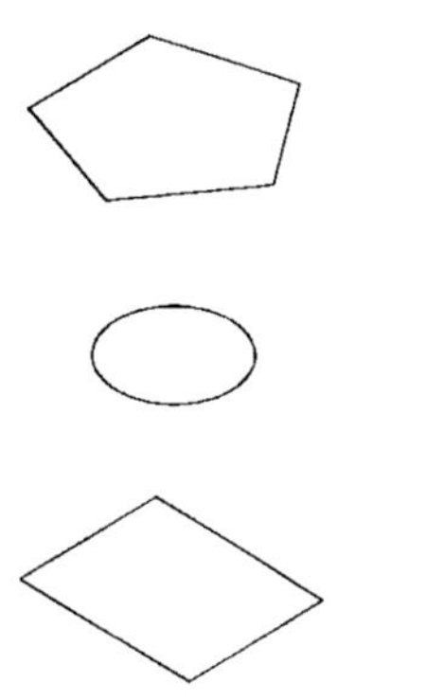

Figure 8-21 *Geometries selected for lofting*

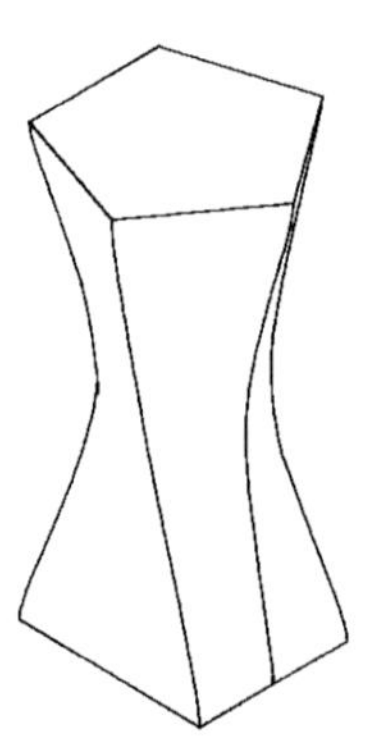

Figure 8-22 *Loft created from the selected geometries*

Closed Cubic

This option is used to create a gradual closed loop lofted feature in which the first selected profile or plane will also be the last, thus forming a closed loop. Figure 8-23 shows the three geometries selected to create the closed cubic loft and Figure 8-24 shows the closed cubic loft created using the selected geometries. In this type of loft, you can control only the starting section of the loft.

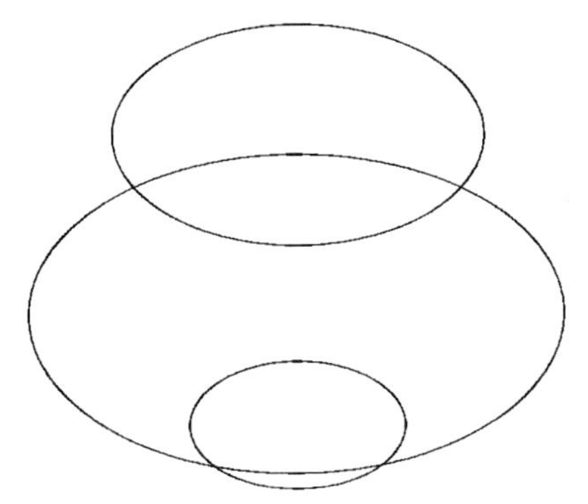

Figure 8-23 *Geometries selected for lofting*

Figure 8-24 *Loft created from the selected geometries*

Minimize Twist

If this check box is selected, the start point of the loft on each selected profile or the planar face will be adjusted to minimize the twist during the blending of the geometries.

Sections to Loft Area

The options provided under this area are:

Redefine

The **Redefine** button is chosen to redefine the sequence of the profiles or the planar faces selected. When you choose this button the **Loft** dialog box will be temporarily closed and you will be asked to redefine the sequence of selecting the geometries.

Start Point

The **Start Point** button is used to manually define the default position of the start point of the loft on each selected profile or planar face. The start points are defined to create twist in the lofted feature.

Reorder

The **Reorder** button is used to reorder the profiles or planar faces selected to loft.

Delete

The **Delete** button is used to delete the profiles or the planar faces selected to loft.

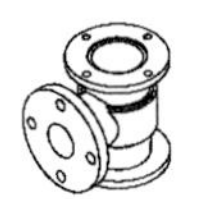

Note

The ***Reorder*** *and the* ***Delete*** *buttons will be available only during the editing of the loft feature.*

Start Section Area

The options provided under this area are:

Tangent to Adjacent Face

This check box will be available only when the start section of the loft is the planar face of an existing feature. If this check box is selected, the new loft created will be tangent to adjacent faces of the planar face selected as the start section of the loft.

Angle

This spinner is used to specify the start angle of the loft, see Figure 8-25. This spinner is not available when the **Tangent to Adjacent Face** check box is selected.

Weight

This spinner is used to specify the distance up to which the loft section will maintain the start angle or the starting face tangency before it blends with the next section, see Figure 8-25.

End Section Area

The options provided under this area are:

Tangent to Adjacent Face

This check box is available only when the last section of the loft is a planar face of an existing feature. If this check box is selected, the new loft feature will be tangent to the faces that are adjacent to the planar face selected as the end section of the loft.

Angle

The **Angle** spinner is used to specify the angle of the end section of the loft, see Figure 8-25. This spinner will not be available when the **Tangent to Adjacent Face** check box is selected.

Weight

The **Weight** spinner is used to specify the distance up to which the loft feature maintain the angle or the tangency with the adjacent faces of the planar face at the end section before it blends with the other loft sections, see Figure 8-25.

Preview

The **Preview** button is used to preview the loft feature before it is actually created.

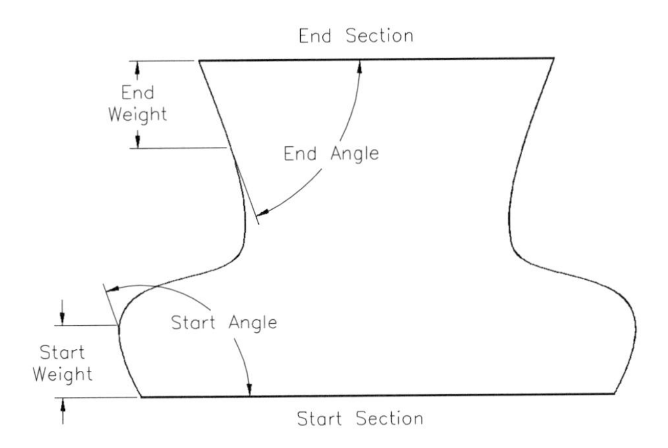

Figure 8-25 Various parameters associated with loft

TUTORIALS

Tutorial 1

In this tutorial you will create the designer model shown in Figure 8-26a. The dimensions to be used are given in Figures 8-26b, 8-26c and 8-26d. Assume the missing dimensions.

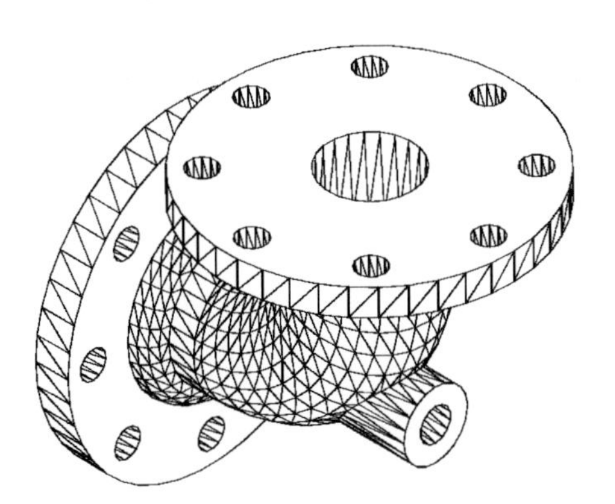

Figure 8-26a Designer model for Tutorial 1

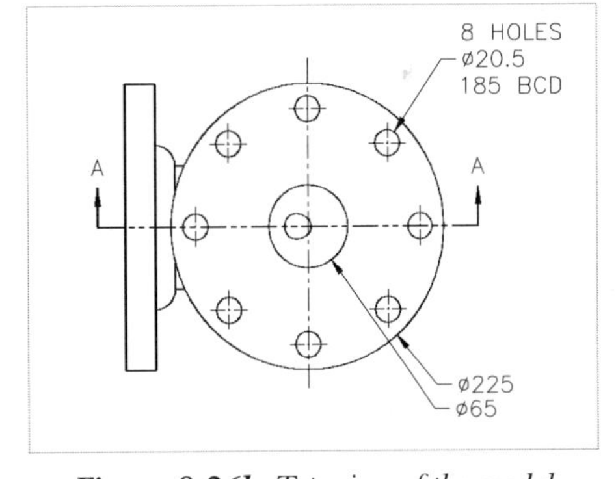

Figure 8-26b Top view of the model

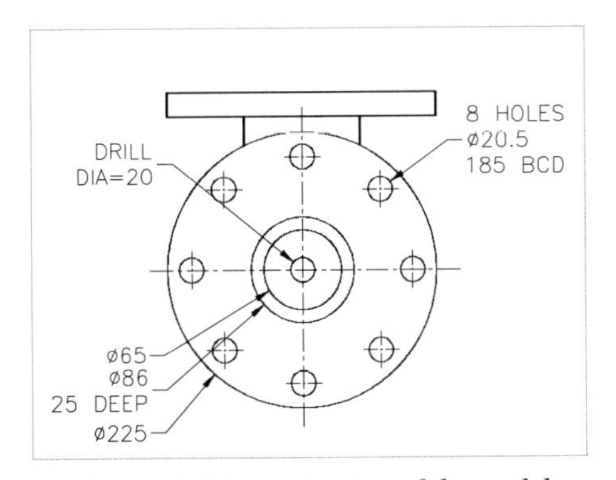

Figure 8-26c Side view of the model

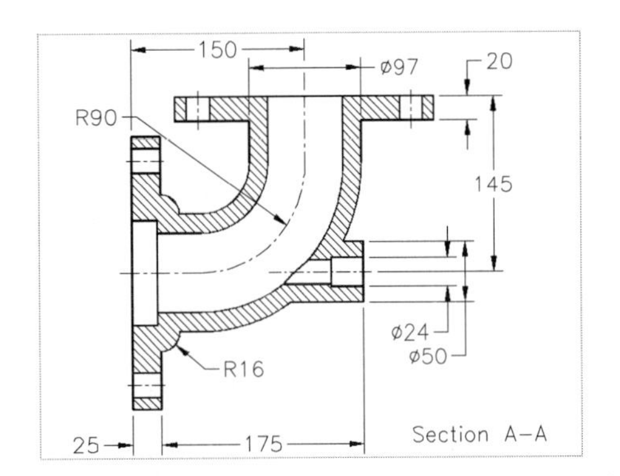

Figure 8-26d Sectioned front view of the model

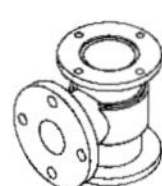

Note

In the top view and the side view of the model, the hidden lines are suppressed for clarity.

1. Increase the limits according to the limits of the model. Define a work plane on world ZX plane. Create rough sketch of the 2D path.

2. Choose the **2D Path** button from the **Profile a Sketch** flyout in the **Part Modeling** toolbar. The prompt sequence is as follows:

Select objects: *Select the path.*
Select objects: Enter
Select start point of the path: *Specify the start point of the path.*
Solved under constrained sketch requiring 3 dimensions or constraints.
Create a profile plane perpendicular to the path? [Yes/No] <Yes>: Enter
Select edge to align X axis or [Z-flip/Rotate] <Accept>: *Align the X axis and then press ENTER.*

3. Dimension the path using the **New Dimension** button from the **Power Dimensioning** flyout in the **Part Modeling** toolbar. The path after dimensioning is shown in Figure 8-27.

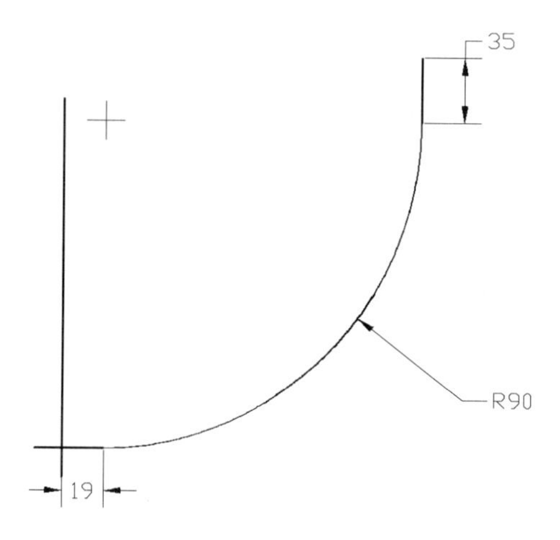

Figure 8-27 *Fully dimensioned 2D path*

4. Enter **9** at the Command prompt.

5. Draw a circle of **97** diameter. Profile the sketch and then dimension it.

6. Choose the **Sweep** button from the **Sketched Features-Extrude** flyout in the **Part Modeling** toolbar to display the **Sweep** dialog box, see Figure 8-28.

7. Select **Normal** from the **Body type** drop-down list. Set the value of the **Draft Angle** spinner to **0**. Select **Path-Only** from the **Termination** drop-down list and then press **OK**. The swept model should look similar to the one shown in Figure 8-29.

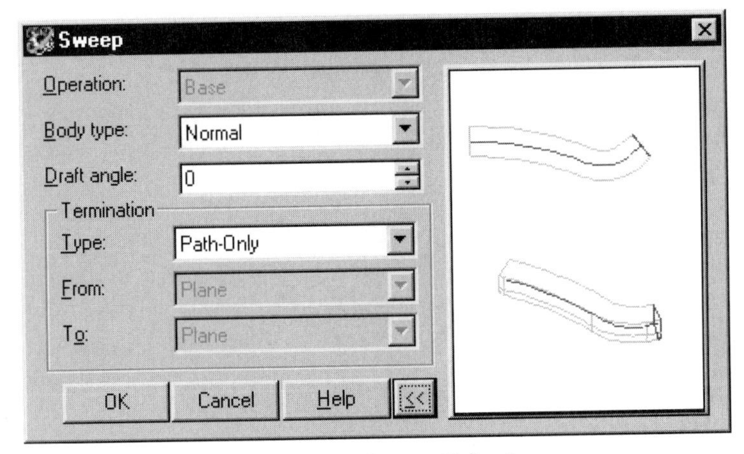

Figure 8-28 *Sweep dialog box*

8. Create the shell feature with 16 wall thickness.

9. Create the next feature on the swept feature. Now, place the counter bore in this extruded feature. The model should now look similar to the one shown in Figure 8-30.

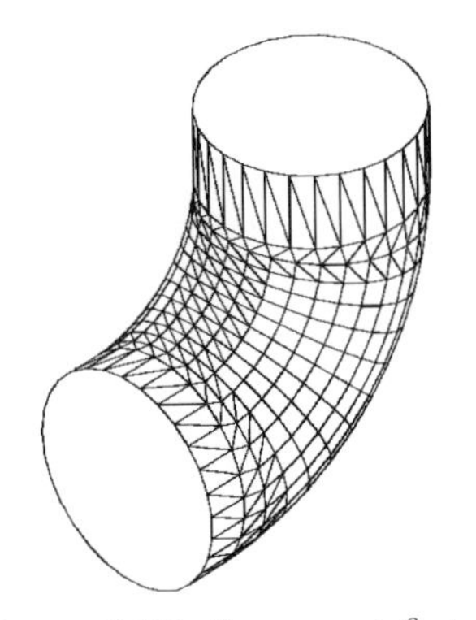

Figure 8-29 *Base sweep feature*

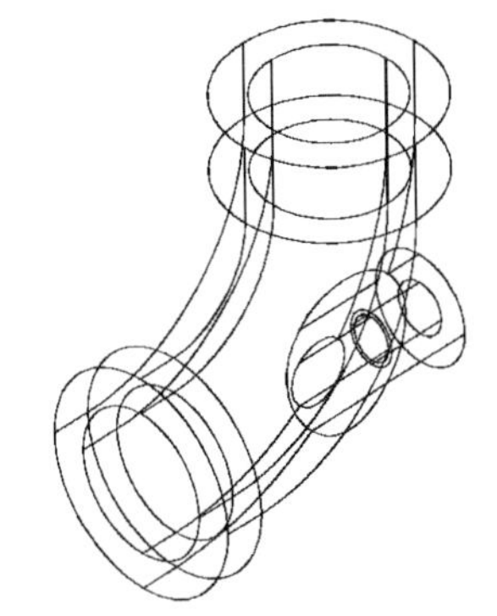

Figure 8-30 *Designer model after creating the shell, extrusion and hole features*

10. Create the remaining features on the base sweep. The final designer model should be similar to the one shown in Figure 8-31.

11. Save this drawing with the name given below:

\MDT Tut\Ch-8**Tut1.dwg**

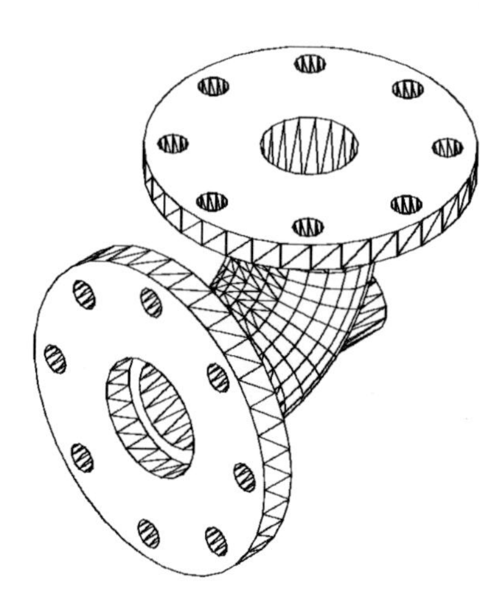

Figure 8-31 Final designer model for Tutorial 1

Tutorial 2

In this tutorial you will create the threaded bolt as shown in Figure 8-32a. The dimensions to be used are shown in Figure 8-32b. Assume the missing dimension.

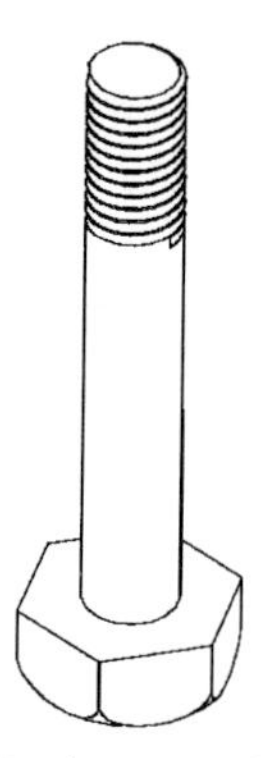

Figure 8-32a Designer model for Tutorial 2

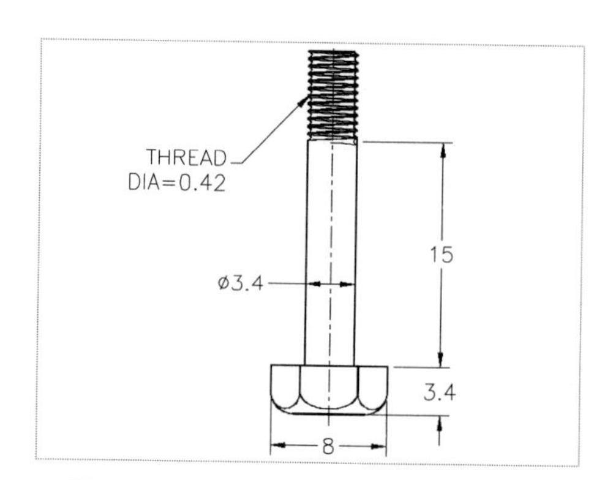

Figure 8-32b Front view of the model

1. Create the base of the bolt as shown in Figure 8-33.

2. Create the next joined feature of 3.4 units diameter and height as 15 units. Define the sketch plane on the top face of the new feature and create another joined feature of 3.4 units diameter and height as 6 units.

3. Right-click in the drawing area to display the context menu. Choose **Sketch Solving > 3D Helix Path**. The prompt sequence is as follows:

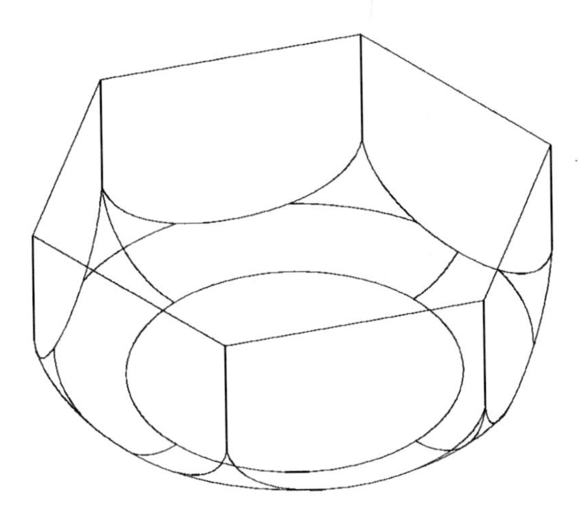

Figure 8-33 *The base of the bolt*

Select work axis, circular edge, or circular face for helical center: *Select the new feature as shown in Figure 8-34.*

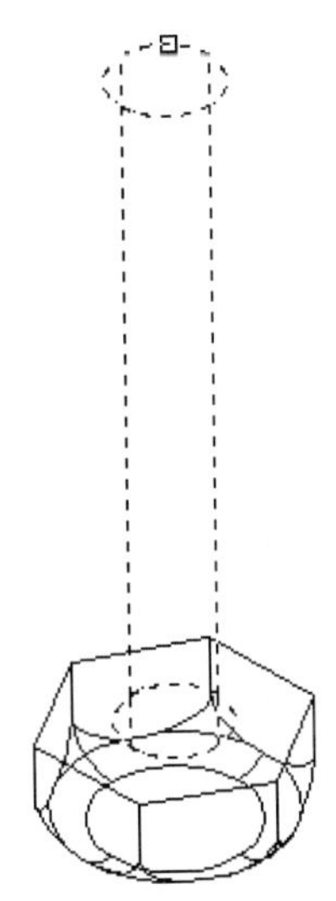

Figure 8-34 *Selecting the feature for helical path*

4. When you select the feature, the **Helix** dialog box will be displayed as shown in Figure 8-35.

5. Select **Pitch and Revolution** from the **Type** area. Set the value of the **Revolution** spinner to **13**.

6. Set the value of the **Pitch** spinner to **0.5**.

7. Set the value of the **Diameter** spinner under the **Shape** area to **3.4**.

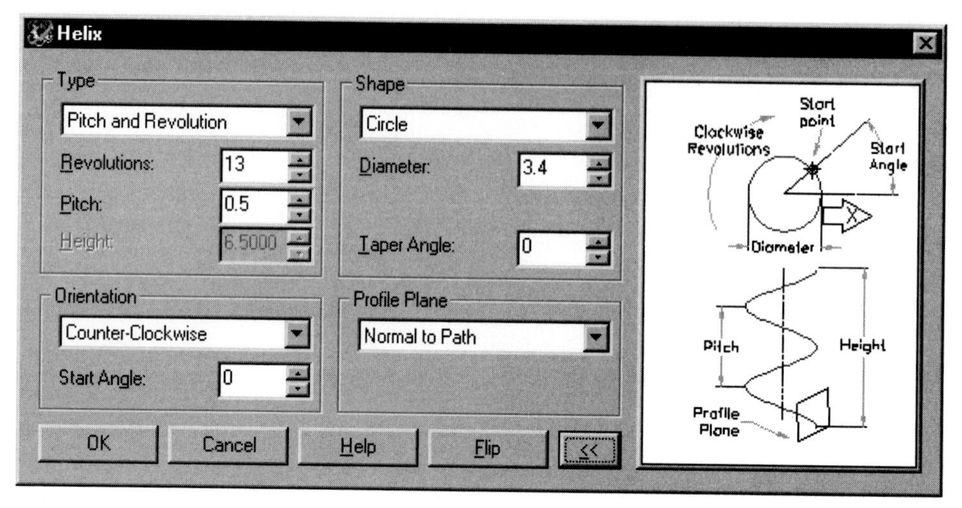

Figure 8-35 *Helix dialog box*

8. Select **Normal to Path** from the drop-down list provided under the **Profile Plane** area. Choose **OK**. A helical path will be created on the model as shown in Figure 8-36.

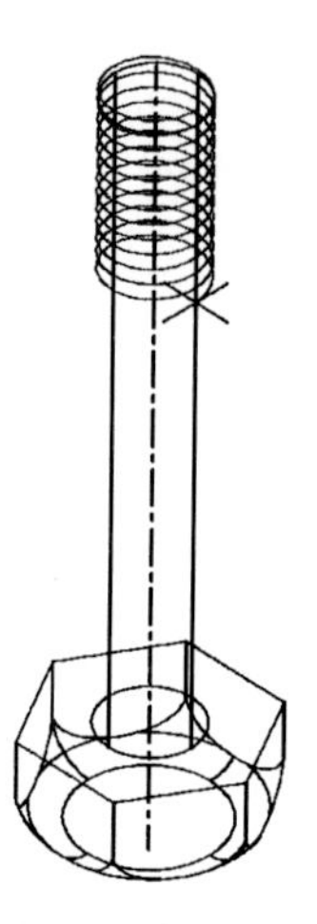

Figure 8-36 *The helical path*

9. Create a circle of 0.42 radius at the start point of the path. Profile it and then dimension it.

10. Choose the **Sweep** button from the **Sketched Features-Extrude** flyout in the **Part Modeling** toolbar to display the **Sweep** dialog box.

11. Select **Cut** from the **Operation** drop-down list and **Path-Only** from the **Termination** drop-down list. Choose **OK**. The final designer model should be similar to the one shown in Figure 8-37.

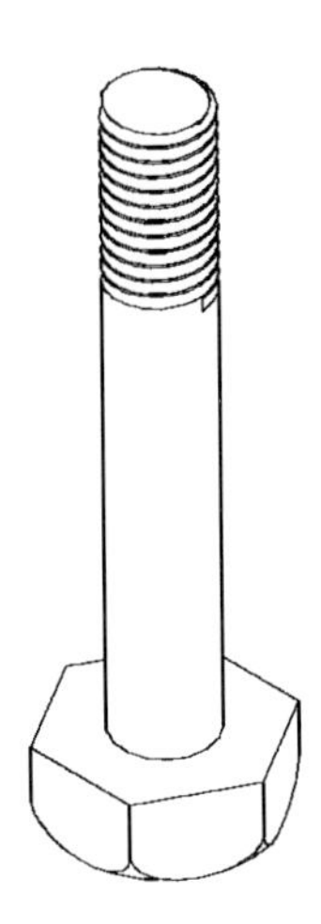

Figure 8-37 *Final designer model for Tutorial 2*

12. Save this drawing with the name given below:

 \MDT Tut\Ch-8**Tut2.dwg**

Tip: *The number of revolutions should be one more than the actual required number of revolutions. The reason for this is that the threads will be created coming out of the model only if the revolutions are one more than the required number.*

Tutorial 3

In this tutorial you will create the designer model shown in Figure 8-38a. The dimensions to be used are given in Figures 8-38b and 8-38c. Assume the missing dimensions.

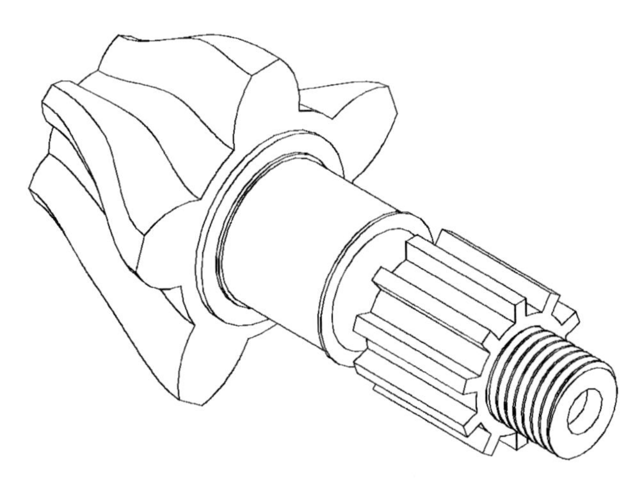

Figure 8-38a *Designer model for Tutorial 3*

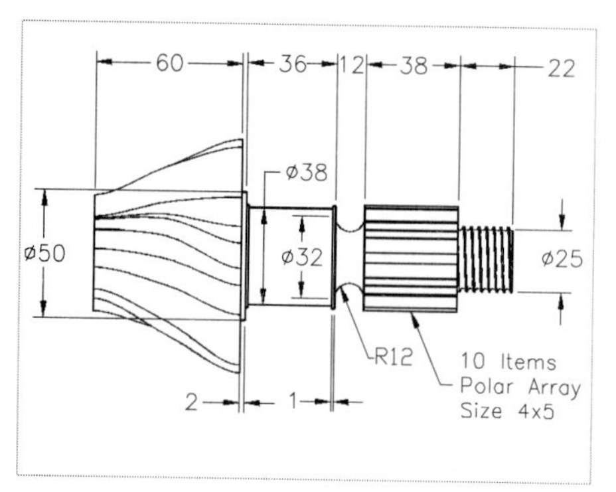

Figure 8-38b *Dimensions for Tutorial 3*

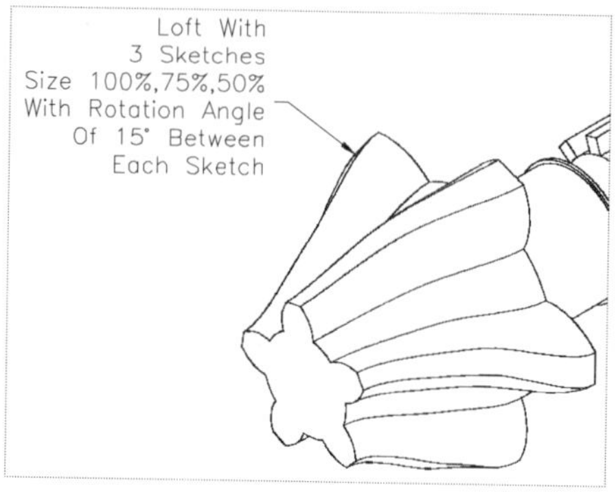

Figure 8-38c *Dimensions for Tutorial 3*

1. Create the base cylindrical feature of 25 units diameter and with 22 units extrusion height. Right-click to display the context menu and choose **Sketch Solving > 3D Helix Path**. The prompt sequence is as follows:

   Select work axis, circular edge, or circular face for helical center: *Select the cylindrical feature.*

2. Select **Pitch and Revolution** from the drop-down list.

3. Set the value of the **Revolution** spinner to **9** and the value of the **Pitch** spinner to **3**.

4. Set the value of the **Diameter** spinner to **25**. Select **Normal to Path** from the drop-down list provided under the **Profile Plane** area. Choose **OK**. The prompt sequence is as follows:

   Select edge to align X axis or [Z-flip/Rotate] <Accept>: *Align the X axis of the plane.*

5. Create the threading using the helical path, see Figure 8-39.

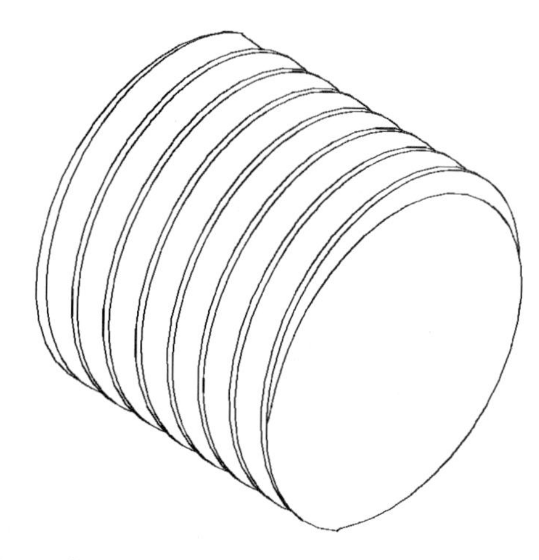

Figure 8-39 *Base feature after creating the threads*

6. Create the remaining features except the loft feature as shown in Figure 8-40.

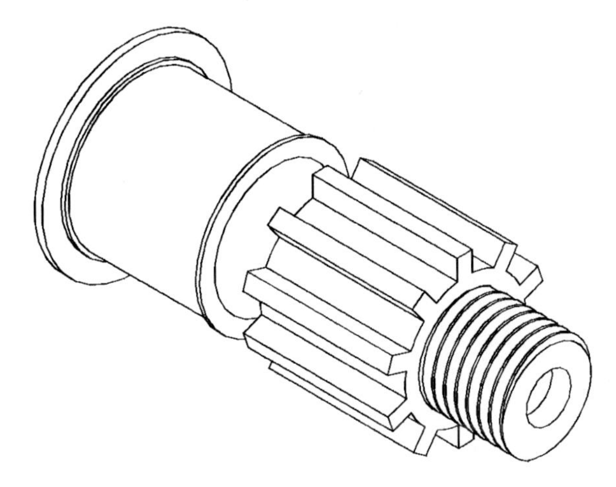

Figure 8-40 *Designer model with various features*

7. Create three sketches on different work planes placed at an offset of 30. Assume some relative dimensions for all three sketches. The rotation angle between the adjacent sketches should be 15 degrees, see Figure 8-41.

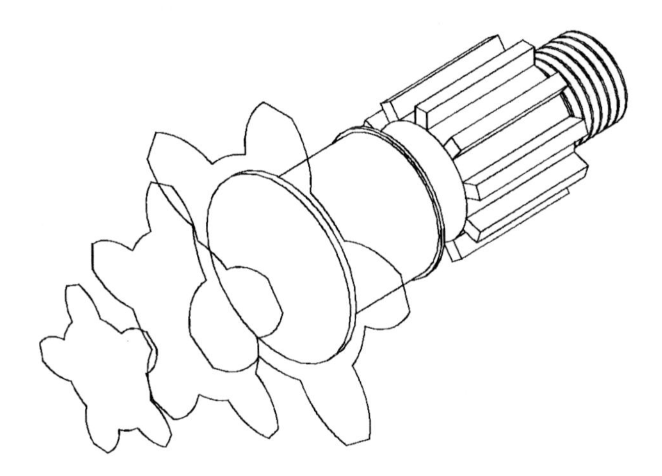

Figure 8-41 *The sketches for lofting*

8. Choose the **Loft** button from the **Sketched Features-Extrude** flyout in the **Part Modeling** toolbar. The prompt sequence is as follows:

Select profiles or planar faces to loft: *Select the first sketch.*
Select profiles or planar faces to loft or [Redefine sections]: *Select the second sketch.*
Select profiles or planar faces to loft or [Redefine sections]: *Select the third sketch.*
Select profiles or planar faces to loft or [Redefine sections]: Enter

9. When you press ENTER after selecting the sketches to loft, the **Loft** dialog box will be displayed as shown in Figure 8-42. Select **Join** from the **Operation** drop-down list.

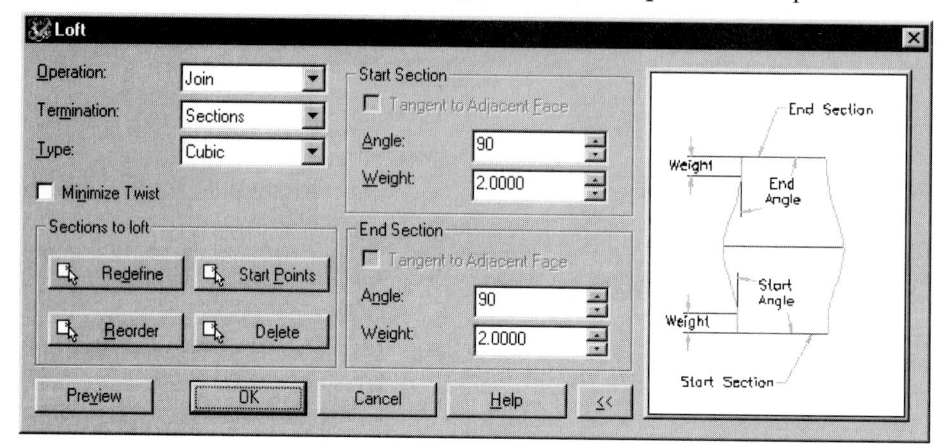

Figure 8-42 *Loft dialog box*

10. Select **Section** from the **Termination** drop-down list. Select **Cubic** from the **Type** drop down list. Clear the **Minimize Twist** check box.

11. Set the value of the **Weight** spinners under the **Start Section** and the **End Section** area to **2**. Choose the **Start Points** button under the **Section to Loft** area. The prompt sequence is as follows:

Specify start points or [preView]: *Specify the start point on sketch 1 as shown in Figure 8-43.*
Specify start points or [preView/Restore default]: *Specify the start point on sketch 2 as shown in Figure 8-43.*
Specify start points or [preView/Restore default]: *Specify the start point on sketch 2 as shown in Figure 8-43.*
Specify start points or [preView/Restore default]: Enter

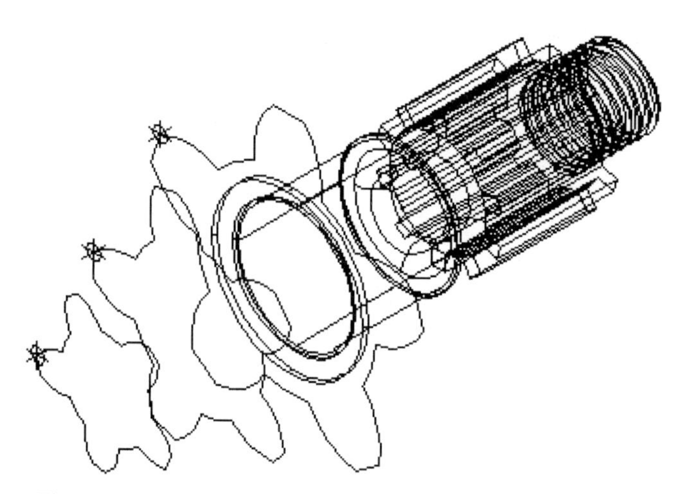

Figure 8-43 *Defining the starting point of loft on the sketches*

14. The **Loft** dialog box will be displayed on the screen again. Choose **OK**.

15. The final designer model should look similar to the one shown in Figure 8-44.

16. Save this drawing with the name given below:

    \MDT Tut\Ch-8**Tut3.dwg**

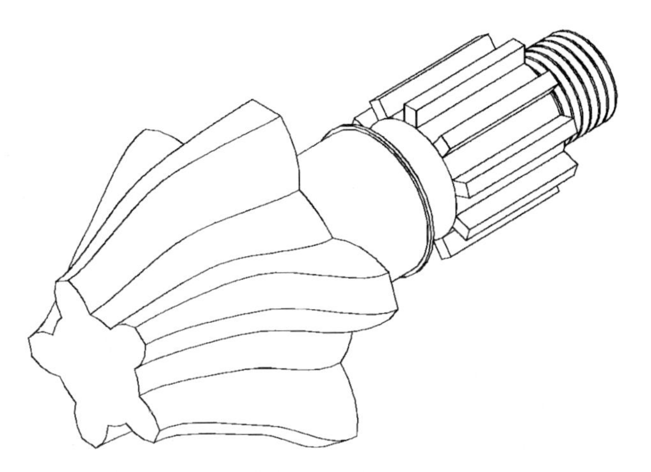

Figure 8-44 *Final designer model for Tutorial 3*

Review Questions

Answer the following questions.

1. Which command is used to convert an open entity into a 2D path?

2. What are the different types of 3D paths that can be created using the **AM3DPATH** command?

3. What are the two types of helical paths that can be created using the **Helix** dialog box?

4. What is the difference between the pipe path and the spline path?

5. Which command is used to sweep the profiles along the 3D paths?

6. What is the difference between sweeping the profile parallel to path and normal to path?

7. Different types of sketches can be blended together using the _________ command.

8. What is the difference between cubic loft and closed cubic loft?

Exercise

Exercise 1

In this exercise you will create the designer model given in Figure 8-45a. The dimensions to be used are given in Figures 8-45b, 8-45c and 8-45d. Assume the missing dimensions.

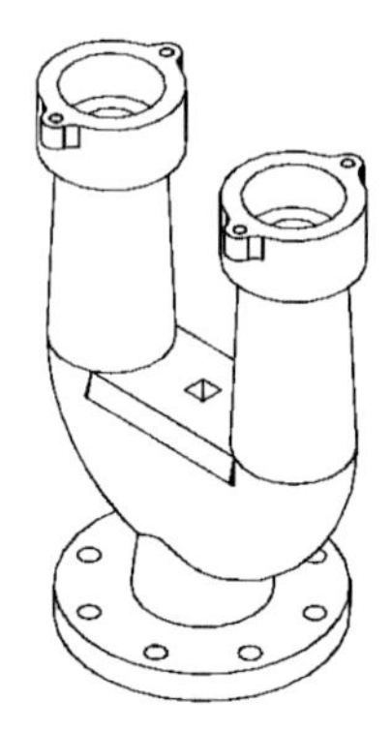

Figure 8-45a *Model for Exercise 1*

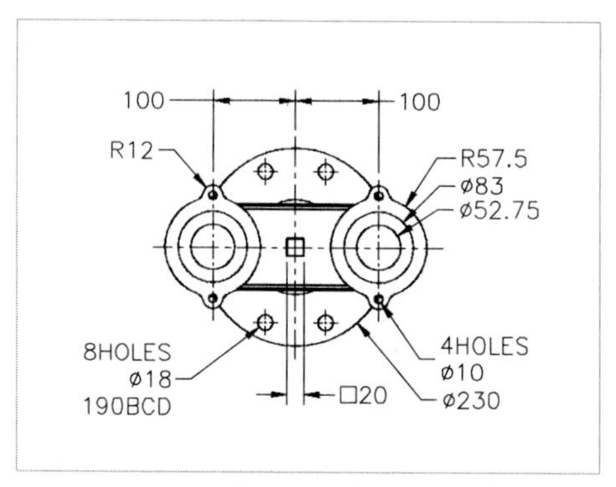

Figure 8-45b *Top view of the model*

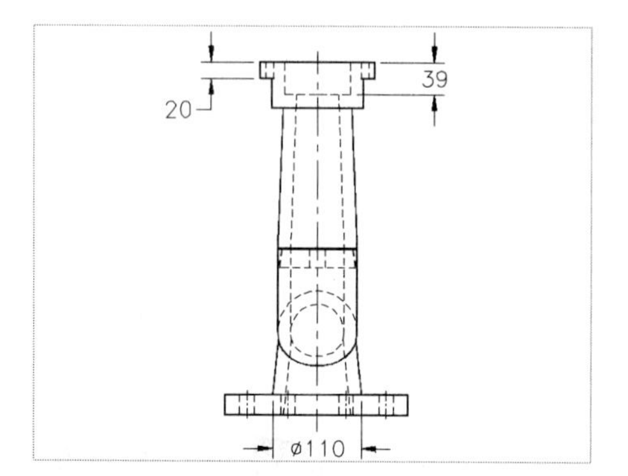

Figure 8-45c *Side view of the model*

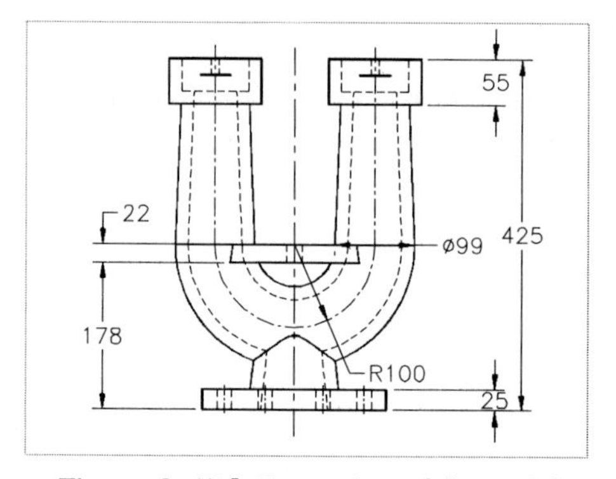

Figure 8-45d *Front view of the model*

Chapter 9

Creating New Parts, Activating the Part, and Mirroring and Combining the Parts

Learning Objectives

After completing this chapter, you will be able to:

- *Create an instance of an existing part.*
- *Convert a 3D solid into a designer model.*
- *Create a new designer model.*
- *Activate the designer model.*
- *Mirror the designer models.*
- *Perform boolean operations on different designer models.*

Commands Covered

- *AMNEW*
- *AMACTIVATE*
- *AMMIRROR*
- *AMCOMBINE*

AMNEW COMMAND

The **AMNEW** command is a very versatile command. This command has the following uses:

1. Creating a copy of an existing designer model.
2. Converting a 3D solid into a designer model.
3. Creating a new designer model.

Creating Copies Of An Existing Designer Part Using The AMNEW Command

Toolbar:	Assembly Modeling > Instance
Menu:	Part > Part > Instance
Context Menu:	Part > Instance
Command:	AMNEW

You can create a replica or an instance of an existing part by selecting the **Instance** option of the **AMNEW** command. The new instance created will be a separate part. However, the initial part will still be the active part. When you select the **Instance** option, you will be prompted to specify the insertion point for placing the new part.

Creating Copies Of An Existing Part Using The Desktop Browser

You can also create the copies of the existing part using the desktop browser. Right-click on the part you want to copy in the tree view of the desktop browser to display the shortcut menu and choose **Copy** as shown in Figure 9-1. You will be prompted to specify the insertion point for the new part.

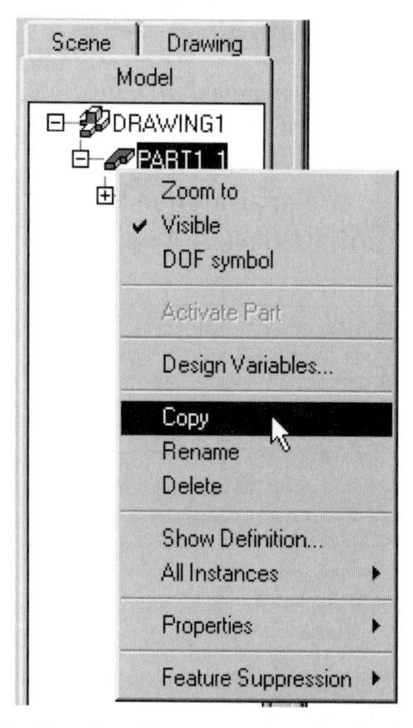

Figure 9-1 *Copying the part using the desktop browser*

Converting A 3D Solid Model Into A Designer Model Using The AMNEW Command

Toolbar: Part Modeling > New Part
Menu: Part > Part > Convert Solid
Command: AMNEW

A solid model created in other software like AutoCAD can be converted into a designer model. This can be done by using the **Part** option of the **AMNEW** command. When you select the **Part** option, you will be prompted to select the 3D model or enter the name of the new part. Instead of specifying the name of the new part, select the AutoCAD 3D solid to be converted into designer model. Once you have selected the 3D solid, you will be prompted to specify the name of the designer part created using the 3D part.

Converting A 3D Solid Model Into A Designer Model Using The Desktop Browser

You can also convert a 3D model into a designer part using the desktop browser. Right-click in the desktop browser to display the shortcut menu and choose **New Part**, see Figure 9-2. You will be prompted to select the 3D model or specify the name of the part. Instead of specifying the name of the part, select the 3D model to be converted into designer part.

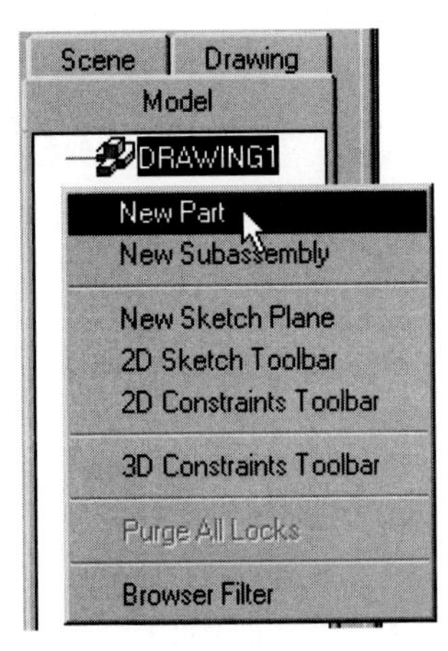

Figure 9-2 *Converting a 3D model into designer model using the desktop browser*

Creating A New Part Using The AMNEW Command

Toolbar: Part Modeling > New Part
Menu: Part > Part > New Part
Context Menu: Part > New Part
Command: AMNEW

Sometimes, you may need to create more than one part in a single drawing. This is done using the **Part** option of the **AMNEW** command. When you invoke this option, you will be prompted to specify the name of the new part.

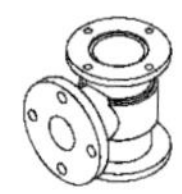

Note

The latest part created will be the active designer part.

Tip: *To create a new part using the desktop browser, follow the same procedure as for converting the 3D model into a designer model. The only difference is that you have to specify the name of the new part in the* ***Select an object or enter new part name*** *prompt.*

ACTIVATING THE PART

If the drawing consists of more then one part than the last part created will be the active part. You can work only on the active part, which, in this case will be the last part. However, if you want to work on some other part, you will have to make that part active. This can be done using the **AMACTIVATE** command or by using the desktop browser.

Activating The Part Using The AMACTIVATE Command

Toolbar:	Part Modeling > New Part > Activate Part
Menu:	Part > Part > Activate Part
Context Menu:	Part > Activate Part
Command:	AMACTIVATE

This command is used to activate the part if the drawing consists of more than one part. You can either select the part to be made active or specify the name of that part at the Command prompt.

Activating The Part Using The Desktop Browser

To activate a part using the desktop browser, right-click on the part to be made active to display the shortcut menu. Choose **Activate Part**, as shown in Figure 9-3.

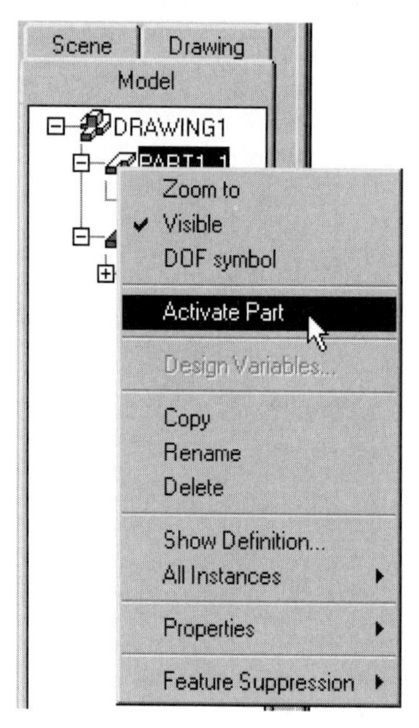

Figure 9-3 *Activating part using the desktop browser*

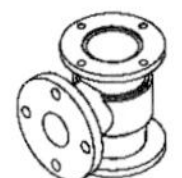

Note

*The **Scene** and the **subAssembly** options of the **AMNEW** command will be discussed in later chapters.*

MIRRORING THE PARTS

You can also create a new part by creating mirror image of an existing part. This is done using the **AMMIRROR** command that is discussed below.

AMMIRROR Command

Toolbar:	Part Modeling > New Part > Mirror Part
Menu:	Part > Part > Mirror Part
Context Menu:	Part > Mirror Part
Command:	AMMIRROR

This command is used to create a mirror image of an existing designer part. However, the mirror image created will be a separate designer part consisting of all the features available in the initial part. You can mirror the part about a planar face, a work plane, or a line. The line can be defined by specifying two points but both the points of the line should lie in the XY plane of the current sketch plane. The new part created by mirroring will be the active part. The options provided under this command are:

Create new part

This option is used to create a new part by mirroring the existing part as shown in Figure 9-4. You will be asked to specify the name of the new part.

Replace instances

This option is used to create a new part that is a mirror image of an existing part. But in this case the original part will be deleted.

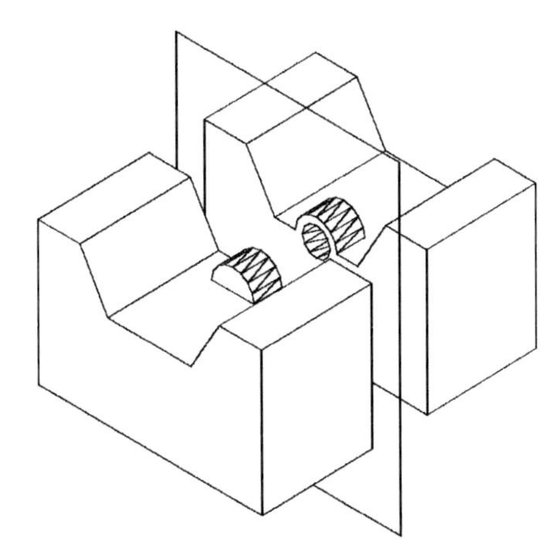

Figure 9-4 *New designer model created by mirroring the existing part about a work plane*

Tip: *If the new part is created by mirroring an existing part about a work plane, then in that case even the work plane will get mirrored. Therefore, now there will be two work planes, one above the other.*

PERFORMING BOOLEAN OPERATIONS ON THE DESIGNER MODELS

You can perform the boolean operations like union, subtract, or intersect on the designer models. This is done using the **AMCOMBINE** command discussed below.

AMCOMBINE Command

Toolbar:	Part Modeling > Placed Features-Hole > Combine
Menu:	Part > Placed Features > Combine
Context Menu:	Placed Features > Combine
Command:	AMCOMBINE

This command is used to perform boolean operations on different designer models. The valid boolean operations that can be performed are union, subtract and intersect. All these boolean operations performed on the designer models will be parametric in nature and can be edited at any point of time. The options provided under this command are:

Cut

This option is similar to the subtract boolean operation. In this operation the material common to the active part and selected parts are removed from the active part. Figure 9-5 shows the models before applying boolean operations and Figure 9-6 shows the model after applying the boolean operation.

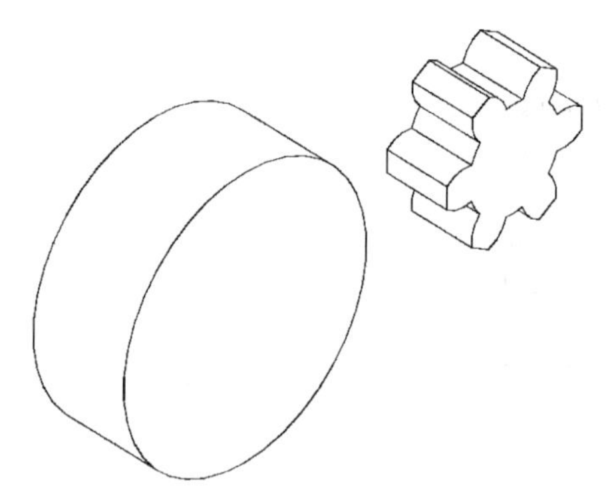

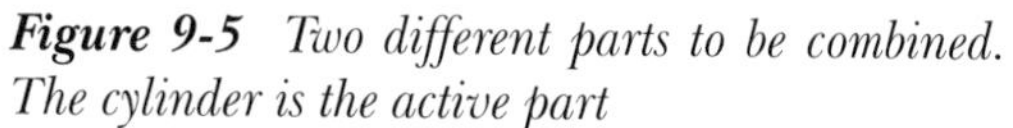

Figure 9-5 *Two different parts to be combined. The cylinder is the active part*

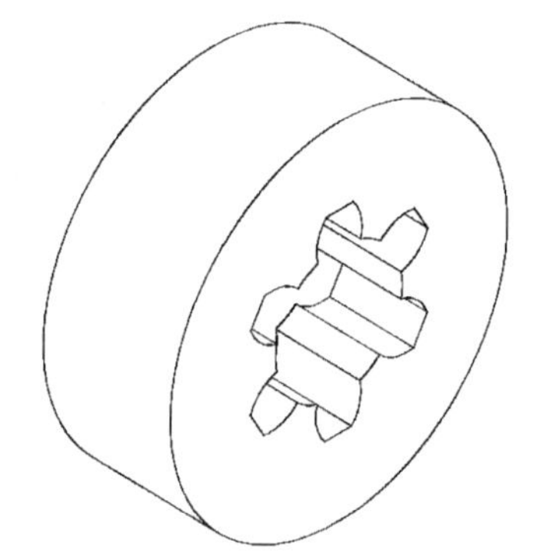

Figure 9-6 *Designer model created after combining both the parts using the Cut option*

Intersect

This option is used to create a new part by retaining the material common to the active part and the selected part. In this case the resultant part does not depend upon the active part and therefore, any part among the selected parts can be active, see Figures 9-7 and 9-8.

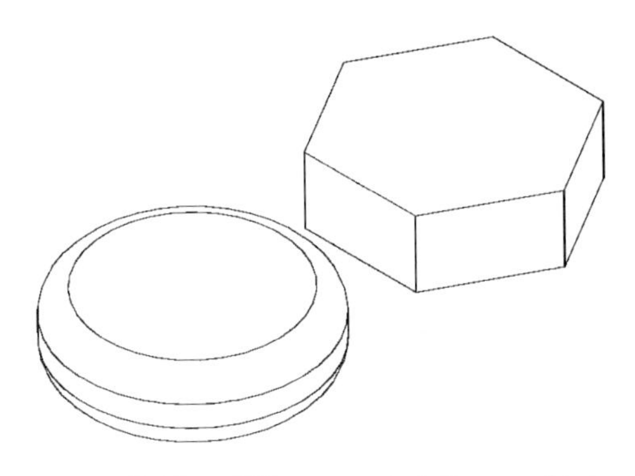

Figure 9-7 *Two different parts before combining*

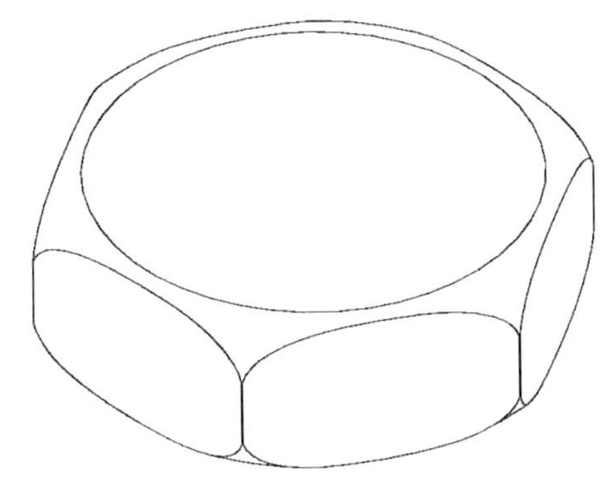

Figure 9-8 *Designer part created by combining the two selected parts using the Intersect option*

Join

This operation is similar to the union boolean operation. In this operation the selected parts are joined with the active part thus converting all of them into a single part. However, this has to be clarified here that only one part can be joined with the active part at a time. Figure 9-9 shows the designer models before joining and Figure 9-10 shows the designer model created after joining.

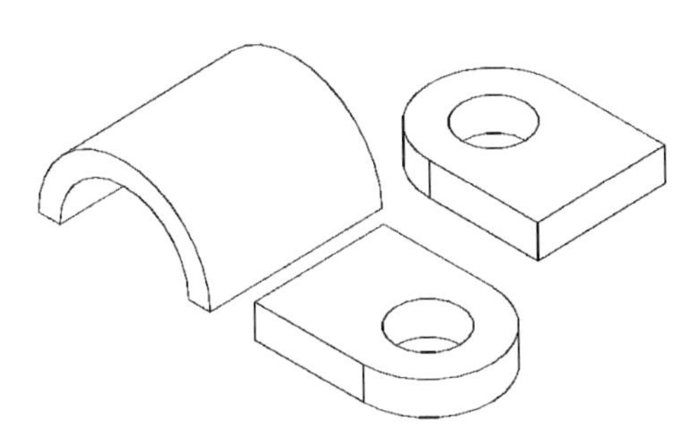

Figure 9-9 *The different parts to be combined*

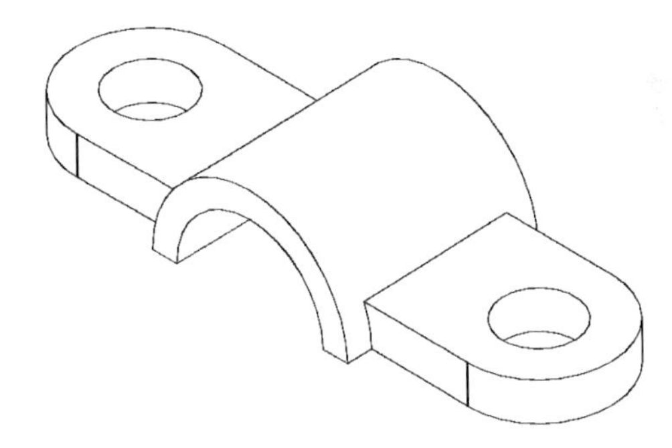

Figure 9-10 *The new designer model created by combining different parts using the join option*

TUTORIALS

Tutorial 1

In this tutorial you will create the bolt, nut and the lock nut for the Plummer Block as shown in the following figures. All three of them should be separate parts. The dimensions to be used are given in Figures 9-11a, 9-11b, 9-11c and 9-11d.

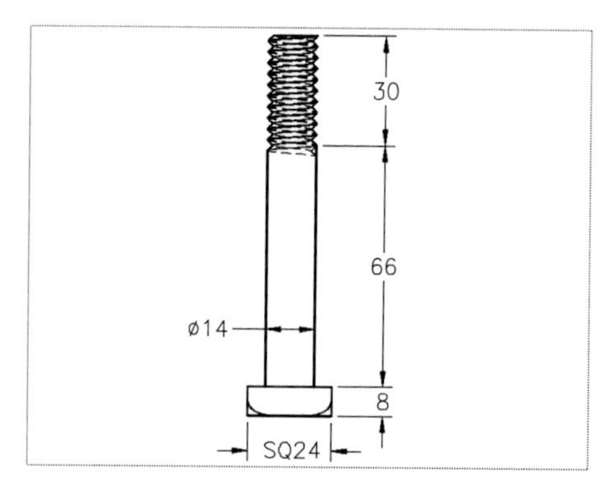

Figure 9-11a The bolt for Tutorial 1

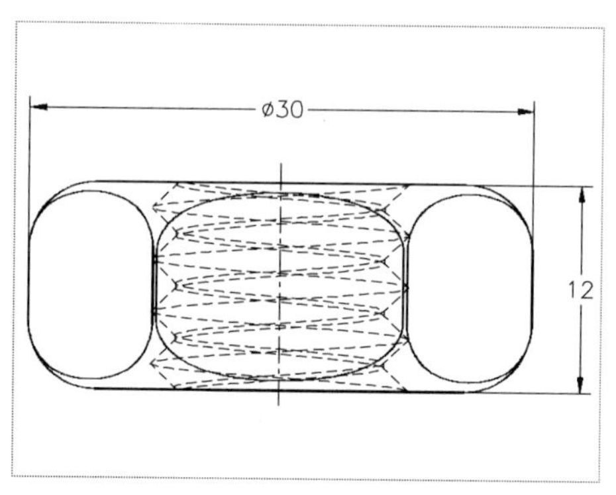

Figure 9-11b The nut for Tutorial 1

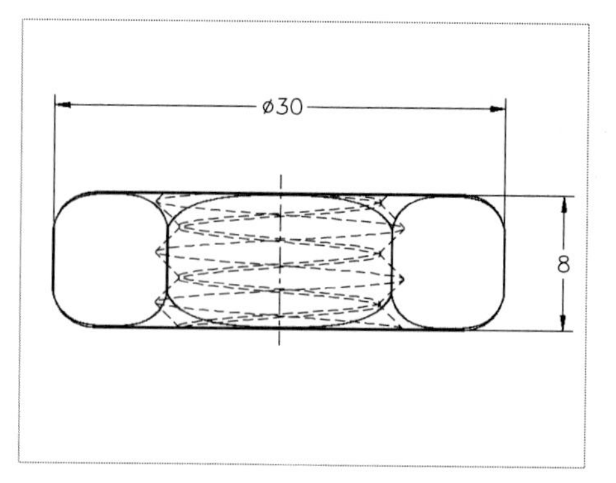

Figure 9-11c The lock nut for Tutorial 1

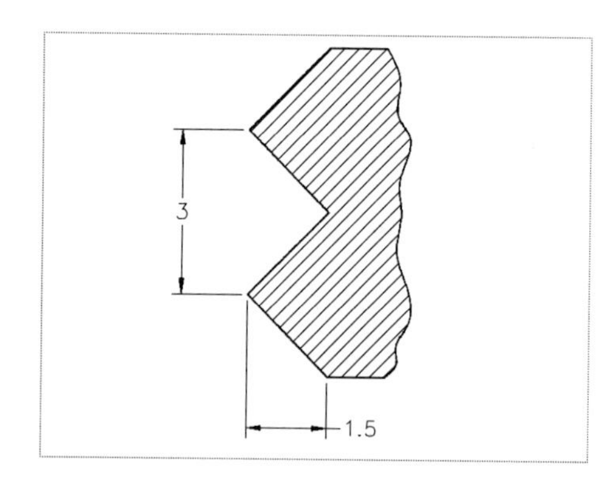

Figure 9-11d The specifications for the threading

1. Create a square with the length of 24 units. Extrude it up to a height of 8 units.

2. Choose the **New Part** button from the **Part Modeling** toolbar. The prompt sequence is as follows:

   Select an object or enter new part name <default name>: **Part 2**

3. Create a cylindrical part with the height of 8 units. Fillet lower circular face of the cylindrical feature. Move it inside the square part, see Figure 9-12.

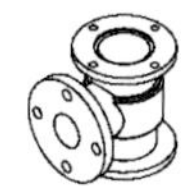

Note

The diameter of the cylindrical feature should be equal to the length of the diagonal of the square of the first part.

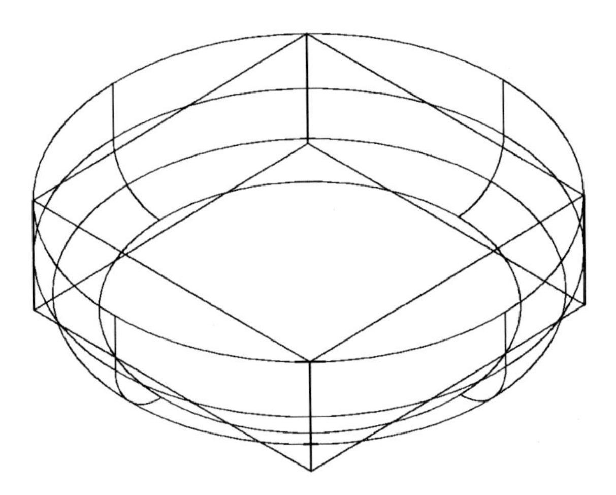

Figure 9-12 *After aligning both the parts*

4. Choose the **Combine** button from the **Placed Features-Hole** flyout in the **Part Modeling** toolbar. The prompt sequence is as follows:

   Enter parametric boolean operation [Cut/Intersect/Join] <Cut>: **I**
   Select part (toolbody) to use for intersecting: *Select the square part.*

5. The part created after combining is shown in Figure 9-13.

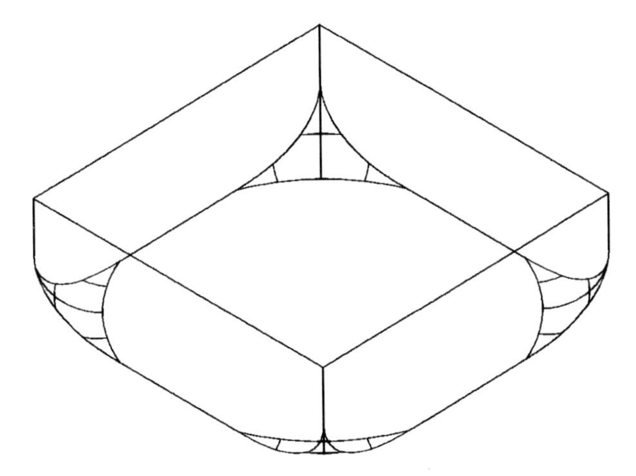

Figure 9-13 *The part created after combining*

6. Create the remaining features on the base feature to create the bolt as shown in Figure 9-14.

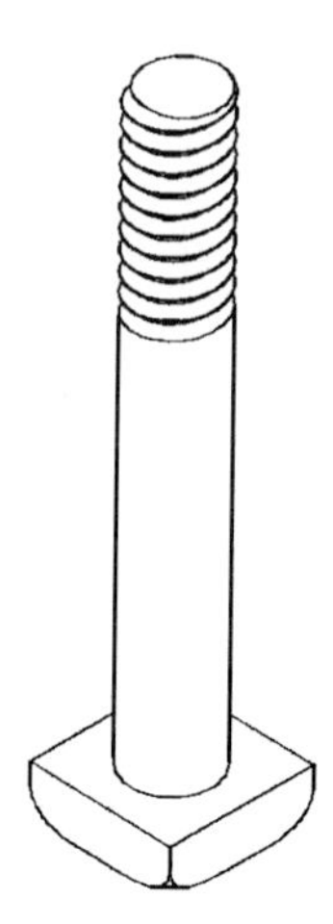

Figure 9-14 The bolt for the Plummer Block

7. Choose the **New Part** button from the **Part Modeling** toolbar. The prompt sequence is as follows:

   Select an object or enter new part name <default name>: **Part 3**

8. Create a cylindrical part for the nut and fillet both of its circular faces.

9. Choose the **New Part** button from the **Part Modeling** toolbar. The prompt sequence is as follows:

   Select an object or enter new part name <default name>: **Part 4**

10. Create a hexagonal part. The radius for the circle to inscribe the hexagon should be same as that of the cylindrical part.

11. Move the cylindrical part inside the hexagonal part.

12. Choose the **Combine** button from the **Placed Features-Hole** flyout in the **Part Modeling** toolbar. The prompt sequence is as follows:

   Enter parametric boolean operation [Cut/Intersect/Join] <Cut>: **I**
   Select part (toolbody) to use for intersecting: *Select the cylindrical part.*

13. Create the internal threading in the nut using the helical path see Figure 9-15.

Tip: *To create internal threading, sketch the work axis using the* ***Sketch*** *option of the* ***AMWORKAXIS*** *command. The length of the work axis should be more than the length of hole and the start point of the axis should be above the start point of hole. Use this work axis for defining the helical path.*

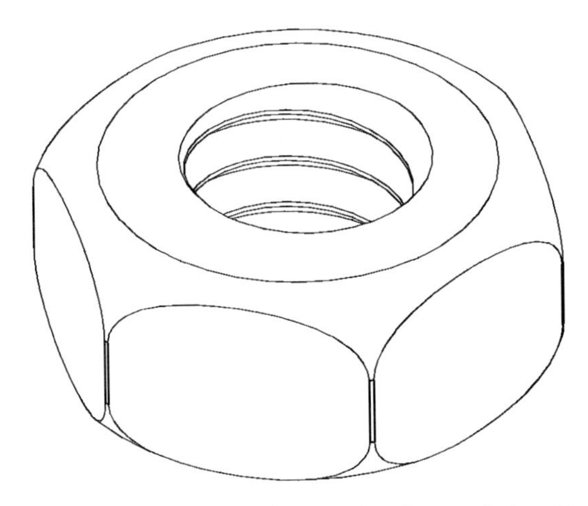

Figure 9-15 *The nut after creating internal threading*

14. Similarly, create the lock nut as shown in Figure 9-16.

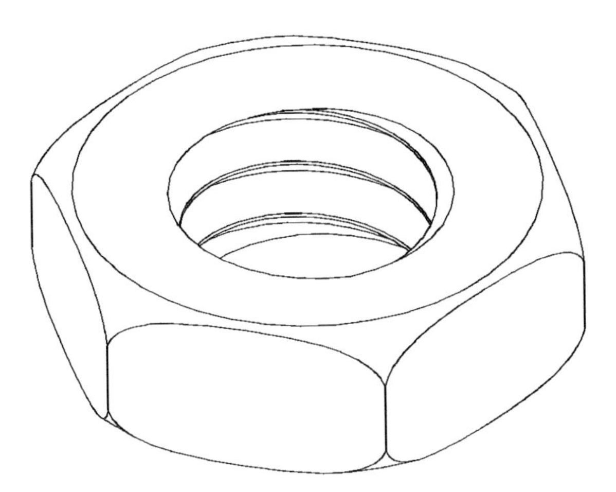

Figure 9-16 *The lock nut*

15. Save this drawing with the name given below:

    MDT Tut\Ch-9**Tut1.dwg**

Tutorial 2

In this tutorial you will create the model shown in Figure 9-17a. The dimensions to be used are given in Figures 9-17b, 9-17c and 9-17d.

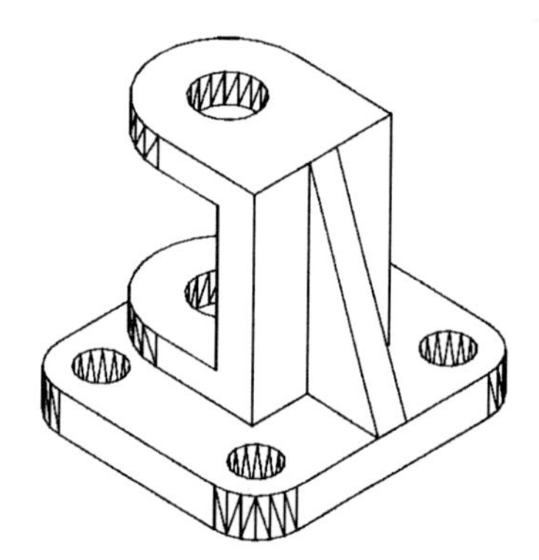

Figure 9-17a Model for Tutorial 2

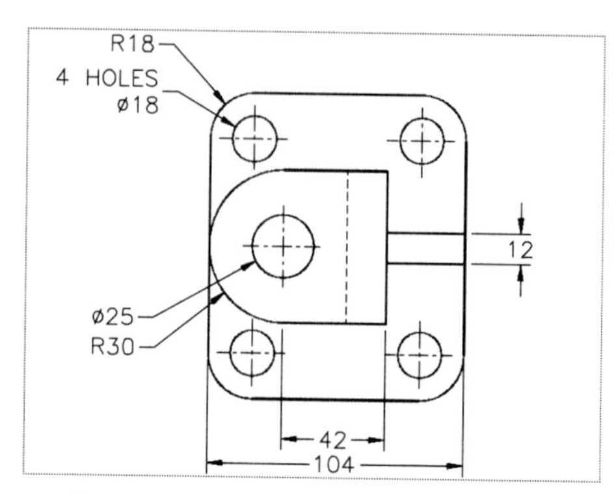

Figure 9-17b Top view of the model

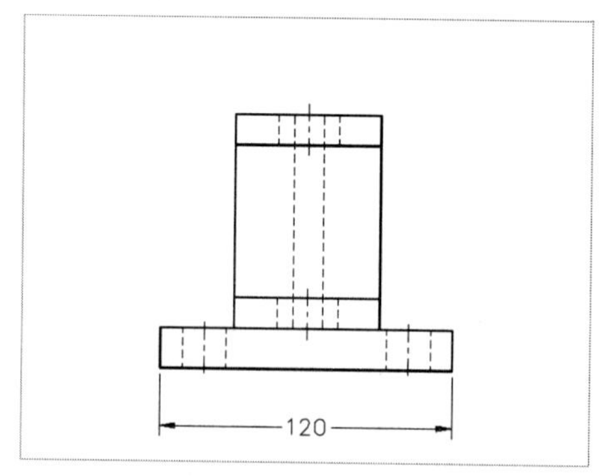

Figure 9-17c Side view of the model

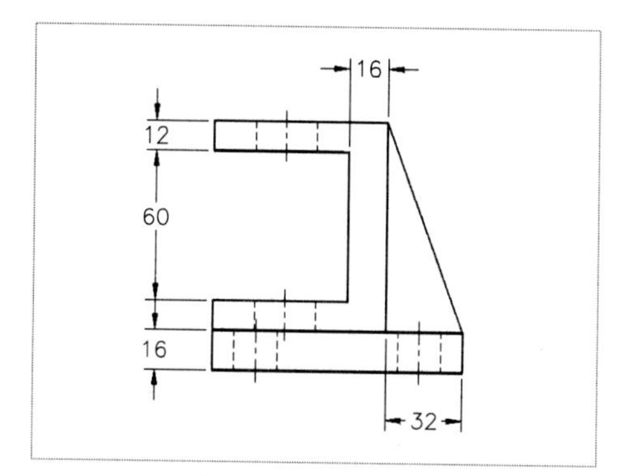

Figure 9-17d Front view of the model

1. Create the base feature for the model and then place the holes as shown in Figure 9-18.

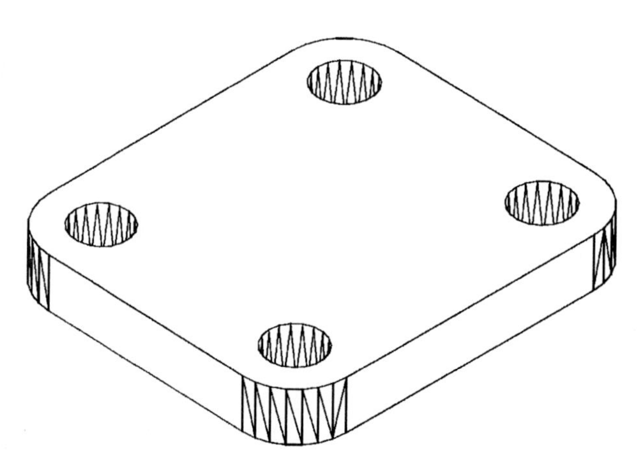

Figure 9-18 Base feature for Tutorial 2

2. Choose the **New Part** button from the **part Modeling** toolbar. The prompt sequence is as follows:

Select an object or enter new part name <default name>: **Part 2**

3. Create the next part and then place a drilled hole in it as shown in Figure 9-19.

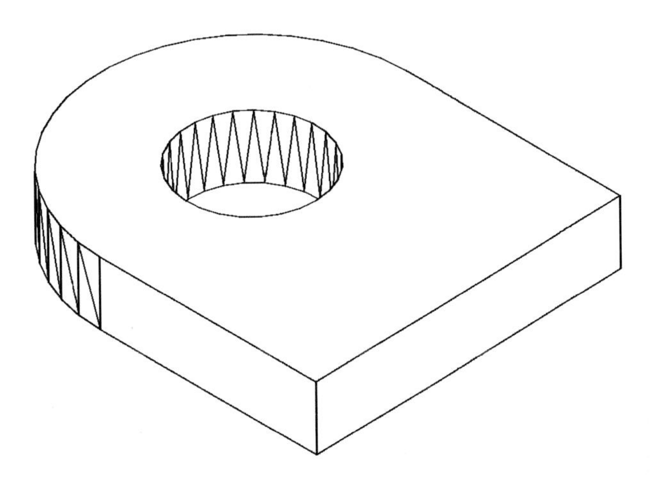

Figure 9-19 *Part 2 after creating the drilled hole*

4. Choose the **Mirror Part** button from the **New Part** flyout in the **Part Modeling** toolbar. The prompt sequence is as follows:

Select part to mirror: *Select Part 2.*
Select planar face to mirror about or [Line]: *Select the planar face as shown in Figure 9-20.*
Enter an option [Next/Accept] <Accept>: Enter
Enter an option [Create new part/Replace instances] <Create new part>: Enter
Enter new part name <default name>: **Part 3**

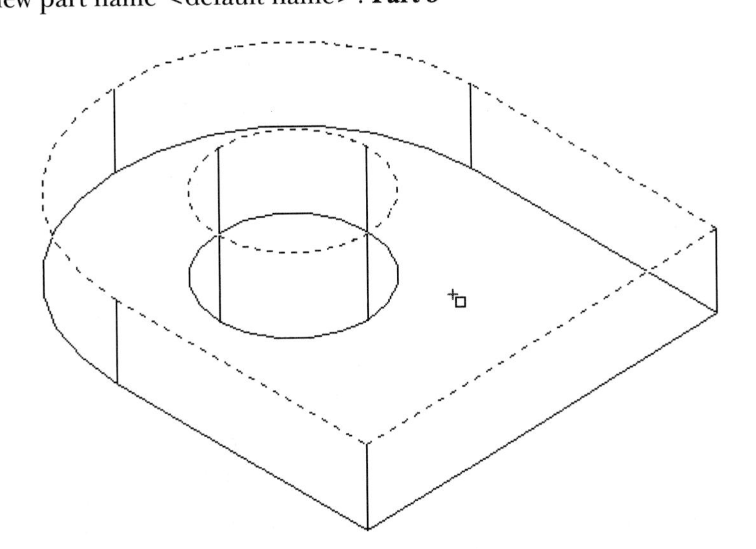

Figure 9-20 *Selecting the face for mirroring the part*

5. Choose the **New Part** button from the **Part Modeling** toolbar. The prompt sequence is as follows:

Select an object or enter new part name <default name>: **Part 4**

6. Create **Part 4** and then align **Part 2** and **Part 3** with **Part 4**, see Figure 9-21.

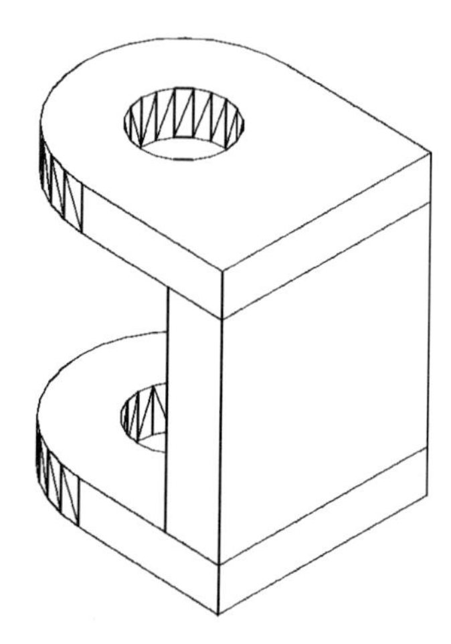

Figure 9-21 *Part 2 and 3 aligned with Part 4*

7. Choose the **Combine Part** button from the **Placed Features-Hole** flyout in the **Part Modeling** toolbar. The prompt sequence is as follows:

Enter parametric boolean operation [Cut/Intersect/Join] <Cut>: **J**
Select part (toolbody) to be joined: *Select Part 2*
Computing ...

8. Similarly, combine **Part 3**. Now combine this part with Part 1.

9. Choose the **New Part** button from the **Part Modeling** toolbar. The prompt sequence is as follows:

Select an object or enter new part name <default name>: **Part 5**

10. Create **Part 5** which is a wedge.

11. Choose the **Combine** button from the **Placed Features-Hole** flyout in the **Part Modeling** toolbar. The prompt sequence is as follows:

Enter parametric boolean operation [Cut/Intersect/Join] <Cut>: **J**
Select part (toolbody) to be joined: *Select Part 5.*
Computing ...

12. Now combine **Part 1** with this part. The final designer model should look similar to the one shown in Figure 9-22.

13. Save this drawing with the name Given below:

    \MDT Tut\Ch-9**Tut2.dwg**

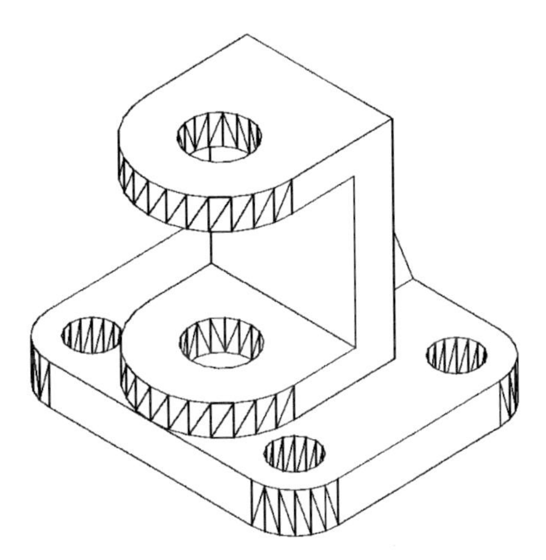

Figure 9-22 *Final designer model for Tutorial 2*

Tutorial 3

In this tutorial you will create the designer model shown in Figure 9-23a. The dimensions to be used are given in Figures 9-23b and 9-23c.

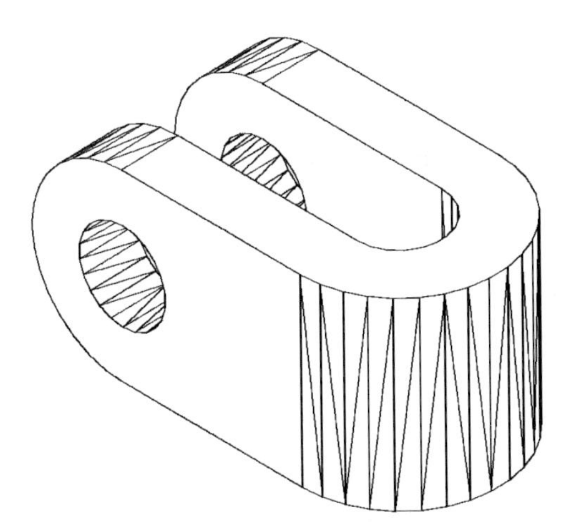

Figure 9-23a *Designer model for Tutorial 3*

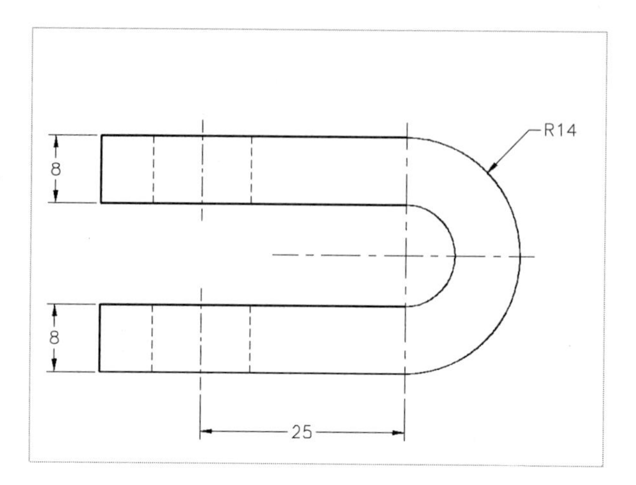

Figure 9-23b *Top view of the model*

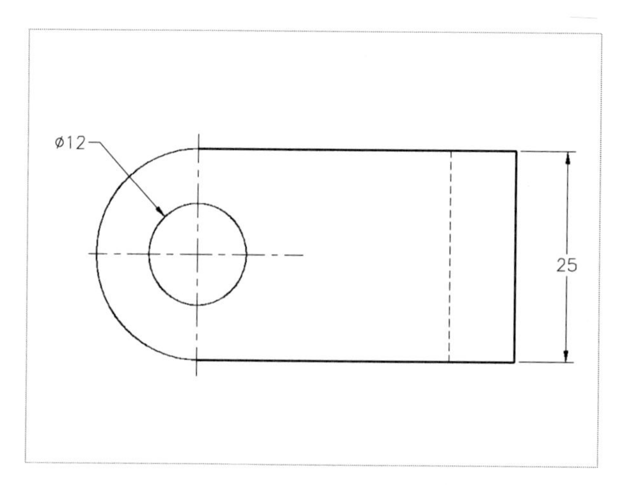

Figure 9-23c *Front view of the model*

1. Create the base feature for the model as shown in Figure 9-24.

2. Choose the **New Part** button from the **Part Modeling** toolbar. The prompt sequence is as follows:

   Select an object or enter new part name <default name>: **Part 2**

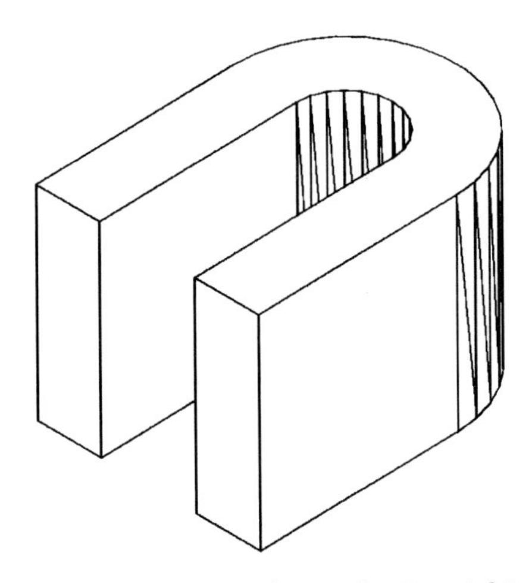

Figure 9-24 *Base feature for Tutorial 3*

3. Create the next part as shown in Figure 9-25.

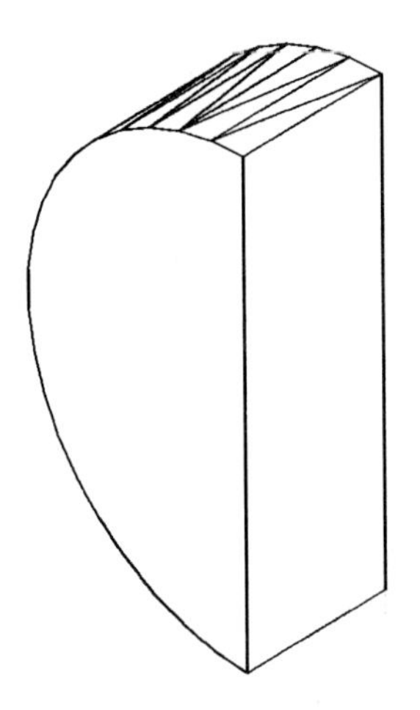

Figure 9-25 *Second part for Tutorial 3*

4. Choose the **Mirror Part** button from the **New Part** flyout in the **Part Modeling** toolbar. The prompt sequence is as follows:

Select part to mirror: *Select the second part.*
Select planar face to mirror about or [Line]: *Select the planar face as shown in Figure 9-26.*
Enter an option [Create new part/Replace instances] <Create new part>: Enter
Enter new part name <default name>: **Part 3**

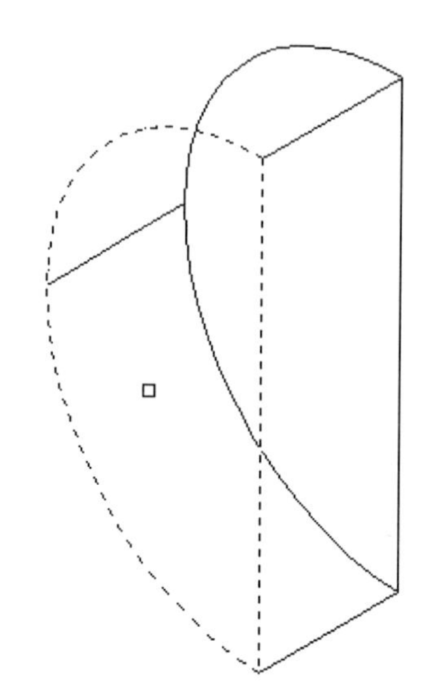

Figure 9-26 *Selecting the face to mirror the part*

5. Align **Part 2** and **Part 3** with **Part 1** as shown in Figure 9-27.

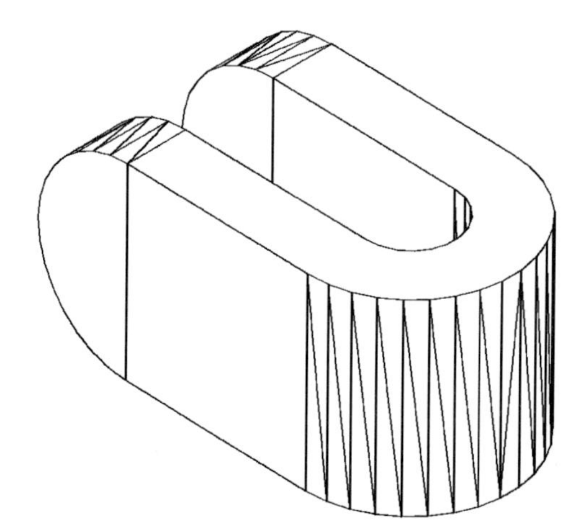

Figure 9-27 *Various parts for Tutorial 3 after aligning*

6. Choose the **Combine** button from the **Placed Features-Hole** flyout in the **Part Modeling** toolbar. The prompt sequence is as follows:

Enter parametric boolean operation [Cut/Intersect/Join] <Cut>: **J**
Select part (toolbody) to be joined: *Select Part 1.*
Computing ...

7. Similarly, combine **Part 2** with this part.

8. Create the drilled through hole in it. The final designer model should look similar to the one shown in Figure 9-28.

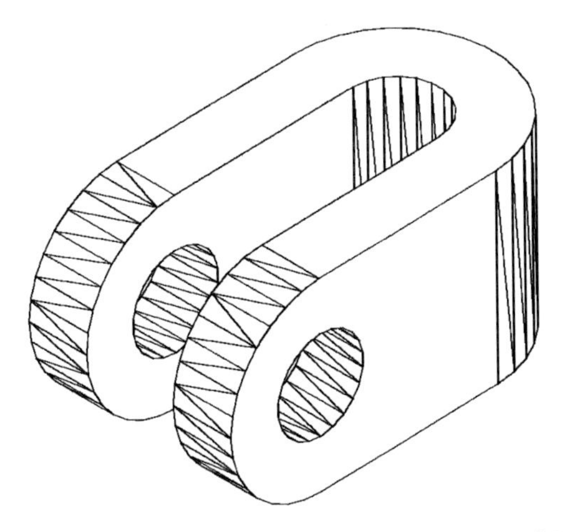

Figure 9-28 *Final designer model for Tutorial 3*

9. Save this drawing with the name given below:

   \MDT Tut\Ch-9**Tut3.dwg**

Review Questions

Answer the following questions.

1. Which option of the **AMNEW** command is used to create a copy of an existing part?

2. Which option of the **AMNEW** command is used to create a new part?

3. The solid models created in AutoCAD software can be converted into designer models. (T/F)

4. The last part created is the active part. (T/F)

5. What is the difference between the **Create new part** and **Replace instances** options of the **AMMIRROR** command?

Exercise

Exercise 1

Create the designer model shown in Figure 9-29a using the combination of **AMCOMBINE** and **AMMIRROR** commands. The dimensions to be used are given in Figures 9-29b and 9-29c. Assume the missing dimensions.

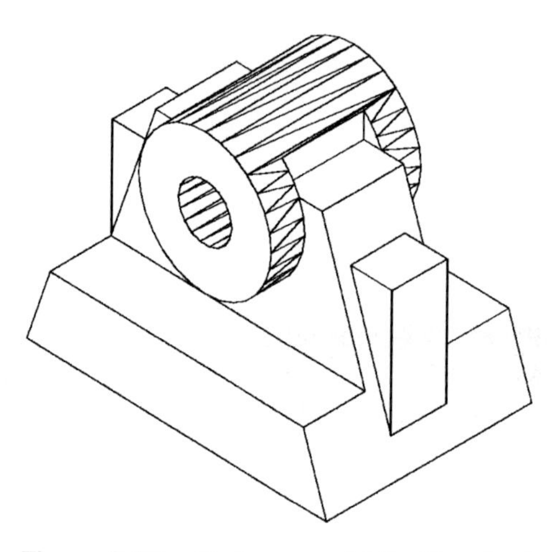

Figure 9-29a *Designer model for Exercise 1*

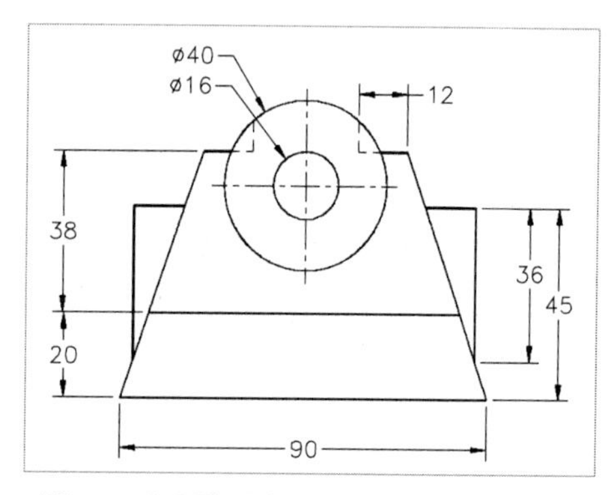

Figure 9-29b *Dimensions for Exercise 1*

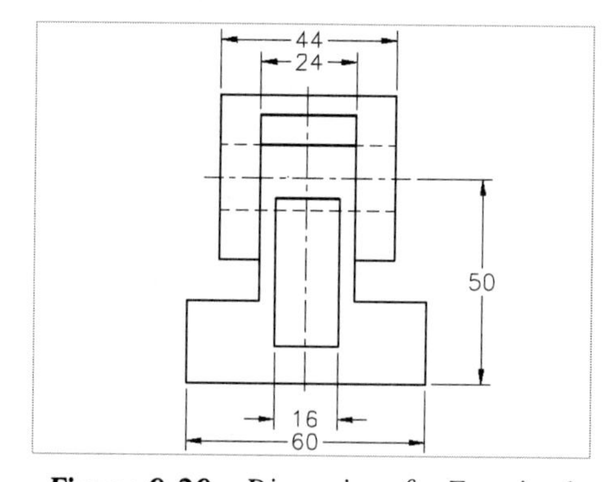

Figure 9-29c *Dimensions for Exercise 1*

Chapter 10

Assembly Modeling I

Learning Objectives

After completing this chapter, you will be able to:

- *Create various assembly parts in same drawing or in different drawings.*
- *Attach various assembly parts created in other drawings to the current drawing.*
- *Assemble different parts by applying the assembly constraints.*
- *Edit the assembly constraints.*

Commands Covered

- ***AMCOPYOUT***
- ***AMCOPYIN***
- ***AMCATALOG***
- ***AMMATE***
- ***AMFLUSH***
- ***AMANGLE***
- ***AMINSERT***
- ***AMEDITCONST***

ASSEMBLY MODELING

The assemblies are the combination of a number of parametric parts placed at their actual desired locations in the assembly using the assembly constraints. These parts (also called components of assembly) can be created in same drawing or in different drawings. The assemblies created in Mechanical Desktop are dynamic in nature. This means that if any part of the assembly is modified, the changes are reflected in the assembly also. By default you work in the part modeling environment. To shift to assembly modeling environment, choose the **Assembly Modeling** button from the **Desktop Main** toolbar.

Tip: *If any part of the assembly, assembled using the assembly constraints, is moved from its original position without deleting the constraint, it is moved back to its original position when the assembly is modified.*

In Mechanical Desktop, the assemblies can be created in the following ways:

1. Creating all of the components of the assembly in same drawing.
2. Creating all of the components of the assembly in same drawing and then copying them out.
3. Creating all of the components of the assembly as separate drawings.

CREATING ALL THE COMPONENTS OF THE ASSEMBLY IN SAME DRAWING

This is the method of creating an assembly whose components are created in the same drawing as different subassemblies. These components are then assembled in the same drawing using the assembly constraints. This type of assemblies are also called the **top-down** assemblies. However, in case of large assemblies, this increases the size of the drawing, and thus requires very large space on the hard disk. The other drawback of this kind of assembly is that the complete assembly is lost if an error occurs in the drawing.

Different subassemblies are created using the **subAssembly** option of the **AMNEW** command. The subassemblies created are the part of the main assembly displayed at the top of the hierarchy in the tree view of the desktop browser. Generally, the name of the main assembly is same as that of the current drawing. Once you have created the subassembly, you need to activate it using the desktop browser. Now, using the **AMNEW** command create a new part which will be the part of the new subassembly.

Tip: *Once you activate a subassembly, all the other subassemblies become inactive. To assemble the components, you will have to activate all the subassemblies individually. You can also activate the main assembly which in turn will activate all the subassemblies.*

Activate the main assembly before creating a new subassembly. The reason for this is that if the main assembly is not activated then the new subassembly will be a part of the last active subassembly and the parts of the same subassembly are not allowed to be assembled.

CREATING ALL THE COMPONENTS IN SAME DRAWING AND THEN COPYING THEM OUT

In this method all the components of the assembly are created in same drawing but they are copied out one by one as a separate drawing. Once you have created all the components and have copied them out, you can open a new drawing and then copy the components of the assembly in and assemble all of them using the assembly constraints. The actual drawing in which all of the components were created may not be saved as all the components have been saved as separate drawings. The components can be copied out using the **AMCOPYOUT** command and can be copied in using the **AMCOPYIN** command. Both of these commands are discussed next.

AMCOPYOUT Command

Toolbar:	Assembly Modeling > Assembly Catalog > Output Part Definition
Menu:	Part > Part > Copy Out
Context Menu:	Part > Copy Out
Command:	AMCOPYOUT

This command is used to save an existing part or a subassembly as a new drawing. When you invoke this command, the **Part/Subassembly Out** dialog box is displayed as shown in Figure 10-1.

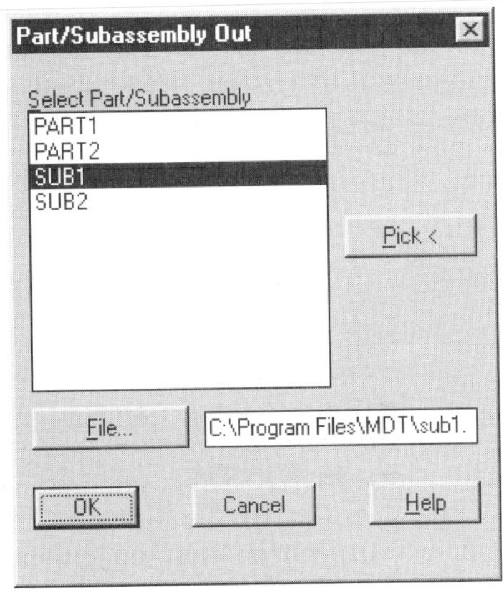

Figure 10-1 Part/Subassembly Out dialog box

Part/Subassembly Out Dialog Box Options

The options provided under this dialog box are:

Select Part/Subassembly Area

This area displays all the parts or subassemblies in the current drawing that can be copied out as a new drawing.

Pick

This button is chosen to select a part or a subassembly to be converted into a new drawing. When you choose this button, the dialog box is temporarily closed and you will be prompted to select the part from the drawing area. Once you have selected the part or the subassembly, the dialog box appears on the screen again.

File

This button is used to specify the name and the path for the new drawing. You can specify the name of the new drawing along with the path in which you want to save it in the **File** text box. You can also directly choose the **File** button to display the **Output File** dialog box. The name and the location of the new drawing can be specified in the **Output File** dialog box, see Figure 10-2.

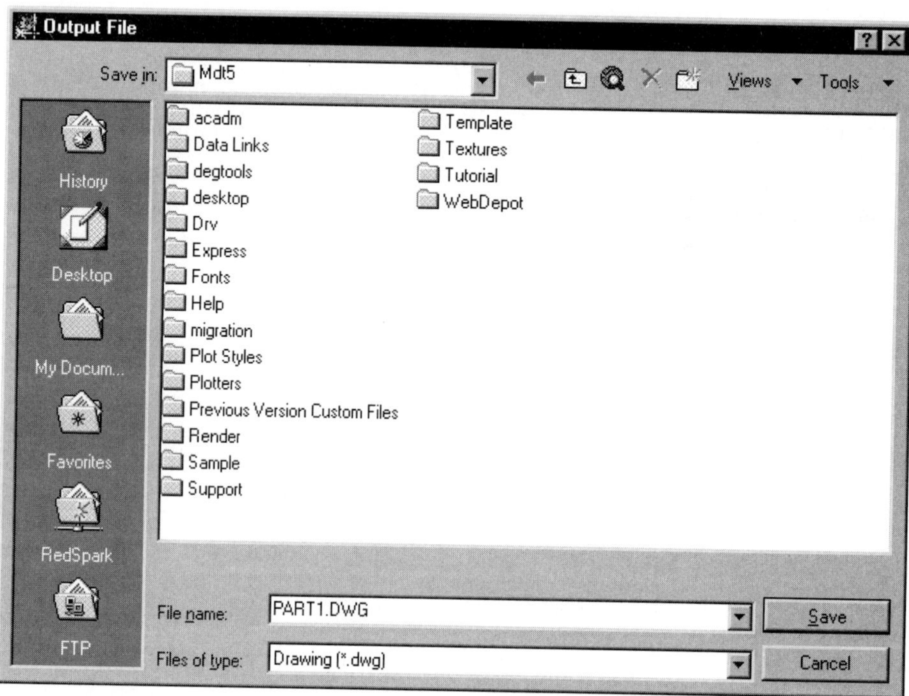

***Figure 10-2** Output File dialog box displayed upon choosing the File button*

After you have selected the part or the subassembly and specified the path and the name, choose **OK**. The **Wblock Preview** window will be displayed for a short duration. This window displays the component that is converted into a new drawing.

AMCOPYIN Command

Toolbar:	Assembly Modeling > Assembly Catalog > Input Part Definition
Menu:	Part > Part > Copy In
Context Menu:	Part > Copy In
Command:	AMCOPYIN

This command is used to copy in the parts or the subassemblies. Any drawing file saved using the **SAVE**, **SAVEAS**, **WBLOCK**, or the **AMCOPYOUT** command can be copied in using this command. When you invoke this command, the **File to Load** dialog box is displayed as shown in Figure 10-3. All the option of this dialog box are similar to those of the **Select File** dialog box displayed upon invoking the **OPEN** command.

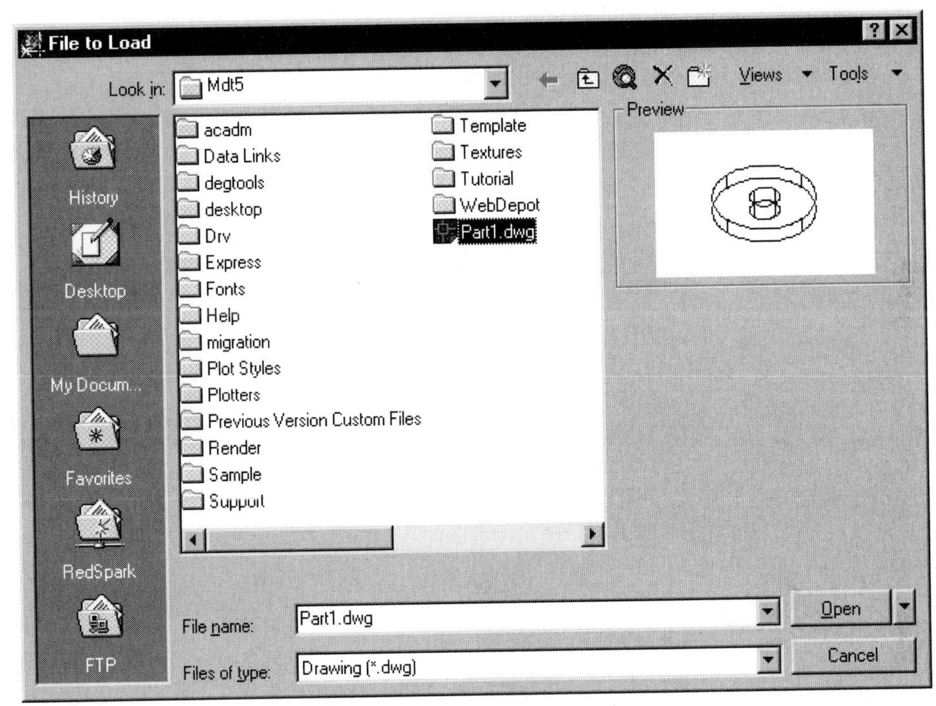

Figure 10-3 *File to Load dialog box*

All of the parts copied in using the **AMCOPYIN** command are converted into local parts. Therefore, you can make any kind of modification in the component in this particular file instead of again opening the drawing of the component and making the modification there.

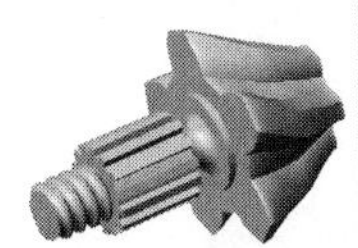

Tip: *When you copy out a part or a subassembly using the* ***AMCOPYOUT*** *command then the drawing views associated with it will not be stored in the new file. You will have to generate the drawing views again in the new file.*

CREATING ALL THE COMPONENTS OF THE ASSEMBLY AS SEPARATE DRAWINGS

This is the method of creating an assembly whose components are created separately as individual drawings. Once all the components are created, you can open a new drawing, recall all the components in this new drawing, and assemble them using the assembly constraints. These types of assemblies are also called **bottom-up** assemblies. The components can be recalled in the current drawing using the **AMCATALOG** command. This command is discussed next.

AMCATALOG Command

Toolbar:	Assembly Modeling > Assembly Catalog
Menu:	Assembly > Catalog
Context Menu:	Assembly Menu > Catalog
Command:	AMCATALOG

This command is used to recall, attach or detach the external parts or subassemblies. The parts or the subassemblies can be localized or can be kept as external parts. However, it is very important to clarify here that the dynamic property of the assemblies are retained only if the part or the subassembly is kept as an external part. When you invoke this command, the **Assembly Catalog** dialog box will be displayed as shown in Figure 10-4.

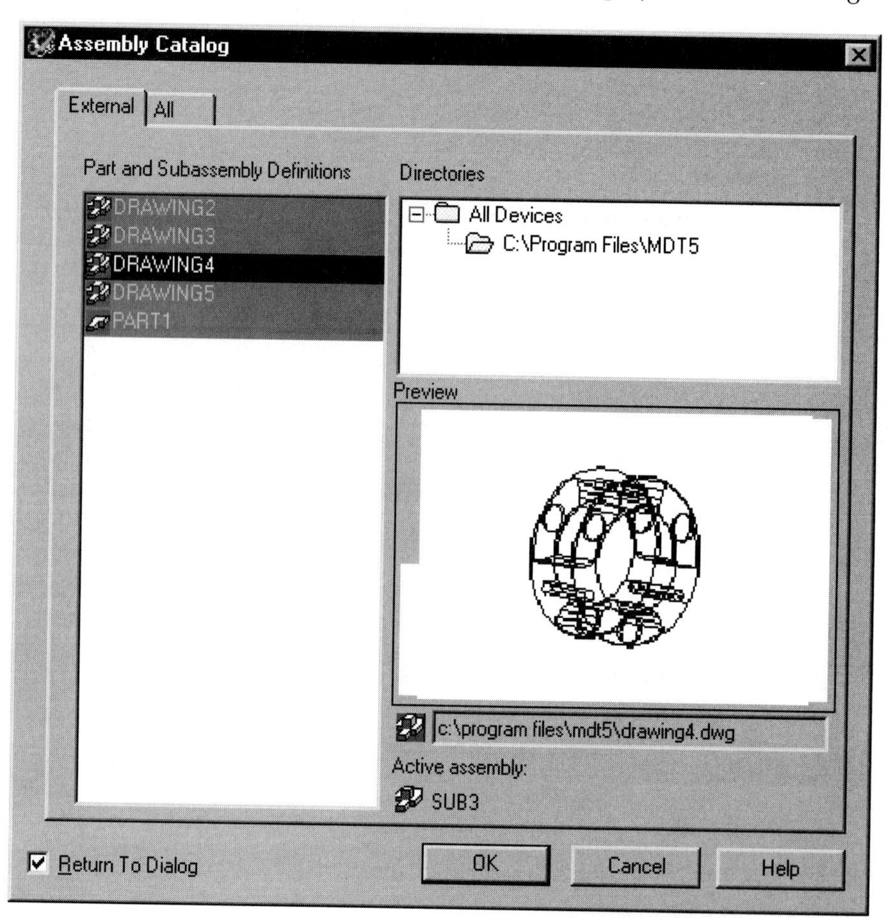

Figure 10-4 *External tab of the Assembly Catalog dialog box*

Assembly Catalog Dialog Box Options (External Tab)

This tab displays all the external parts or subassemblies in the selected path or the directory that can be attached, see Figure 10-4. This tab is divided into three areas. All three areas are discussed next individually.

Part and Subassembly Definitions Area

This area displays all the parts or the subassemblies in the selected path that can be attached. All the parts that are not attached are displayed with grey background and the parts or the subassemblies that are attached are displayed with the white background. To attach a part or a subassembly, double-click on the it. You can also attach a part or a subassembly by choosing the **Attach** option from the shortcut menu displayed on right-clicking on it, see Figure 10-5.

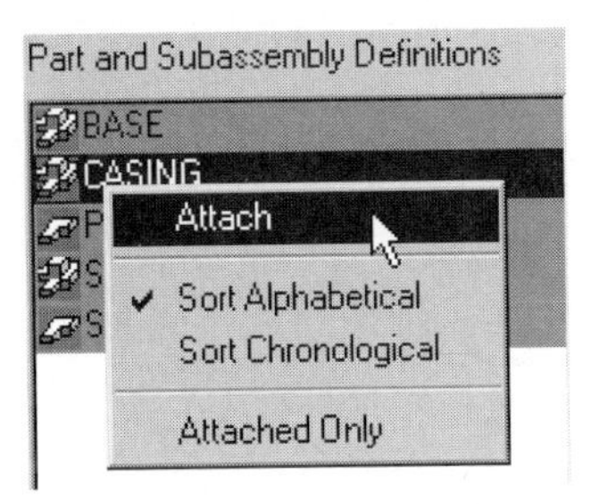

Figure 10-5 *Shortcut menu displayed upon right-clicking on the unreferenced part*

The options provided under this shortcut menu are:

Attach. This option is used to attach the selected part or subassembly.

Sort Alphabetical. This option is used to sort all the available parts or subassemblies in the alphabetical order.

Sort Chronological. This option is used to sort all the available parts or subassemblies in the order of their occurrence.

Attached Only. If this option is chosen, the next time you invoke the **Assembly Catalog** dialog box, only the attached parts or subassemblies will be displayed under the **Part and Subassembly Definition** area.

If you right-click on the attached part or subassembly, the shortcut menu displayed will be different, see Figure 10-6. The options provided under this shortcut menu are:

Instance. This option is used to create an instance of the attached component in the current drawing. The new object will be a separate component.

Rename Definition. This option is used to rename the part or the subassembly attached in the current drawing. The **Rename Definition** dialog box will be

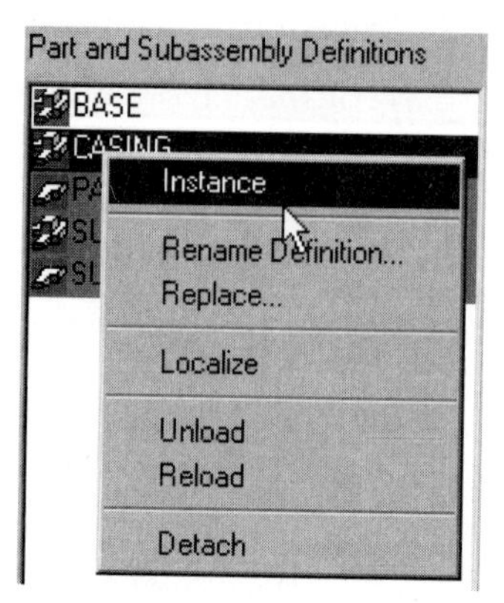

Figure 10-6 *Shortcut menu displayed upon right-clicking on the referenced part*

displayed when you choose this option, see Figure 10-7. The new name of the component will be displayed in the brackets along with the original name of the component under the **Part and Subassembly Definition** area. The options provided under this dialog box are discussed next.

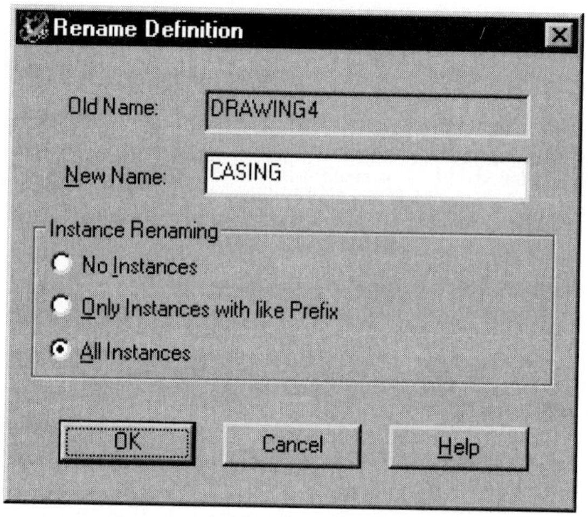

Figure 10-7 *Rename Definition dialog box*

Old Name. This text box displays the original name of the component.

New Name. This text box is used to specify new name for the component.

No Instance. This radio button is selected to disable the renaming of the component.

Only Instances with like Prefix. This radio button is selected to rename all the components in the current drawing that have similar prefix.

All Instances. This radio button is used to rename all the instances of the component in the current drawing.

Replace. The **Replace** option is used to replace the current component with some other component. The **Replace Definition** dialog box will be displayed when you choose this option, see Figure 10-8. This dialog box has a drop-down list from which you can select the new component to replace the existing one.

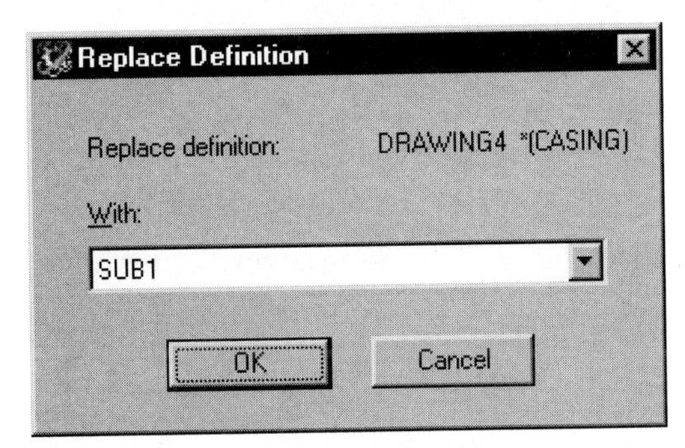

Figure 10-8 Replace Definition dialog box

Localize. The **Localize** option is used to convert the external component into a local part. The link with the original drawing will be lost if the part is localized.

Unload. The **Unload** option will unload the component from the current drawing. However, the database of the component will still remain in the memory of the current drawing.

Reload. The **Reload** option will reload the component. This option works only if the database of the component is still there in the memory of the current drawing.

Detach. The **Detach** option is used to detach the component from the current drawing. No information regarding the component remain in the memory of the current drawing.

Directories Area

The **Directories** area displays the directories in which the parts or the subassemblies are saved. By default the directory in which Mechanical Desktop was loaded will be available. You can add a directory by using the shortcut menu displayed by right-clicking in this area, see Figure 10-9.

The options provided under this shortcut menu are:

Add Directory. The **Add Directory** option is used to add a directory to the

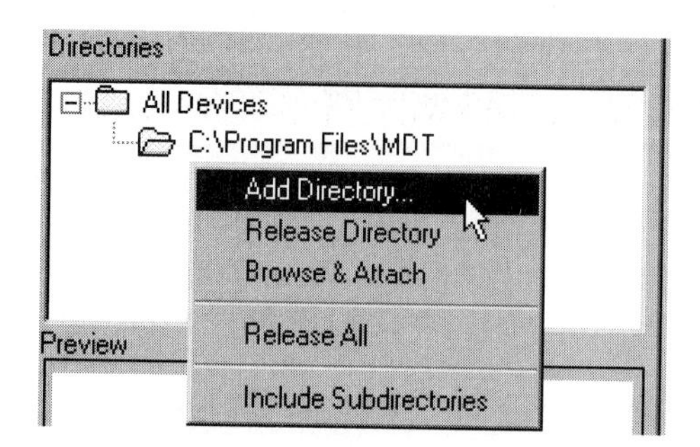

Figure 10-9 *Shortcut menu displayed upon right-clicking in the Directories area*

existing list of directories. The **Browse for Folder** dialog box will be displayed as shown in Figure 10-10 when you choose this option. You can select the required directory from this dialog box. All the parts or subassemblies available in the new directory will be displayed under the **Part and Subassembly Definition** area.

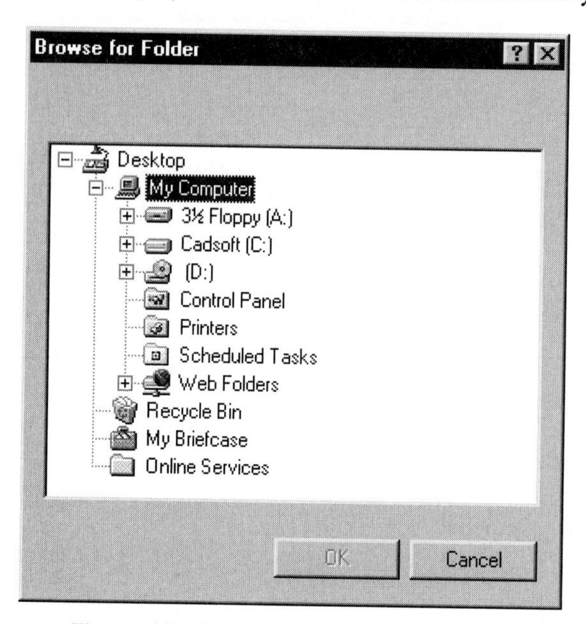

Figure 10-10 *Browse for Folder dialog box*

Release Directory. The **Release Directory** option is used to remove the last added directory from the list of directories.

Browse & Attach. The **Browse & Attach** option is used to attach a single part to the current drawing. When you choose this option, the **External file to attach** dialog box will be displayed, see Figure 10-11.

Release All. The **Release All** option is used to remove all the selected and the

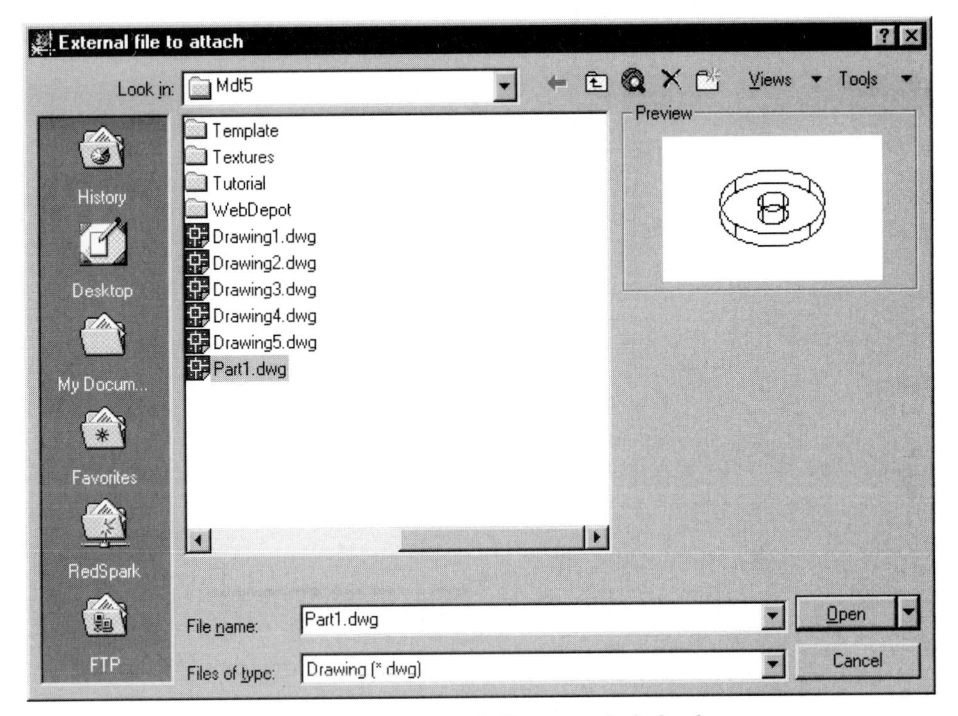

Figure 10-11 *External file to attach dialog box*

default directories from the assembly catalog.

Include Subdirectories. The Include Subdirectories option is used to include the subdirectories along with the directories in the assembly catalog. It is very important to clarify here that this option has to be chosen before adding the directory containing the subdirectories.

Preview Area

This area displays the preview of the selected part or subassembly.

All Tab

The **All** tab of the **Assembly Catalog dialog** box displays all the external or local parts in the current drawing, see Figure 10-12. This tab is also divided into three areas.

External Assembly Definition Area

This area displays all the external parts attached to the current drawing. When you right-click on the external part in this area, the shortcut menu is displayed as shown

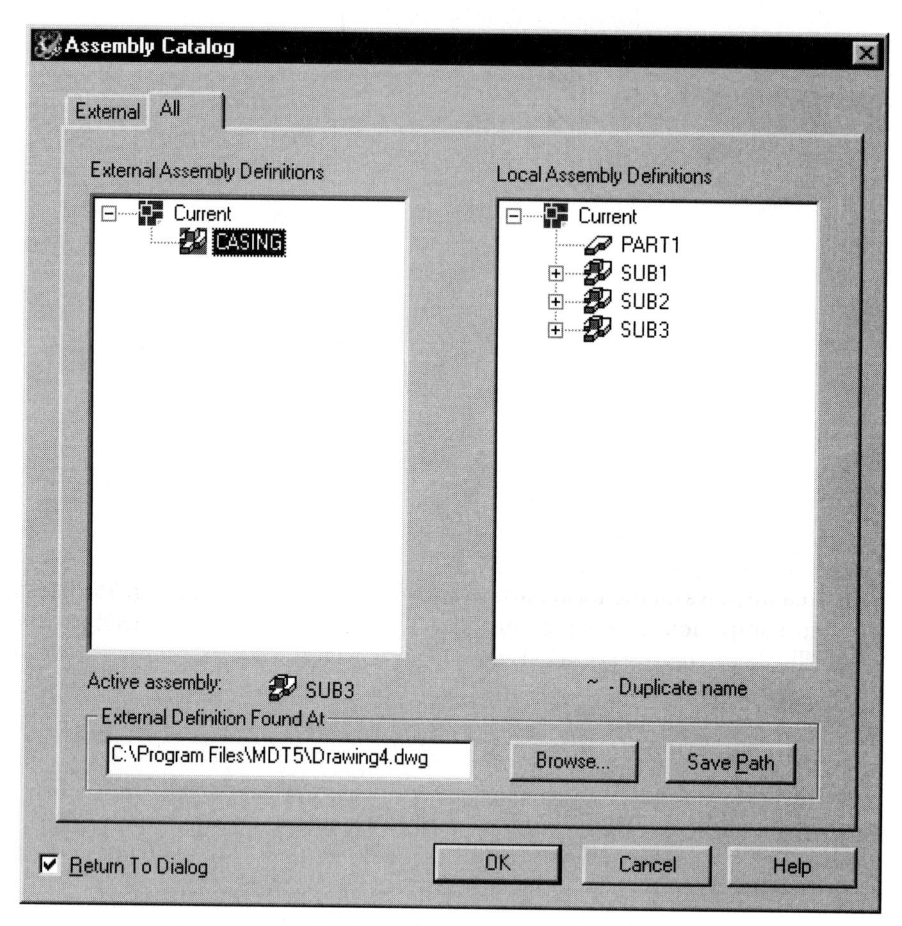

Figure 10-12 *All tab of the Assembly Catalog dialog box*

in Figure 10-13. All of the options of this shortcut menu are similar to those discussed in the **External** tab. The only difference is that this shortcut menu has a new option of **Purge Lock**. This option is discussed below.

Purge Lock. There are two kinds of locks displayed on the external components; the red lock and the grey lock. The red lock indicates that this external part is locked by the current drawing. It generally appears when a local part is externalized. The red lock can be purged. The grey lock generally appears when the external part is unloaded. This lock cannot be purged and the component has to be reloaded. When you choose this option, the red lock is purged and you will be given a warning that the specified file will be saved and you will not be allowed to undo the operations performed up to this point.

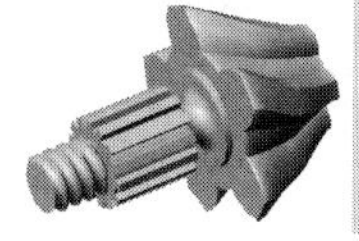

Tip: *You can also purge the lock using the shortcut menu displayed upon right-clicking on the component in the desktop browser. Choose the* ***Purge Lock*** *option to purge the lock.*

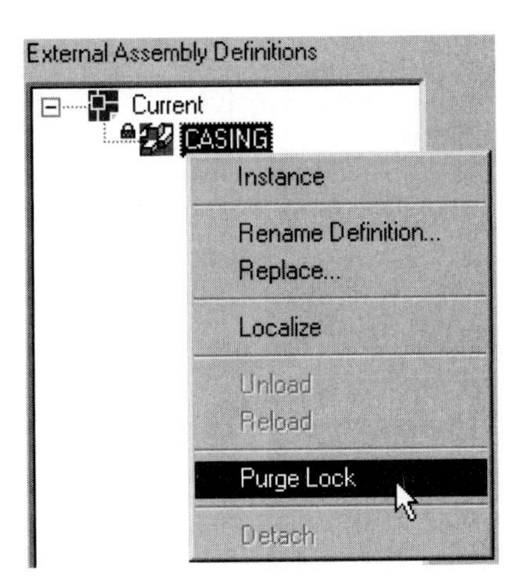

Figure 10-13 *Shortcut menu displayed upon right-clicking on the external component in the External Assembly Definition area*

Local Assembly Definition Area

This area displays all the local parts available in the current drawing. You can copy the selected component or even externalize the selected component using the shortcut menu displayed upon right-clicking on the selected component, see Figure 10-14.

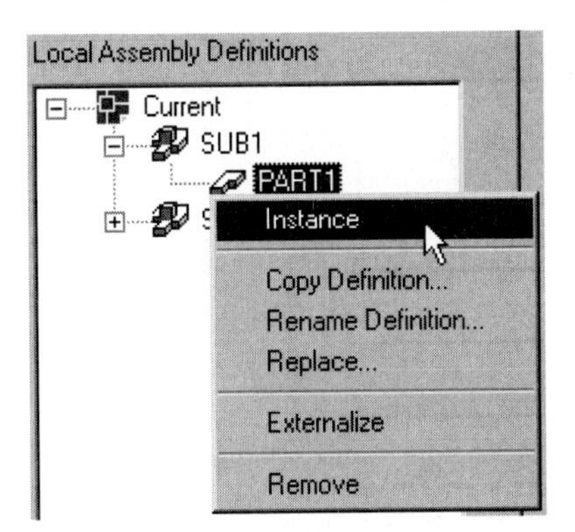

Figure 10-14 *Shortcut menu displayed upon right-clicking on the component in the Local Assembly Definition area*

The options provided under this shortcut menu are:

Instance. The **Instance** option is used to create an instance of the selected part or subassembly.

Copy Definition. The **Copy Definition** option is used to copy the selected part. When you choose this option, the **Copy Definition** dialog box is displayed, see Figure 10-15. However, this option is available only when you select a part and not the complete subassembly.

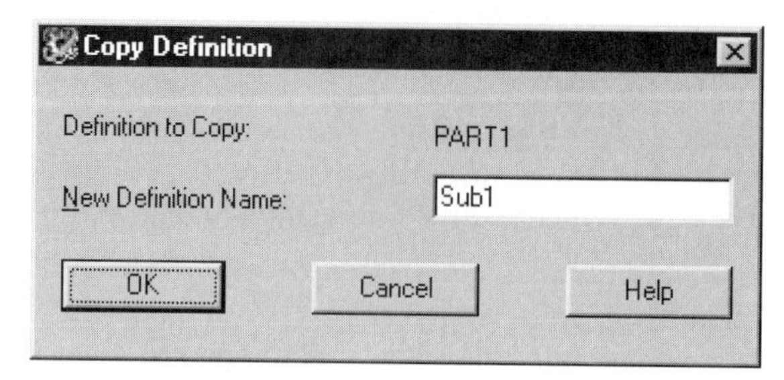

Figure 10-15 *Copy Definition dialog box*

Externalize. This option is used to convert the selected local part or subassembly into an external part. When you choose this option, the **New External File** dialog box will be displayed. You can specify the name of the external part in this dialog box.

Remove. This option is used to delete the selected part or subassembly. When you choose this option, a warning will be displayed and you will be prompted to confirm the deletion.

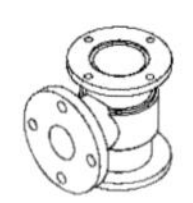

Note

The ***Rename Definition*** *and the* ***Replace*** *options are similar to those discussed under the* ***External*** *tab.*

Tip: *The parts or the subassemblies can also be localized by dragging them and dropping them in the* ***Local Assembly Definition*** *area. Similarly, you can externalize them by dragging and dropping them in the* ***External Assembly Definition*** *area.*

External Definition Found At Area

This area displays the path from which the external component is attached. It is available only when an external component is selected in the **External Assembly Area**.

Browse. This button is chosen to browse for an external component. When you choose this button, the **Locate External Definition** dialog box will be displayed, see Figure 10-16.

Save Path. This button is chosen to save the path in which the current external component is saved.

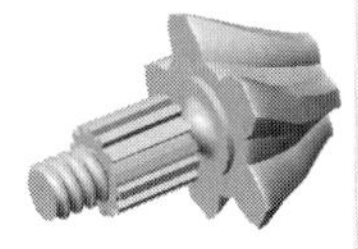

Tip: *To edit an external component, right-click on it in the desktop browser to display the shortcut menu. Select the* ***Open to Edit*** *option and that file will be opened for modification.*

Any part or subassembly saved using the ***AMCOPYOUT*** *command can be attached using the* ***AMCATALOG*** *command.*

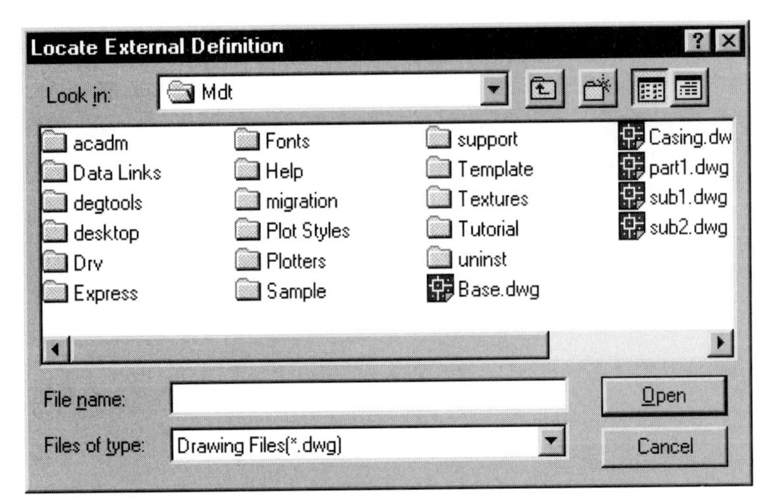

Figure 10-16 *Locate External Definition dialog box*

Return To Dialog

If this check box is selected, the **Assembly Catalog** dialog box will be displayed again after the current task is completed.

Tip: *You have to refresh the external component using the **AMREFRESH** command after making the modifications, if it is attached using the **AMCATALOG** command, to view the effects of modification. However, it is advisable to reload the component using the **All** tab of the **Assembly Catalog** dialog box displayed upon invoking the **AMCATALOG** command.*

ASSEMBLING THE COMPONENTS

You can assemble different components of the assembly using the assembly constraints. Mechanical Desktop provides the following four types of assembly constraints:

- Mate
- Flush
- Angle
- Insert

These constraints can be applied using the following commands.

AMMATE Command

Toolbar:	Assembly Modeling > 3D Assembly Constraints > Mate
Menu:	Assembly > 3D Constraints > Mate
Context Menu:	3D Constraints > Mate
Command:	AMMATE

This command is used to apply the mate constraints to the components of the assembly. The mate constraint makes a point, axis, a planar face or a non planar face of a component coincidental with that of another component. You can specify the offset distance between the selected components. The options provided under this command are:

Point

The **Point** option uses the specified points for applying the mate constraints. Different types of components have different points that can be selected for applying the mate constraint. For example, a cylindrical component has the center points of both the faces that can be selected to apply the mate constraint. A linear edge of a component has three points that can be selected to apply the mate constraint; two at the end points and one at the mid point of the edge, see Figures 10-17 and 10-18.

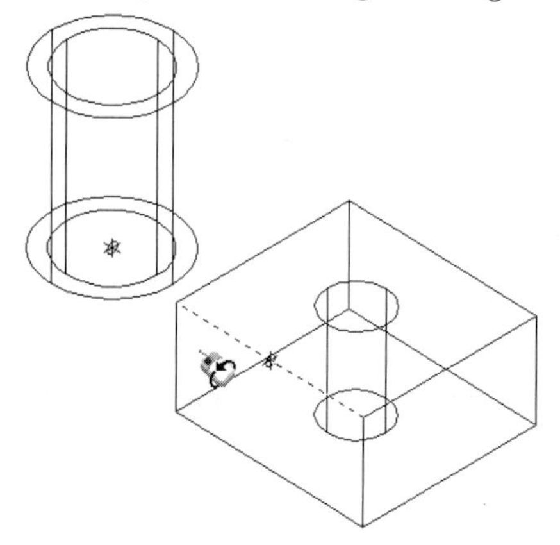

Figure 10-17 *Selecting the points to apply the Mate constraint*

Figure 10-18 *Components after applying the Mate constraint*

aXis

This option uses the axes of the components to apply the mate constraint. In case of the components with planar faces, the linear edge of the face is considered as an axis. The axis of a cylindrical or conical part passes through the center of the circle. Figure 10-19 shows two component with their axes selected for assembling and Figure 10-20 both the components after assembling.

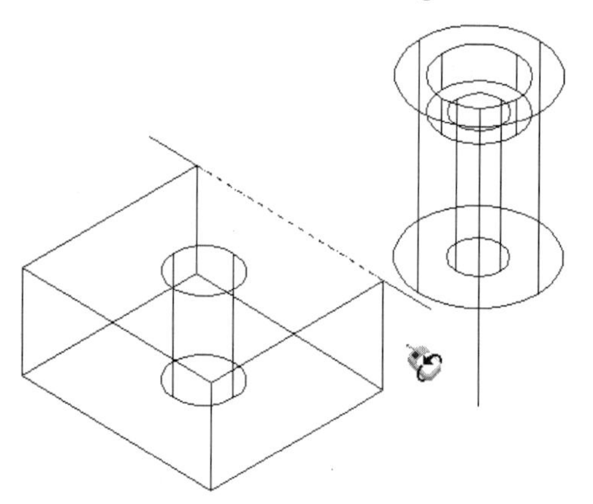

Figure 10-19 *Selecting the axes of components*

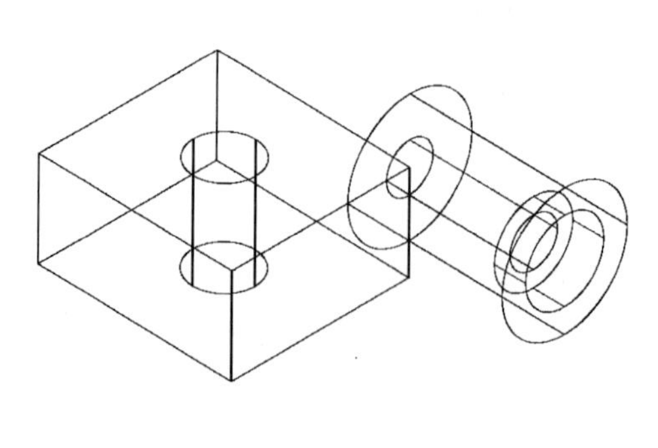

Figure 10-20 *Components after assembly*

fAce

This option uses the planes or the faces of the components to apply the mate constraint. Mechanical Desktop gives the flexibility of selecting the planar faces or even the non planar faces, like the outer wall of a cylinder as shown in Figures 10-21 and 10-22.

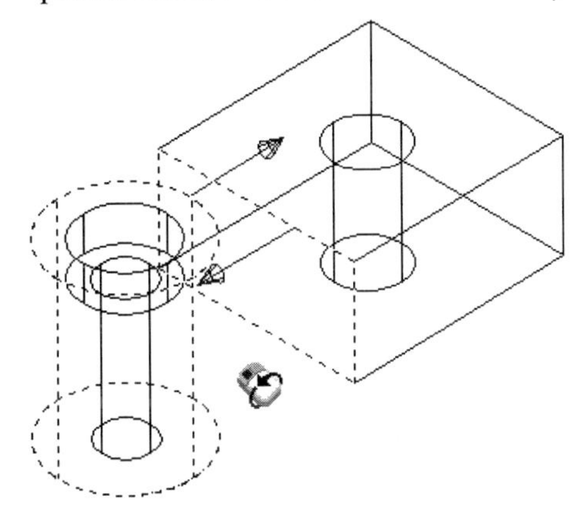

Figure 10-21 *Selecting the planar and non planar faces to apply the mate constraint*

Figure 10-22 *Components after applying the mate constraint*

Clear

This option clears the current selection set.

Cycle

Cycling through will, one by one, display all of the options that can be used for applying the mate constraint like face, axis or a point.

Next

This option displays the next option available for applying the mate constraint.

Flip

This option is used to reverse the direction of the constraint.

Offset

This option is used to apply an offset distance between the assembly components, see Figure 10-23.

AMFLUSH Command

Toolbar:	Assembly Modeling > 3D Assembly Constraints > Flush
Menu:	Assembly > 3D Constraints > Flush
Context Menu:	3D Constraints > Flush
Command:	AMFLUSH

This command is used to apply the flush constraint between the selected components. The flush constraint defines coplanar position between two components. Using this

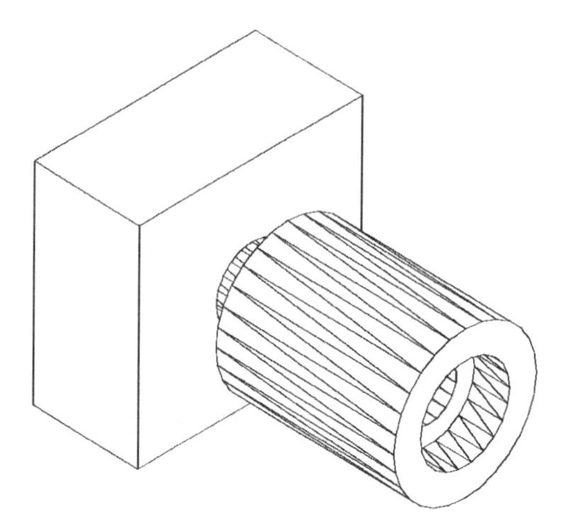

Figure 10-23 Components assembled with an offset distance of 1

constraint you can assemble two different components parallel to each other with the help of their faces. You can also specify offset distance between the selected components. Figure 10-24 shows two components with their faces selected to apply the flush constraint and Figure 10-25 shows both the components after assembling.

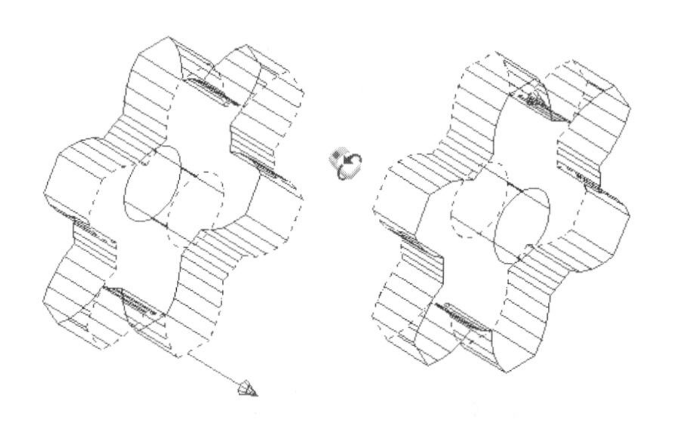

Figure 10-24 Selecting the faces to apply the flush constraint

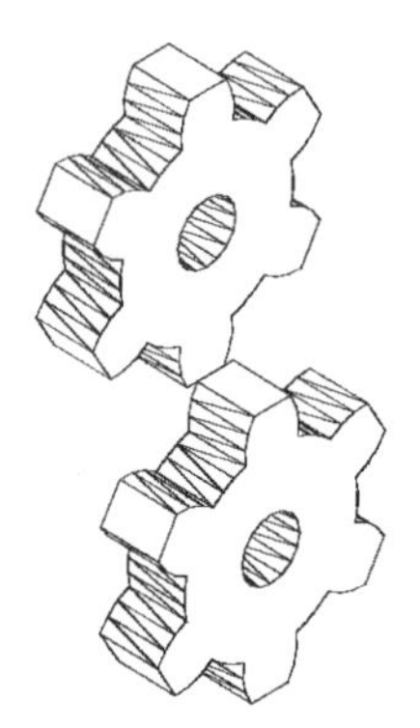

Figure 10-25 Components after assembling

AMANGLE Command

Toolbar: Assembly Modeling > 3D Assembly Constraints > Angle
Menu: Assembly > 3D Constraints > Angle
Context Menu: 3D Constraints > Angle
Command: AMANGLE

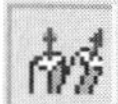

This command is used to apply the angle constraint to the selected objects. The angle constraint is used to assemble two components at a certain angle to each other. The angle can be defined between two planes, vectors (axis), points or a combination of these. The options provided under this command are:

Point

The **Point** option uses specified points for applying the angle constraints. Different types of components have different points that can be selected for applying the angle constraint. For example, a cylindrical component has the center points of both the faces that can be selected to apply the angle constraint. A linear edge of a component has three points that can be selected to apply the angle constraint; two at the end points and one at the mid point of the edge. The other points that can be used for applying this constraint are a work point or an AutoCAD point.

aXis

The axis is also termed as vector in this case. This option uses the axes of the components to apply the angle constraint. In case of the components with planar faces, the linear edge of the face is considered as an axis. The axis of a cylindrical or conical part passes through the center of the circle. You can also directly select a work axis for applying the angle constraint.

fAce

This option uses work planes or planar faces of the components to apply the angle constraint. Figure 10-26 shows the faces of the components selected to apply the angle constraint and Figure 10-27 shows the components after applying the constraint.

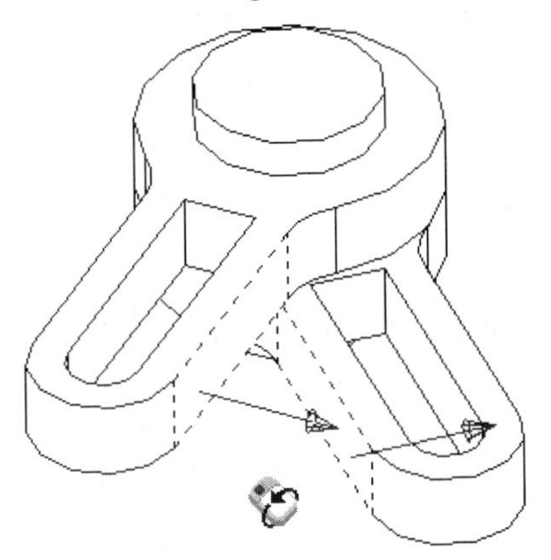

Figure 10-26 *Selecting the faces of components to apply the angle constraint*

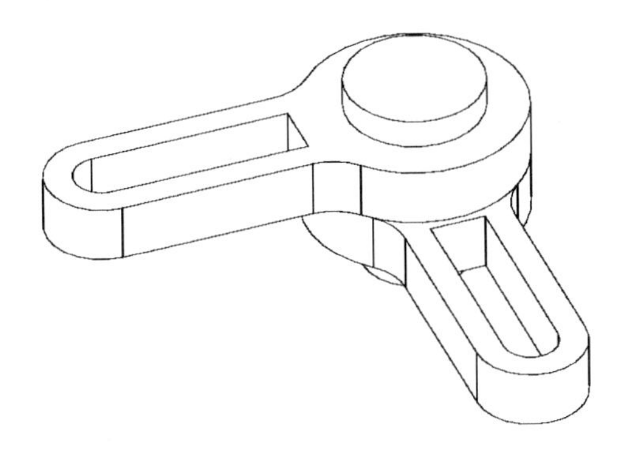

Figure 10-27 *Components after applying the angle constraint of 90 degrees*

Angle

This option is used to specify the angle between the components.

Next

This option is selected to display the next option that is available for applying the angle constraint.

cYcle

Cycling through will one by one display all of the options that can be used for applying the angle constraint like face, axis or a point.

Clear

This option is used to clear the current selection set.

AMINSERT Command

Toolbar:	Assembly Modeling > 3D Assembly Constraints > Insert
Menu:	Assembly > 3D Constraints > Insert
Context Menu:	3D Constraints > Insert
Command:	AMINSERT

This command is used to apply the insert constraint. The insert constraint is defined in terms of a central axis between two cylindrical or conical components, two holes or a combination of both. Applying this constraint allows two different components to share same orientation of the central axis and at the same time makes the selected faces coplanar. You can also specify offset distance between the two faces. Figure 10-28 shows two components displaying the faces selected for applying the insert constraint and Figure 10-29 shows the assembled components after the constraint is applied.

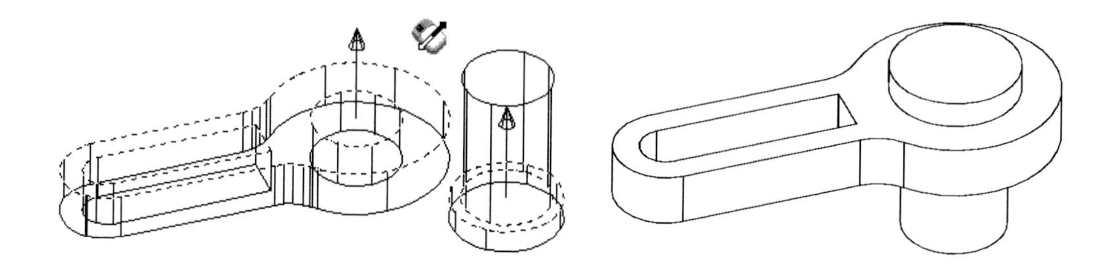

Figure 10-28 *Selecting the faces to apply the insert constraint*

Figure 10-29 *Components after assembling using the insert constraint*

EDITING THE ASSEMBLY CONSTRAINTS

All the assembly constraints applied to the components of the assembly can be edited or deleted using the **AMEDITCONST** command or by using the desktop browser.

Editing The Constraints Using The AMEDITCONST Command

Toolbar:	Assembly Modeling > 3D Assembly Constraints > Edit Constraints
Menu:	Assembly > 3D Constraints > Edit
Context Menu:	Assembly Menu > 3D Constraints > Edit
Command:	AMEDITCONST

This command is used to edit the assembly constraints applied to the components. When you invoke this command, the **Edit 3D Constraints** dialog box will be displayed as shown in Figure 10-30.

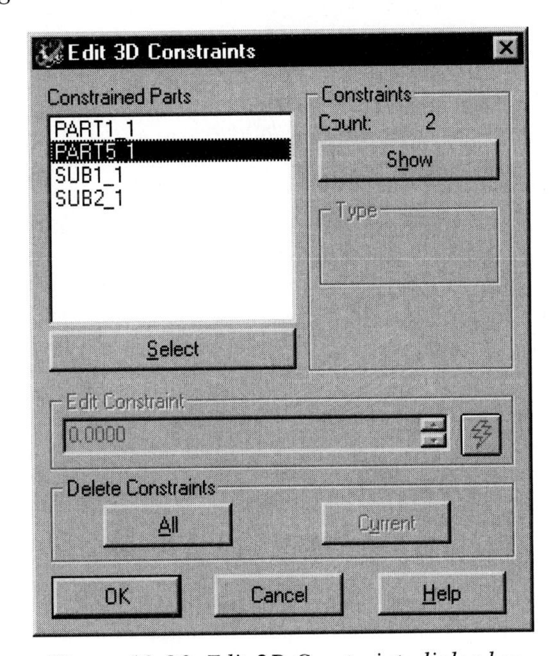

Figure 10-30 Edit 3D Constraints dialog box

Edit 3D Constraints Dialog Box Options

The options provided under this dialog box are:

Constrained Parts Area

This area displays all the components on which the assembly constraints have been applied.

Constraints Area

The options under this area will be available only when you select a component from the **Constrained Parts** area. The **Count** option under this area displays the total number of assembly constraints that are applied to the selected part. By default this area has only one button which is the **Show** button.

Show. The **Show** button is used to show all the assembly constraints that are applied to the selected component. If the component has more than one constraint, then this area displays the **Next** and **Previous** buttons to cycle through all the constraints see Figure 10-31.

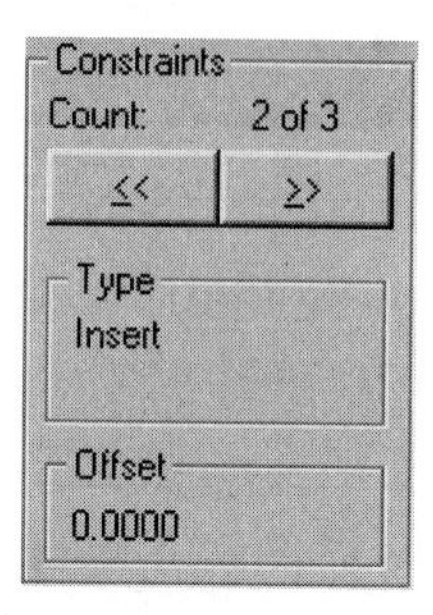

Figure 10-31 *Constraints area displaying the Next and Previous buttons*

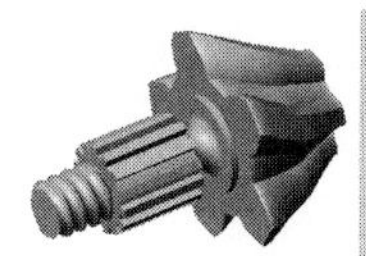

Tip: *You can move the* ***Edit 3D Constraints*** *dialog box in one of the corners of the screen and view all the constraints that are applied on the components as they will be displayed in the background.*

Type Area

This area displays the type of constraints that are applied to the selected component.

Offset Area

This area displays the offset distance specified for the current constraint.

Select

This button is chosen to select the component to edit from the graphics screen. The dialog box will not close but the cursor will be moved to the graphics screen to allow you to select the component.

Edit Constraint Area

The spinner provided in this area is used to specify new offset value for the assembly constraint.

Update Constraint. The **Update Constraint** button is available on the right side of the spinner in the **Edit Constraint** area. This button is chosen to update the current constraint to the new values that are set in the spinner.

Delete Constraints Area

This area provides you with the following two buttons:

All. The **All** button is chosen to delete all of the assembly constraints that are applied to the selected component.

Current. The **Current** button is chosen to delete the current constraint from the selected component.

Editing The Constraints Using The Desktop Browser

The assembly constraints that are applied to the components can also be edited from the desktop browser using the shortcut menu displayed upon right-clicking on the constraint to edit. Choose the **Edit** option to edit the constraint and **Delete** option to delete the constraint from the shortcut menu. When you choose the **Edit** option, the **Edit 3D Constraint** dialog box (for desktop browser) will be displayed as shown in Figure 10-32.

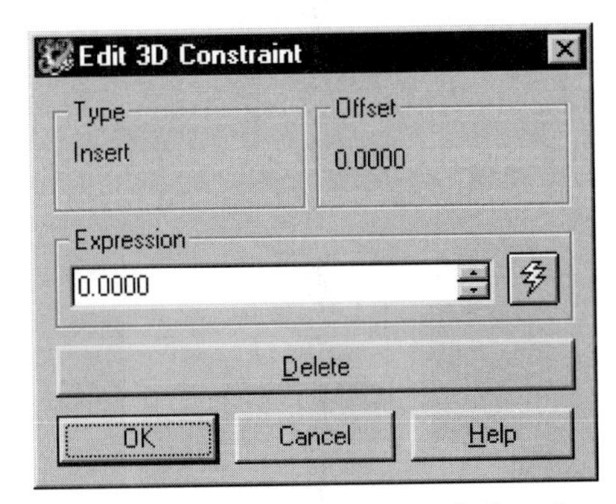

Figure 10-32 Edit 3D Constraint (desktop browser)

Edit 3D Constraint Dialog Box (For Desktop Browser) Options

The options provided under this dialog box are:

Type

This area displays the type of the constraint selected to edit.

Offset

This area displays the offset value for the selected constraint.

Expression

This spinner is used to specify the offset value for the selected constraint.

Update Constraint

This button is chosen to update the selected constraint to the new values.

Delete

This button is chosen to delete the selected constraint.

TUTORIALS

Tutorial 1

In this tutorial you will create and assemble the components of the Plummer Block. The Plummer Block assembly is shown in Figure 10-33 and Figure 10-34. Figures 10-35a through 10-38 displays the dimensions of all components of the Plummer block assembly.

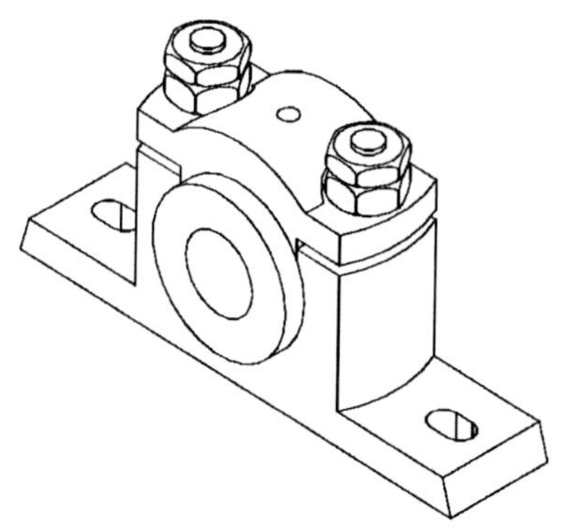

Figure 10-33 Assembly for Tutorial 1

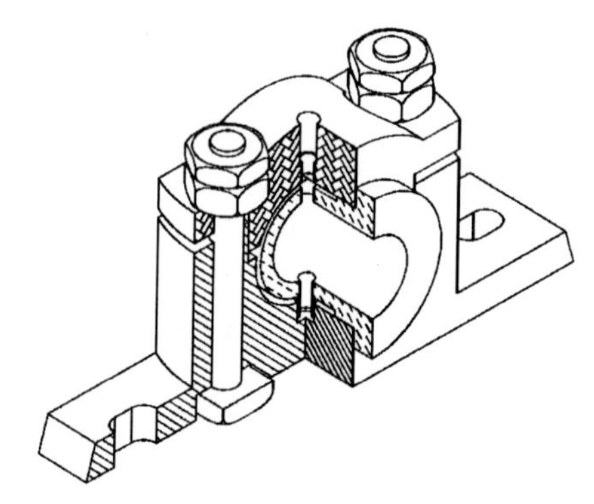

Figure 10-34 Assembly for Tutorial 1

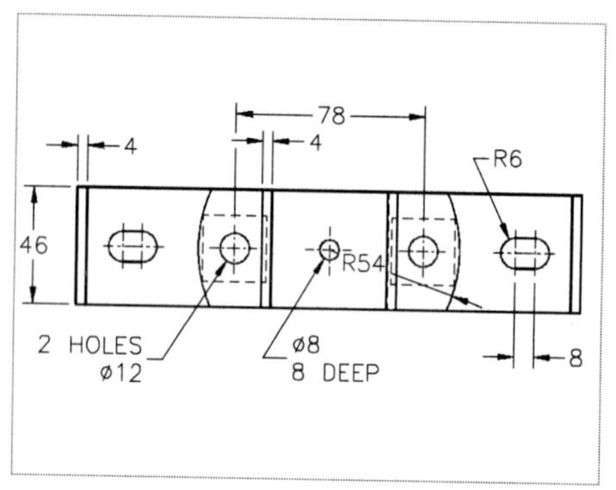

Figure 10-35a Top view of CASTING

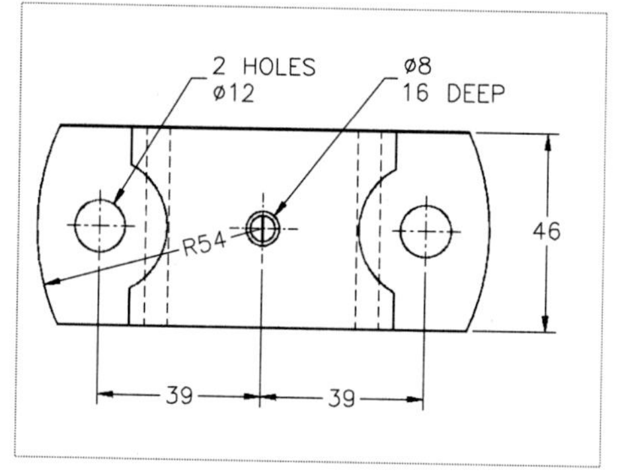

Figure 10-36a Top view of CAP

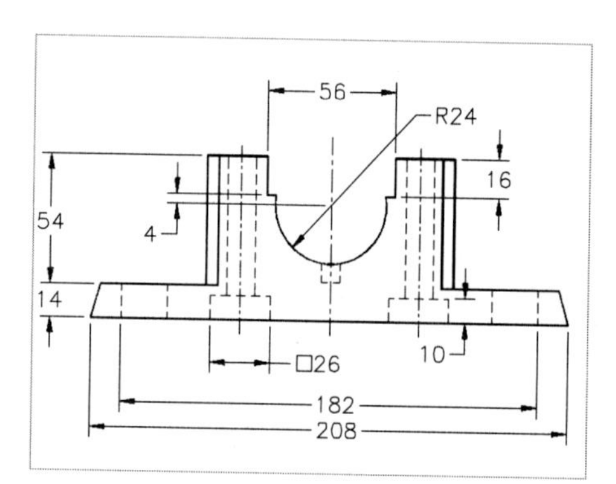

Figure 10-35b Front view of CASTING

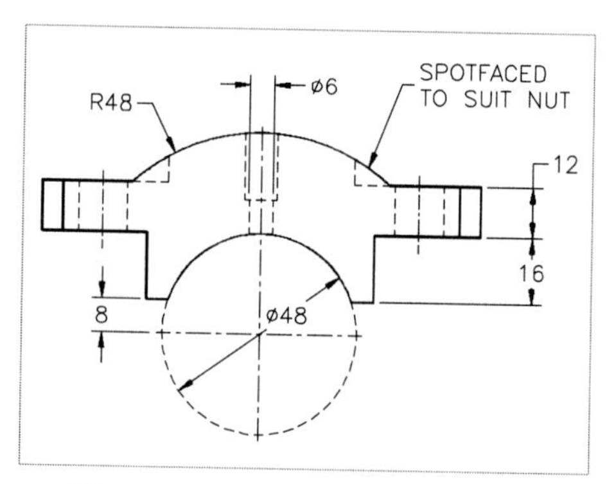

Figure 10-36b Front view of CAP

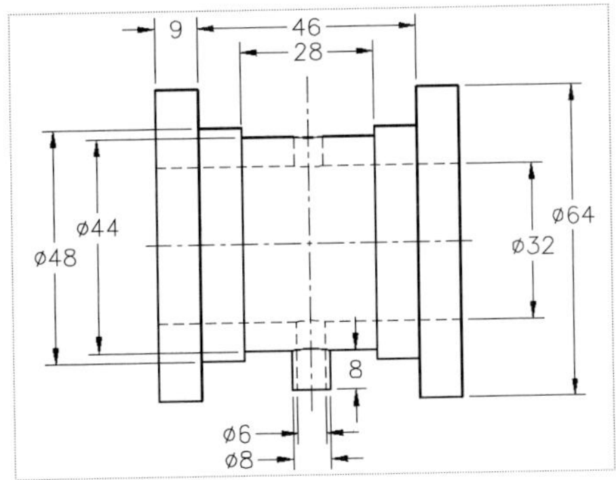

Figure 10-37 *Details of BRASSES*

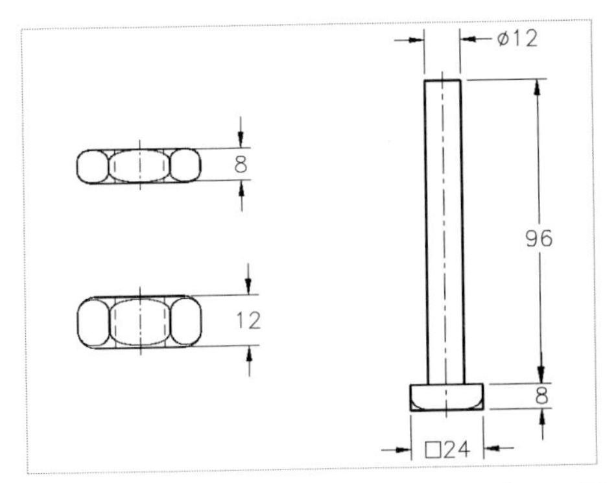

Figure 10-38 *Details of NUTS and BOLT*

1. Create the CASTING.

2. Choose the **Out Part Definition** button from the **Assembly Catalog** flyout in the **Assembly Modeling** toolbar to display the **Part/Subassembly Out** dialog box, see Figure 10-39.

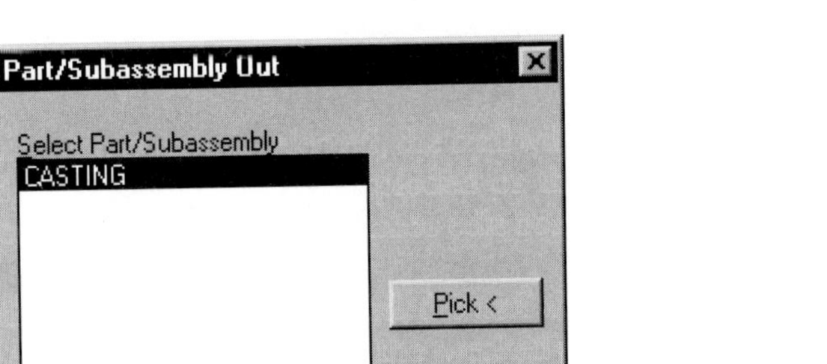

Figure 10-39 *Part/Subassembly Out dialog box*

3. Choose the **Pick** button. The dialog box will disappear and the prompt sequence is as follows:

   Select part or subassembly: *Select the component.*

4. Choose the **File** button to display the **Output File** dialog box. Select the directory and save it with the name C:\MDT Tut\Ch-10\Assembly**CASTING.dwg**.

5. Similarly, create the other components and copy them out using the **AMCOPYOUT** command.

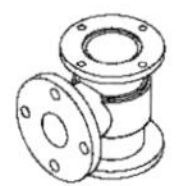

Note

Delete the component once it has been copied out and then create the next component. Repeat this procedure until all of the components are created and copied out.

6. Open a new file.

7. Choose the **Assembly Modeling** button from the **Desktop Main** toolbar to proceed to the assembly mode.

8. Choose the **Assembly Catalog** button from the **Assembly Modeling** toolbar to display the **Assembly Catalog** dialog box, see Figure 10-40.

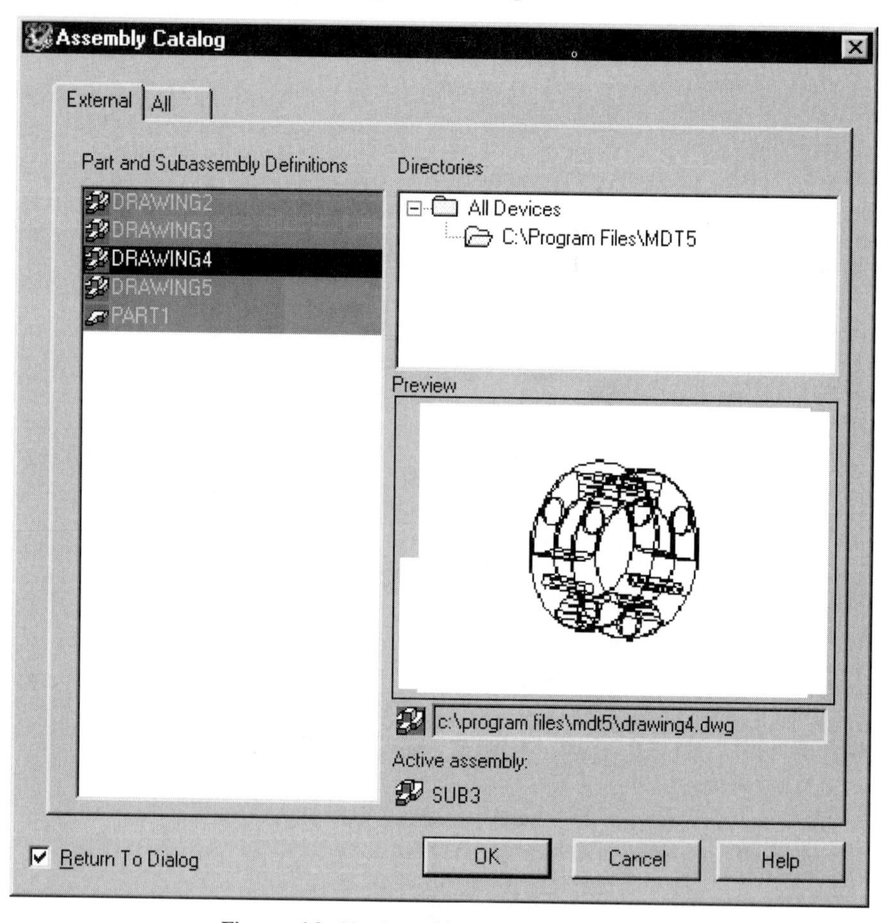

Figure 10-40 *Assembly Catalog dialog box*

9. Right-click under the **Directories** area to display the shortcut menu. Choose **Add Directory** to display the **Browse for Folder** dialog box. Select the directory C:\MDT Tut\Ch-10\Assembly and add it to the current directory. All the components created in this assembly will be displayed under the **Part and Subassembly Definition** area.

10. Double-click on the **CASTING** under the **Part and Subassembly Definition** area. The prompt sequence is as follows:

    Number of errors found: 0   Number of errors fixed: 0
    Auditing Mechanical Desktop Data complete.
    Specify new insertion point: *Specify the insertion point for the component.*
    Specify insertion point for another instance or <continue>: Enter

11. Double-click on the **CAP** under the **Part and Subassembly Definition** area. The prompt sequence is as follows:

    Number of errors found: 0   Number of errors fixed: 0
    Auditing Mechanical Desktop Data complete.
    Specify new insertion point: *Specify the insertion point for the component.*
    Specify insertion point for another instance or <continue>: Enter

12. Choose **OK**.

13. Choose the **Insert** button from the **3D Assembly Constraints** flyout in the **Assembly Modeling** toolbar. The prompt sequence is as follows:

    Select first circular edge: *Select the circular edge of the **CASTING** as shown in Figure 10-41.*
    First set = Plane/Axis
    Enter an option [Clear/Flip] <accEpt>: Enter

    Select second circular edge: *Select the circular edge of the **CAP** as shown in Figure 10-41.*
    Second set = Plane/Axis
    Enter an option [Clear/Flip] <accEpt>: Enter
    Enter offset <0.0000>: **4**

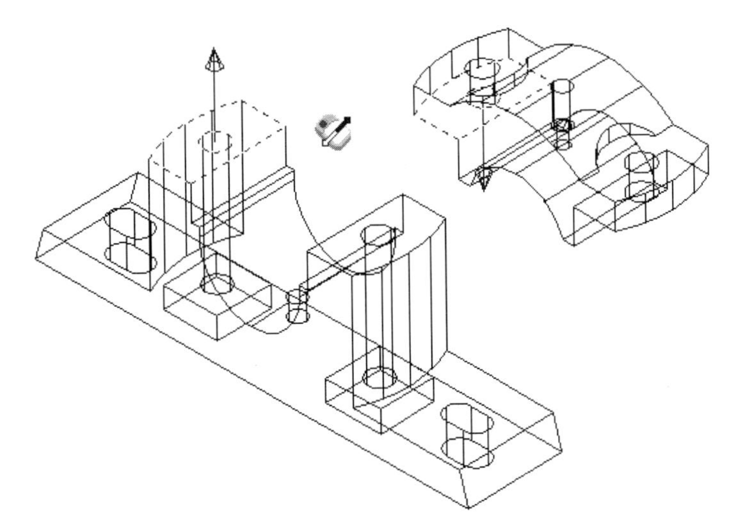

Figure 10-41 *Selecting the edges to apply the Insert constraint*

14. Choose the **Assembly Catalog** button to display the **Assembly Catalog** dialog box.

15. Double-click on the **BRASSES** under the **Parts and Subassembly Definition** area. The prompt sequence is as follows:

    Number of errors found: 0   Number of errors fixed: 0
    Auditing Mechanical Desktop Data complete.
    Specify new insertion point: *Specify the insertion point for the component.*
    Specify insertion point for another instance or <continue>: Enter

16. Choose **OK**.

17. Choose the **Insert** button from the **3D Assembly Constraints** flyout in the **Assembly Modeling** toolbar. The prompt sequence is as follows:

    Select first circular edge: *Select the circular edge of the **BRASSES** as shown in Figure 10-42.*
    First set = Plane/Axis
    Enter an option [Clear/Flip] <accEpt>: Enter

    Select second circular edge: *Select the circular edge of the **CASTING** as shown in Figure 10-42.*
    Second set = Plane/Axis
    Enter an option [Clear/Flip] <accEpt>: Enter
    Enter offset <0.0000>: **0**

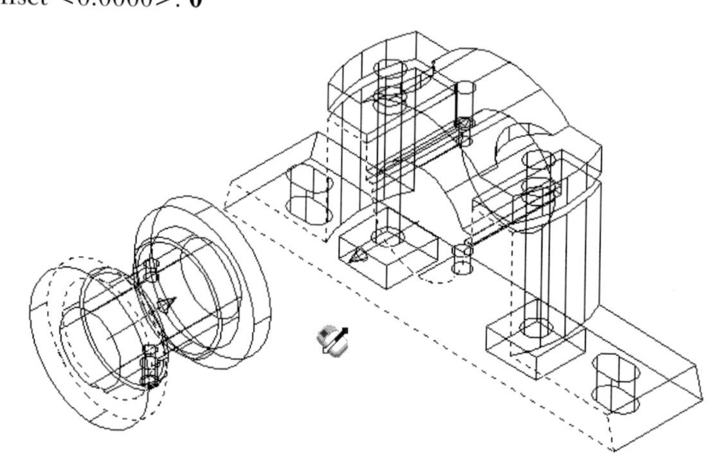

Figure 10-42 *Selecting the faces to apply the Insert constraint*

18. Make the **CAP** and the **BRASSES** invisible by right-clicking on them in the desktop browser and choosing **Invisible** from the shortcut menu.

19. Choose the **Assembly Catalog** button from the **Assembly Modeling** toolbar to display the **Assembly Catalog** dialog box.

20. Double-click on the **BOLT** under the **Part and Subassembly Definition** area. The prompt sequence is as follows:

    Number of errors found: 0 Number of errors fixed: 0
    Auditing Mechanical Desktop Data complete.
    Specify new insertion point: *Specify the insertion point for the component.*
    Specify insertion point for another instance or <continue>: *Specify the insertion point for the instance of the component.*
    Specify insertion point for another instance or <continue>: Enter

21. Choose **OK**.

22. Choose the **Insert** button from the **3D Assembly Constraints** flyout in the **Assembly Modeling** toolbar. The prompt sequence is as follows:

    Select first circular edge: *Select the circular edge of the **BOLT** as shown in Figure 10-43.*
    First set = Plane/Axis
    Enter an option [Clear/Flip] <accEpt>: Enter

    Select second circular edge: *Select the circular edge of the **CASTING** as shown in Figure 10-43.*
    Sccond set = Plane/Axis
    Enter an option [Clear/Flip] <accEpt>: Enter
    Enter offset <0.0000>: **0**

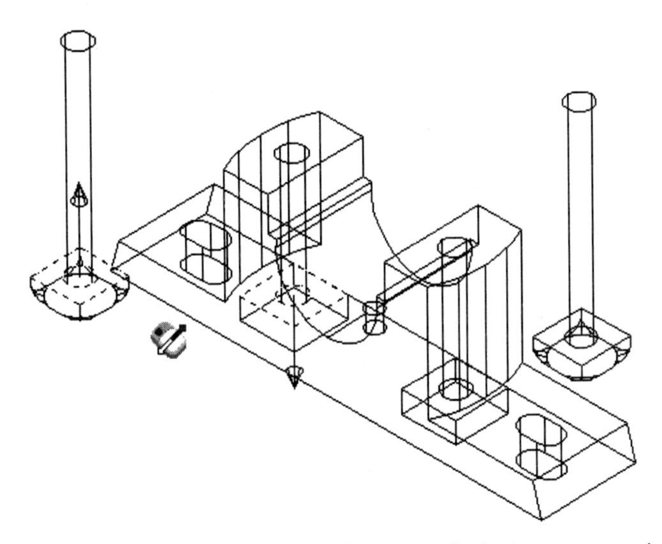

Figure 10-43 *Selecting the edges to apply the insert constraint*

23. Similarly, assemble the other bolt.

24. Make the **CAP** visible and the **CASTING** invisible using the desktop browser.

25. Choose the **Assembly Catalog** button to display the **Assembly Catalog** dialog box.

26. Double-click on the **NUT** under the **Part and Subassembly Definition** area. The prompt sequence is as follows:

    Number of errors found: 0 Number of errors fixed: 0
    Auditing Mechanical Desktop Data complete.
    Specify new insertion point: *Specify the insertion point for the component.*
    Specify insertion point for another instance or <continue>: *Specify the insertion point for the instance of the component.*
    Specify insertion point for another instance or <continue>: Enter

27. Double-click on the **LOCK NUT** under the **Part and Subassembly Definition** area. The prompt sequence is as follows:

    Number of errors found: 0 Number of errors fixed: 0
    Auditing Mechanical Desktop Data complete.
    Specify new insertion point: *Specify the insertion point for the component.*
    Specify insertion point for another instance or <continue>: *Specify the insertion point for the instance of the component.*
    Specify insertion point for another instance or <continue>: Enter

28. Choose **OK**.

29. Choose the **Insert** button from the **3D Assembly Constraints** flyout in the **Assembly Modeling** toolbar. The prompt sequence is as follows:

    Select first circular edge: *Select the circular edge of the* ***NUT*** *as shown in Figure 10-44.*
    First set = Plane/Axis
    Enter an option [Clear/Flip] <accEpt>: Enter

    Select second circular edge: *Select the circular edge of the* ***CAP*** *as shown in Figure 10-44.*
    Second set = Plane/Axis
    Enter an option [Clear/Flip] <accEpt>: Enter

    Enter offset <0.0000>: **0**

30. Similarly, assemble the other nut.

31. Choose the **Insert** button from the **3D Assembly Constraints** flyout in the **Assembly Modeling** toolbar. The prompt sequence is as follows:

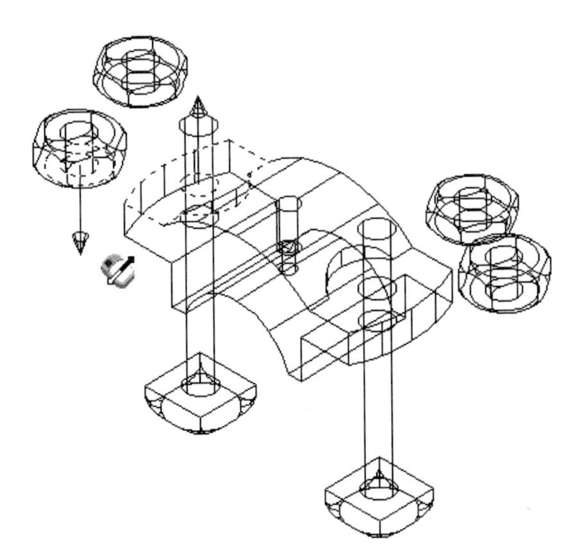

Figure 10-44 *Selecting the edges to apply the insert constraint*

Select first circular edge: *Select the circular edge of the* ***LOCK NUT*** *as shown in Figure 10-45.*
First set = Plane/Axis
Enter an option [Clear/Flip] <accEpt>: Enter

Select second circular edge: *Select the circular edge of the* ***NUT*** *as shown in Figure 10-45.*
Second set = Plane/Axis
Enter an option [Clear/Flip] <accEpt>: Enter

Enter offset <0.0000>: **0**

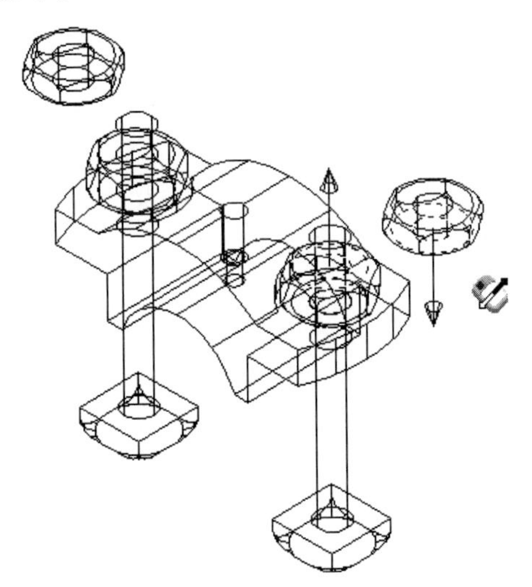

Figure 10-45 *Selecting the edges to apply the insert constraint*

32. Make all the components visible using the desktop browser. The final assembly should be

similar to the one shown in Figure 10-46.

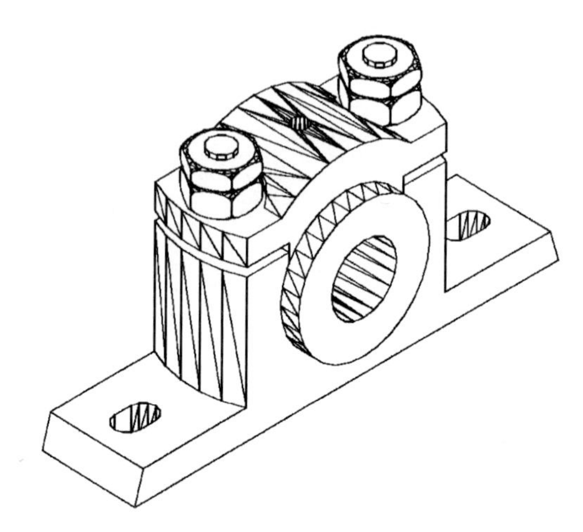

Figure 10-46 *Final assembly for Tutorial 1*

33. Save this assembly with the name given below:

    \MDT Tut\Ch-10\Assembly**PBAssembly.dwg**

Tutorial 2

In this tutorial you will create all the components of Drill Press Vice and then assemble them as shown in Figure 10-47 and Figure 10-48. Figures 10-49a through 10-52 displays the dimensions of all the components of the Drill Press Vice assembly. Assume the missing dimensions.

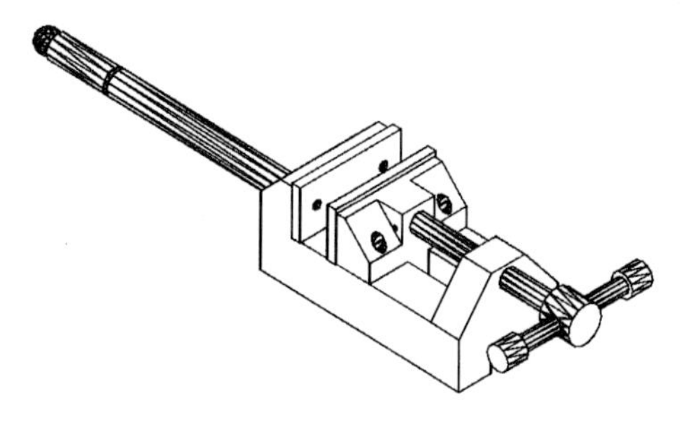

Figure 10-47 *Assembly for Tutorial 2*

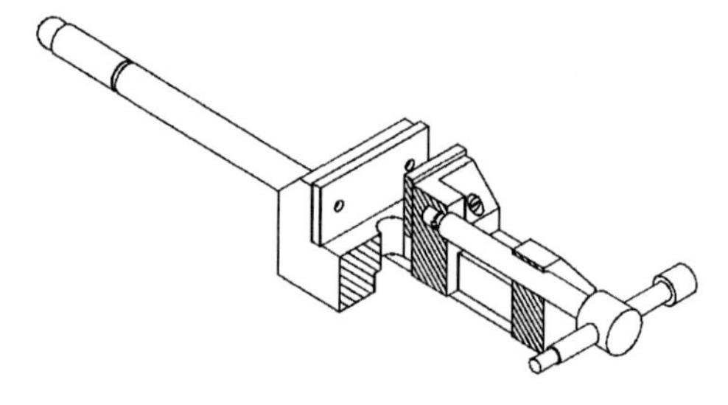

Figure 10-48 *Assembly for Tutorial 2*

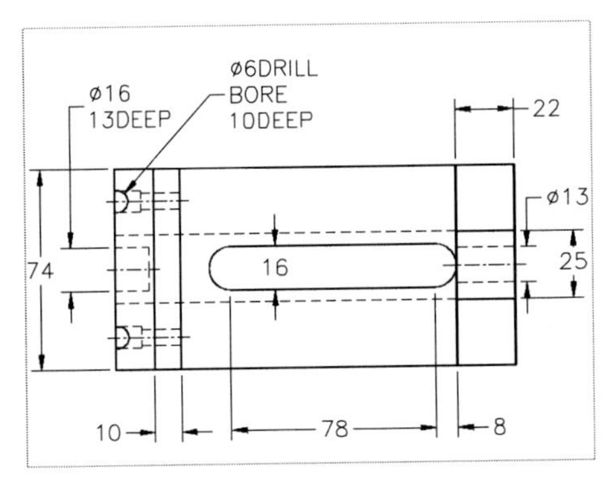

Figure 10-49a *Top view of the BASE*

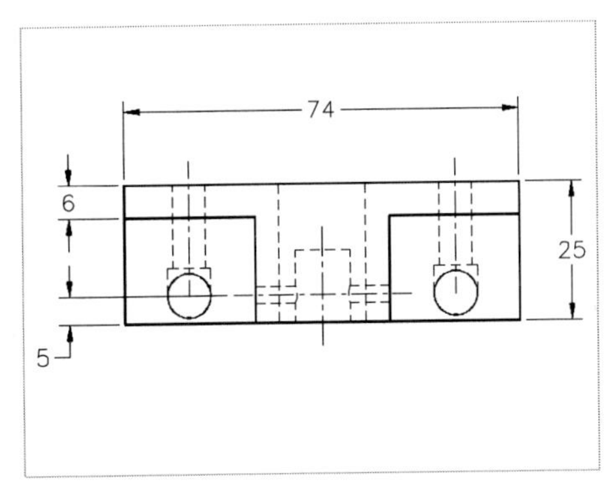

Figure 10-50a *Top view of the MOVABLE JAW*

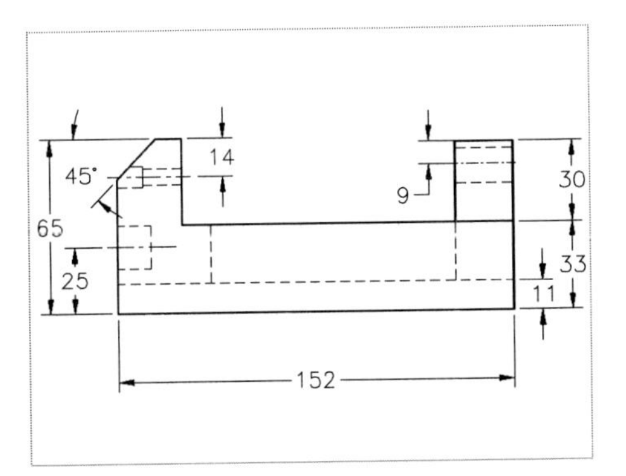

Figure 10-49b *Front view of the BASE*

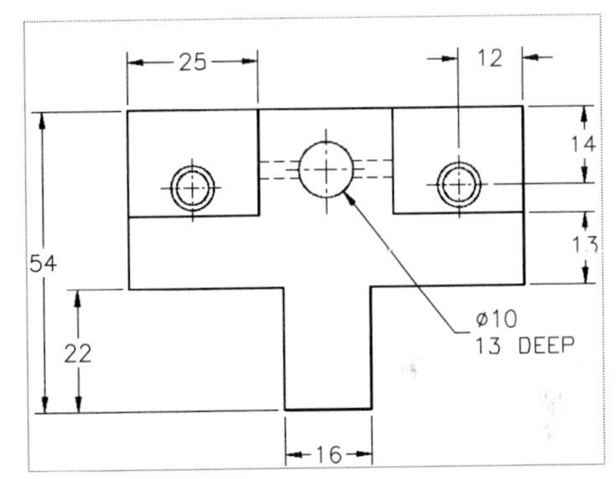

Figure 10-50b *Front view of the MOVABLE JAW*

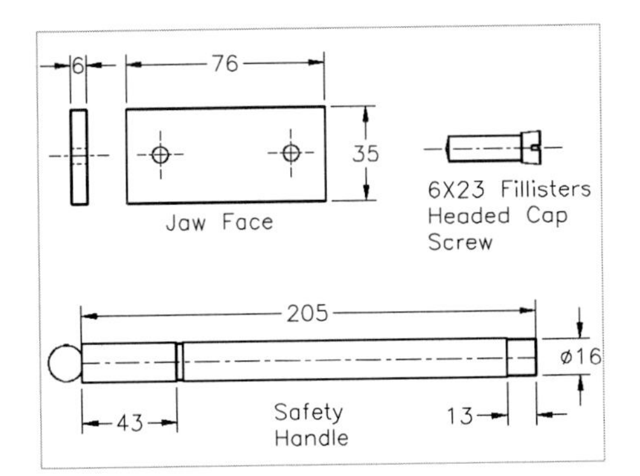

Figure 10-51 *Dimensions for Jaw Face, Cap Screw and Safety Handle*

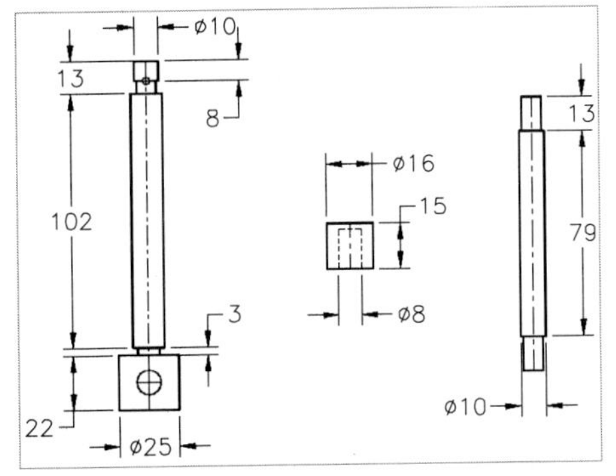

Figure 10-52 *Dimensions for Clamp Screw, Handle Stop and Clamp Screw Handle*

1. Create all components of the assembly in different drawings and save them in the directory C:\MDT Tut\Ch-10\Assembly.

2. Open a new drawing.

3. Choose the **Assembly Catalog** button to display the **Assembly Catalog** dialog box.

4. Right-click under the **Directories** area to display the shortcut menu. Choose **Add Directory** to display the **Browse for Folder** dialog box. Select the directory C:\MDT Tut\Ch-10\Assembly and add it to the current list of directories. All of the components saved in this directory will be displayed under the **Part and Subassembly Definition** area.

5. Double-click on the **BASE** under the **Part and Subassembly Definition** area. The prompt sequence is as follows:

   Number of errors found: 0   Number of errors fixed: 0
   Auditing Mechanical Desktop Data complete.
   Specify new insertion point: *Specify the insertion point of the component.*
   Specify insertion point for another instance or <continue>: Enter

6. Double-click on the **MOVABLE JAW** under the **Part and Subassembly Definition** area. The prompt sequence is as follows:

   Number of errors found: 0   Number of errors fixed: 0
   Auditing Mechanical Desktop Data complete.
   Specify new insertion point: *Specify the insertion point of the component.*
   Specify insertion point for another instance or <continue>: Enter

7. Choose **OK**.

8. Choose the **Mate** button from the **3D Assembly Constraints** flyout in the **Assembly Modeling** toolbar. The prompt sequence is as follows:

   Select first set of geometry: *Select the edge of the **MOVABLE JAW** as shown in Figure 10-53.*
   First set = Axis, (line)
   Select first set or [Clear/fAce/Point/cYcle] <accEpt>: Enter

   Select second set of geometry: *Select the edge on the **BASE** as shown in Figure 10-53.*
   Second set = Axis, (line)
   Select second set or [Clear/fAce/Point/cYcle] <accEpt>: Enter
   Enter offset <0.0000>: Enter

9. Choose the **Assembly Catalog** button to display the **Assembly Catalog** dialog box.

10. Double-click on the **JAW FACES** under the **Part and Subassembly Definition** area. The prompt sequence is as follows:

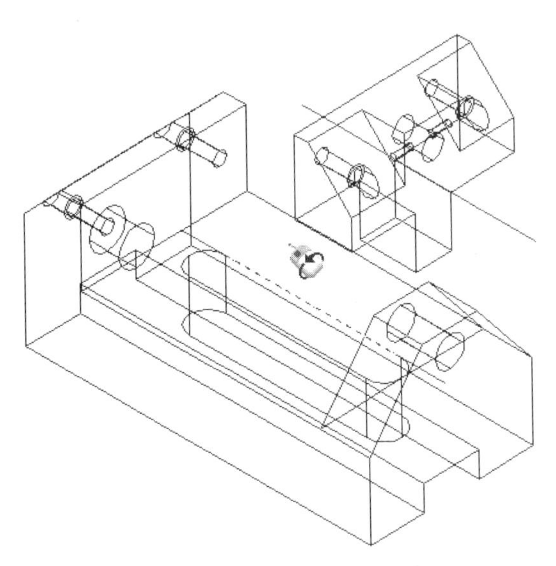

Figure 10-53 *Selecting the edges to apply the Mate constraint*

Number of errors found: 0 Number of errors fixed: 0
Auditing Mechanical Desktop Data complete.
Specify new insertion point: *Specify the insertion point of the component.*
Specify insertion point for another instance or <continue>: *Specify the insertion point for the instance of the component.*
Specify insertion point for another instance or <continue>: Enter

11. Choose **OK**.

12. Choose the **Insert** button from the **3D Assembly Constraints** flyout in the **Assembly Modeling** toolbar. The prompt sequence is as follows:

Select first circular edge: *Select the circular edge of the **JAW FACES** as shown in Figure 10-54.*
First set = Plane/Axis
Enter an option [Clear/Flip] <accEpt>: Enter

Select second circular edge: *Select the circular edge of the **BASE** as shown in Figure 10-54.*
Second set = Plane/Axis
Enter an option [Clear/Flip] <accEpt>: Enter
Enter offset <0.0000>: Enter

13. Similarly, assemble the other jaw face with the movable jaw.

14. Choose the **Assembly Catalog** button to display the **Assembly Catalog** dialog box.

15. Double-click on the **CLAMP SCREW** under the **Part and Subassembly Definition** area. The prompt sequence is as follows:

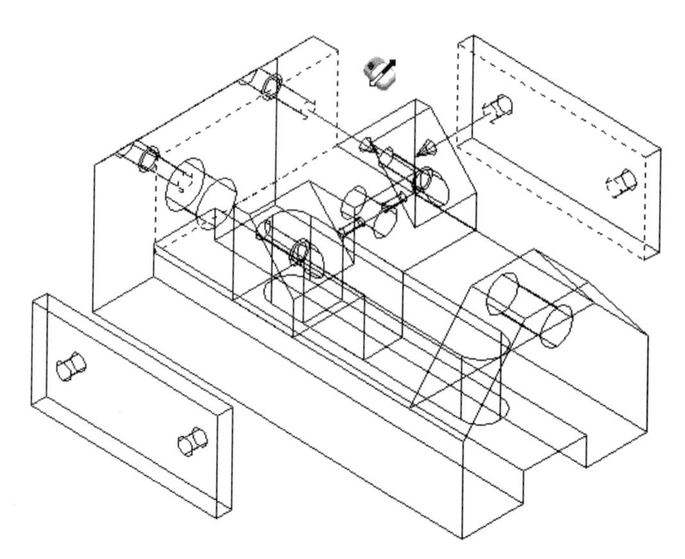

Figure 10-54 Selecting the edges to apply the Insert constraint

Number of errors found: 0 Number of errors fixed: 0
Auditing Mechanical Desktop Data complete.
Specify new insertion point: *Specify the insertion point of the component.*
Specify insertion point for another instance or <continue>: Enter

16. Choose **OK**. Make both the jaw faces invisible using the desktop browser.

17. Choose the **Insert** button from the **3D Assembly Constraints** flyout in the **Assembly Modeling** toolbar. The prompt sequence is as follows:

Select first circular edge: *Select the circular edge of the* ***CLAMP SCREW*** *as shown in Figure 10-55.*
First set = Plane/Axis
Enter an option [Clear/Flip] <accEpt>: Enter

Select second circular edge: *Select the circular edge of the* ***MOVABLE JAW*** *as shown in Figure 10-55.*
Second set = Plane/Axis
Enter an option [Clear/Flip] <accEpt>: Enter

Enter offset <0.0000>: Enter

18. Make the movable jaw and the base invisible using the desktop browser.

19. Choose the **Assembly Catalog** button to display the **Assembly Catalog** dialog box.

20. Double-click on the **CLAMP SCREW HANDLE** under the **Part and Subassembly Definition** area. The prompt sequence is as follows:

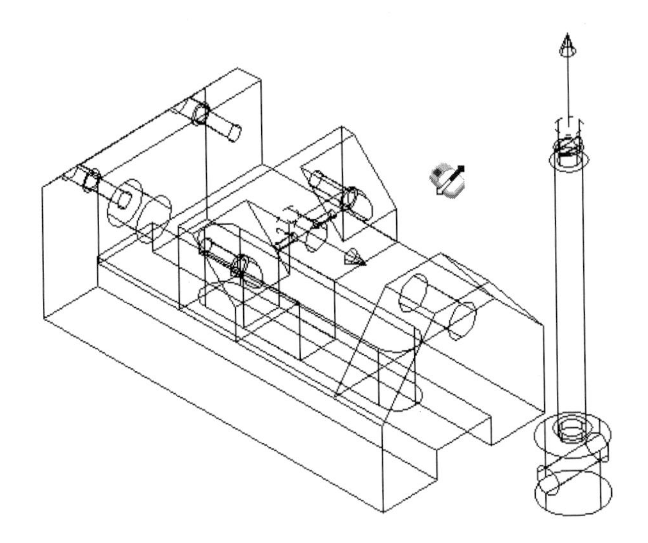

Figure 10-55 *Selecting the edges to apply the Insert constraint*

Number of errors found: 0 Number of errors fixed: 0
Auditing Mechanical Desktop Data complete.
Specify new insertion point: *Specify the insertion point of the component.*
Specify insertion point for another instance or <continue>: Enter

21. Double-click on the **HANDLE STOP** under the **Part and Subassembly Definition** area. The prompt sequence is as follows:

 Number of errors found: 0 Number of errors fixed: 0
 Auditing Mechanical Desktop Data complete.
 Specify new insertion point: *Specify the insertion point of the component.*
 Specify insertion point for another instance or <continue>: *Specify the insertion point for the instance of the component.*
 Specify insertion point for another instance or <continue>: *Press* Enter

22. Choose **OK**.

23. Choose the **Mate** button from the **3D Assembly Constraints** flyout in the **Assembly Modeling** toolbar. The prompt sequence is as follows:

 Select first set of geometry: *Select the axis of the* ***CLAMP SCREW HANDLE*** *as shown in Figure 10-56.*
 First set = Axis, (arc)
 Select first set or [Clear/fAce/Point/cYcle] <accEpt>: Enter

 Select second set of geometry: *Select the axis of the* ***CLAMP SCREW*** *as shown in Figure 10-56.*
 Second set = Axis, (cylinder)
 Select second set or [Clear/Point/Next/cYcle] <accEpt>: Enter

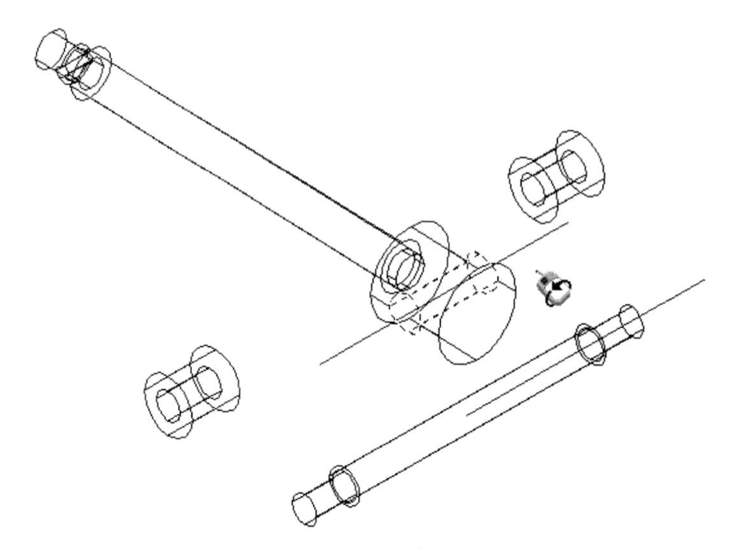

Figure 10-56 *Selecting the axes to apply the Mate constraint.*

Enter offset <0.0000>: Enter

24. Again choose the **Mate** button from the **3D Assembly Constraints** flyout in the **Assembly Modeling** toolbar. The prompt sequence is as follows:

Select first set of geometry: *Select the point on the **CLAMP SCREW HANDLE** as shown in Figure 10-57.*
First set = Point, (arc)
Select first set or [Clear/aXis/fAce/cYcle] <accEpt>: Enter

Select second set of geometry: *Select the point on the **CLAMP SCREW** as shown in Figure 10-57.*
Second set = Point, (spline)
Select second set or [Clear/aXis/cYcle] <accEpt>: Enter

Enter offset <0.0000>: **27**

Tip: *Due to the Mate constraint (Axis) applied to the **CLAMP SCREW HANDLE** and **CLAMP SCREW**, the central axis orientation will not change. The next Mate constraint (Point) will only specify the offset distance.*

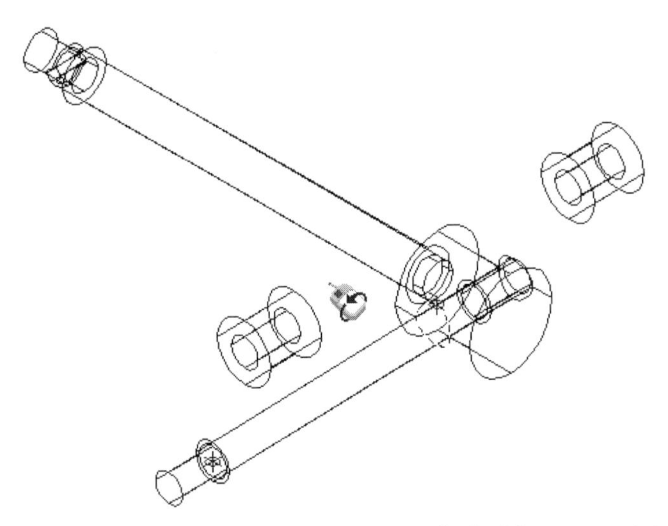

Figure 10-57 *Selecting the points to apply the Mate constraint*

25. Choose the **Insert** button from the **3D Assembly Constraints** flyout in the **Assembly Modeling** toolbar. The prompt sequence is as follows:

Select first circular edge: *Select the circular edge of the **HANDLE STOP** as shown in Figure 10-58.*
First set = Plane/Axis
Enter an option [Clear/Flip] <accEpt>: Enter

Select second circular edge: *Select the circular edge of the **CLAMP SCREW HANDLE** as shown in Figure 10-58.*
Second set = Plane/Axis
Enter an option [Clear/Flip] <accEpt>: Enter

Enter offset <0.0000>: Enter

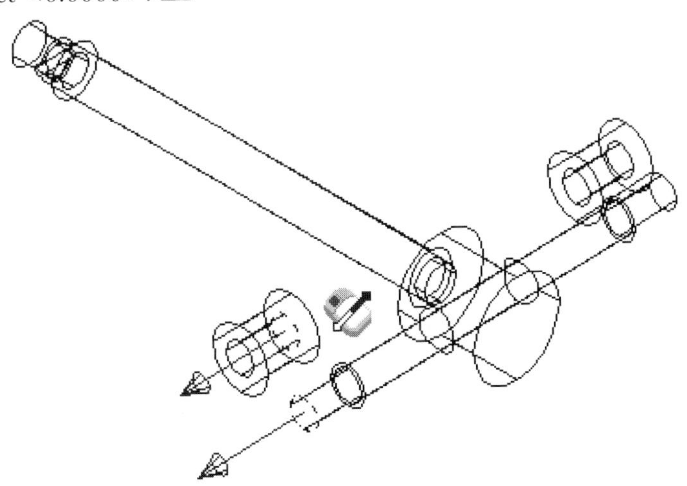

Figure 10-58 *Selecting the edges to apply the Insert constraint*

26. Similarly, assemble the other handle stop.

27. Make the base and movable jaw visible and the rest of the components invisible.

28. Choose the **Assembly Catalog** button to display the **Assembly Catalog** dialog box.

29. Double-click on the **HEADED SCREW** under the **Part and Subassembly Definition** area. The prompt sequence is as follows:

    Number of errors found: 0 Number of errors fixed: 0
    Auditing Mechanical Desktop Data complete.
    Specify new insertion point: *Specify the insertion point of the component.*
    Specify insertion point for another instance or <continue>: *Specify the insertion point for the instance of the component.*
    Specify insertion point for another instance or <continue>: *Specify the insertion point for the next instance of the component.*
    Specify insertion point for another instance or <continue>: *Specify the insertion point for the next instance of the component.*

30. Choose **OK**.

31. Choose the **Insert** button from the **3D Assembly Constraints** flyout in the **Assembly Modeling** toolbar. The prompt sequence is as follows:

    Select first circular edge: *Select the circular edge of the **HEADED SCREW** as shown in Figure 10-59.*
    First set = Plane/Axis
    Enter an option [Clear/Flip] <accEpt>: Enter

    Select second circular edge: *Select the circular edge of the **MOVABLE JAW** as shown in Figure 10-59.*
    Second set = Plane/Axis
    Enter an option [Clear/Flip] <accEpt>: Enter

    Enter offset <0.0000>: Enter

32. Similarly, assemble the remaining headed screws.

33. Choose the **Assembly Catalog** button to display the **Assembly Catalog** dialog box.

34. Double-click on the **SAFETY HANDLE** under the **Part and Subassembly Definition** area. The prompt sequence is as follows:

    Number of errors found: 0 Number of errors fixed: 0
    Auditing Mechanical Desktop Data complete.
    Specify new insertion point: *Specify the insertion point of the component.*
    Specify insertion point for another instance or <continue>: Enter

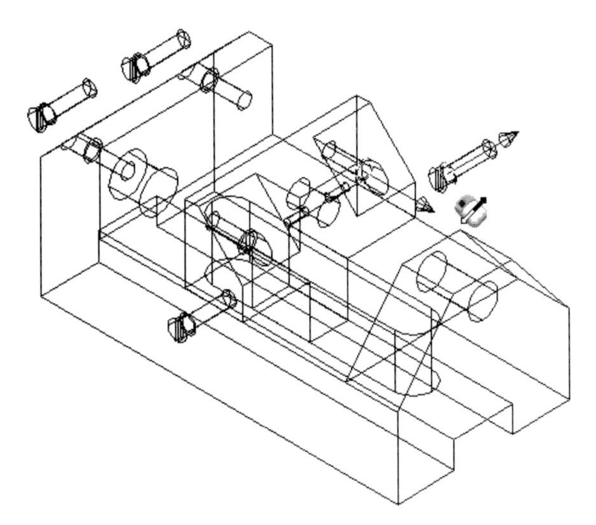

Figure 10-59 Selecting the edges to apply the Insert constraint

35. Choose **OK**.

36. Choose the **Insert** button from the **3D Assembly Constraints** flyout in the **Assembly Modeling** toolbar. The prompt sequence is as follows:

Select first circular edge: *Select the circular edge of the **SAFETY HANDLE** as shown in Figure 10-60.*
First set = Plane/Axis
Enter an option [Clear/Flip] <accEpt>: Enter

Select second circular edge: *Select the circular edge of the **BASE** as shown in Figure 10-60.*
Second set = Plane/Axis
Enter an option [Clear/Flip] <accEpt>: Enter
Enter offset <0.0000>: Enter

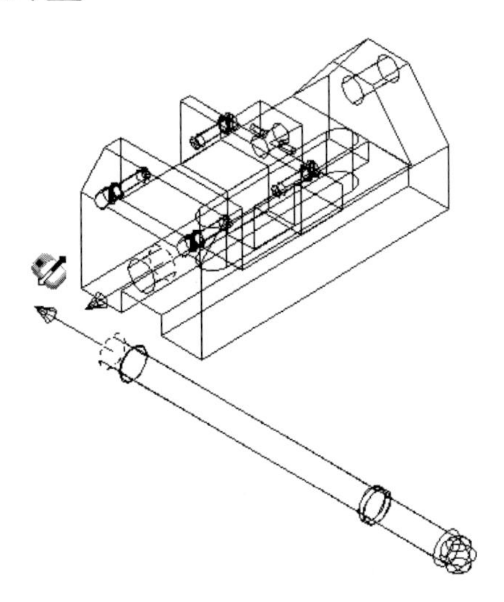

Figure 10-60 Selecting the edges to apply the Insert constraint

37. The final assembly should be similar to the one shown in Figure 10-61.

38. Save this drawing with the name given below:

    \MDT Tut\Ch-10\Assembly**Tut2.dwg**

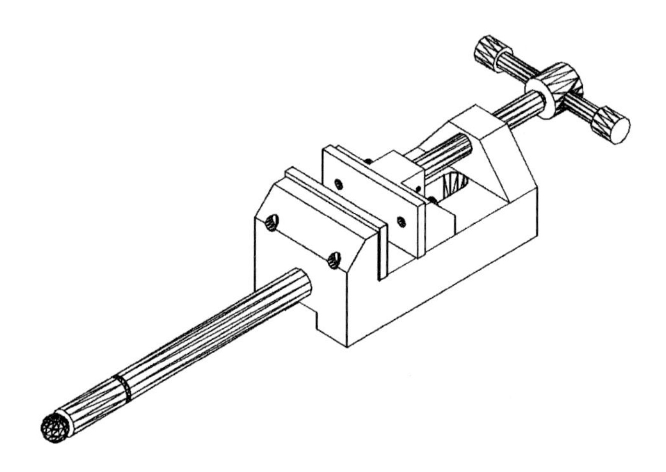

Figure 10-61 *Drill Press Vice Assembly*

Review Questions

Answer the following questions.

1. What are the various options for creating the assemblies?

2. The drawing views of the component can be copied to another drawing using the **AMCOPYOUT** command. (T/F)

3. The components copied in using the **AMCOPYIN** command are local parts. (T/F)

4. Which command is used to copy the components as external parts?

5. What are the various types of assembly constraints?

Exercise

Exercise 1

Create all components of the Screw Jack assembly and then assemble them as shown in Figure 10-62 and 10-63. Figures 10-64 through 10-67 give the dimensions of all components of Screw Jack assembly.

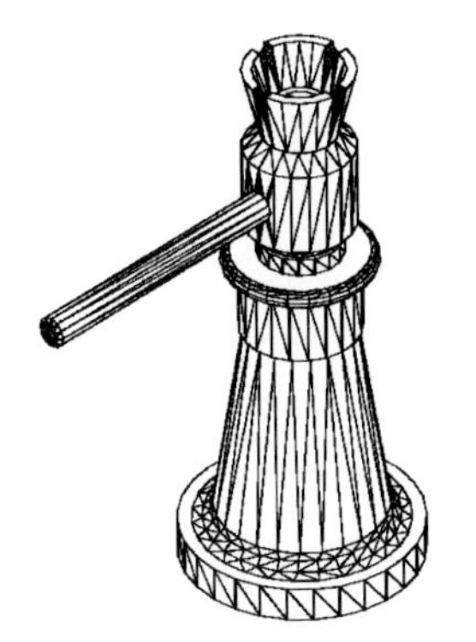

Figure 10-62 Screw Jack assembly

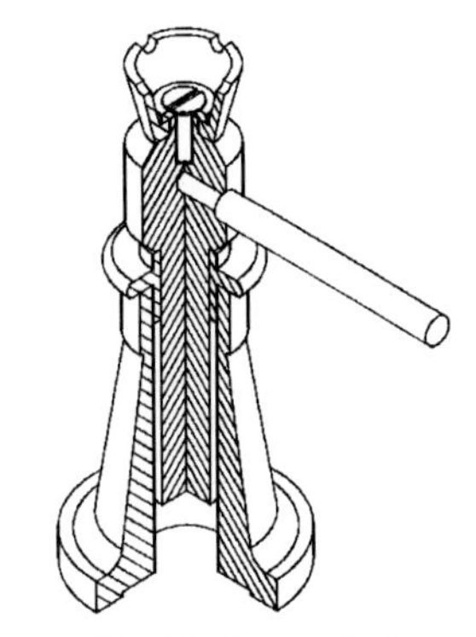

Figure 10-63 Screw Jack assembly

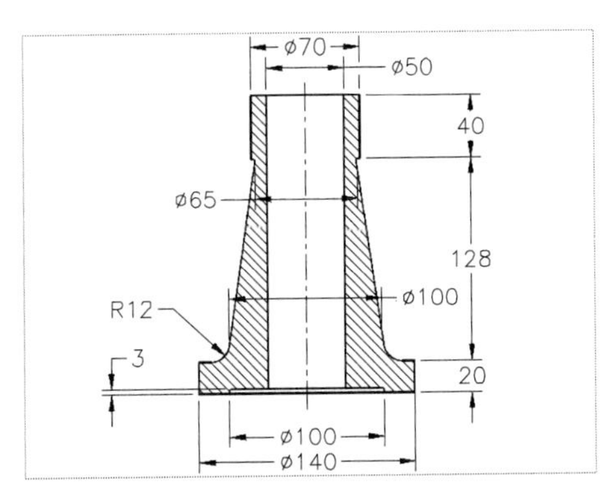

Figure 10-64 Details of CASTING

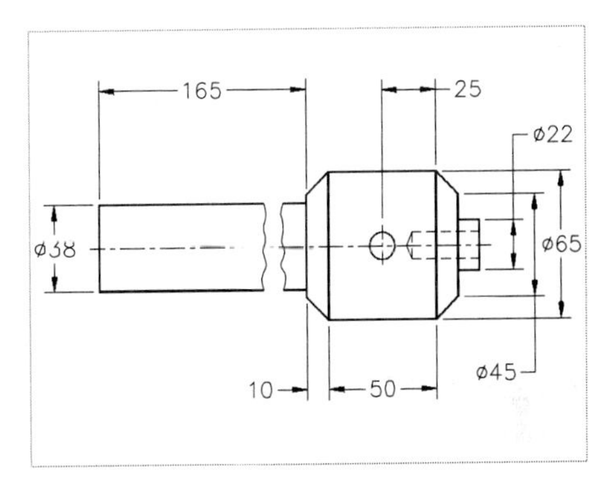

Figure 10-65 Details of SCREW

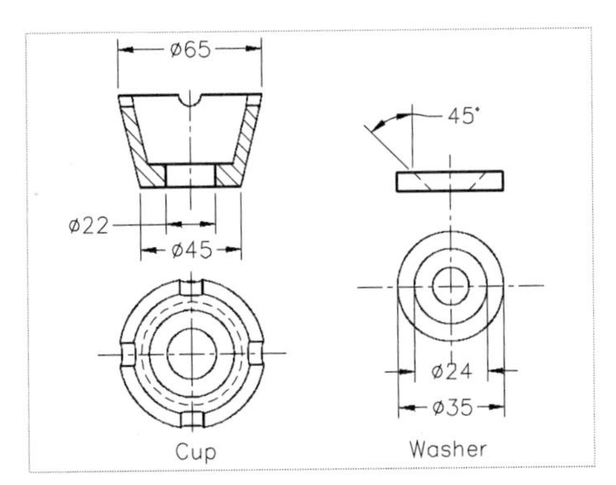

Figure 10-66 Dimensions for Exercise 1

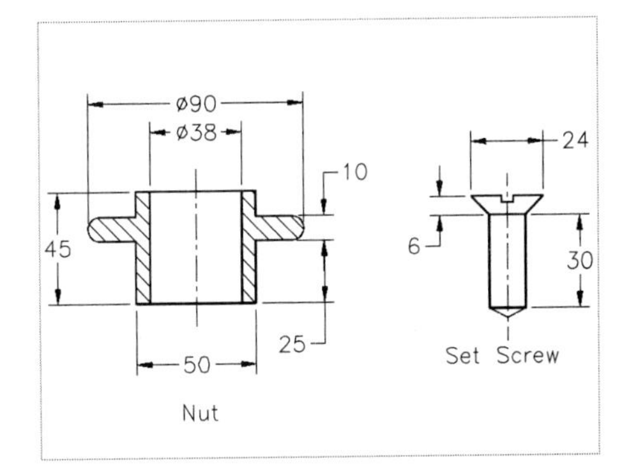

Figure 10-67 Dimensions for Exercise 1

Chapter 11

Assembly Modeling II

Learning Objectives

After completing this chapter, you will be able to:

- *Analyze the assemblies.*
- *Create scenes.*
- *Create exploded views of the assembly in the scenes.*
- *Edit the exploded views.*
- *Add tweaks and trails to the exploded assembly.*
- *Edit or delete the tweaks and trails.*

Commands Covered

- ***AMASSMPROP***
- ***AMINTERFERE***
- ***AMDIST***
- ***AMNEW***
- ***AMXFACTOR***
- ***AMTWEAK***
- ***AMDELTWEAKS***
- ***AMTRAIL***
- ***AMEDITTRAIL***
- ***AMDELTRAIL***

ANALYZING THE ASSEMBLIES

Once you have assembled the components and created the assembly, you need to analyze it by calculating the mass property of the components, check them for interference or calculate minimum distance between two selected points. This allows you to increase the efficiency of the assembly and reduce the material loss during manufacturing.

Calculating The Mass Properties Of The Components Using The AMASSMPROP Command

Toolbar:	Assembly Modeling > Mass Property
Menu:	Assembly > Analysis > Mass Properties
Context Menu:	Assembly Menu > Analysis > Mass Properties
Command:	AMASSMPROP

This command is used to calculate the mass properties of the components of assembly. You can assign a desired material to the component and then calculate the properties based on that material. The calculated mass property can be written to a .MPR file. When you invoke this command, you will be prompted to select the part or subassembly whose mass properties you wish to calculate. You can directly select the part or subassembly from the drawing area or display the names of the parts that can be selected. As soon as you select the part or subassembly, the **Assembly Mass Properties** dialog box will be displayed, see Figure 11-1.

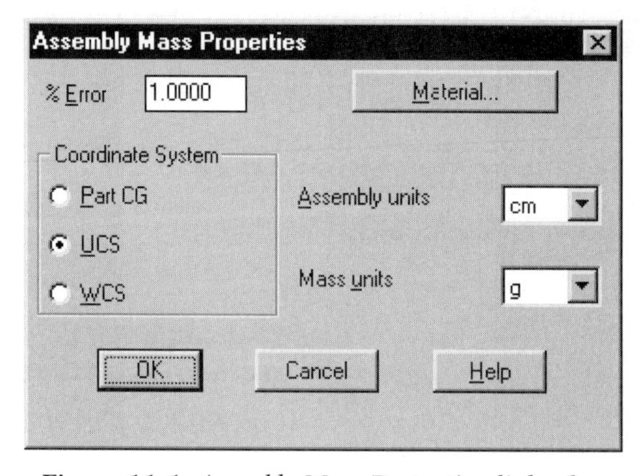

Figure 11-1 *Assembly Mass Properties dialog box*

Assembly Mass Properties Dialog Box Options

The options provided in this dialog box are:

%Error

This edit box displays the maximum error allowed, in the component, in terms of percentage.

Material

This button is chosen to assign the desired material to the component. When you choose this button, the **Select Material** dialog box will be displayed as shown in Figure 11-2. You can select the desired material from the drop-down list provided in this dialog box. When you select a particular material, the specifications related to that material are displayed in this dialog box. If initially you have selected more than one part, then to assign the material to a particular part or subassembly, select the material from the drop-down list and then select the part or subassembly from the **Part/Subassemblies definition** area. Now, choose the **Assign** button to assign the desired material to the selected part or subassembly.

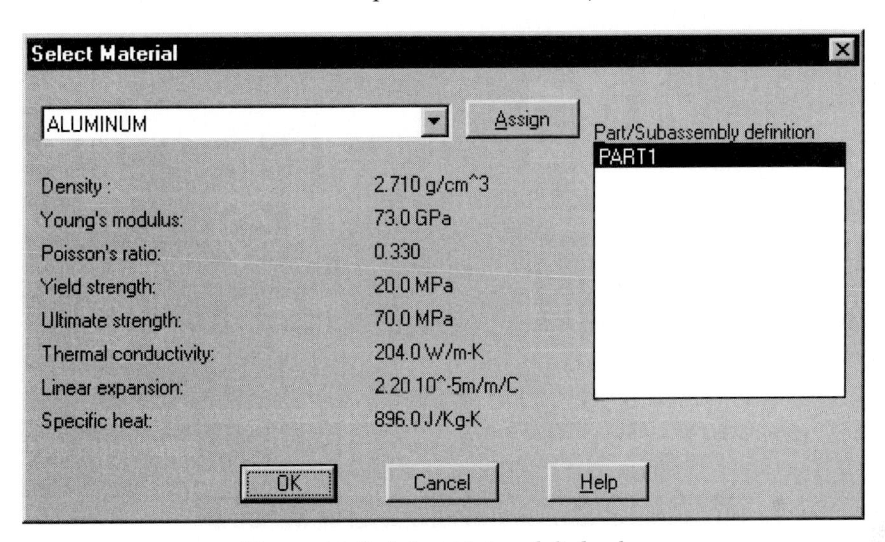

Figure 11-2 *Select Material dialog box*

Coordinate System Area

This area provides the option of calculating the mass property based on the center of gravity of the part or subassembly you have selected, the current user coordinate system or the world coordinate system.

Assembly units

This drop-down list provides the units for measurement of length in a part or a subassembly. The measurements can be made in terms of inches, centimeters or millimeters.

Mass units

This drop-down list provides the units in which the mass property of the part or subassembly has to be calculated. The units for the mass can be selected in the terms of pounds, grams or kilograms.

Once you have assigned the material to the component and selected the coordinate system and units, choose **OK**. The **Assembly Mass Property Results** dialog box will be displayed as shown in Figure 11-3.

You can write the results of the assembly mass property in the .MPR file by choosing the **File**

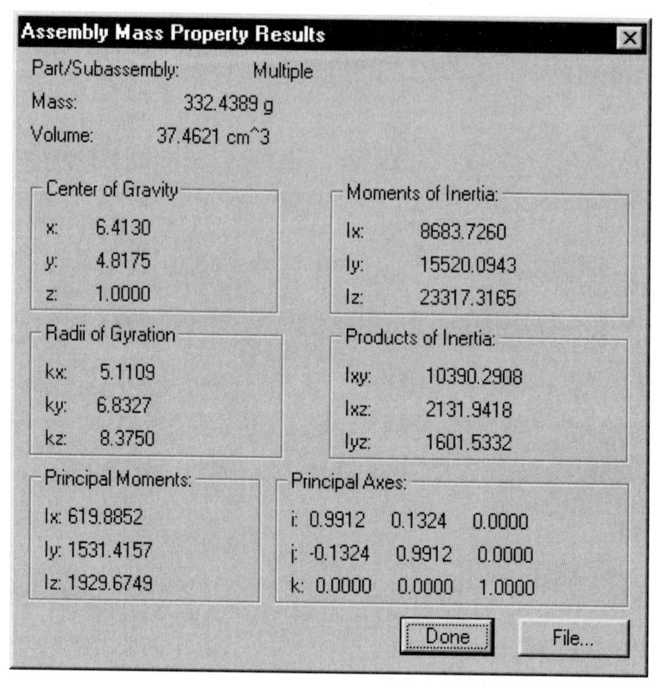

Figure 11-3 *Assembly Mass property Results dialog box*

button from the **Assembly Mass Property Results** dialog box. You can specify the name of the file in the **New Massprop File** dialog box, see Figure 11-4.

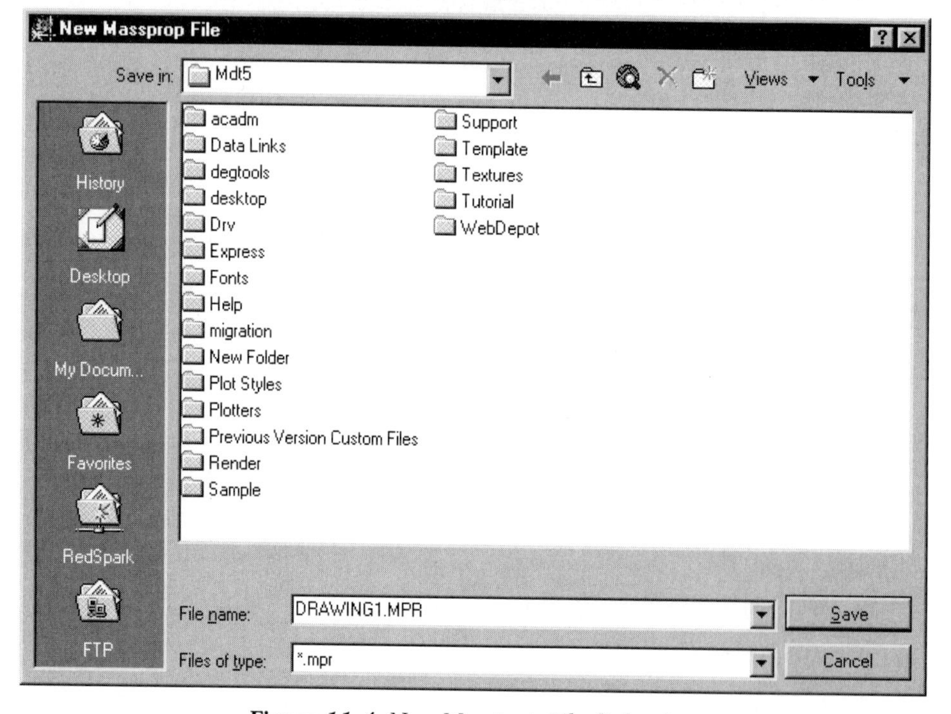

Figure 11-4 *New Massprop File dialog box*

Checking The Assemblies For Interference Using The AMINTERFERE Command

Toolbar: Assembly Modeling > Mass Property > Check 3D Interference
Menu: Assembly > Analysis > Check Interference
Context Menu: Assembly Menu > Analysis > Check Interference
Command: AMINTERFERE

This command is used to check the interference between the selected set of components by analyzing their position with respect to each other. You will be given an option of creating the interference solid that can be used to analyze the interference. You can also select the nested parts or subassemblies to check the interference along with the main assembly. The options provided under this command are:

Nested part or subassembly selection?

This option is used if you wish to select the nested parts or subassemblies.

Yes

If you select **Yes** in the **Nested part or subassembly selection** prompt all the nested parts or the subassemblies inside the selected components will be selected to be checked for interference. All subassemblies or parts will be displayed, one by one, to be checked for interference.

No

If you select **No** in the **Nested part or subassembly selection** prompt all the nested parts or the subassemblies inside the selected components will be neglected and only the selected components will be checked for interference.

Create interference solids?

This option is used if you wish to create the interference solid.

Yes

If you select **Yes** in the **Create interference solid** prompt then Mechanical Desktop will create the interference solid. This solid can be used to calculate the amount of interference between the selected components of assembly. However, the new interference solid created will be a 3D solid part and not a designer part.

No

If you select **No** in the **Create interference solid** prompt there will be no interference solid created.

Highlight pairs of interfering parts/subassemblies?

This option allows you to highlight the interfering parts.

Yes

If you select **Yes** in the **Highlight pairs of interfering parts/subassemblies** prompt, all the components that are interfering will be highlighted one by one.

No

If you select **No** in the **Highlight pairs of interfering parts/subassemblies** prompt the interfering components will not be highlighted.

Calculating The Minimum 3D Distance Between The Selected Components Using The AMDIST Command

Toolbar: Assembly Modeling > Mass Property > Minimum 3D Distance
Menu: Assembly > Analysis > Minimum 3D Distance
Context Menu: Assembly Menu > Analysis > Minimum 3D Distance
Command: AMDIST

This command is used to calculate minimum distance between the selected components in the 3D space. You will be prompted to select the first set of components and the second set of components to calculate the distance. The result of this command can be viewed either as a numeric value in the command line using the **Display** option or can be displayed as a line between the selected components along with the numeric value displayed in the command line using the **Line** option.

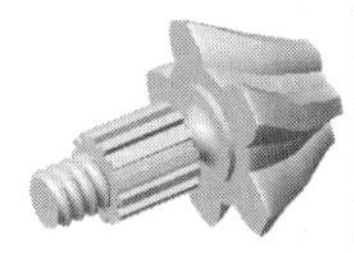

Tip: *When you wish to view the result of the **AMDIST** command as a numeric value in the command line, an imaginary line displaying the minimum distance is created. This line can be removed by any command that leads to the regeneration of drawing.*

CREATING THE SCENES

The scenes are created to explode the assemblies thus providing a better visualization of the mating components and their location in the assembly. The scenes can be created using the **Scene** option of the **AMNEW** command. You can also create a scene by selecting **Assembly Menu > New Scene** from the context menu. New scene can also be created by choosing the **New Scene** button from the **Scenes** toolbar.

EXPLODING THE ASSEMBLIES

Exploding the assembly makes the components of the assembly move by the specified value along the direction of constraint. However, only the components assembled using the mate or the insert constraint can be exploded. If the components are assembled using more than one mate or insert constraint on one component then the components do not explode. The reason for this is that in this case the components will have more than one direction to move. Thus, the components prefer to stay at their original location.

Exploding The Assembly Using The Command Line

The assemblies can be exploded using the command line with the help of the **AMNEW** or the **AMXFACTOR** command. Use the **Scene** option of the **AMNEW** command to create a scene and also to define the explosion factor for the scene. All of the components of the assembly will be exploded with equal explosion factor that is specified in the **AMNEW** command. However, you cannot specify different explosion factors for different components using this command. To specify different explosion factors for different components you have to use the **AMXFACTOR** command.

AMXFACTOR Command (Scene Explosion Factor)

Toolbar: Scenes > Scene Explosion Factor
Menu: Assembly > Exploded Views > Scene Explosion Factor
Context Menu: Scene Explosion Factor
Command: AMXFACTOR

This option of the **AMXFACTOR** command is used to specify the overall explosion factor for the scene. Figure 11-5 shows an unexploded assembly and Figure 11-6 shows an assembly with the scene explosion factor of 10.

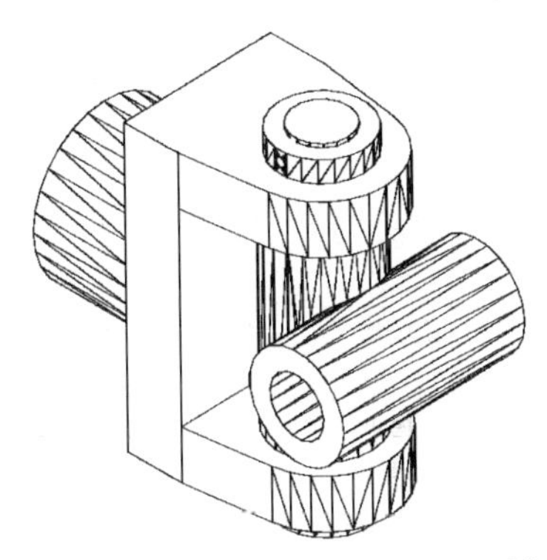

Figure 11-5 *Unexploded assembly*

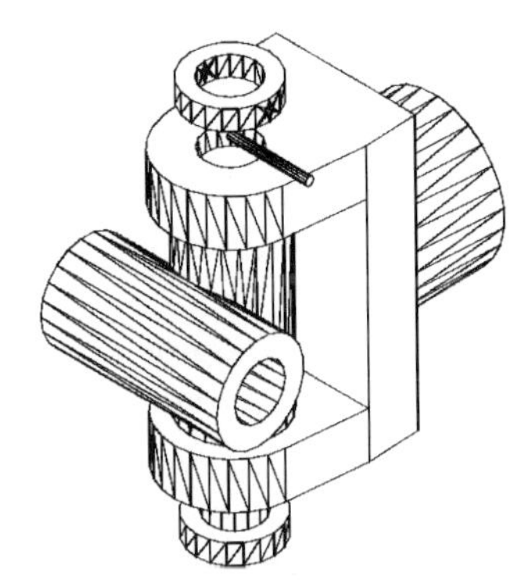

Figure 11-6 *Assembly with 25 explosion factor*

AMXFACTOR Command (Part or Subassembly Explosion Factor)

Toolbar: Scenes > Part or Subassembly Explosion Factor
Menu: Assembly > Exploded Views > Part Explosion Factor
Context Menu: Part Explosion Factor
Command: AMXFACTOR

This option of the **AMXFACTOR** command is used to specify different explosion factor for each part or subassembly. When you invoke this option, you will be prompted to select the part or subassembly. You can specify a new explosion factor or can reset the predefined explosion factor, see Figures 11-7 and 11-8.

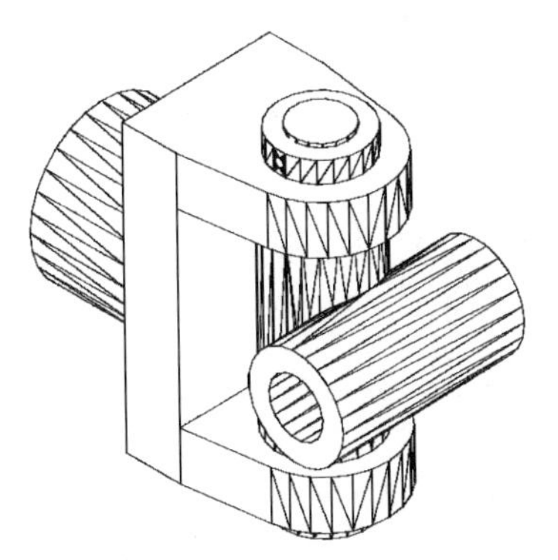

Figure 11-7 *Unexploded assemblies*

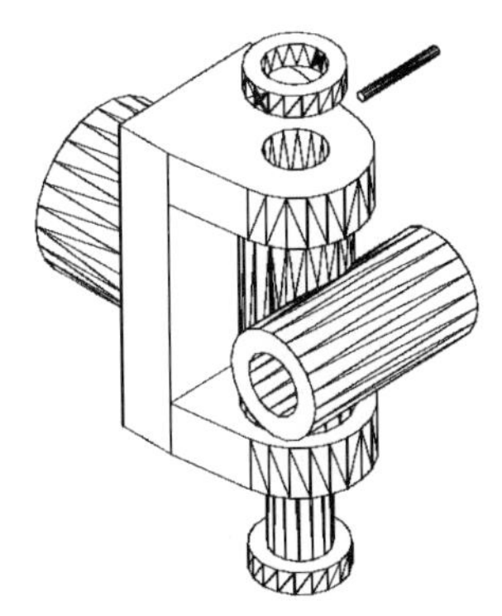

Figure 11-8 *Different explosion factor*

Exploding The Assembly Using The Desktop Browser

You can specify the scene explosion factor or the part explosion factor using the desktop browser. To specify the scene explosion factor, right-click on the scene to be exploded in the tree view hierarchy of the desktop browser to display the shortcut menu, see Figure 11-9.

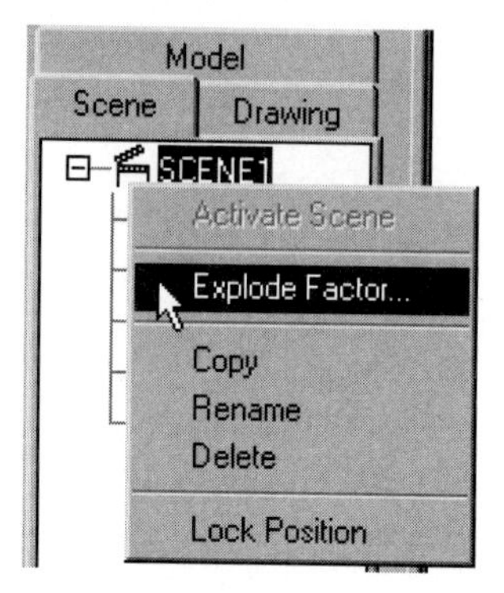

Figure 11-9 *Specifying the scene explosion factor using the desktop browser*

Choose the **Explode Factor** option to display the **Explode Factor** dialog box as shown in Figure 11-10. You can specify the desired explosion factor in this dialog box.

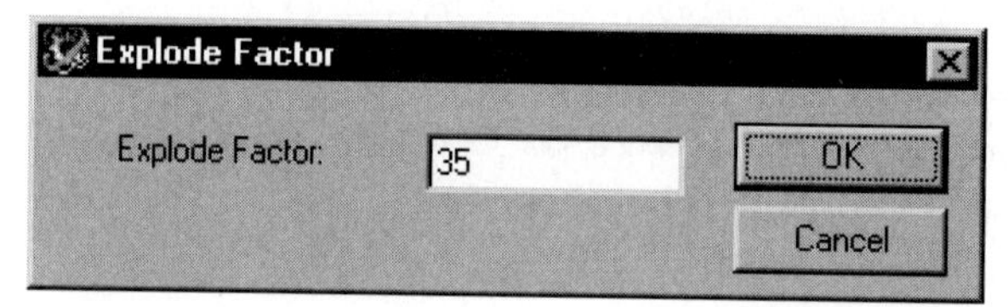

Figure 11-10 *Explode Factor dialog box for specifying the scene explosion factor*

To specify the explosion factor for the part or the subassembly, right-click on the selected part in the desktop browser to display the shortcut menu, see Figure 11-11.

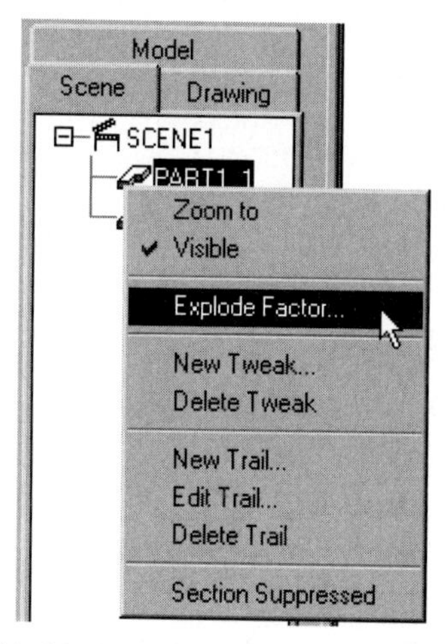

Figure 11-11 *Specifying the part explosion factor*

Choose the **Explode Factor** option to display the **Explode Factor** dialog box as shown in Figure 11-12. Specify the part explosion factor in this dialog box.

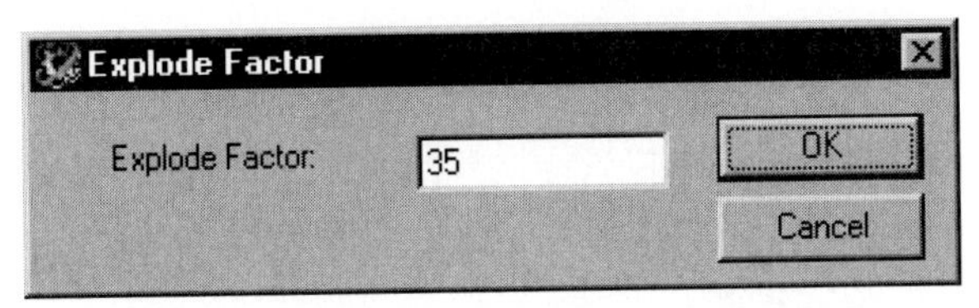

Figure 11-12 *Explode Factor dialog box for specifying the part explosion factor*

ADDING TWEAKS TO THE COMPONENTS OF ASSEMBLY

As you know that exploding the assembly makes the components of the assembly move in the direction of the assembly constraint applied to the components. Therefore, sometimes exploding of the assembly gives undesirable results as shown in Figure 11-13.

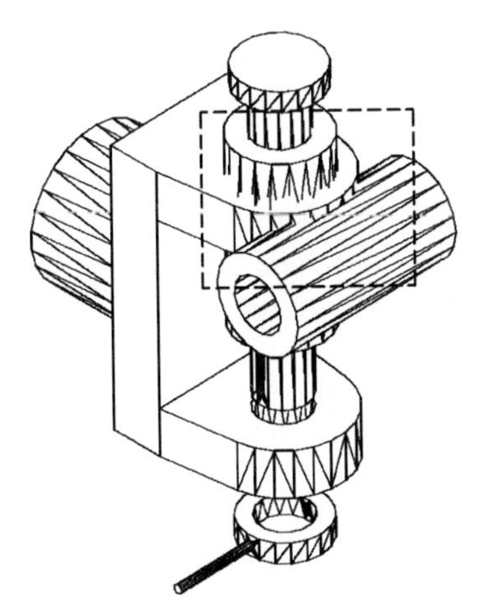

Figure 11-13 *The exploded assembly*

In such cases you need to tweak the components causing the problem instead of exploding them. Tweaking is defined as the method of redefining the position of the components in the scenes. Tweaking gives you the flexibility of moving the constrained component in any specified direction. You can even rotate the component in any specified direction. You can tweak the component using the **AMTWEAK** command or by using the desktop browser.

Tweaking The Components Using The AMTWEAK Command

Toolbar:	Scenes > New Tweak
Menu:	Assembly > Exploded Views > New Tweak
Context Menu:	New Tweak
Command:	AMTWEAK

This command is used to add tweaks by moving or rotating the selected component in the scene. When you invoke this command, you will be asked to select the component to be tweaked. One you have selected the component, the **Tweak Part/Subassembly** dialog box will be displayed as shown in Figure 11-14.

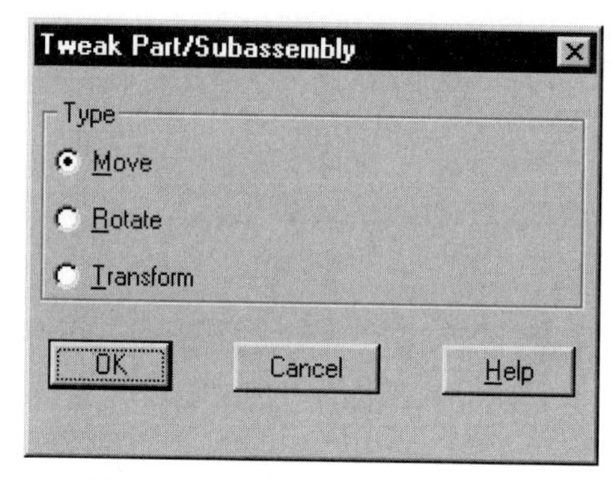

Figure 11-14 *Tweak Part/Subassembly dialog box*

Tweak Part/Subassembly Dialog Box Options

The options provided under this dialog box are:

Move

The **Move** option is used to move the selected component taking the reference of another component. When you select this option and choose **OK**, all components that can be used for reference remain visible along with the selected component and the rest of them are made invisible. You will be prompted to select a reference component and the direction in which the component will be moved is displayed with an arrow. To move the component in the opposite direction, enter a negative value.

Rotate

This option is used to rotate the selected component taking the reference of another component. When you select this option and choose **OK**, all components that can be used for reference remain visible along with the selected component and the rest of them are made invisible. You will be prompted to select a reference component and the direction in which the component will be rotated is displayed with an arrow. To rotate the component in the opposite direction, enter a negative angle value.

Transform

This option is a combination of the **Move** and the **Rotate** options. You can move and rotate the selected component by selecting this option. The main advantage of this option over the other two options is that you do not require a reference object in this case. This option has the following two sub-options:

Move. This sub-option is used to move the selected component. The component can be moved using various methods. The first method is the **Specify start point** method. Using this method you can sketch the direction for movement by selecting two points in the drawing area. The second method is the **View Direction** method. This method uses the current view to define the direction for movement. The third method is the **Wire** method. This method uses 2D entities like line, polyline or arc to define the view direction. Apart from these methods you can also move the selected components along the X, Y or the Z direction using the **X**, **Y** or the **Z** method.

Rotate. This sub-option is used to rotate selected component about a selected center.

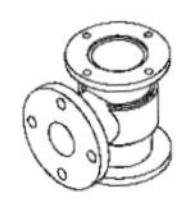

Note

*The methods of rotation are same as those of the **Move** option discussed above.*

Tweaking The Components Using The Desktop Browser

You can also tweak the components using the desktop browser. Right-click on the component to be tweaked in the tree view of the desktop browser to display the shortcut menu as shown in Figure 11-15.

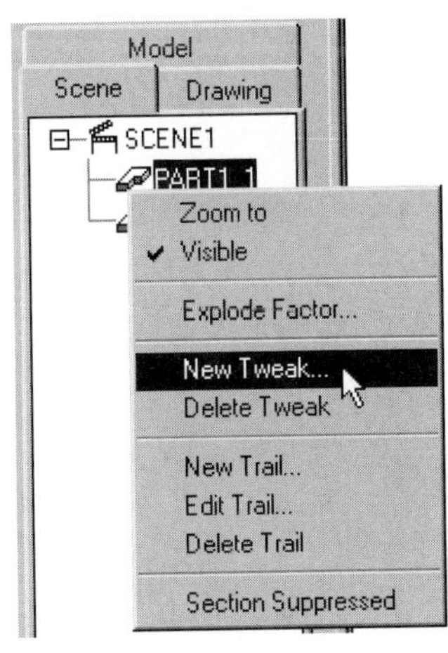

***Figure 11-15** Tweaking the component using the desktop browser*

Choose the **New Tweak** option to display the **Tweak Part/Subassembly** dialog box. The options provided in this dialog box are similar to those discussed in the **AMTWEAK** command.

DELETING THE TWEAKS

The tweaks added to the components can be deleted either using the **AMDELTWEAKS** command or by using the desktop browser.

Deleting The Tweaks Using The AMDELTWEAKS Command

Toolbar:	Scenes > Delete Tweak
Menu:	Assembly > Exploded Views > Delete Tweak
Context Menu:	Delete Tweak
Command:	AMDELTWEAKS

This command is used to delete the tweaks added to the components in the scenes. Once the tweaks are deleted, the selected component is moved back to its original location.

Deleting The Tweaks Using The Desktop Browser

You can also delete the tweaks using the desktop browser. Right-click on the component in the tree view of the desktop browser to display the shortcut menu, see Figure 11-16. Choose the **Delete Tweak** option to delete the tweaks.

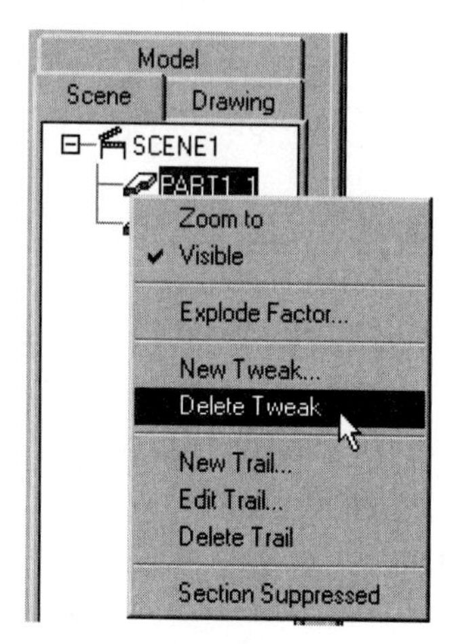

***Figure 11-16** Deleting the tweaks using the desktop browser*

ADDING TRAILS TO THE EXPLODED OR TWEAKED COMPONENTS

The trails are the parametric lines displaying the path and the direction of the mating components. This makes the visualization of the exploded or the tweaked component easier. The trails can be added to the component either by using the **AMTRAIL** command or by using the desktop browser.

Adding The Trails Using The AMTRAIL Command

Toolbar:	Scenes > New Trail
Menu:	Assembly > Exploded Views > New Trail
Context Menu:	New Trail
Command:	AMTRAIL

When you invoke this command, you will be prompted to select the component to which the trail has to be added. Once you select the component, the **Trail Offsets** dialog box will be displayed as shown in Figure 11-17.

Trail Offsets Dialog Box Options

The options provided under this dialog box are:

Offset at Current Position Area

This area controls the creation of the trails at the exploded or the tweaked position.

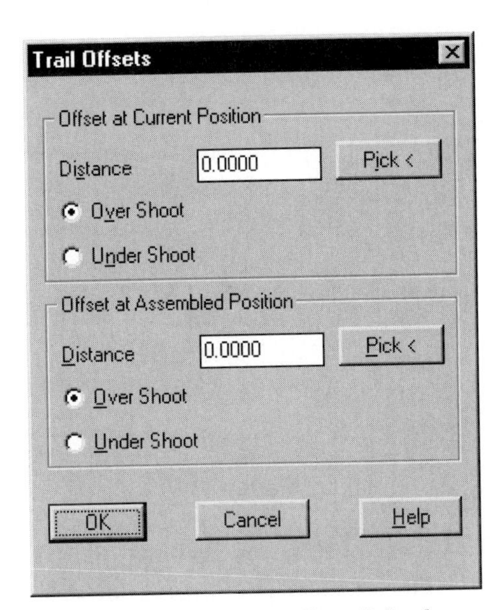

Figure 11-17 *Trail Offsets dialog box*

Distance. The **Distance** edit box is used to specify the length of the trail at the exploded or the tweaked position.

Over Shoot. If this radio button is selected, the trail will be extended beyond the mating point, edge or the face at the exploded or the tweaked position. The overshoot distance is specified in the **Distance** edit box.

Under Shoot. If this radio button is selected, the trail will fall short of the mating point, edge or the face at the exploded or the tweaked position. The undershoot distance is specified in the **Distance** edit box.

Offset at Assembled Position Area

This area controls the creation of the trails at the original assembled position.

Distance. The **Distance** edit box is used to specify the length of the trail at the assembled position.

Over Shoot. If this radio button is selected, the trail will be extended beyond the mating point, edge or the face at the assembled position. The overshoot distance is specified in the **Distance** edit box.

Under Shoot. If this radio button is selected, the trail will fall short of the mating point, edge or the face at the assembled position. The undershoot distance is specified in the **Distance** edit box.

Figure 11-18 shows an unexploded assembly and Figure 11-19 shows an exploded assembly displaying the trails.

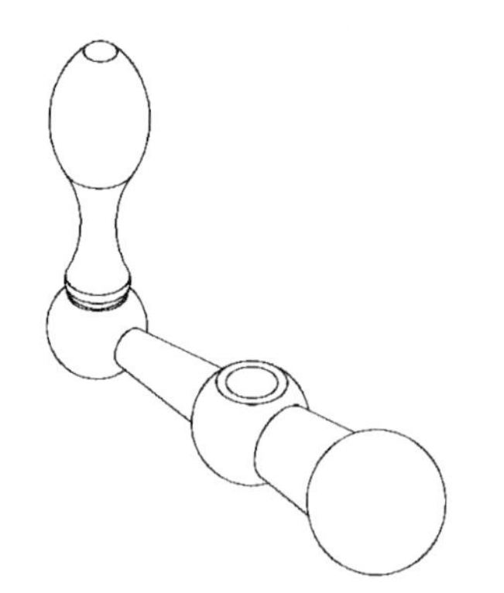

Figure 11-18 *Unexploded assembly*

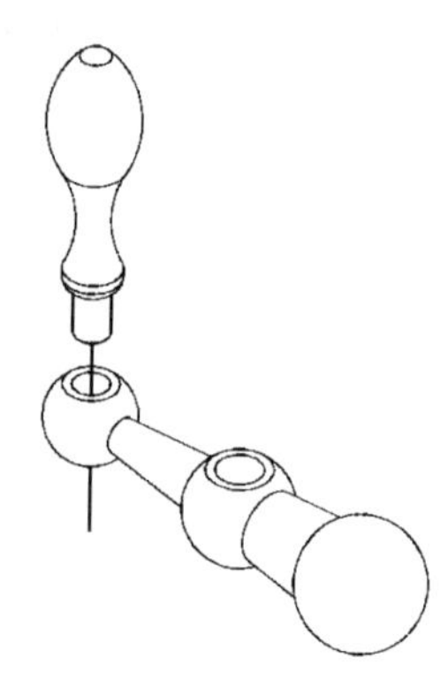

Figure 11-19 *Exploded assembly with trails displaying the path*

Adding The Trails Using The Desktop Browser

You can also add trails to the selected component using the desktop browser. Right-click on the component to display the shortcut menu, see Figure 11-20.

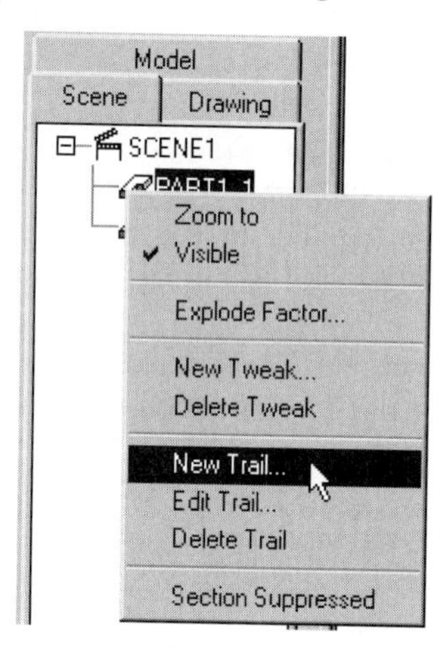

Figure 11-20 *Creating the trails using the desktop browser*

Choose the **New Trail** option and select the component. The **Trails Offsets** dialog box will be displayed and you can specify the required values in this dialog box.

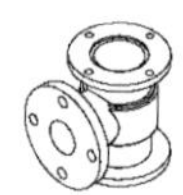

Note

*The **Trail Offsets** dialog box displayed using the desktop browser is the same as that displayed using the **AMTRAIL** command.*

EDITING THE TRAILS

The trails added to the components can be edited either using the **AMEDITTRAIL** command or by using the desktop browser.

Editing The Trail Using The AMEDITTRAIL Command

Toolbar:	Scenes > Edit Trail
Menu:	Assembly > Exploded Views > Edit Trail
Context Menu:	Edit Trail
Command:	AMEDITTRAIL

When you invoke this command, you will be asked to select the trail to be edited. Once you have selected the trail, the **Trail Offsets** dialog box will be displayed with the original settings. You can edit the desired settings from this dialog box.

Editing The Trail Using The Desktop Browser

To edit the trail using the desktop browser, right-click on the component in the desktop browser to display the shortcut menu. Choose the **Edit Trail** option and select the trail to be edited. The **Trail Offsets** dialog box will be displayed and you can make the desired modification in this dialog box.

DELETING THE TRAILS

The trails added to the components can be deleted either by using the **AMDELTRAIL** command or by using the desktop browser.

Deleting The Trail Using The AMDELTRAIL Command

Toolbar:	Scenes > Delete Trail
Menu:	Assembly Exploded Views > Delete Trail
Context Menu:	Delete Trail
Command:	AMDELTRAIL

When you invoke this command, you will be asked to select the trail to be deleted. As soon as you select the trail, it will be deleted. No warning will be displayed before deleting the trails.

Deleting The Trail Using The Desktop Browser

To delete the trail using the desktop browser, right-click on the component in the desktop browser to display the shortcut menu. Choose the **Delete Trail** option and select the trail to be deleted.

TUTORIALS

Tutorial 1

In this tutorial you will analyze and explode the Plummer Block assembly created in Chapter 10. Add tweaks and trails to the exploded view.

1. Open the file C:\MDT Tut\Ch-10\Assembly**Tut1.dwg**.

2. Choose the **Check 3D Interference** button from the **Mass Properties** flyout in the **Assembly Modeling** toolbar. The prompt sequence is as follows:

Nested part or subassembly selection? [Yes/No] <No>: Enter
Select first set of parts or subassemblies: *Select* ***CASTING***.
Select first set of parts or subassemblies: Enter

Select second set of parts or subassemblies: *Select* ***CAP***.
Select second set of parts or subassemblies: *Select* ***BRASSES***.
Select second set of parts or subassemblies: *Select* ***NUT 1***.
Select second set of parts or subassemblies: *Select* ***LOCK NUT 1***.
Select second set of parts or subassemblies: *Select* ***BOLT 1***.
Select second set of parts or subassemblies: *Select* ***NUT 2***.
Select second set of parts or subassemblies: *Select* ***LOCK NUT 2***.
Select second set of parts or subassemblies: *Select* ***BOLT 2***.
Select second set of parts or subassemblies: Enter

Parts/subassemblies do not interfere.

As the parts do not interfere, therefore, the assembly is error free.

Tip: *If the parts or the subassemblies interfere, create the interference solid and then remove that much material from the interfering components. Right-click on the interfering component in desktop browser and choose the* ***Open to Edit*** *option for modification.*

3. Choose the **Scene** tab from the desktop browser to shift to the scene mode.

4. Choose the **New Scene** button from the **Scenes** toolbar to create a new scene. The prompt sequence is as follows:

Enter new scene name of the active assembly (PBASS) <SCENE1>: Enter
Enter overall explosion factor <0.0000>: **20**
Activate new scene? [Yes/No] <Yes>: Enter

5. The exploded assembly should look similar to the one shown in Figure 11-21.

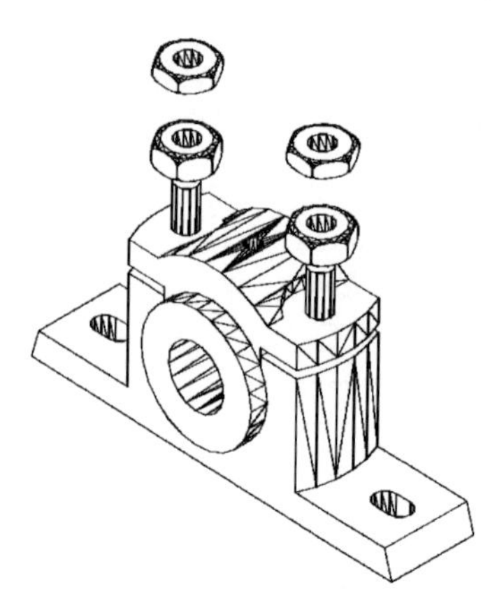

Figure 11-21 *Exploded assembly*

6. Choose the **New Tweak** button from the **Scenes** toolbar and select the **CAP** to display the **Tweak Part/Subassembly** dialog box as shown in Figure 11-22.

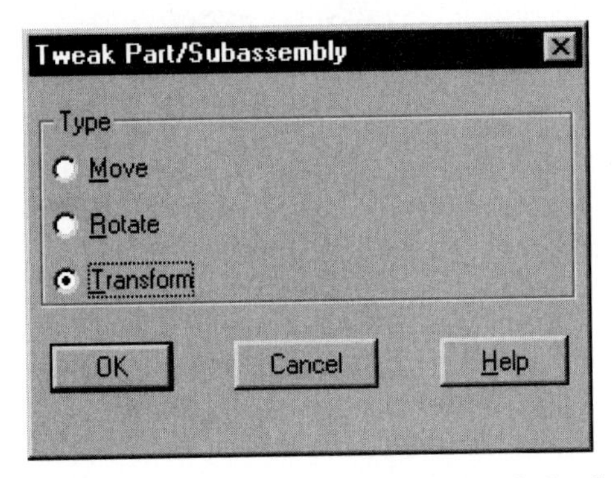

Figure 11-22 *Tweak Part/Subassembly dialog box*

7. Select the **Transform** radio button and choose **OK**. The prompt sequence is as follows:

   Enter an option [eXit/Move/Rotate] <Move>: Enter
   Define direction and length:
   Specify start point or [Viewdir/Wire/X/Y/Z]: **Z**
   Enter length <1.0000>: **25**
   Enter an option [Flip/Accept] <Accept>: *Make sure the direction of movement is positive Z.*
   Enter an option [eXit/Move/Rotate] <Move>: **X**

8. Choose the **New Tweak** button from the **Scenes** toolbar and select the **BOLT** to display the **Tweak Part/Subassembly** dialog box.

9. Select the **Transform** radio button and choose **OK**. The prompt sequence is as follows:

   Enter an option [eXit/Move/Rotate] <Move>: Enter
   Define direction and length:
   Specify start point or [Viewdir/Wire/X/Y/Z]: **Z**
   Enter length <1.0000>: **50**
   Enter an option [Flip/Accept] <Accept>: *Make sure the direction of movement is negative Z.*
   Enter an option [eXit/Move/Rotate] <Move>: **X**

10. Similarly, tweak the other bolt through the same distance.

11. The assembly after tweaking should look similar to the one shown in Figure 11-23.

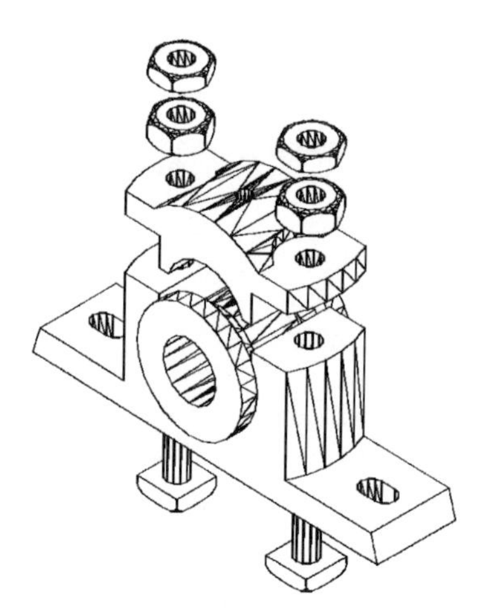

Figure 11-23 *Exploded and tweaked assembly*

12. Choose the **New Trail** button from the **Scenes** toolbar and select the hole on the left side of the **CAP** to display the **Trail Offsets** dialog box as shown in Figure 11-24.

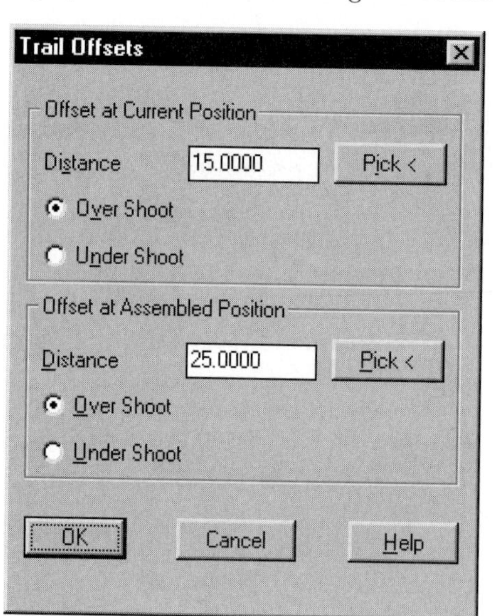

Figure 11-24 *Trail Offsets dialog box*

13. Enter **15** in the **Distance** edit box under the **Offset at Current Position** area.

14. Enter **25** in the **Distance** edit box under the **Offset at Assembled Position** area. Choose **OK**.

15. Similarly, add trails to the remaining components.

16. The assembly after adding the trails to all components should look similar to the one shown in Figure 11-25.

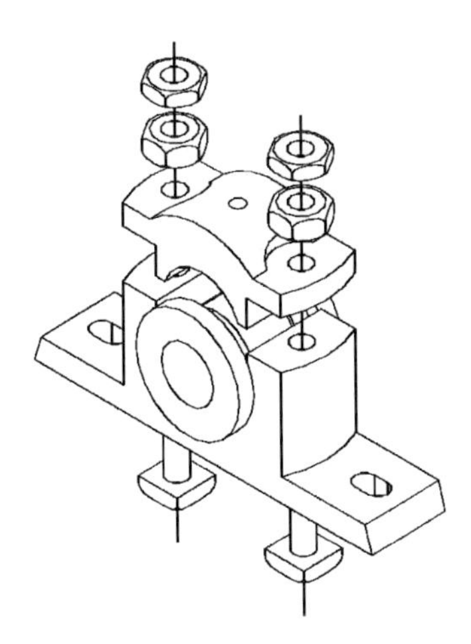

Figure 11-25 *Assembly after adding trails*

17. Save this assembly with the name given below:

    \MDT Tut\Ch-11**Tut1.dwg**

Tutorial 2

In this tutorial you will create an exploded view of the Screw Jack assembly created in Exercise 1 of Chapter 10. Add tweaks and trails to the exploded view.

1. Open the file C:\MDT Tut\Ch-10\Assembly**Exr1.dwg**.

2. Choose the **Scene** tab from the desktop browser to shift to the scene mode.

3. Choose the **New Scene** button from the **Scenes** toolbar to create a new scene. The prompt sequence is as follows:

   Enter new scene name of the active assembly (SJASS) <SCENE1>: Enter
   Enter overall explosion factor <0.0000>: **25**
   Activate new scene? [Yes/No] <Yes>: Enter

4. The exploded assembly should look similar to the one shown in Figure 11-26.

5. Choose the **New Tweak** button from the **Scenes** toolbar and select **SET SCREW** to display the **Tweak Part/Subassembly** dialog box.

6. Select the **Transform** radio button and choose **OK**. The prompt sequence is as follows:

   Enter an option [eXit/Move/Rotate] <Move>: Enter
   Define direction and length:
   Specify start point or [Viewdir/Wire/X/Y/Z]: **Z**

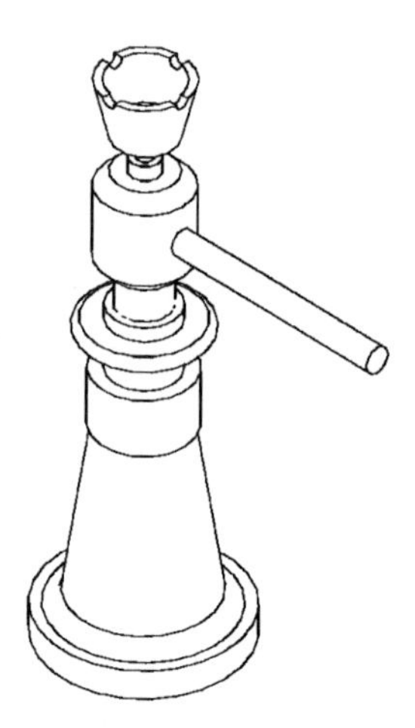

Figure 11-26 *Exploded assembly*

Enter length <1.0000>: **20**
Enter an option [Flip/Accept] <Accept>: *Make sure the direction of movement is positive Z.*
Enter an option [eXit/Move/Rotate] <Move>: **X**

7. Choose the **New Tweak** button from the **Scenes** toolbar and select **WASHER** to display the **Tweak Part/Subassembly** dialog box.

8. Select the **Transform** radio button and choose **OK**. The prompt sequence is as follows:

   Enter an option [eXit/Move/Rotate] <Move>: Enter
   Define direction and length:
   Specify start point or [Viewdir/Wire/X/Y/Z]: **Z**
   Enter length <1.0000>: **125**
   Enter an option [Flip/Accept] <Accept>: *Make sure the direction of movement is positive Z.*
   Enter an option [eXit/Move/Rotate] <Move>: **X**

9. Choose the **New Tweak** button from the **Scenes** toolbar and select **TOMMY BAR** to display the **Tweak Part/Subassembly** dialog box.

10. Select the **Transform** radio button and choose **OK**. The prompt sequence is as follows:

    Enter an option [eXit/Move/Rotate] <Move>: Enter
    Define direction and length:
    Specify start point or [Viewdir/Wire/X/Y/Z]: **X**
    Enter length <1.0000>: **60**
    Enter an option [Flip/Accept] <Accept>: *Make sure the direction of movement is positive X.*
    Enter an option [eXit/Move/Rotate] <Move>: **X**

11. The assembly after tweaking should look similar to the one shown in Figure 11-27.

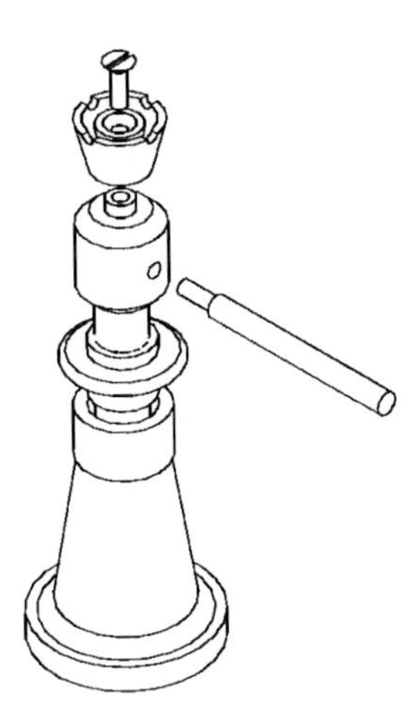

Figure 11-27 *Exploded and tweaked assembly*

12. Choose the **New Trail** button from the **Scenes** toolbar and select the **CUP** to display the **Trail Offsets** dialog box.

13. Enter **15** in the **Distance** edit box under the **Offset at Current Position** area.

14. Enter **25** in the **Distance** edit box under the **Offset at Assembled Position** area. Choose **OK**.

15. Similarly, add trails to the remaining components.

16. The final assembly after adding the trails should look similar to the one shown in Figure 11-28.

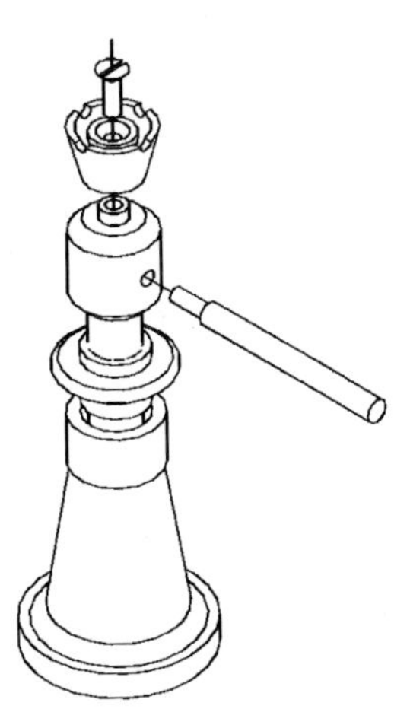

Figure 11-28 *Final assembly after adding the trails*

17. Save this assembly with the name given below:

    \MDT Tut\Ch-11**Tut2.dwg**

Review Questions

Answer the following questions.

1. What is the need for analyzing the assembly?
2. What are the various commands used to analyze the assembly?
3. Define tweaking.
4. What is the difference between tweaking and exploding?
5. What are trails?

Exercise

Exercise 1

In this exercise you will create the components of the Screw Clamp (Medium Size) and then assemble them as shown in Figure 11-29 and Figure 11-30. Figures 11-31a through 11-33 displays dimensions of all the components of the Screw Clamp assembly. After creating the assembly, analyze it and then explode it. Add tweaks and trails to the exploded assembly.

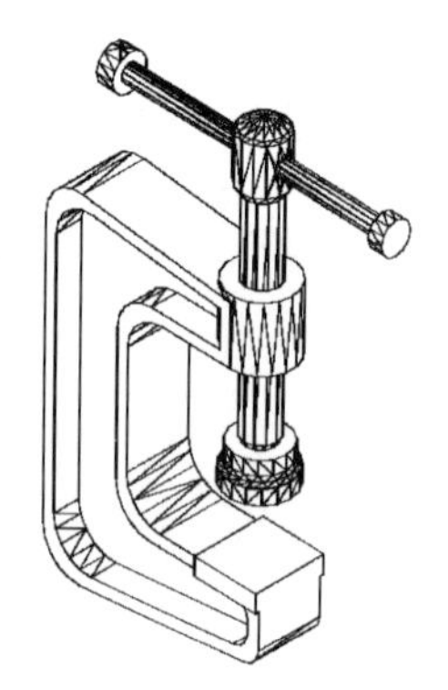

Figure 11-29 *Assembly for Exercise 1*

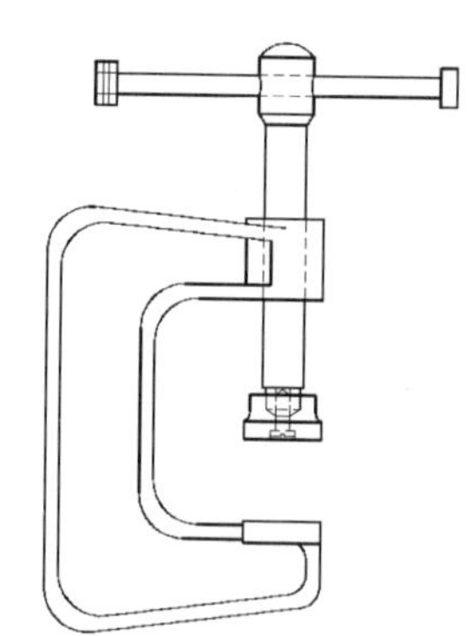

Figure 11-30 *Assembly for Exercise 1*

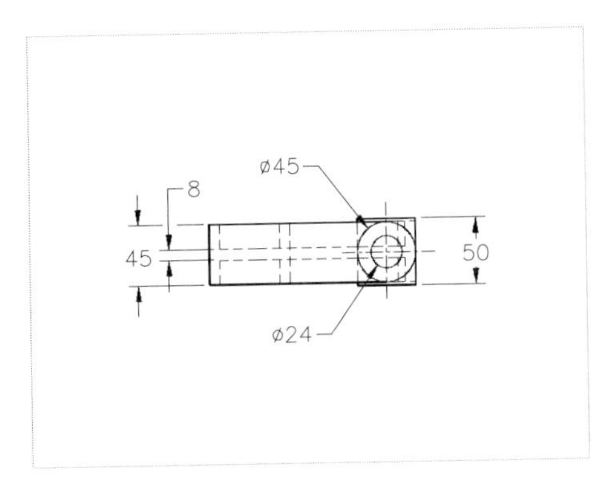

Figure 11-31a Top view of the body

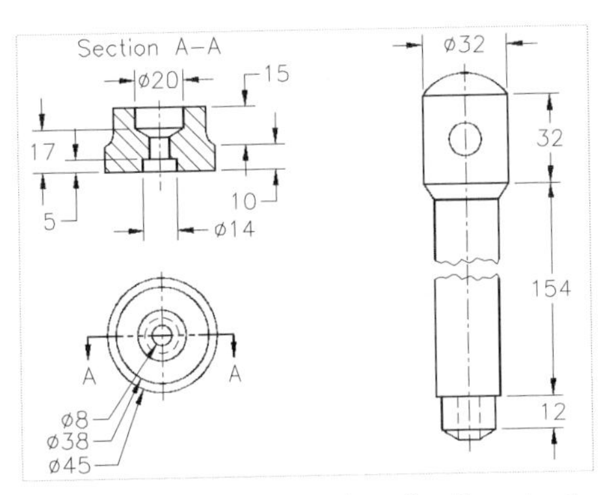

Figure 11-32 Dimensions for Exercise 1

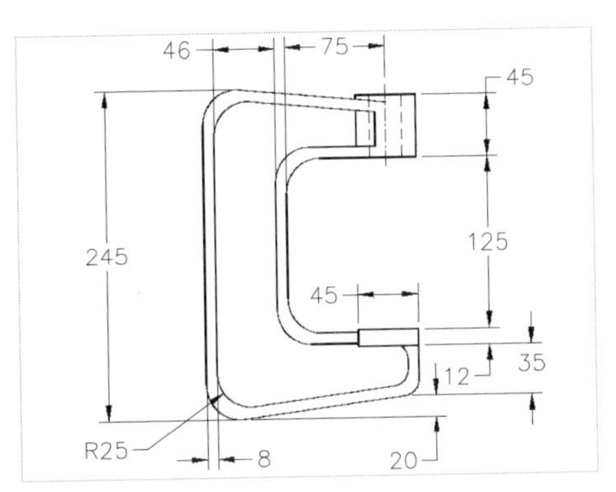

Figure 11-31b Front view of the body

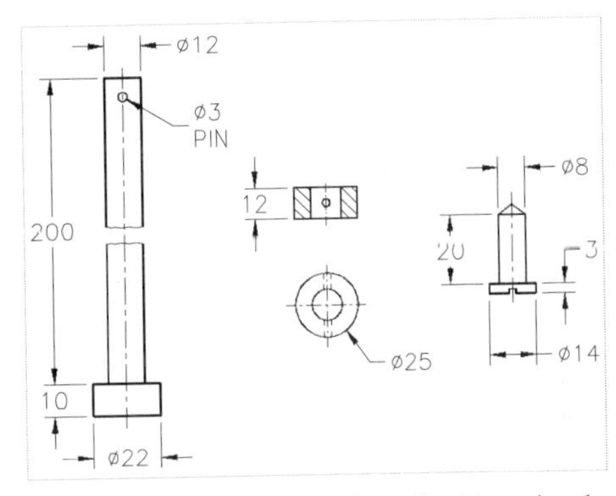

Figure 11-33 Dimensions for Exercise 1

Chapter 12

Creating and Modifying the Drawing Views

Learning Objectives

After completing this chapter, you will be able to:

- *Create drawing views of the assemblies or individual components.*
- *Modify the drawing view.*
- *Obtain information about the drawing views.*
- *Export the drawing views as 2D entities.*
- *Adjust drawing options.*

Commands Covered

- *AMDWGVIEW*
- *AMEDITVIEW*
- *AMCOPYVIEW*
- *AMMOVEVIEW*
- *AMDELVIEW*
- *AMLISTVIEW*
- *AMVIEWOUT*
- *AMOPTIONS*

CREATING THE DRAWING VIEWS

Once the final assembly is created, its drawing views have to be created in the layouts. You can create 2D orthographic or isometric views from the assemblies or individual components. The bidirectional associative nature of Mechanical Desktop ensures that the drawing views are updated once the assembly or the component is modified and the assembly or the component is updated once the dimensions in the drawing views are modified. The views can be created in the layouts using the **AMDWGVIEW** command or by using the desktop browser.

Creating The Drawing Views Using The AMDWGVIEW Command

Toolbar:	Drawing Layout > New View
Menu:	Drawing > New View
Context Menu:	New View
Command:	AMDWGVIEW

This command is used to create 2D drawing views, from the designer parts or assemblies, in the layouts. Before invoking this command you must have a designer part that can be selected for creating the drawing view. If you are working with the assemblies, you need to create a scene of the assembly before creating the drawing views. When you invoke this command, the **Create Drawing View** dialog box will be displayed as shown in Figure 12-1.

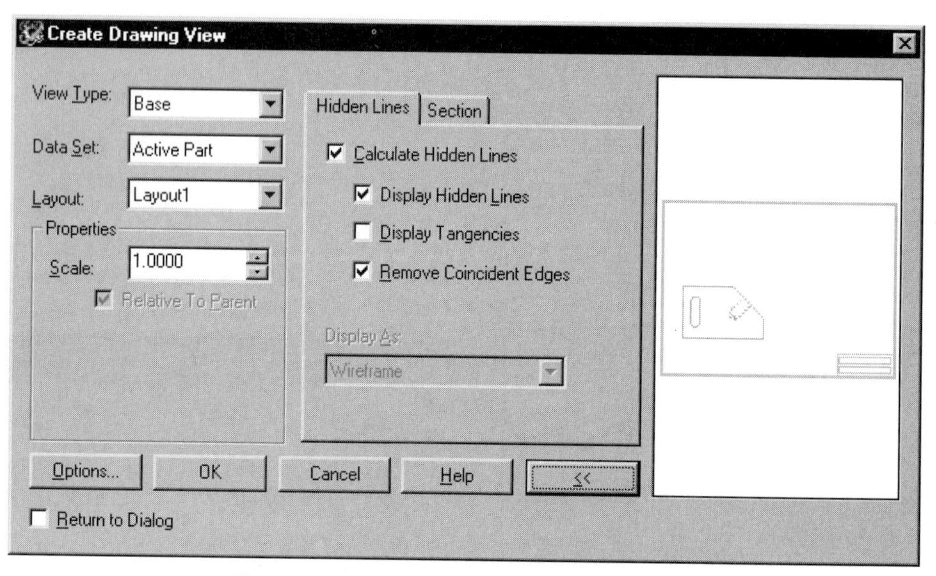

Figure 12-1 Create Drawing View dialog box

Create Drawing View Dialog Box Options

The options provided under this dialog box are:

View Type

The **View Type** drop-down list provides the type of view that can be generated. If it is the first view then this drop-down list provides the following three options:

Base. The **Base** option is used to create the base view of the selected designer part or scene. You can select the view plane and the X, Y, Z axes direction of the view. This view can be used for generating other views, see Figure 12-2. This view is called the parent view and all of the views created using this view are modified if any modification is made in this view.

Multiple. This option is used for creating multiple projected views. This option first creates the base view and then generates the projected views from the base view.

Broken. This option is used to create broken orthographic view of the selected part or assembly. You can specify any break gap for the view, see Figure 12-2.

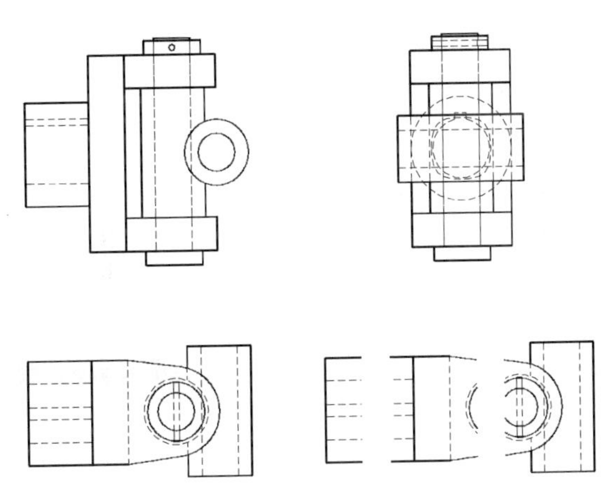

Figure 12-2 *Different views created using the Multiple option and the broken view using the Broken option*

Once you have created the base view, the **View Type** drop-down list will provide the following options in addition to the three options discussed above.

Ortho. The **Ortho** option is used to generate the orthographic view from the selected parent view. The orthographic view is created with same scale factor as that of the parent view.

Auxiliary. This option is used to generate the auxiliary view from the selected parent view. When you create the auxiliary view, you will be asked to specify an auxiliary plane. This plane defined by you will be taken as the viewing plane and the 2D drawing view will be created normal to the specified plane, see Figure 12-3.

Iso. The **Iso** option is used to generate the isometric view from the selected parent view. Isometric view is defined as a 2D representation of the 3D object. This view allows the visualization of all three axes of the designer model or the assembly as shown in Figure 12-3. Any view can be selected as the parent view for generating the isometric view. The isometric view can have a scale factor different from that of the parent view.

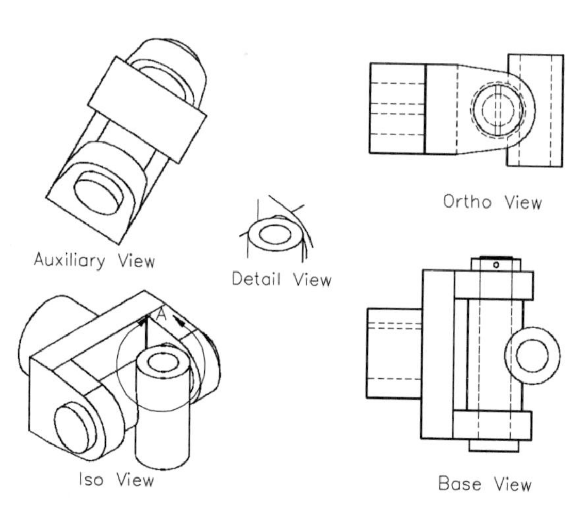

Figure 12-3 *Different type of drawing views*

Detail. The **Detail** option is used to generate the detailed view from the selected parent view. You will be asked to specify a point and then draw a rectangle about that point to generate the detailed view.

Data Set

This drop-down list provides the options that can be used for selecting the data for creating the drawing views. The options provided under this drop-down list are:

Active Part. The **Active Part** option creates the drawing views of the active part. This option is not available when you are working with assembly in which components are attached using the **AMCATALOG** command.

Scene. The **Scene** option creates the drawing view of the specified scene. This is used for creating the drawing views of exploded or unexploded assemblies.

Group. This option uses the group of objects for creating the drawing views. This option will be available only if you have created a group using the **GROUP** command.

Select. This option is used to select the component for creating the drawing views.

Select Part. This option uses the specified part for creating the drawing views.

Layout

The **Layout** drop-down list displays the available layouts in which you can create the drawing views. You can select the desired layout from this drop-down list.

Properties Area

The options in this area changes depending upon the type of view selected from the **View Type** drop-down list. If you select **Base** from the **View Type** drop-down list, this area will provide the following option:

Scale. The **Scale** spinner is used for specifying the view scale. This spinner is not available while generating the orthographic or auxiliary views.

If you select **Multiple** from the **View Type** drop-down list, the options will change as shown in Figure 12-4. This area will then provide the following options:

Scale. The **Scale** spinner is used to specify the view scale for the base view.

Iso Scale. This spinner is used to specify the view scale for the isometric view.

Relative To Parent. If this check box is selected, the view scale of the isometric view will be calculated relative to the parent view.

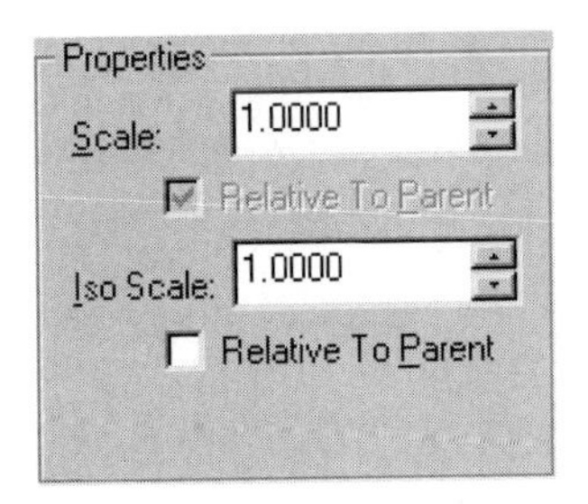

Figure 12-4 Options provided under the properties area for creating multiple views

If you select **Ortho** or **Auxiliary** from the **View Type** drop-down list, no option in this area will be available. If you select **Iso** from the **View Type** drop-down list, this area will provide the following option along with the **Scale** option:

Relative To Parent. If this check box is selected, the view scale of the isometric view will be calculated relative to the parent view.

If you select **Detail** from the **View Type** drop-down list, the options provided under this area will change as shown in Figure 12-5. The options that will be provided are:

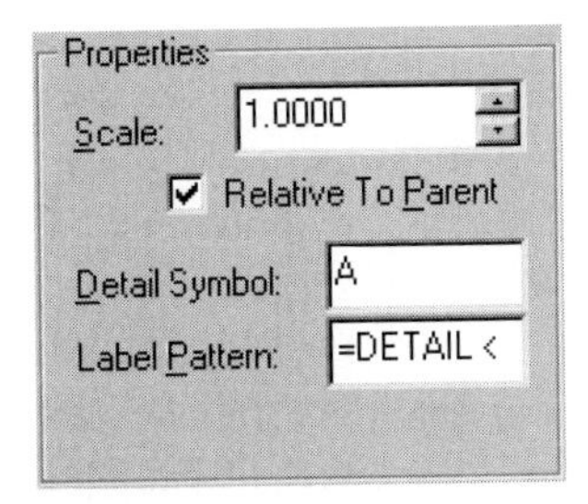

Figure 12-5 Options provided under the properties area for creating the Detail view

Scale. The **Scale** spinner is used to specify the view scale for the detail view.

Relative To Parent. If this check box is selected, the view scale for the detail view will be calculated relative to the parent view.

Detail Symbol. This text box will display the symbol that will be displayed on the parent view of the detail view.

Label Pattern. This text box displays the view label of the detail view.

If you select **Broken** from the **View Type** drop-down list, this area will provide the following option along with the **Scale** option:

Break Gap. The **Break Gap** spinner is used to specify the break gap between the sub views of the broken view.

The other part of the **Create Drawing View** dialog box has two tabs; **Hidden Lines** tab and the **Section** tab. The options provided under these are discussed below:

Hidden Lines Tab

This tab provides the options related to the hidden lines in the views.

Calculate Hidden Lines. This check box is selected to calculate the hidden lines in the view. If this check box is cleared, the hidden lines will not be displayed in the views. The views can displayed only as wireframes or as wireframes with silhouettes (Figures 12-6 and 12-7) by selecting these options from the **Display As** drop-down list provided near the bottom of this tab .

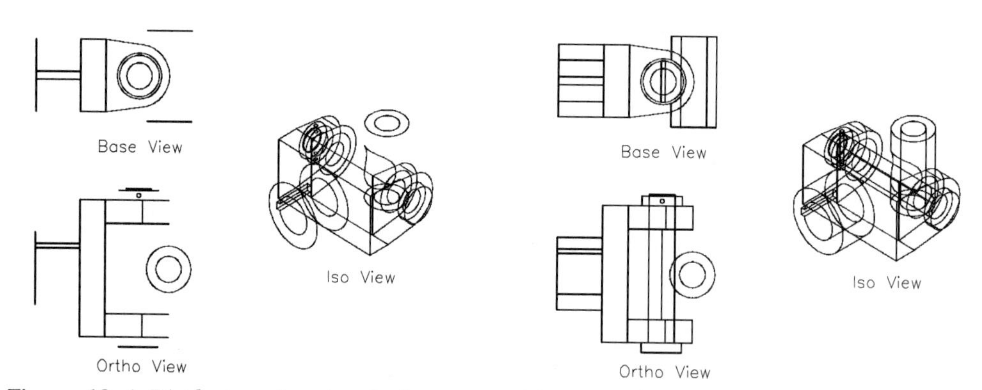

Figure 12-6 *Displaying views in wireframes*

Figure 12-7 *Displaying views in wireframe with silhouettes*

Display Hidden Lines. This check box will be available only if the **Calculate Hidden Lines** check box is selected. This check box is selected to display the hidden lines in the drawing views. If this check box is cleared, the hidden lines will not be displayed in the drawing views. Figure 12-8 shows the drawing views displaying the hidden lines and Figure 12-9 shows the drawing views without the hidden lines.

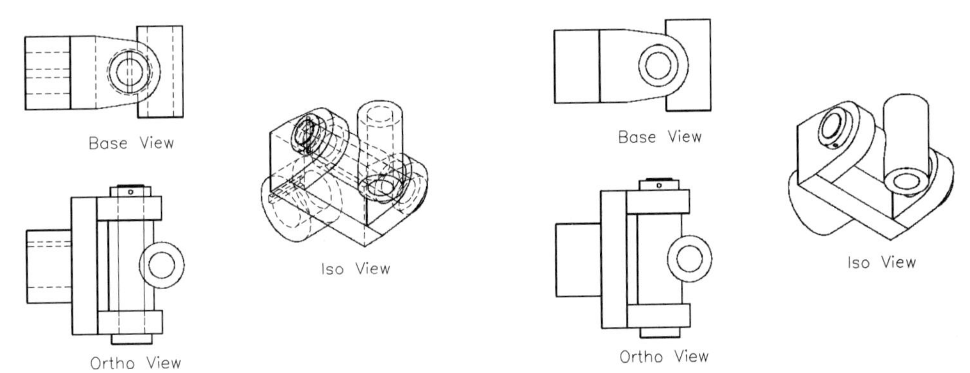

Figure 12-8 Drawing views displaying the hidden lines

Figure 12-9 Drawing views without the hidden lines

Display Tangencies. If this check box is selected, the tangent silhouette edges of the cylindrical part will be displayed in the views.

Remove Coincident Edges. If this check box is selected, the coincidental edges will be removed from the drawing views. Only one of the coincidental edges will be displayed in this case.

Section Tab

The options provided under this tab are used to generate the orthographic or auxiliary section views from the selected parent view. You need to first create a base view and then generate an orthographic or auxiliary section view from that base view. The sectioning properties can be specified using the options provided under this tab. This tab provides the following options:

Type. This drop-down list provides the options for the type of section view. Depending upon the type of view selected from the **View Type** drop-down list, the options under this tab changes. The **Section** tab area will be available only when you select an option from this drop-down list. The various options provided under this tab are:

Full. This option when selected generates a fully sectioned view from the selected parent view. This option is available for the base, orthographic, auxiliary, or the broken views. The section plane can be defined by a work plane or a point. The point can be an imaginary point specified in the parent view. Figure 12-10 shows the base view, the full sectioned orthographic view and the isometric view of the sectioned view.

Half. This option is selected to generate a half sectioned view from the selected parent view. The section plane can be specified using a work plane or a point. This option is available only for the orthographic and the auxiliary views and not for the broken views. Figure 12-11 shows the base view, half sectioned orthographic view and the isometric view of the sectioned view.

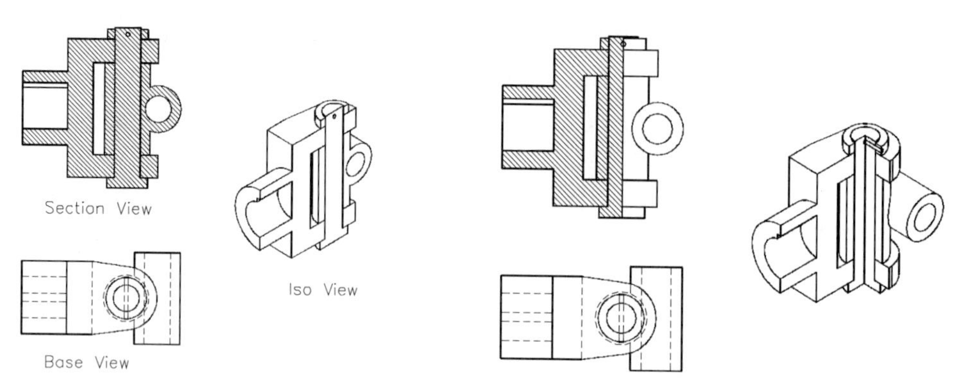

Figure 12-10 *The base view, full section view and the isometric view of section view*

Figure 12-11 *The base view, half section view and the isometric view of section view*

Offset. This option is used for generating an offset sectioned view using a cut line. This option is available for base, orthographic, auxiliary or broken views.

Aligned. This option is used to create an aligned section view using a cut line. However, for generating this view the cut line should have a maximum of two segments.

Breakout. This option is used for generating the section view using the break line sketch. Figure 12-12 shows the assembly with the breakline sketch and Figure 12-13 shows the drawing views created using the breakline sketch.

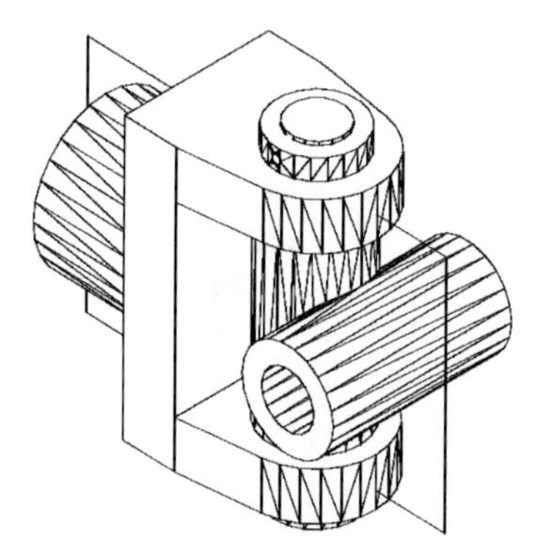

Figure 12-12 *The breakline sketch in the assembly*

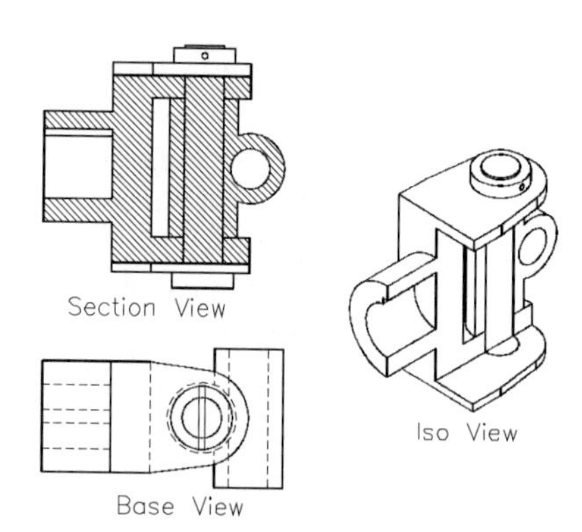

Figure 12-13 *Drawing views created using the breakline sketch*

Radial. This option is used to create radial section view from the selected base view. When you select the parent view, it displays all of the work planes associated with it. You can select any of this work plane as the cutting plane for the radial section view.

Label. This text box displays the section symbol for the section view.

Label Pattern. This text box displays the name of the section view.

Hatch. This check box is used to hatch the sectioned portion of the section view. If this check box is selected, the sectioned portion of the section view will be hatched.

Pattern. This button is chosen to specify the hatch pattern, scale factor and other parameters for the hatching in the sectioned view. When you choose this button, the **Hatch Pattern** dialog box will be displayed as shown in Figure 12-14. The options under this dialog box are similar to those under the **Boundary Hatch** dialog box displayed upon invoking the **BHATCH** command.

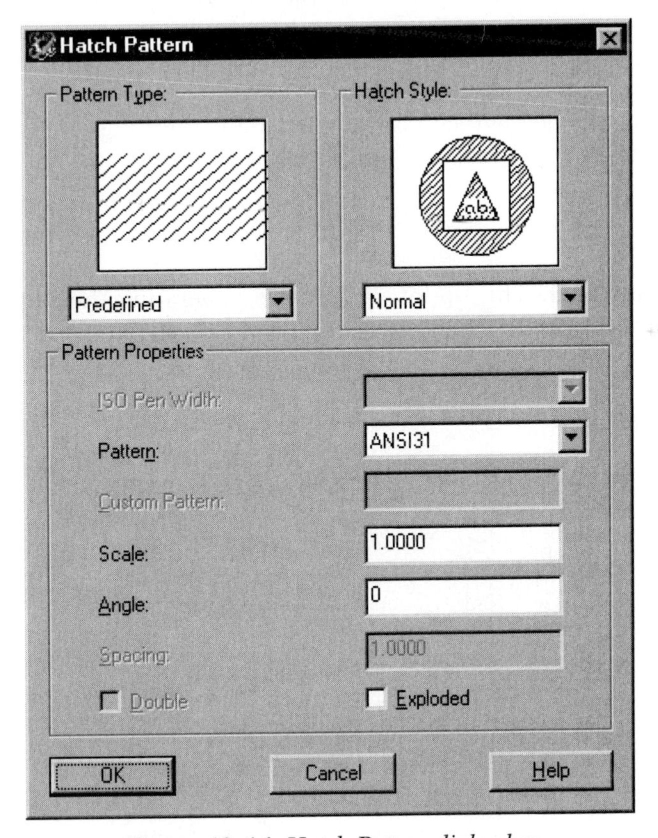

Figure 12-14 Hatch Pattern dialog box

Hide Obscured Hatch. This check box is selected to hide the vague hatching in the section view. Generally, the hatch pattern coming out of the view is hidden by selecting this check box.

Options

This button is chosen to adjust the drawing options using the **Drawing** tab of the

Desktop Options dialog box. The options under this tab are discussed later in this chapter.

Return to Dialog

This check box is selected to display this dialog box again after the current view is created.

Tip: *You can toggle between different types of views by clicking in the preview window of the* ***Create Drawing View*** *dialog box. Different types of views available in the* ***View Type*** *drop-down list will be displayed one by one.*

Creating The Drawing Views Using The Desktop Browser

The drawing views of a designer part or the assembly can also be created using the desktop browser. If the current view is the first view or the base view, right-click on the desired layout under the **Drawing** tab of the desktop browser to display the shortcut menu as shown in Figure 12-15. Choose the **New View** option to display the **Create Drawing Views** dialog box. Once you have created the base view, you can create the remaining views by right-clicking on the base view in the tree view of the desktop browser and selecting the **New View** option from the shortcut menu.

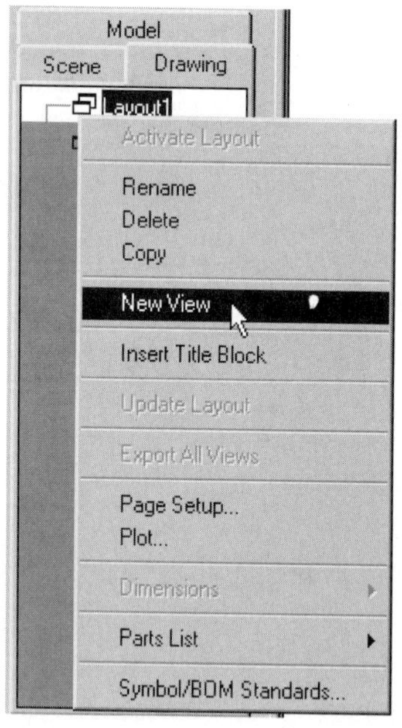

Figure 12-15 *Creating the drawing views using the desktop browser*

SUPPRESSING THE COMPONENTS IN THE SECTION VIEWS

Generally, when a section view is generated, the components like nuts, bolts, lock nuts and so on are not sectioned. If you do not want any component to section while creating the section views, you will have to suppress them in the **Scene** mode. The components are suppressed in the scene when the section view is created. To suppress the desired component, right-click on it in the **Scene** tab of the desktop browser. Choose the **Sectioned Suppressed** option from the shortcut menu. Now if you create the sectioned drawing views of the assembly, the components that were suppressed will not be sectioned. Figure 12-16 shows the drawing views of an assembly without suppressing the components and Figure 12-17 shows the drawing views of the same assembly after suppressing the components.

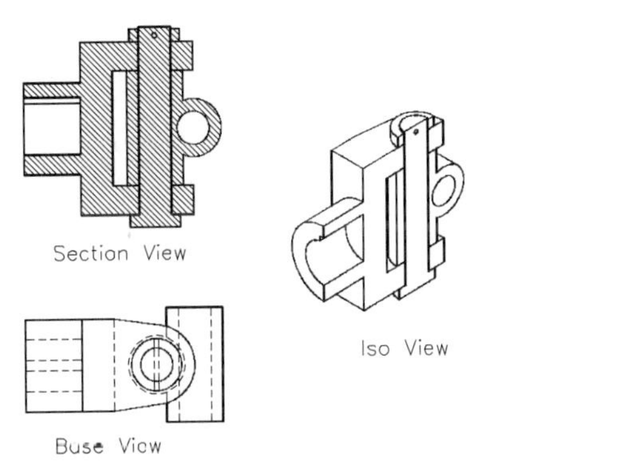

Figure 12-16 Creating drawing views without suppressing the components

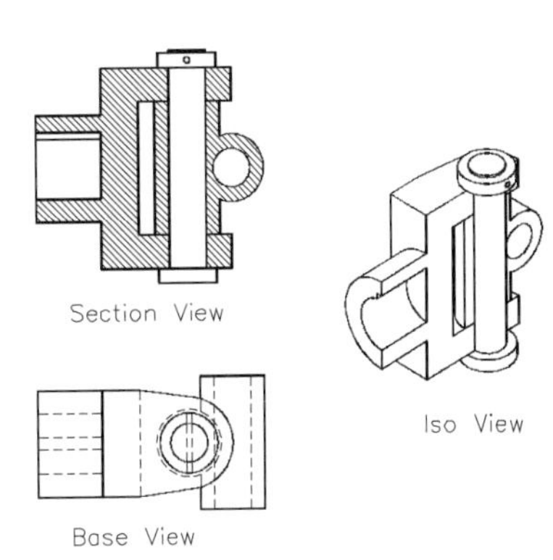

Figure 12-17 Creating drawing views with the components suppressed

ASSIGNING DIFFERENT HATCH PATTERNS TO ALL COMPONENTS IN THE SECTION VIEWS

Generally, section views of huge assemblies are very complicated. The reason for this is that all of the components are hatched with similar hatch pattern and different components cannot be distinguished. This makes understanding of the section views very difficult. This situation can be avoided by using different hatch patterns for different components in lieu of the similar hatch patterns for all components. This is done with the help of the **AMPATTERNDEF** command discussed below.

AMPATTERNDEF Command

Toolbar:	Assembly Modeling > Assign Attributes > Hatch Pattern
Menu:	Assembly > Assembly > Hatch Patterns
Command:	AMPATTERNDEF

This command is used to assign a different hatch pattern to the various components of the assembly, so that each component can be easily distinguished in the section view. Figure 12-18 shows a section view of the assembly with similar hatch pattern for all the components and Figure 12-19 shows a section view of the same assembly with a different hatch pattern for all the components. However, it is important to clarify here that you should assign different hatch patterns to the components before creating the drawing views. When you invoke this command, the **Hatch Pattern** dialog box will be displayed as shown in Figure 12-20.

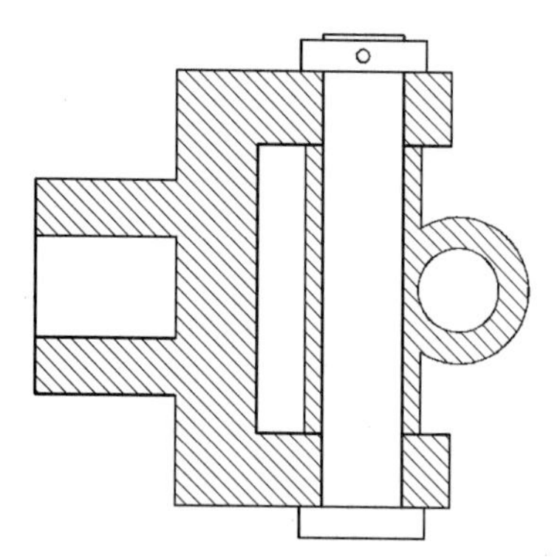

Figure 12-18 Section views showing the components with similar hatch patterns

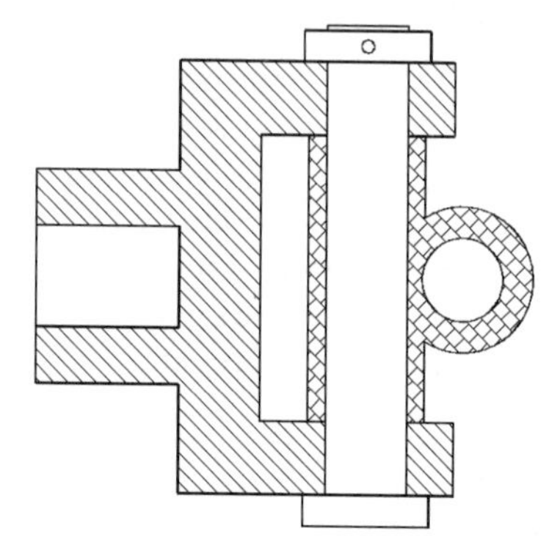

Figure 12-19 Section views showing the components with different hatch patterns

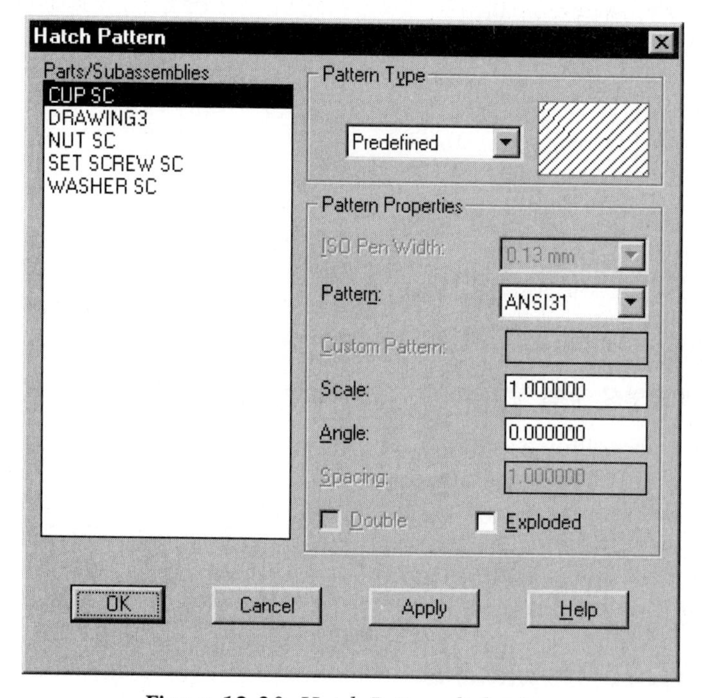

Figure 12-20 Hatch Pattern dialog box

Hatch Pattern Dialog Box Options

Parts/Subassemblies Area

This area displays main assembly along with the components of the assembly. The main assembly and the other components will be arranged in alphabetic order. The default hatch pattern specified to the assembly will be for all components. To specify a different hatch pattern to the components, select the desired component from this area and assign the hatch pattern by selecting it from the **Pattern** drop-down list under the **Pattern Properties** area.

Tip: *The spacing defined using the **Pattern** button under the **Section** tab of the **Create Drawing View** dialog box is valid only for the default hatch pattern. For the other hatch patterns, use the **Scale** edit box of the **Hatch Pattern** dialog box displayed upon invoking the **AMPATTERNDEF** command.*

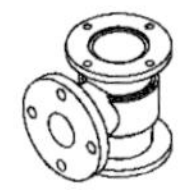

Note

*The rest of the options under the **Hatch Pattern** dialog box are similar to those under the **Boundary Hatch** dialog box displayed upon invoking the **BHATCH** command.*

EDITING THE DRAWING VIEWS

The drawing views created in the layouts can be edited either using the **AMEDITVIEW** command or by using the desktop browser.

Editing The Drawing Views Using The AMEDITVIEW Command

Toolbar:	Drawing Layout > Edit View
Menu:	Drawing > Edit View
Context Menu:	Edit View
Command:	AMEDITVIEW

This command is used for editing the drawing views created in the layouts. When you invoke this command, you will be asked to select the view to be edited. Depending upon whether the view selected for editing is a section view or not a section view, the dialog box changes. For example, if the view selected for editing is a base view, the dialog box that will be displayed will be as shown in Figure 12-21.

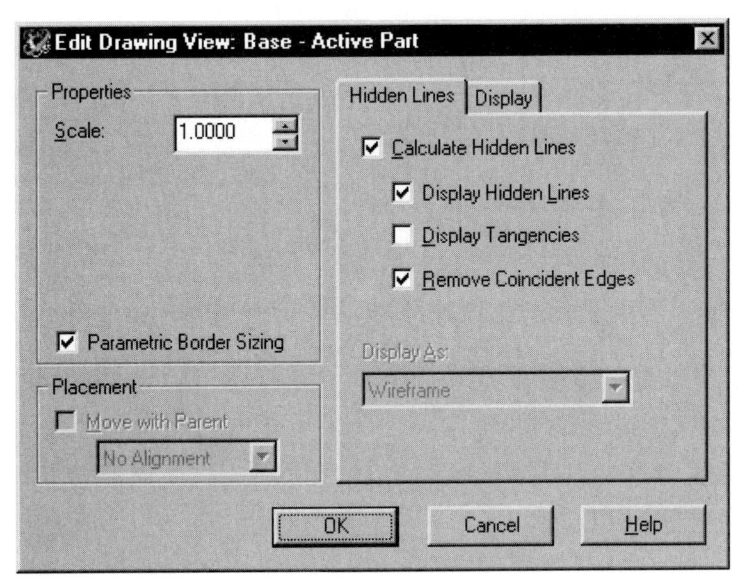

Figure 12-21 *Edit Drawing View dialog box*

Edit Drawing View Dialog Box Options

The options provided under this area are:

Properties Area

The options under this area changes depending upon the type of view selected for editing. All the options for various views are discussed below:

Scale. This spinner is used to modify the view scale. This spinner is available only for the base, detail, broken and the isometric views.

Parametric Border Sizing. This check box will be available only for the base, ortho and auxiliary views. If this check box is cleared, the size of the view or the view border will not change upon changing the scale of the view.

Relative to Parent. If this check box is selected, the scale of the current view will be calculated in relation to the parent view.

Detail Symbol. This text box displays the symbol for the detail view.

Label Pattern. This text box displays the name of the detail view.

Redefine Boundary. This check box is selected for redefining the boundary of the detail view.

Break Gap. This spinner is used to modify the break gap in the broken views.

Modify Sub Views. This check box is selected to modify the sub views in the broken view. You can redefine the boundary of a sub view, delete the sub view or add a sub view using this option.

Placement Area

The options provided under this area are:

Move with Parent. This check box is available only if the current view is not the parent view. If this check box is selected, the view will move along with the parent view from which it is generated. This check box has a drop-down list associated with it. This drop-down list is available only if the **Move with Parent** check box is cleared. The options provided under this drop-down list are:

No Alignment. If this option is selected, the view will not be aligned with the parent view.

Horizontal. This option is selected to horizontally align the selected view with its parent view.

Vertical. This option is selected to vertically align the selected view with its parent view.

Hidden Lines Tab

The options provided under this tab are similar to those under the **Hidden Lines** tab of the **Create Drawing View** dialog box.

Display Tab

The options provided under the **Display** tab are:

Tapped Holes Thread Lines. This check box is selected to display the tapped holes created using the **AMHOLE** command in the drawing views.

Parametric Dimensions. This check box is selected to display the parametric dimensions in the drawing views.

Surface UV Flow Lines. This check box is selected to display the UV flow lines of the surfaces in their drawing views.

Automatic Centerlines. This check box is selected to display the centerlines in the drawing views.

Centerline Settings. This button is chosen to control the settings of the centerline in the drawing views. When you choose this button, the **Automatic Centerlines** dialog box will be displayed as shown in Figure 12-22.

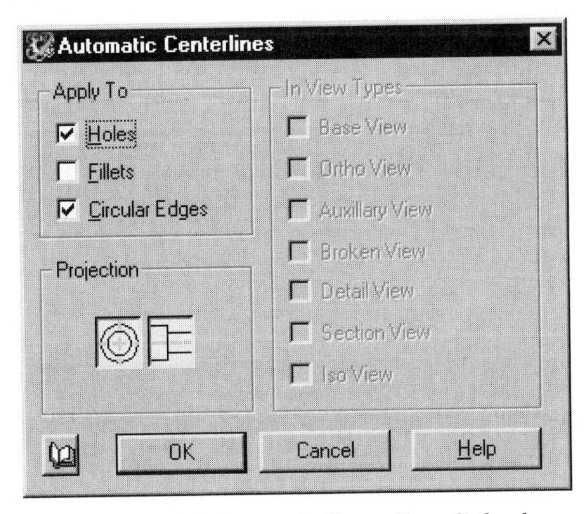

Figure 12-22 *Automatic Centerlines dialog box*

Edge Properties. This button is chosen to modify the edge properties in the selected view. When you choose this button, the dialog box is closed and you will be provided the following options:

Select. This option is used to modify the properties like color, layer, linetype, linetype scale of the selected edges using the **Edge Properties** dialog box, see Figure 12-23. You can also hide the selected edges using this dialog box.

Remove all. This option is used to clear all the edge properties modifications made to the edges using the **Edge Properties** dialog box.

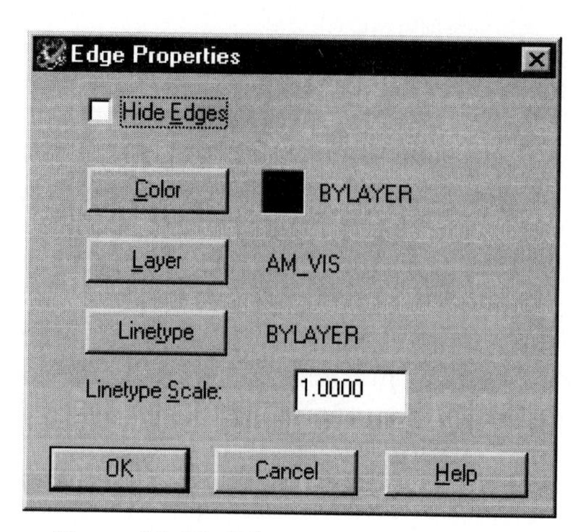

Figure 12-23 *Edge Properties dialog box*

Unhide all. This option is used to unhide all of the hidden edges in the selected view.

If the view selected for editing is a section view, the dialog box will have **Section** tab in addition to the **Hidden Lines** and the **Display** tab. The options provided under the **Section** tab are similar to those discussed under the **Section** tab of the **Create Drawing View** dialog box.

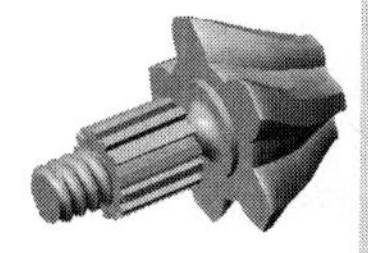

Tip: *To display the hatching in the isometric view of the section view, first create the isometric view of the section view. Now, invoke the* ***AMEDITVIEW*** *command and select the isometric view. Select the* ***Section*** *tab of the* ***Edit Drawing View*** *dialog box and select the* ***Hatch*** *check box. You can also modify the hatch pattern using the* ***Pattern*** *button under the same tab.*

You can also edit a drawing view by double-clicking on it.

Editing The Drawing Views Using The Desktop Browser

Editing of the drawing views is also possible using the desktop browser. Right-click on the view to be edited to display the shortcut menu and choose the **Edit** option, see Figure 12-24. Depending upon the view selected, the related dialog box will be displayed.

COPYING THE DRAWING VIEWS*

The drawing views created in the layouts using the **AMDWGVIEW** command cannot be copied using the simple **COPY** command. These views can be copied either using the **AMCOPYVIEW** command or the desktop browser.

Copying The Drawing Views Using The AMCOPYVIEW Command

Toolbar:	Drawing Layout > Copy View
Menu:	Drawing > Copy View
Context Menu:	Copy View
Command:	AMCOPYVIEW

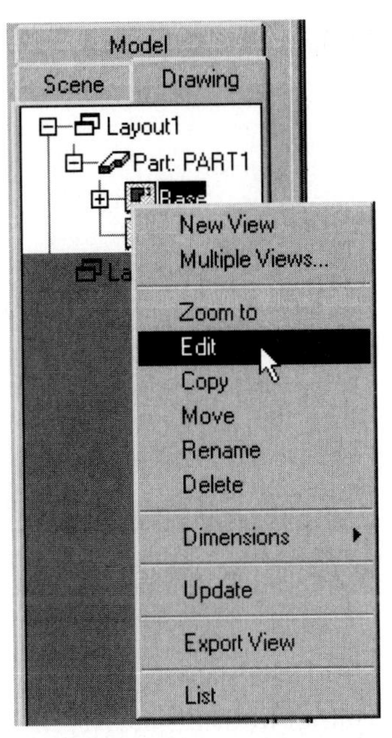

Figure 12-24 *Editing the drawing views using the desktop browser*

This command is used to copy the selected drawing views to another location in same layout or in the other layouts. When you invoke this command, you will be prompted to select the view to be copied. If you select a view that has dependent views, then you will be prompted to specify whether or not you want to copy the dependent views also. However, it has to be kept in mind that the dependent view can be copied to a new location in the same layout and not in the other layouts. In case you want to copy the view to another layout, you can directly specify the name of the layout in the **Enter destination layout name or [?] <Layout1>** prompt. You can also select the layout using the **Layouts** dialog box displayed upon entering **?** in the **Enter destination layout name or [?] <Layout1>** prompt.

Copying The Drawing Views Using The Desktop Browser

To copy the drawing view using the desktop browser, right-click on it to display the shortcut menu and choose the **Copy** option. You will be prompted to specify the new location of the view.

MOVING THE DRAWING VIEWS

Similar to copying the views, the drawing views created using the **AMDWGVIEW** command also cannot be moved using the simple **MOVE** command. These views can be moved from their original location using either the **AMMOVEVIEW** command or the desktop browser.

Moving The Drawing Views Using The AMMOVEVIEW Command

Toolbar: Drawing Layout > Move View
Menu: Drawing > Move View
Context Menu: Move View
Command: AMMOVEVIEW

This command is used to move the drawing views in the layouts. All views created taking the current view as the parent view will also move with this view until this option is cleared using the **Edit Drawing View** dialog box. The orthographic and the auxiliary views remain aligned with the parent view even after the parent view is moved. When you invoke this command you will be asked to select the view to be moved and its new location.

Moving The Drawing Views Using The Desktop Browser

To move the drawing view using the desktop browser, right-click on it to display the shortcut menu and choose the **Move** option. You will be asked to specify the new location of the view.

DELETING THE DRAWING VIEWS

The drawing views can be deleted either by using the **AMDELVIEW** command or by using the desktop browser.

Deleting The Drawing Views Using The AMDELVIEW Command

Toolbar: Drawing Layout > Delete View
Menu: Drawing > Delete View
Context Menu: Delete View
Command: AMDELVIEW

The drawing views created in the layouts cannot be deleted by the simple erasing methods. They can only be deleted using the **AMDELVIEW** command. If the selected view has some dependent views then you will be prompted to confirm the deletion of the dependent views using the **Delete Dependent Views** dialog box, see Figure 12-25.

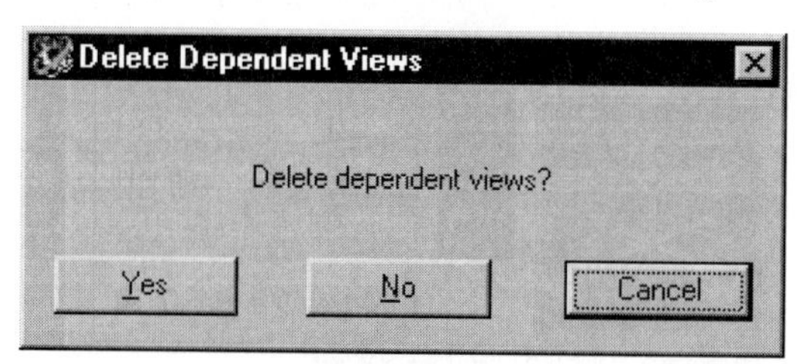

Figure 12-25 *Delete Dependent Views dialog box*

Deleting The Views Using The Desktop Browser

To delete a drawing view using the desktop browser, right-click on the view and choose the **Delete** option from the shortcut menu. If the selected view has some dependent views, the **Delete Dependent Views** dialog box will be displayed and you will be prompted to confirm the deletion of the dependent views.

LISTING OF VIEWS

You can extract some information regarding the drawing views like the type of view, view scale, view direction, center point of the view and so on created in the layouts. This is done using the **AMLISTVIEW** command or by using the desktop browser.

Listing Of Drawing Views Using The AMLISTVIEW Command

Toolbar:	Drawing Layout > List Drawing
Menu:	Drawing > List Drawing
Command:	AMLISTVIEW

When you invoke this command, you will be asked to select the view to be listed. As soon as you select the view, the information regarding the view will be displayed in the AutoCAD Text Window.

Listing Of Drawing Views Using The Desktop Browser

To list a view using the desktop browser, right-click on it and choose **List** from the shortcut menu. The information regarding the selected view will be displayed in the AutoCAD Text Window.

EXPORTING THE DRAWING VIEWS AS 2D ENTITIES

The drawing views created in the layouts can be exported as 2D entities into a new file. This is done using the **AMVIEWOUT** command or by using the desktop browser.

Exporting The Drawing Views Using The AMVIEWOUT Command

Menu:	Drawing > Export View
Command:	AMVIEWOUT

The drawing views created using the **AMDWGVIEW** command can be exported as 2D entities into a new file using this command. These views become a combination of basic 2D AutoCAD entities like lines, circles, arcs and so on. They can be easily modified using the AutoCAD commands. When you invoke this command, the **Export drawing views to AutoCAD 2D** dialog box will be displayed to specify the name of new file, see Figure 12-26.

Exporting The Drawing Views Using The Desktop Browser

Right-click on the view to be exported and choose the **Export View** option from the shortcut menu. The **Export drawing views to AutoCAD 2D** dialog box will be displayed. You can specify the name of the new file in this dialog box and then select the views to be exported in the specified file.

ADJUSTING THE DRAWING OPTIONS

The adjusting of drawing options is a very important aspect of creating the drawing view. This is done using the **AMOPTIONS** command discussed next.

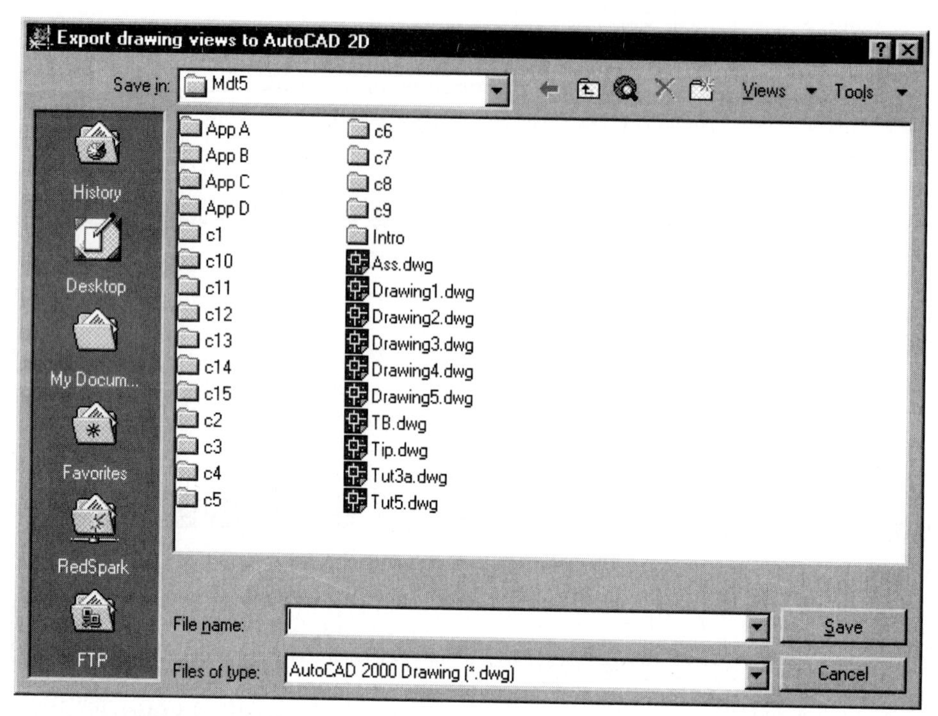

Figure 12-26 *Export drawing views to AutoCAD 2D dialog box*

AMOPTIONS Command*

Toolbar:	Drawing Layout > Drawing Options
Menu:	Drawing > Drawing Options
Command:	AMOPTIONS

This command is used for adjusting various drawing options. When you invoke this command in the **Drawing** mode, the **Mechanical Options** dialog box will be displayed with the **Drawing** tab as shown in Figure 12-27.

Mechanical Options (Drawing Tab) Dialog Box Options

The options provided under this dialog box are:

Suppress Area

The **Suppress** area provides you with the following check boxes:

Hidden Line Calculations. If this check box is selected then the hidden lines in all views will not be recalculated every time you enter the **Drawing** mode.

Automatic View Updates. If this check box is cleared then all drawing views will be updated every time you enter the **Drawing** mode. This check box is generally not cleared so that the time wastage is reduced.

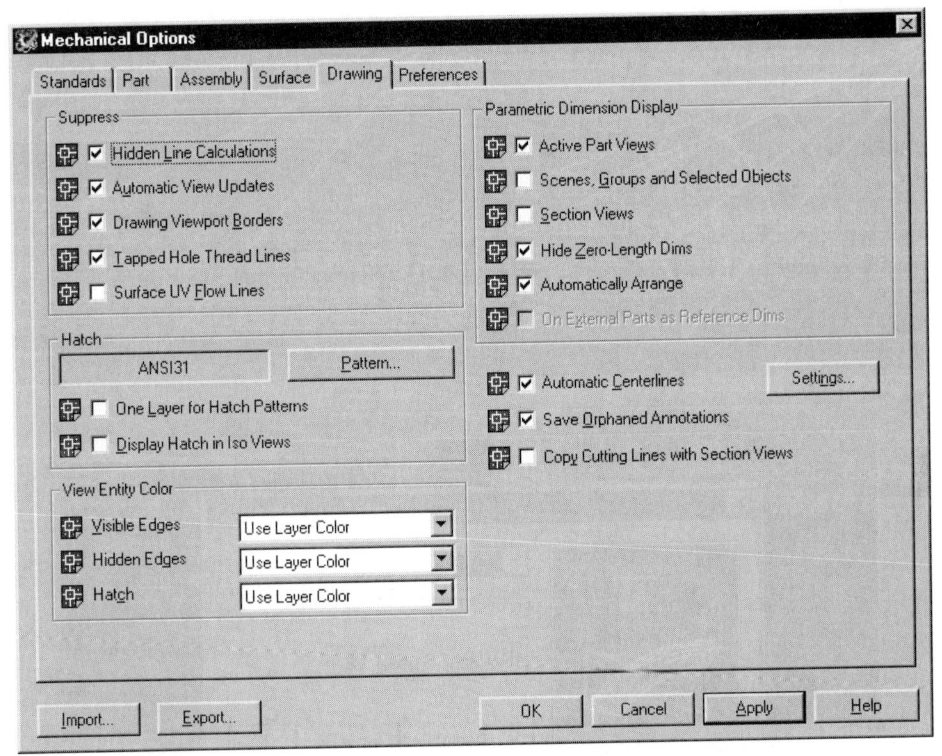

Figure 12-27 Drawing tab of the Mechanical Options dialog box

Drawing Viewport Borders. This check box is selected to suppress the display of the border of viewports.

Tapped Holes Thread Lines. This check box is selected to suppress the tapped holes lines.

Surface UV Flow Lines. This check box is selected to suppress the UV flow lines in the drawing views of the surfaces.

Hatch Area

The options provided under the **Hatch** area are:

Pattern. This button is used to change the default hatch pattern for the hatching in the section views. When you choose this button, the **Hatch Pattern** dialog box is displayed. You can select any hatch pattern to be made the default hatch pattern.

One Layer for Hatch Patterns. If this check box is selected then all hatch patterns will be placed on a single layer.

Display Hatch in Iso Views. If this check box is selected then the hatching will be automatically displayed when you generate isometric view of the section view.

View Entity Color Area

The options provided under this area are used to control the color of the visible edges, hidden edges and the hatch patterns. You can use the part color, the layer color or the part and feature color for them.

Automatic Centerlines

This check box is selected to automatically display the centerlines when the view is generated.

Settings

This button is chosen to modify the centerlines settings. When you choose this button, the **Centerlines** dialog box will be displayed.

Copy Cutting Lines with Section Views

This check box is selected to copy the cutting lines along with the view when the section views are copied.

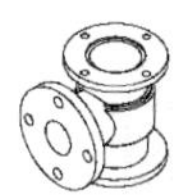

Note

*The remaining options of the **Drawing** tab of the **Mechanical Options** dialog box will be discussed in the later chapters.*

Tip: *To change the projection mode for the drawing views from third angle to first angle and vice-versa, use the **AMPROJTYPE** system variable. The value 1 is for the third angle projection and the value 0 is for the first angle projection.*

TUTORIALS

Tutorial 1

In this tutorial you will create the plan, left half sectioned front elevation, and the isometric view of the section view of the exploded Plummer Block assembly created in the Tutorial 1 of Chapter 11. Make sure that the nuts and bolts are not sectioned and that all components are hatched with different hatch patterns. Display the hatching in the isometric view also.

1. Open the file \MDT Tut\Ch-11\Tut1.dwg.

2. Left-click on the + sign available in the left of the **SCENE 1** in the desktop browser to ensure all of the components of the assembly are displayed.

3. Right-click on **Bolt_1** and choose the **Section Suppressed** option from the shortcut menu. Choosing this option will suppress the sectioning of these components in the section view.

4. Similarly, suppress the remaining bolts, nuts and the lock nuts.

5. Choose the **Model** tab of the desktop browser to shift to the model mode.

6. Choose the **Hatch Patterns** button from the **Assign Attributes** flyout in the **Assembly Modeling** toolbar to display the **Hatch Pattern** dialog box.

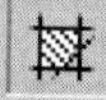

7. Select **BRASSES** from the **Part/Subassemblies** area. Enter **25** in the **Scale** edit box.

8. Select **CAP** from the **Part/Subassemblies** area and select **AR-HBONE** from the **Pattern** drop-down list. Enter **0.75** in the **Scale** edit box.

9. Select **CASTING** from the **Part/Subassemblies** area and select **DASH** from the **Pattern** drop-down list. Enter **25** in the **Scale** edit box and **-45** in the **Angle** edit box. Choose **Apply**.

11. Choose the **Drawing** tab of the desktop browser to switch to the drawing mode and set the value of the **AMPROJTYP** variable to **1**.

12. Choose the **New View** button from the **Drawing Layout** toolbar to display the **Create Drawing View** dialog box as shown in Figure 12-28.

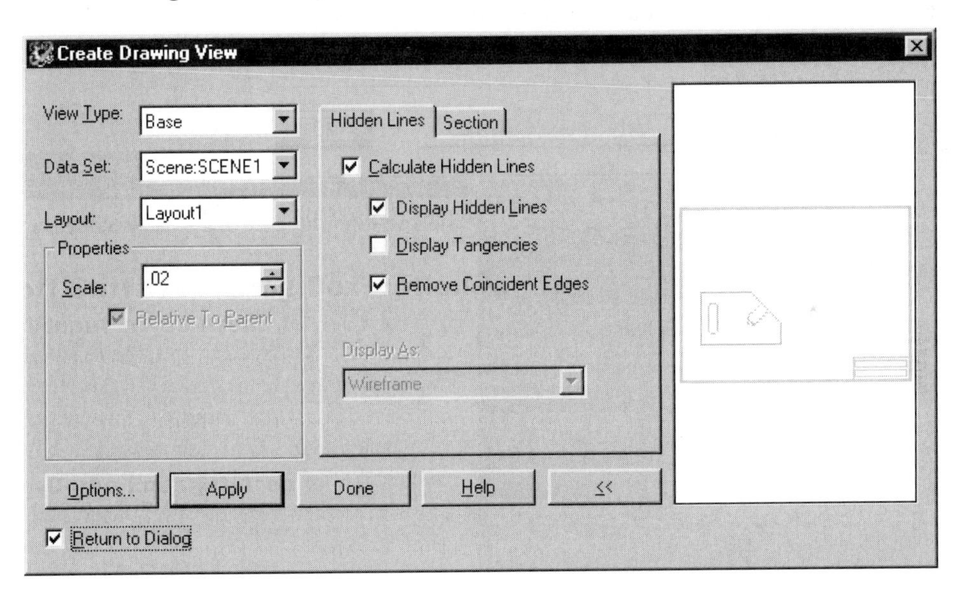

Figure 12-28 Create Drawing View dialog box

13. Select **Base** from the **View Type** drop-down list and **Scene:SCENE1** from the **Data Set** drop-down list.

14. Select **Layout1** from the **Layout** drop-down list and set the value of the **Scale** spinner under the **Properties** area to **0.02**.

15. Choose the **Options** button to display the **Drawing** tab of the **Desktop Options** dialog box. Select the **One Layer for Hatch Patterns** and the **Display Hatch in Iso View** check boxes.

15. Choose **OK** to exit this dialog box. The **Create Drawing View** dialog box will be redisplayed.

16. Select the **Return to Dialog** check box. Choose **Apply**. The prompt sequence is as follows:

    Select planar face, work plane or [Ucs/View/worldXy/worldYz/worldZx]: **X**
    Define X axis direction:
    Select work axis, straight edge or [worldX/worldY/worldZ]: **X**
    Adjust orientation [Flip/Rotate] <Accept>: *Make sure the Z axis direction is positive Z.*
    Regenerating layout.
    Specify location of base view: *Place the view close to the top right corner of the drawing window.*
    Specify location of base view: Enter

17. Select **Ortho** from the **View Type** drop-down list.

18. Choose the **Section** tab and select **Half** from the **Type** drop-down list.

19. Select the **Hatch** and **Hide Obscured Hatch** check boxes.

20. Clear the **Return to Dialog** check box. Choose **OK**. The prompt sequence is as follows:

    Select parent view: *Select the previous view*
    Specify location for orthogonal view: *Specify the placement point below the previous view.*
    Enter section through type [Point/Work plane] <Work plane>: **P**
    Specify point in parent view for depth of section: *Specify the point as the center point of the oil hole on the **CAP**.*
    Side of half section [Flip/Accept] <Accept>: *Make sure the section portion is on the left side.*

21. The two views created should be similar to those shown in Figure 12-29.

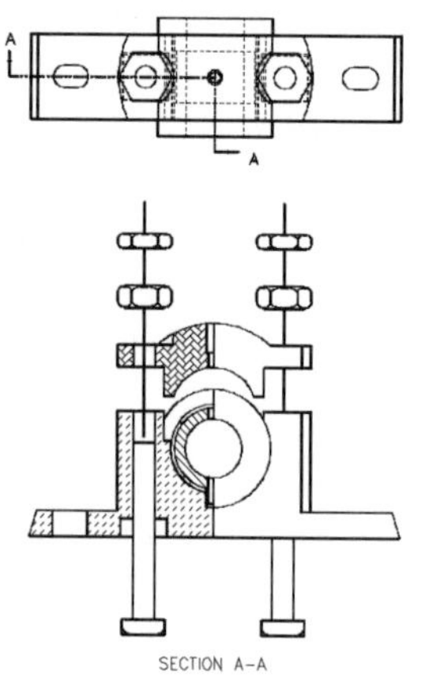

Figure 12-29 *Plan and sectioned front elevation views*

22. Choose the **New View** button from the **Drawing Layout** toolbar to display the **Create Drawing View** dialog box.

23. Select **Iso** from the **View Type** drop-down list. Choose **OK**. The prompt sequence is:

Select parent view: *Select the section view.*
Specify location for isometric view: *Specify the placement point for the view on the left side of the previous views.*

24. The layout after creating the views is shown in Figure 12-30. Save this drawing with the name given below:

\MDT Tut\Ch-12**Tut1.dwg**

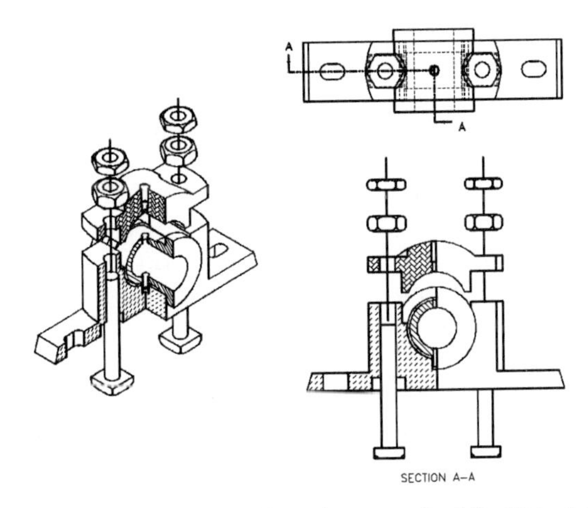

Figure 12-30 *The drawing views required for Tutorial 1*

Tutorial 2

In this tutorial you will create the plan view, full sectioned front elevation view and the isometric view of the section view of the Screw Jack assembly created in Tutorial 2 of Chapter 11. All components should have different hatch patterns. Make sure that the set screw and the tommy bar are not sectioned. Do not display the hatching in the isometric view.

1. Open the file \MDT Tut\Ch-11\Tut2.dwg.
2. Create a new scene named **SCENE 2** with explosion factor of 0.
3. Right-click on **SET SCREW** and choose the **Section Suppressed** option from the shortcut menu.
4. Similarly, suppress the **TOMMY BAR**.
5. Choose the **Model** tab of the desktop browser to enter the model mode.
6. Choose the **Hatch Patterns** button from the **Assign Attributes** flyout in the **Assembly Modeling** toolbar to display the **Hatch Pattern** dialog box.

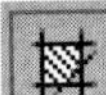

7. Select **CASTING** from the **Part/Subassemblies** area. Enter **25** in the **Scale** edit box.

8. Select **CUP** from the **Part/Subassemblies** area and select **CROSS** from the **Pattern** drop-down list. Enter **20** in the **Scale** edit box.

9. Select **NUT** from the **Part/Subassemblies** area and select **DASH** from the **Pattern** drop-down list. Enter **20** in the **Scale** edit box and **-45** in the **Angle** edit box. Choose **Apply**.

10. Select **SCREW** from the **Part/Subassemblies** area and select **AR-HBONE** from the **Pattern** drop-down list. Enter **1.5** in the **Scale** edit box.

11. Select **WASHER** from the **Part/Subassemblies** area and select **ZIGZAG** from the **Pattern** drop-down list. Enter **10** in the **Scale** edit box.

12. Choose the **Drawing** tab of the desktop browser to proceed to the drawing mode.

13. Set the value of the **AMPROJTYPE** system variable to **1**.

14. Choose the **New View** button from the **Drawing Layout** toolbar to display the **Create Drawing View** dialog box.

15. Select **Base** from the **View Type** drop-down list and **Scene:SCENE1** from the **Data Set** drop-down list.

16. Select **Layout1** from the **Layout** drop-down list and set the value of the **Scale** spinner under the **Properties** area to **0.02**.

17. Choose the **Options** button to display the **Mechanical Options** dialog box. Choose the **Drawing** tab. Select the **One Layer for Hatch Patterns** and the **Display Hatch in Iso View** check boxes. Choose **OK** to return to **Create Drawing View** dialog box.

18. Select the **Return to Dialog** check box. Choose **Apply**. The prompt sequence is as follows:

    Select planar face, work plane or [Ucs/View/worldXy/worldYz/worldZx]: **X**
    Define X axis direction:
    Select work axis, straight edge or [worldX/worldY/worldZ]: **X**
    Adjust orientation [Flip/Rotate] <Accept>: *Make sure the Z axis direction is positive Z.*
    Regenerating layout.
    Specify location of base view: *Place the view close to the top of the drawing window.*
    Specify location of base view: Enter

19. Select **Ortho** from the **View Type** drop-down list.

20. Choose the **Section** tab and select **Half** from the **Type** drop-down list.

21. Make sure the **Hatch** and **Hide Obscured Hatch** check boxes are selected.

22. Clear the **Return to Dialog** check box. Choose **OK**. The prompt sequence is as follows:

    Select parent view: *Select the previous view*
    Specify location for orthogonal view: *Specify the placement point below the previous view.*
    Enter section through type [Point/Work plane] <Work plane>: **P**
    Specify point in parent view for depth of section: *Specify the point as the center point of the **CASTING** in the plan view.*

23. Right-click on the **Section** view in the desktop browser. Select the **New View** option from the shortcut menu to display the **Create Drawing View** dialog box.

24. Select **Iso** from the **View Type** drop-down list. Clear the **Return to Dialog** check box. Choose **OK**. The prompt sequence is as follows:

    Specify location for isometric view: *Specify the placement point for the isometric view on the right side of the previous views.*
    Specify location for isometric view: Enter

25. The layout after generating all the views is shown in Figure 12-31.

26. Save this drawing with the name given below:

    \MDT Tut\Ch-12**Tut2.dwg**

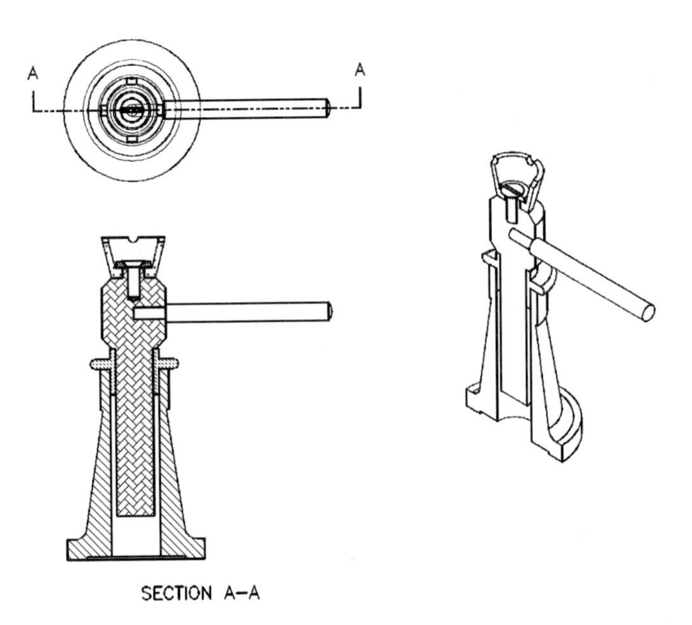

Figure 12-31 *Plan view, full section front view and isometric view*

Review Questions

Answer the following questions.

1. Define drawing views.

2. You can create the drawing views for a single part as well as for the assemblies. (T/F)

3. What are the various types of views that can be created using the **AMDWGVIEW** command?

4. What is the difference between the auxiliary views and the isometric views?

5. What are the obscured hatches in case of the section views?

6. You cannot specify different hatch pattern for different components of assembly in the section views. (T/F)

7. What is suppressing of sections in the section views?

8. What is the difference between first angle projection and third angle projection?

Exercise

Exercise 1

In this exercise you will create plan, full sectioned front elevation and the isometric view of the section view of the Drill Press Vise assembly created in Tutorial 2 of Chapter 10. The components should have different hatch patterns. Do not display the hatching in the isometric view. Also create a detail view giving the details of the HEADED SCREW.

Tip: *If you assign hatch pattern to the components after creating the views, you will have to update the views by choosing the **Update Drawing View** button from the **Drawing Layout** toolbar. You can also update the drawing view by right-clicking on it in the desktop browser and choosing **Update** from the shortcut menu.*

Chapter 13

Dimensioning the Drawing Views

Learning Objectives

After completing this chapter, you will be able to:

- *Dimension the drawing views.*
- *Edit the dimensions in the drawing views.*
- *Add a Bill Of Material to the views.*
- *Add Balloons to the components in views.*
- *Control the visibility in the layouts.*

Commands Covered

- ***AMREFDIM***
- ***AMOPTIONS***
- ***AMPOWEREDIT***
- ***AMMODDIM***
- ***AMMOVEDIM***
- ***AMDIMALIGNED***
- ***AMDIMJOIN***
- ***AMDIMINSERT***
- ***AMDIMBREAK***
- ***AMDIMFORMAT***
- ***AMBOM***
- ***AMVISIBLE***

DIMENSIONING THE DRAWING VIEWS

Dimensioning the drawing views is the last step in creating a model or an assembly. One of the advantages of Mechanical Desktop is that it automatically displays all the parametric dimensions, assigned to the designer part during its creation using any of the parametric dimensioning techniques, in the drawing views. You can also add some reference dimensions in the selected view.

While working with parts, the dimensions are automatically displayed in the drawing views. However, the dimensions can also be displayed on the external parts attached in the assembly using the **AMCATALOG** command. The bidirectional associative nature of Mechanical Desktop ensures that the drawing views are automatically updated once the designer part is modified and the designer part is updated once the dimensions in the drawing views are modified. It is very important here to mention that this property does not work in the case of the assemblies created by attaching the external components.

ADDING REFERENCE DIMENSIONS IN THE DRAWING VIEWS

As we know that the parametric dimensions are automatically displayed in the drawing views, but however, sometimes these dimensions are not enough to explain a particular view. Therefore, you need to add some reference dimensions, that otherwise would have been displayed in some other view, in the current view. The reference dimensions can be added using the **AMPOWERDIM**, **AMREFDIM** or the **AMAUTODIM** command. The functioning of the **AMPOWERDIM** and the **AMAUTODIM** command have already been discussed in previous chapters. In this chapter the **AMREFDIM** command will be discussed.

AMREFDIM Command

Toolbar:	Drawing Layout > Power Dimensioning > Reference Dimensions
Menu:	Annotate > Reference Dimensions
Context Menu:	Annotate Menu > Reference Dimensions
Command:	AMREFDIM

This command is used to add reference dimensions in the drawing views as shown in Figures 13-1 and 13-2. The options provided under this command changes depending upon the type of entity selected for dimensioning. If the entity selected for dimensioning is a line segment, the options are:

Undo

The **Undo** option is used to clear the current selection set.

Hor

The **Hor** option is used to place the horizontal dimension.

Ver

The **Ver** option is used to place the vertical dimension.

Align

The **Align** option is used to dimension an aligned entity.

Par

The **Par** option is used to assign parallel dimension to the selected entities.

aNgle

The **aNgle** option is used to assign angular dimension to the selected entity.

Ord

The **Ord** option is used to assign the ordinate dimension to the selected entity.

reF

The **reF** option is used to place the reference dimension text inside brackets.

Basic

If this option is selected, the reference dimension text is placed inside a box.

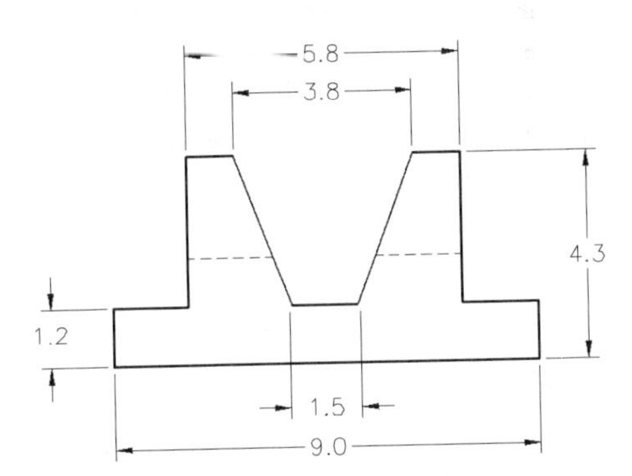

Figure 13-1 *Drawing view showing the parametric dimensions assigned to the component*

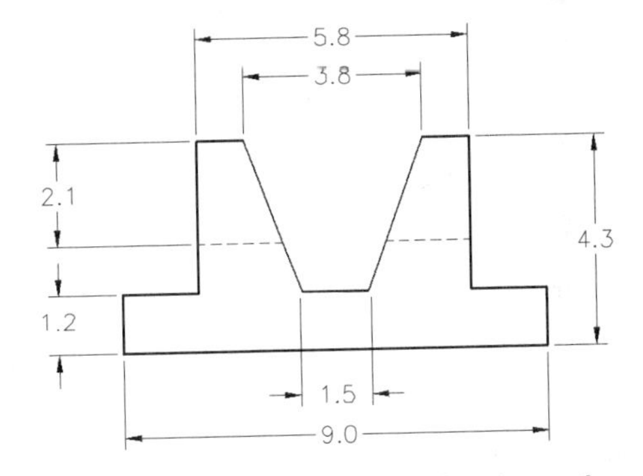

Figure 13-2 *Drawing view showing reference dimension along with the parametric dimensions*

If the entity selected for dimensioning is an arc or a circle, the following options will be provided in addition to the **Undo**, **Ord**, **reF** and **Basic** options:

Diameter

The **Diameter** option is used to place the dimension in terms of diameter.

Radius

The **Radius** option is used to place the dimension in terms of radius.

DISPLAYING DIMENSIONS IN THE ASSEMBLY VIEWS

The parametric dimensions assigned to the components during the creation can also be displayed in the assembly drawing views. This is done by setting some options in the **Drawing** tab of the **Mechanical Options** dialog box (Figure 13-3) displayed upon invoking the **AMOPTIONS** command. You can also invoke this command by selecting the **Annotation Options** from the **Annotate** menu.

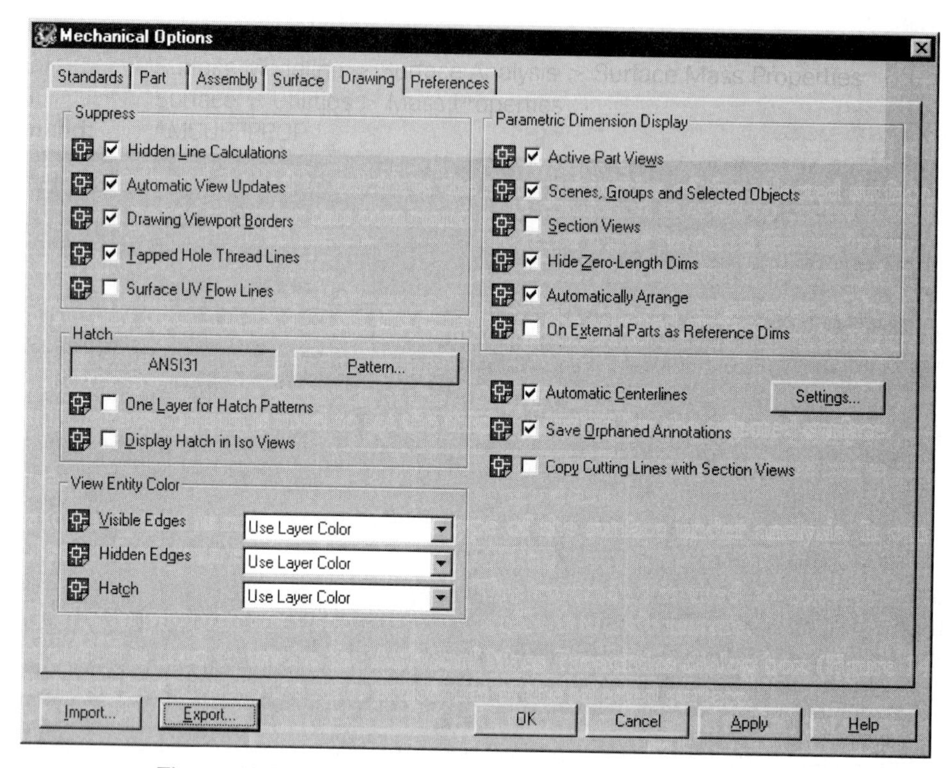

Figure 13-3 Annotation tab of Mechanical Options dialog box

Mechanical Options (Drawing Tab) Dialog Box Options

Parametric Dimension Display Area

This area contains the options related to when and how the parametric dimensions should be displayed in the views.

Active Part Views

This check box is selected to display the parametric dimensions on the drawing views of the active part. This check box is selected by default.

Scenes, Groups and Selected Objects

This check box is selected to display the parametric dimensions on the drawing views created using the **Scene**, **Group** or the **Select** option from the **Data Set** drop-down list of the **Create Drawing Views** dialog box. This option is used to display the parametric dimensions on the drawing views of the assemblies.

Section Views

If this check box is selected, the dimensions will also be displayed on the section views.

Hide Zero-Length Dims

This check box is selected to ensure that if there exists a parametric dimension of zero length, it is not displayed in the drawing views.

Automatically Arrange

This check box is selected to automatically arrange the dimensions in the drawing views.

On External Parts as Reference Dims

This check box is available only when the **Scenes, Groups, Selected Objects** check box is selected. If this check box is selected, the parametric dimensions will be displayed on the external parts as reference dimensions.

Save Orphaned Annotations

This check box is selected to save all the orphaned annotation.

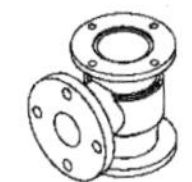

Note

*The remaining options of the **Drawing** tab of the **Mechanical Options** dialog box have been discussed in Chapter 12.*

EDITING THE DIMENSIONS

You can perform any kind of editing on the dimensions in drawing views. However, the result of the editing operation will be different in the case of the automatically displayed parametric dimensions and the forced reference dimensions. The bidirectional associative property is valid only in the case of the parametric dimensions for the active part.

Editing The Parametric Dimensions (AMPOWEREDIT Command)

Toolbar:	Drawing Layout > Power Dimensioning > Power Edit
Menu:	Annotate > Edit Dimensions > Power Edit
Context Menu:	Annotate Menu > Edit Dimensions > Power Edit
Command:	AMPOWEREDIT

The parametric dimensions that automatically appear in the drawing views can be edited using the **AMPOWEREDIT** command. This command is used to edit the dimensions in the drawing views. You can also invoke this command by double-clicking on the dimension you want to edit. When you invoke this command, you will be prompted to select the dimension to be edited. The **Power Dimensioning** dialog box will be displayed upon selecting the dimension for editing as shown in Figure 13-4. It is very important here to mention that the **Expression** edit box will be available only for the parametric dimensions of the active part.

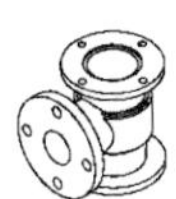

Note

*The options provided in the **Power Dimensioning** dialog box are similar to those discussed in **AMPOWERDIM** command in Chapter 4.*

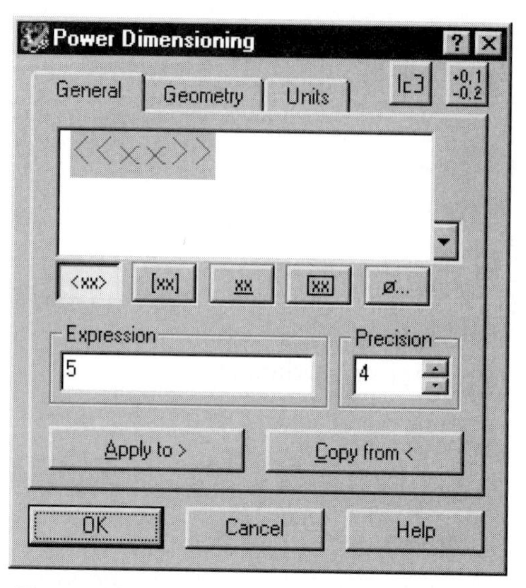

Figure 13-4 Power Dimensioning dialog box

AMMODDIM Command

Toolbar:	Drawing Layout > Power Dimensioning > Edit Dimensions
Menu:	Annotate > Edit Dimensions > Edit Dimension
Context Menu:	Annotate Menu > Edit Dimensions > Edit Dimension
Command:	AMMODDIM

This command is also used for editing the dimensions in the drawing views. When you invoke this command, you will be prompted to select the dimension to be edited. You have to enter the new dimension value in the Command prompt.

Once you have modified the dimensions using any of the above mentioned commands, you will have to update the drawing view using the **AMUPDATE** command to view the affects of modification. In this case you will be asked whether you want to update the designer model also.

Editing The Reference Dimensions

The reference dimensions can be edited using the **AMPOWEREDIT** command. This is similar to the editing of the parametric dimensions and the **Power Dimensioning** dialog box will be displayed when you select the dimension to be edited. However, the **Expression** edit box will not be available in this case.

MODIFYING THE DIMENSION APPEARANCE

Mechanical Desktop allows you to perform various types of modification in the appearance of the dimensions. You can move, align, join, insert or break the dimensions in the drawing views. These modifications are done using the following commands:

Moving The Dimensions Using The AMMOVEDIM Command

Toolbar: Drawing Layout > Power Dimensioning > Move Dimensions
Menu: Annotate > Edit Dimensions > Move Dimension
Context Menu: Annotate Menu > Edit Dimensions > Move Dimension
Command: AMMOVEDIM

This command is used to move the dimensions from one view to other, reattach the dimensions or change the location of dimension text. The options provided under this command are:

Flip

The **Flip** option is used to redefine the location of dimension text in the drawing views.

Move

The **Move** option is used to move the dimension from one place to another in the drawing view. You can also move the selected dimension from one view to the other using this option.

move mUltiple

The **move mUltiple** option is used to move multiple dimensions in a single attempt.

Reattach

The **Reattach** option is used to reattach the selected dimension to a new point using the extension line of the dimension.

Aligning The Dimensions Using The AMDIMALIGN Command

Toolbar: Drawing Layout > Power Dimensioning > Edit Format > Align Dimension
Menu: Annotate > Edit Dimensions > Align Dimension
Context Menu: Annotate Menu > Edit Dimensions > Align Dimension
Command: AMDIMALIGN

This command is used to arrange the dimensions in the drawing views. Generally, when you place the reference dimensions in the views, the views become untidy. This command is then used to align the selected dimensions with a base dimension and arrange them in a proper order. Figure 13-5 shows a drawing view in which the dimensions are not aligned and Figure 13-6 shows a drawing view in which the dimensions are aligned.

Joining The Dimensions Using The AMDIMJOIN Command

Toolbar: Drawing Layout > Power Dimensioning > Edit Format > Join Dimensions
Menu: Annotate > Edit Dimensions > Join Dimension
Context Menu: Annotate Menu > Edit Dimensions > Join Dimension
Command: AMDIMJOIN

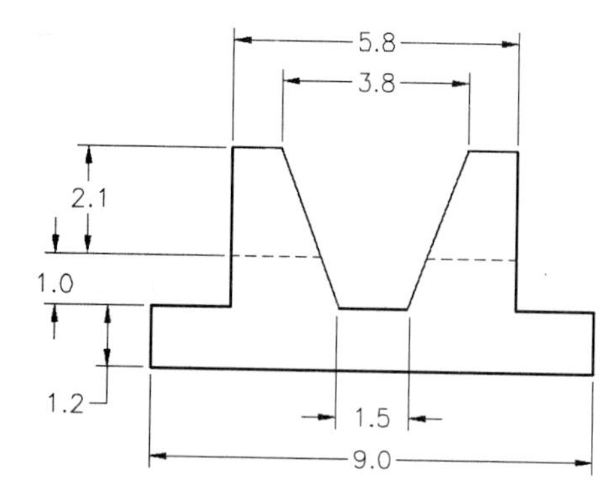

Figure 13-5 *Drawing view before aligning the dimension*

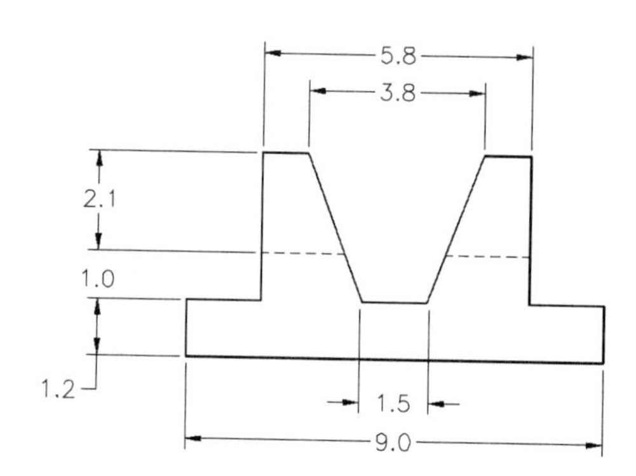

Figure 13-6 *Drawing view after aligning the dimension*

This command is used to join the dimensions in the drawing views. This reduces the number of dimensions in the view, see Figures 13-7 and 13-8. When you invoke this command, you will be prompted to select the base dimension. Once you have selected the base dimension, you will be asked to select the dimension to be joined.

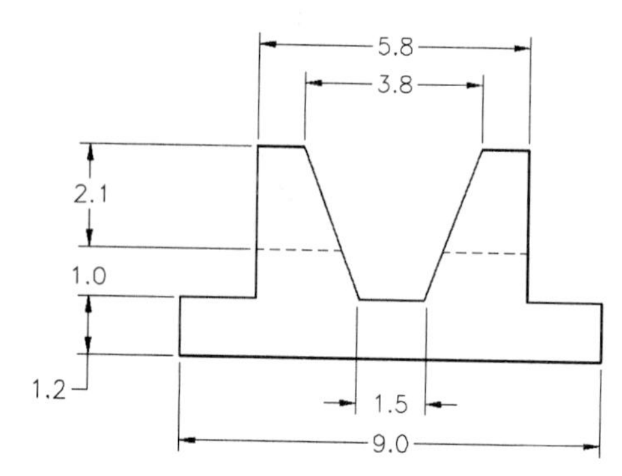

Figure 13-7 *Drawing view before joining the dimension*

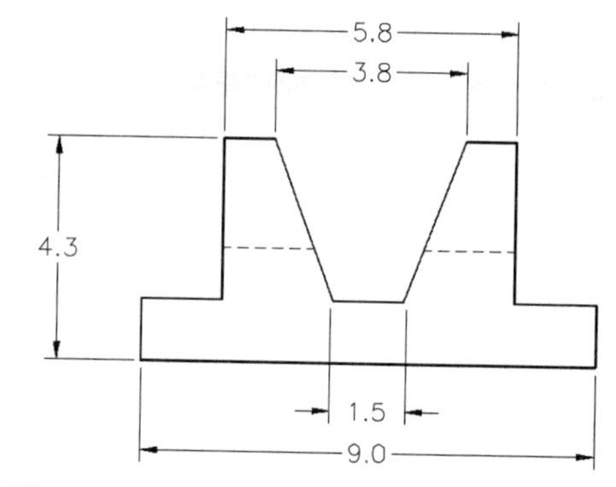

Figure 13-8 *Drawing view after joining the dimension*

Inserting The Dimensions Using The AMDIMINSERT Command

Toolbar: Drawing Layout > Power Dimensioning > Edit Format > Insert Dimension
Menu: Annotate > Edit Dimensions > Insert Dimension
Context Menu: Annotate Menu > Edit Dimensions > Insert Dimension
Command: AMDIMINSERT

This command is used to break a single dimension into more than one individual dimension. The dimension is broken from a selected point up to a specified point in the drawing view and a new dimension is placed between the specified points. However, this option increases the number of dimensions in the drawing views. Figure 13-9 shows the dimensions before breaking and Figure 13-10 shows the drawing view after breaking the dimension into three individual dimensions.

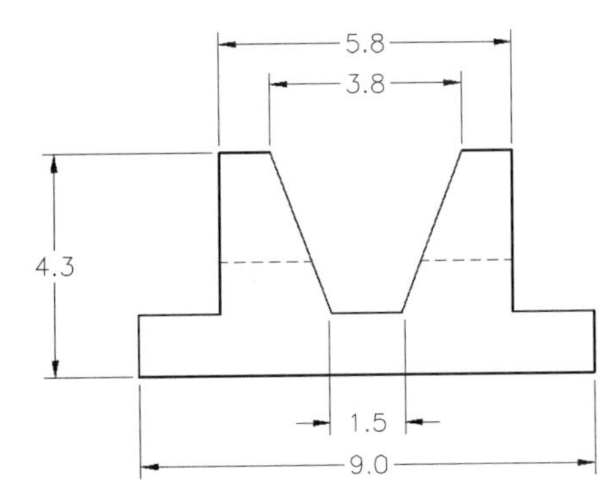

Figure 13-9 Drawing view before inserting the dimension

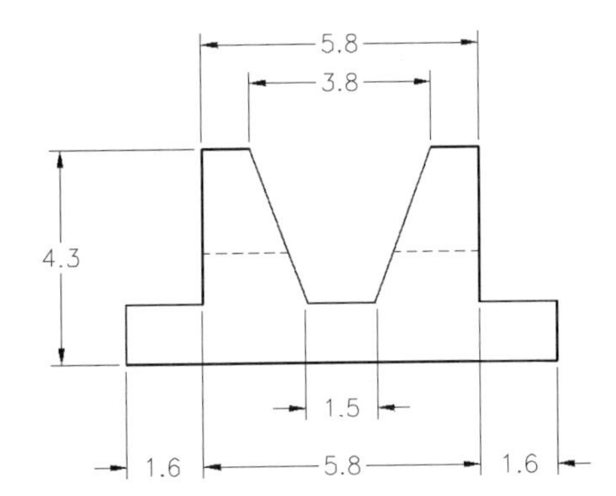

Figure 13-10 Drawing view after inserting the dimension

Breaking The Dimension Or Extension Lines Using The AMDIMBREAK Command

Toolbar:	Drawing Layout > Power Dimensioning > Edit Format > Break Dimension
Menu:	Annotate > Edit Dimensions > Break Dimension
Context Menu:	Annotate Menu > Edit Dimensions > Break Dimension
Command:	AMDIMBREAK

This command is used to break the dimension line or the extension line, see Figures 13-11 and 13-12. This command is similar to the **BREAK** command in AutoCAD.

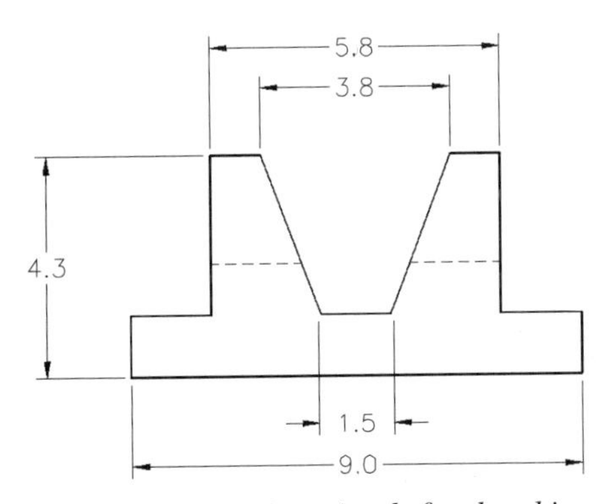

Figure 13-11 Drawing view before breaking the dimension line

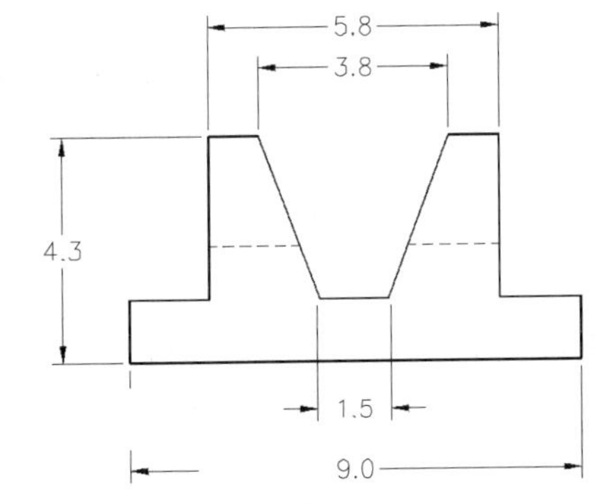

Figure 13-12 Drawing view after breaking the dimension and extension lines

EDITING THE DIMENSION FORMAT

The format of the dimensions in the drawing views can be edited using the **AMDIMFORMAT** command discussed next.

AMDIMFORMAT Command

Command: AMDIMFORMAT

This command is used to edit the dimension format in the drawing views. When you invoke this command, you will be asked to select the dimension whose format has to be edited. The **Dimension Formatter** dialog box will be displayed upon selecting the dimension, see Figure 13-13. Depending upon whether the dimension selected for editing is linear, angular, radius or diameter, the appearance of this dialog box changes. The options provided under various tabs of this dialog box are similar to those under the **Modify Dimension Style** dialog box displayed upon invoking the **DIMSTYLE** command in AutoCAD.

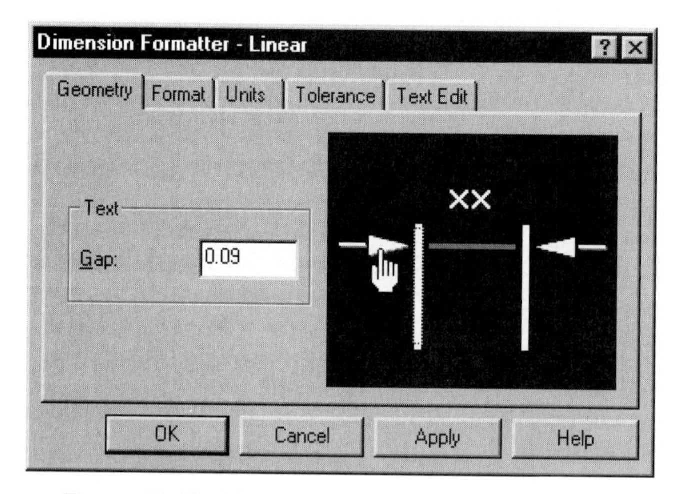

Figure 13-13 *Dimension Formatter-Linear dialog box*

Dimension Formatter Dialog Box Options (Geometry Tab)

The options under this tab are used to modify the geometry of the dimensions. You can modify the arrowheads, dimension lines or the extension lines by selecting them from the preview window using the hand that is displayed in place of the cursor in the preview window, see Figure 13-13.

Format Tab

The options provided under this tab are used to specify the location of the dimension text.

Units Tab

The options provided under this tab are related to the primary and the alternate units.

Tolerance

The options under this tab are used to modify the method and precision of the tolerance in the drawing views.

Text Edit

The options provided under this tab are used to edit the primary and alternate dimension text. You can also add some special symbols using this tab.

ADDING BILL OF MATERIAL TO THE DRAWING VIEWS

The Bill Of Material (BOM) is a tabular representation of all the components in assembly, their quantities, material, manufacturer and so on. However, before defining the BOM database, you need to select the symbol standard to be used by the BOM. The symbol standard can be selected using the **AMSYMSTD** command discussed below:

AMSYMSTD Command*

Command: AMSYMSTD

This command is used to select the symbols and standards for the BOMs to be used in the drawing views. When you invoke this command, the **Symbol Standards** dialog box will be displayed as shown in Figure 13-14. You can select the system standard by right-clicking on it and choosing **Insert standard**, see Figure 13-14.

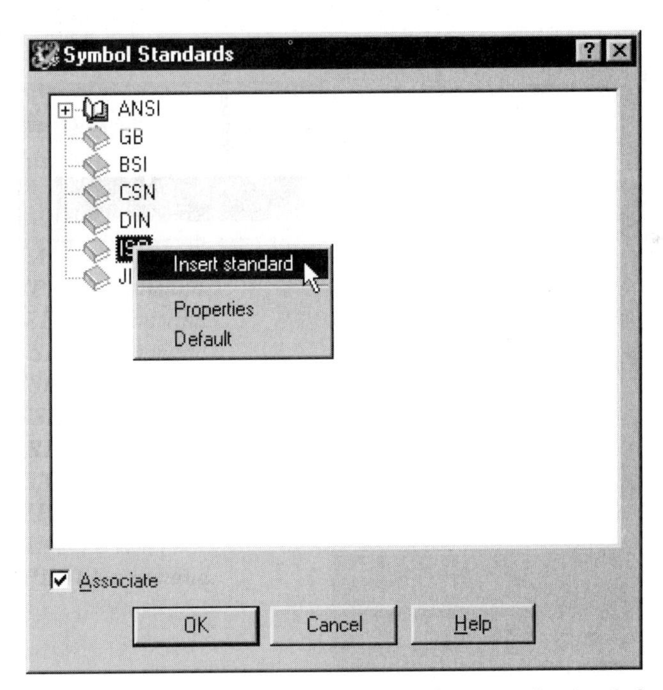

Figure 13-14 *Inserting a standard using the Symbol Standards dialog box*

You can set the standard properties for the selected symbol using the **Standard Properties for...** dialog box displayed by double-clicking on the selected symbol, see Figure 13-15.

The selected symbol will be displayed now with a + sign on the left side. You can open the related directories by clicking on this sign using the pick button of the mouse. The values for leaders, parts list, balloons and so on can be set using this dialog box by double-clicking on them.

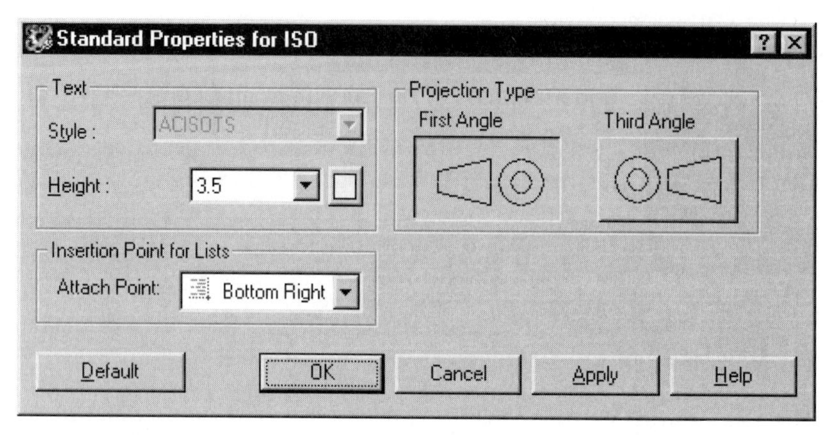

Figure 13-15 *Selecting the properties related to the selected standard*

After selecting the symbols and standards for the BOMs, you can now create the BOM database using the **AMBOM** command discussed below:

AMBOM Command

Toolbar:	Drawing Layout > Power Dimensioning > Edit BOM Database
Menu:	Annotate > Parts List > BOM Database
Command:	AMBOM

This command is used for creating, editing and deleting the BOM database. When you invoke this command for the first time or when you do not have the database for BOM, the **BOM** dialog box will be displayed as shown in Figure 13-16.

BOM Dialog Box Options

The options provided under this dialog box are:

Print

This button is chosen to print the BOM database.

Add parts

This button is chosen to add new parts to the BOM database.

Remove column

This button is chosen to delete the selected column from the BOM database.

Insert column

This button is chosen to insert the a new column at the specified position.

Add item

This button is chosen to add a new row at the end of the BOM database.

Insert item

This button is chosen to add a new row at the specified position.

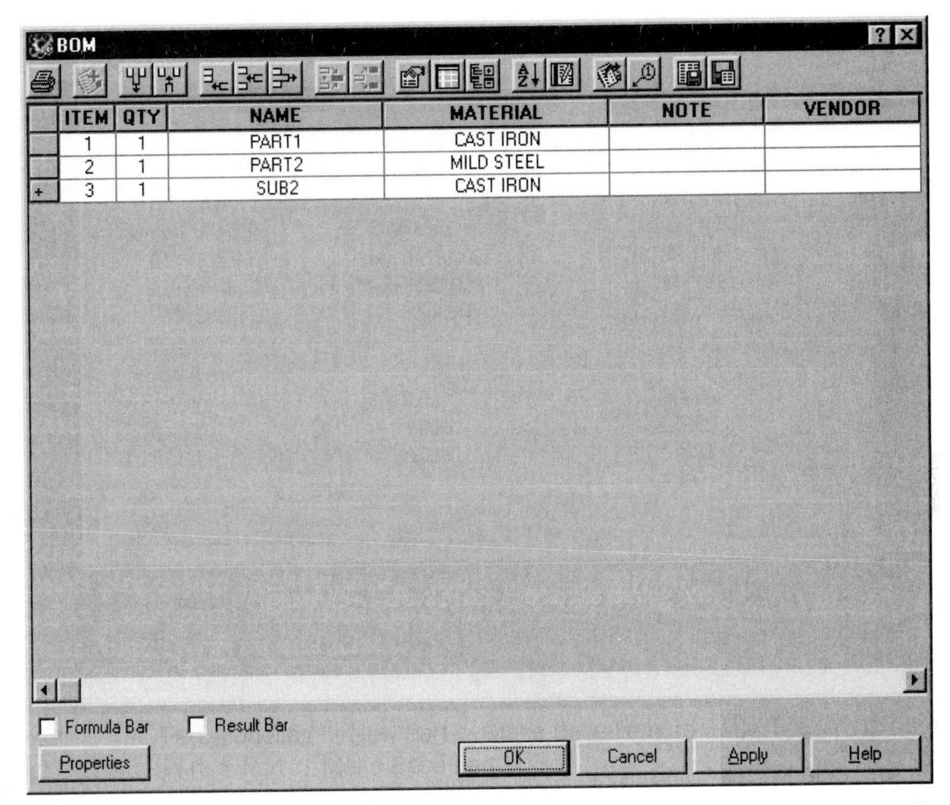

Figure 13-16 BOM dialog box

Delete item
This button is chosen to delete the selected row.

Merge items
This button is chosen to merge the selected rows. However, only the rows that have similar data can be merged.

Split items
This button is chosen to split the selected row.

Assembly Properties
This button is chosen to display the **Assembly Properties** dialog box to edit the assembly properties.

BOM Representation
This button is chosen to display the **BOM Representation** dialog box for editing the external representation of the BOM.

Extend/Collapse all subassemblies
This button is chosen to extend or collapse the subassemblies in the BOM database.

Sort

This button is chosen to display the **Sort** dialog box for sorting the BOM database.

Set values

This button is chosen to set the values for different columns in the BOM database.

Insert parts list

This button is chosen to insert the parts list with the current BOM database in the layouts. When you select this button, the **BOM** dialog box will be temporarily closed to allow you to insert the part list. You can change the default name of the parts list. The part list can be created for all parts, selected parts, parts in specified range, parts displayed in the specified sheet or the parts displayed in the selected view. Figure 13-17 shows drawing views of the Knuckle Joint assembly with the part list (BOM) inserted in the layout.

Ballooning

This button is chosen to add the balloons to the drawing views in the layouts. The BOM in the layout displays various components in the assembly, their quantity and so on. But however, it is very difficult to identify the parts in the drawing views. To avoid this vague situation you need the callouts for the components in the drawing views. These callouts are called the balloons. The symbol standards for the balloons are controlled using the **AMSYMSTD** command. The balloons can be placed automatically or manually. Figure 13-17 shows the drawing views with the balloons.

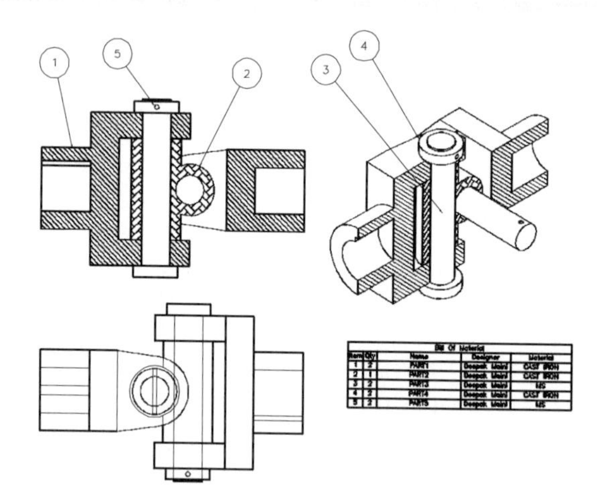

Figure 13-17 *Drawing views showing BOM and balloons*

Export

This button is chosen to export the current database into a database file in the Microsoft Access (mdb) format.

Import

This button is chosen to import an existing database file into the BOM database.

Formula Bar

This check box is selected to display the formula bar in the **BOM** dialog box for using the variables in the BOMs.

Result Bar

This button is chosen to display the result bar in the **BOM** dialog box.

Properties

This button is chosen to display the **BOM Properties** dialog box.

CONTROLLING THE VISIBILITY IN THE LAYOUTS

The visibility in the layouts can be controlled using the **AMVISIBLE** command discussed next.

AMVISIBLE Command

Toolbar:	Drawing Layout > Drawing Visibility
Menu:	Drawing > Drawing Visibility
Command:	AMVISIBLE

This command in layouts is used to control the visibility of the options related to the drawing views. When you invoke this command, the **Desktop Visibility** dialog box will be displayed as shown in Figure 13-18.

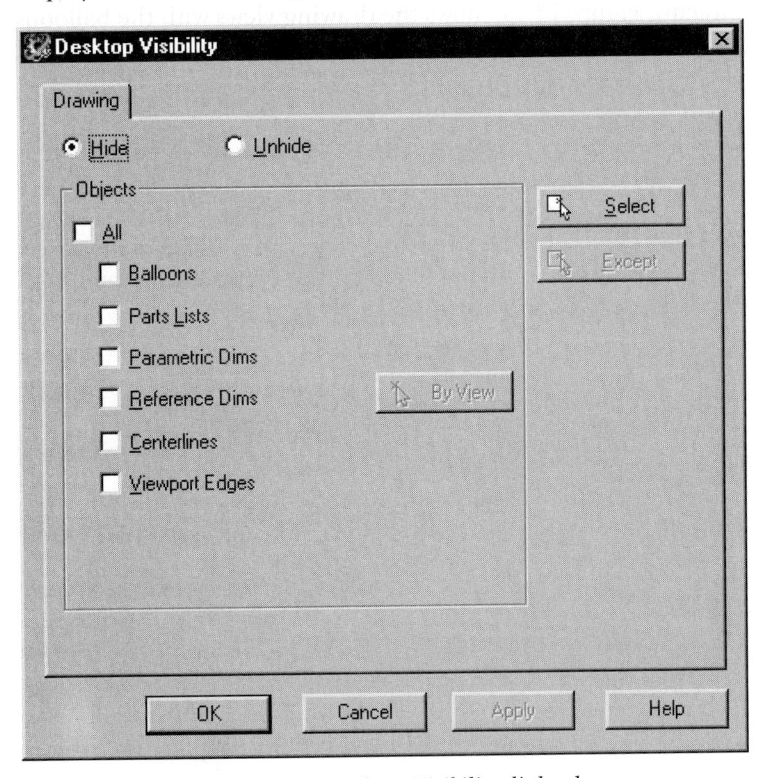

Figure 13-18 *Desktop Visibility dialog box*

Desktop Visibility Dialog Box Options (Drawing Tab)

Hide

This radio button is chosen to hide the selected objects in the drawing views.

Unhide

This radio button is chosen to unhide the selected objects in the drawing views.

Objects Area

All. This check box when selected controls the visibility of all the balloons, part lists, dimensions, centerlines and view port edges in the drawing views.

Balloons. This check box is selected to control the visibility of the balloons in the layouts.

Parts Lists. This check box is selected to control the visibility of the parts list in the layouts.

Parametric Dims. This check box is used to control the visibility of the parametric dimensions in the drawing views.

Reference Dims. This button is chosen to control the visibility of reference dimensions in the drawing views.

Centerlines. This check box is used to control the visibility of the centerlines in the drawing views.

Viewport Edges. This check box is used to control the visibility of the viewport edges in the layouts.

By View. This button is chosen to control the visibility of the objects in the selected views. This button is available only if the **Parametric Dims**, **Reference Dims** or the **Centerlines** check box is selected.

Select

This button is chosen to select the objects to control their visibility.

Except

This button is chosen to remove the specified objects from the selection set created using the **All** check box. This button is available only if the **All** check box is selected.

TUTORIALS

Tutorial 1

In this tutorial you will create the dimensioned drawing views of the designer model created in Tutorial 3 of Chapter 3. The details of the drawing views to be created are:

1. Top View	2. Full Sectioned Front View
3. Left Side View	4. Isometric View of the Section View.

1. Open the file \MDT Tut\Ch-3\Tut3.dwg. Choose the **Drawing** tab in the desktop browser and create the top view.

2. Create the full section front view taking the top view as the parent view.

3. Create the right side view.

4. Create the isometric view of the section view.

5. Double-click on the iso view to display the **Edit Drawing View** dialog box.

6. Choose the **Section** tab. Select the **Hatch** check box and the **Hide Obscured Hatch** check box. Choose **OK**.

7. All the dimensioned drawing views should look similar to the one shown in Figure 13-19. Save this drawing with the name given below:

   \MDT Tut\Ch-13**Tut1.dwg**

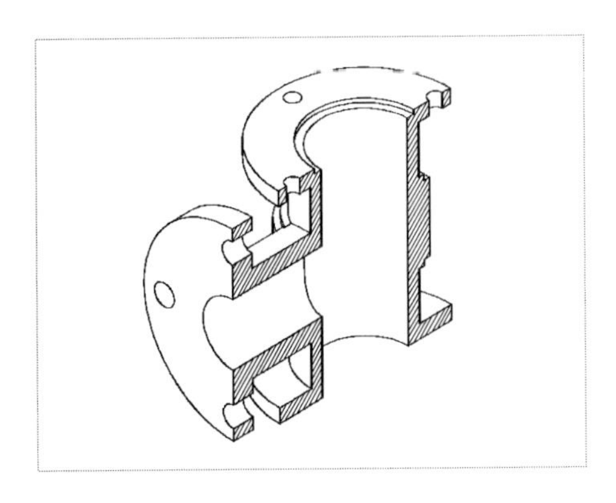

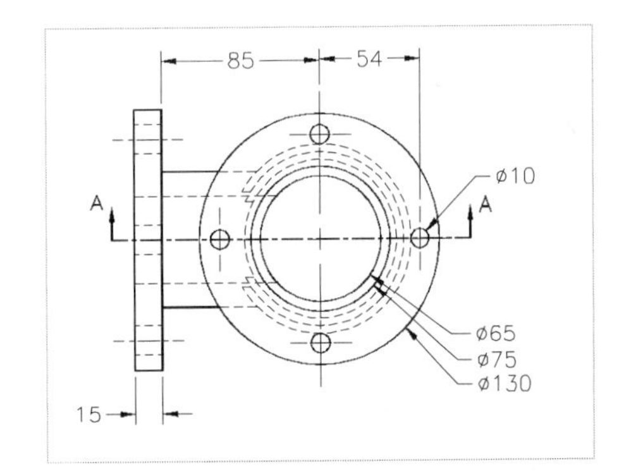

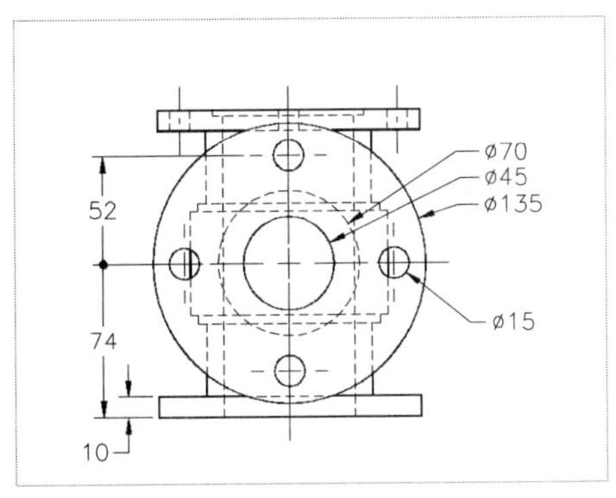

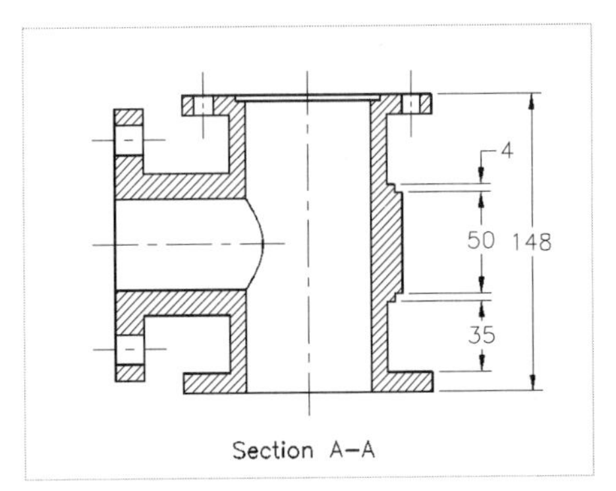

Figure 13-19 *Various views for Tutorial 1*

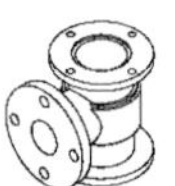

Note

If any dimensions ares missing, you can add them using the ***AMREFDIM*** *or the* ***AMPOWERDIM*** *command. You can also reposition the dimensions by dragging them using the grips.*

Tutorial 2

In this tutorial you will insert a title block of A4 size and then create the drawing views of the Screw Jack assembly created in Exercise 1 of Chapter 10, in the title block. Add the Parts list and balloons to the drawing views. The name of the part list should be Bill Of Material. The views to be created are:

1. Top View
2. Sectioned Front View
3. Left Side View
4. Isometric View of the Section View

The Bill Of Material should include the following columns:

1. Item
2. Quantity
3. Name
4. Material

1. Open the specified assembly and then choose the **Scene** tab and create a new scene with zero explosion factor.

2. Choose the **Drawing** tab to shift to the **Drawing** mode.

3. Right-click on **Layout 1** to display the shortcut menu.

4. Choose the **Insert Title Block** option.

5. Enter **1** in the AutoCAD Text Window. The prompt sequence is as follows:

   Create a drawing named iso_a4.dwg? <Y>: **N**

   Enter an option [Align/Create/Scale viewports/Options/Title block/Undo]: Enter

6. Now, create the required view in this title block as shown in Figure 13-20.

7. Enter **AMSYMSTD** at the Command prompt to display the **Symbol Standards** dialog box.

8. Open the directories under **ANSI**. Now open the **BOM and Balloon Support** directory.

9. Double-click on the **Parts List** to display the **Parts List Properties for ANSI** dialog box

10. Select **Top Left** from the **Attach Point** drop-down list.

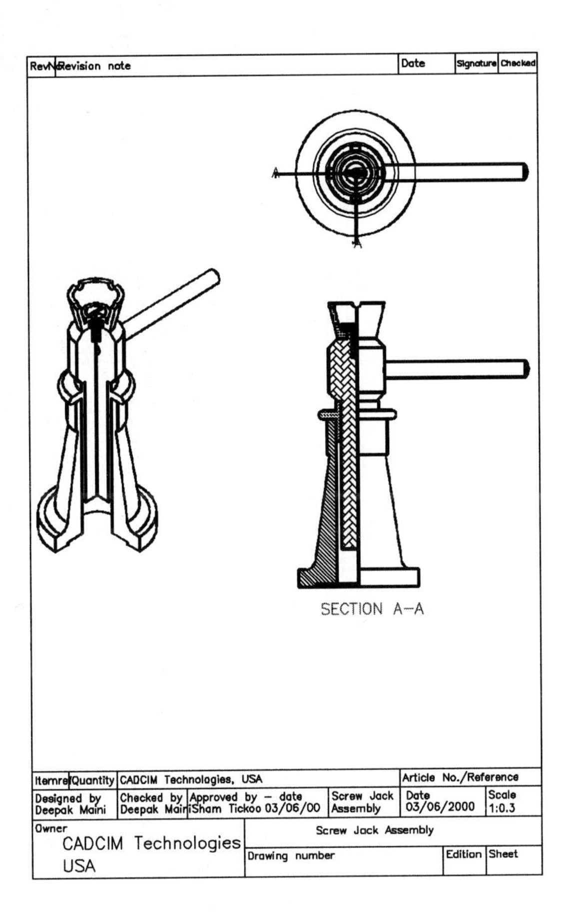

Figure 13-20 *The views created inside the title block*

11. Select **Designer** from the **Parts List** area and then choose the **Remove Item** button.

12. Choose **Apply** and then choose **OK**.

13. Double-click on **Balloon** to display the **Balloon Properties for ANSI** dialog box.

14. Set the value of the **Text Height** spinner to **0.12**. Choose **Apply** and choose **OK**.

15. Choose **OK** in the **Symbol Standards** dialog box.

16. Choose the **Edit BOM Database** button from the **Power Dimensioning** flyout in the **Drawing Layout** toolbar. The prompt sequence is as follows:

Bom table [Delete/Edit] <Edit>: Enter *The **BOM** dialog box will be displayed.*

17. Choose the **Properties** button to display the **BOM Properties for ANSI** dialog box.

18. Select the **NAME** column and then choose the **Center Align Text** from the **Data Alignment** area. Choose **Apply** and then choose **OK**.

19. Double-click on the **Material** column in front of **CASTING SC** and enter **CAST STEEL**.

20. Similarly, enter the material values for the rest of the components, see Figure 13-21.

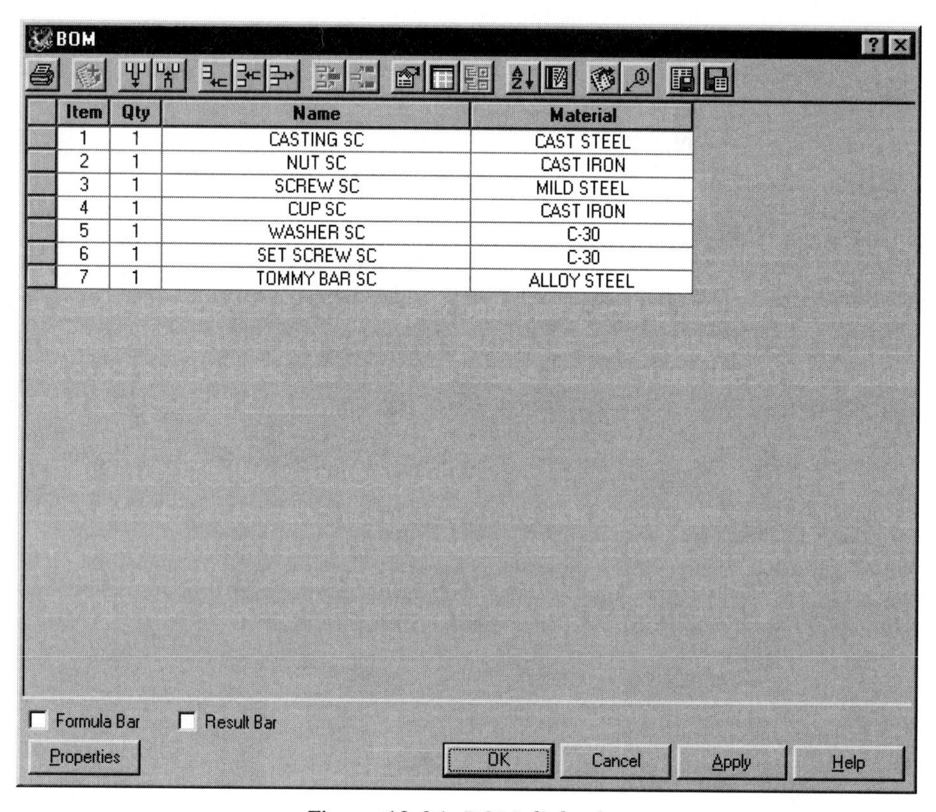

Item	Qty	Name	Material
1	1	CASTING SC	CAST STEEL
2	1	NUT SC	CAST IRON
3	1	SCREW SC	MILD STEEL
4	1	CUP SC	CAST IRON
5	1	WASHER SC	C-30
6	1	SET SCREW SC	C-30
7	1	TOMMY BAR SC	ALLOY STEEL

Figure 13-21 *BOM dialog box*

21. Choose the **Insert parts list** button in the **BOM** dialog box to display the **Parts List** dialog box. Choose **Apply** and then choose **OK**. The prompt sequence is as follows:

    Specify location: *Place the Parts List.*

22. Choose **OK** in the **BOM** dialog box.

23. Choose **Power Dimensioning > Edit BOM Database > Place Balloon** from the **Drawing Layout** toolbar. The prompt sequence is as follows:

    Select part/assembly or [auTo/autoAll/Collect/Manual/One/Renumber/rEorganize]: **O**

Select pick object: *Select the point on the **CASTING SC**.*
Select a start point of balloon: *Select a point on the component.*
Next Point: *Specify the placement point.*
Next Point: Enter

Similarly, add balloons to the remaining components.

24. The layout after creating the required drawing view and after adding BOM and balloons should look similar to the one shown in Figure 13-22.

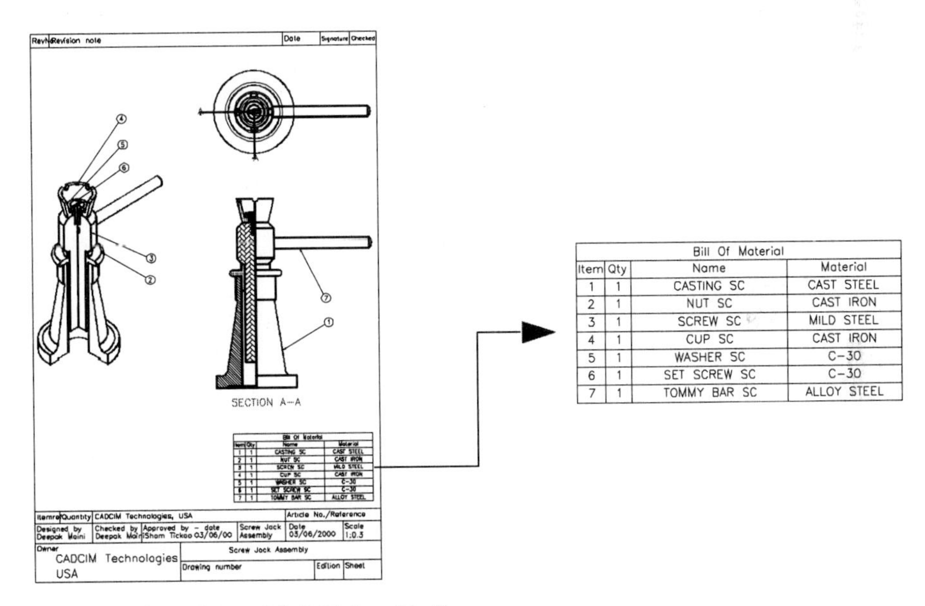

***Figure 13-22** Drawing views with BOM and balloons*

25. Save this drawing with the name given below:

\MDT Tut\Ch-13**Tut3.dwg**

Review Questions

Answer the following questions.

1. What is the difference between the parametric dimensions and the reference dimensions in the drawing views?

2. The **AMPOWEREDIT** command can be used for editing the reference dimensions in the drawing views. (T/F)

3. The bidirectional associative nature of Mechanical Desktop is valid for reference dimensions. (T/F)

4. The dimensions can be shown in the drawing views of the assembly. (T/F)

5. Define BOMs and Balloons.

Exercise

Exercise 1

In this exercise you will place the BOM and balloons in the drawing views of the Plummer Block assembly created in Tutorial 1 of Chapter 12. The BOM should have the following columns:

1. Item
2. Quantity
3. Name
4. Material

Chapter 14

Surface Modeling

Learning Objectives

After completing this chapter, you will be able to:

- *Create different types of surfaces.*
- *Create different types of surface primitives.*
- *Edit surfaces.*
- *Calculate the mass properties of selected surfaces.*

CREATING THE SURFACES

Surfaces are the objects having length and width but no thickness or negligible thickness. Mechanical Desktop allows you to create the surfaces in the **Model** mode using the Non Uniform Rational Bezier Splines. Mechanical Desktop provides you with a number of commands for creating the surfaces. These commands are discussed next.

AMEXTRUDESF Command

Toolbar:	Surface Modeling > Swept Surface > Extruded Surface
Menu:	Surface > Create Surface > Extrude
Command:	AMEXTRUDESF

This command is used to create an extruded surface by extruding the selected wires along the specified path. The extrusion direction can be specified using the X, Y, Z axes, view direction or using a specified object, see Figures 14-1 and 14-2.

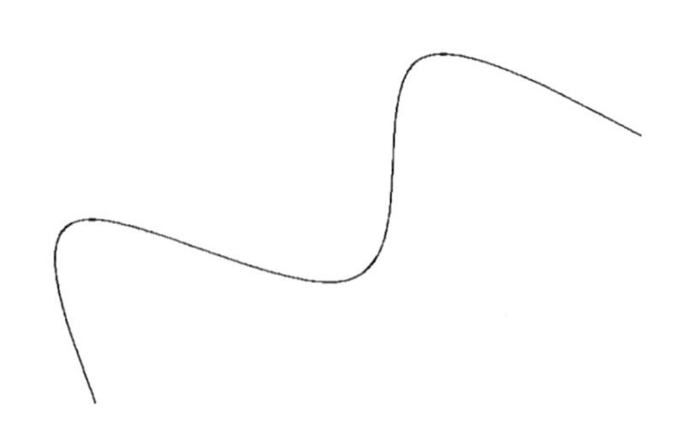

Figure 14-1 *Spline before creating surface*

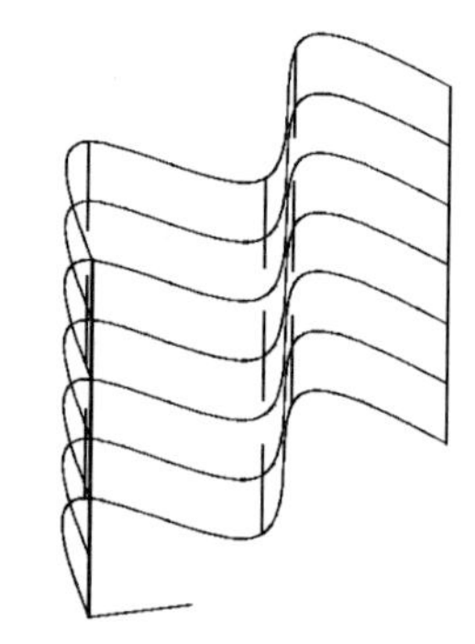

Figure 14-2 *Surface created by extruding the spline*

AMREVOLVESF Command

Toolbar:	Surface Modeling > Swept Surface > Revolved Surface
Menu:	Surface > Create Surface > Revolve
Command:	AMREVOLVESF

This command is used to create a revolved surface by rotating the path curves about the specified axis.

AMSWEEPSF Command

Toolbar:	Surface Modeling > Swept Surface > Swept Surface
Menu:	Surface > Create Surface > Sweep
Command:	AMSWEEPSF

This command is used to create a swept surface. The path curve can be swept along a single wire or more than one wire, see Figures 14-3 and 14-4. When you invoke this command, you will be prompted to select the cross section and then the rails about which the cross section is to be swept. The **Sweep Surface** dialog box will be displayed after

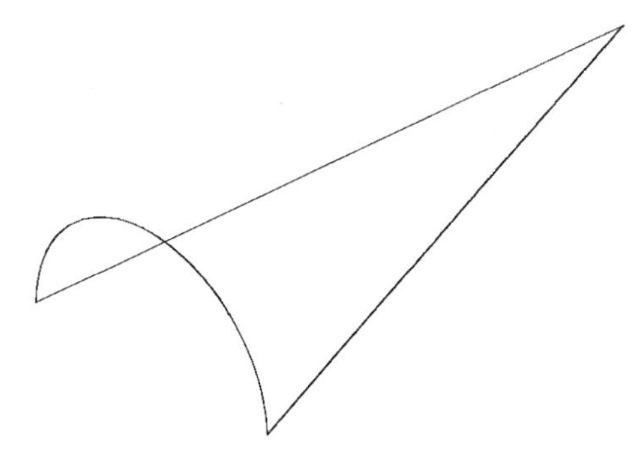

Figure 14-3 *Arc before creating surface*

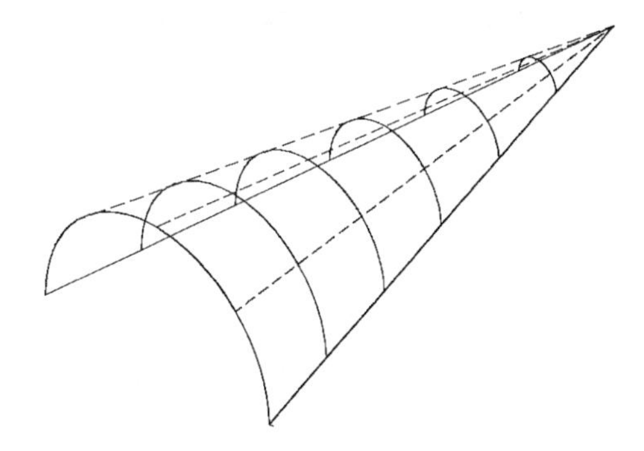

Figure 14-4 *Surface created by sweeping the arc*

you have selected the rails, see Figure 14-5.

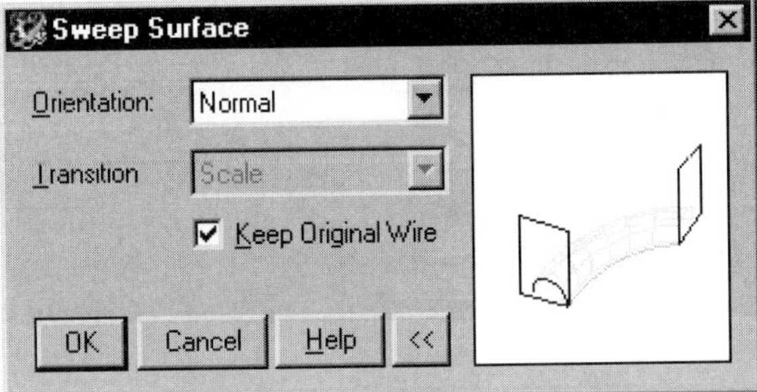

Figure 14-5 *Sweep Surface dialog box*

Sweep Surface Dialog Box Options

The options provided under this dialog box are:

Orientation

This drop-down list provides the options related to the cross section and rails. However, this drop-down list will be available only when a single rail is selected for sweeping the cross section. The options provided under this drop-down list are:

Normal. If this option is selected, the cross section will be held perpendicular to the selected rail when it is being swept along the rail.

Parallel. If this option is selected, the cross section will be held parallel to the selected rail when it is being swept along the rail.

Direction. This option is used to manually specify the direction of the swept surface. You can specify the direction as the view direction, a wire or using a start and an endpoint. You can also use the X, Y or Z axis to specify the direction of the sweep surfaces.

Transition

The options under this drop-down list will be available only when two rails are selected.

Scale. This radio button is selected to uniformly adjust the cross section between the selected rails.

Stretch. This radio button is selected to adjust the cross section only across the selected rails.

Keep Original Wire

If this check box is selected then the original cross section as well as the rails will be retained even after the swept surface is created.

AMTUBE Command

Toolbar:	Surface Modeling > Swept Surface > Tubular Surface
Menu:	Surface > Create Surface > Tubular
Command:	AMTUBE

This command is used to create a tubular surface using a base curve, see Figures 14-6 and 14-7. Depending upon the type of object used as the wire, the surfaces created differ. You will be prompted to specify the overall diameter of the tube and the radius at bends, if any. You can automatically define the radius at all the bends or manually define the radius at each bend.

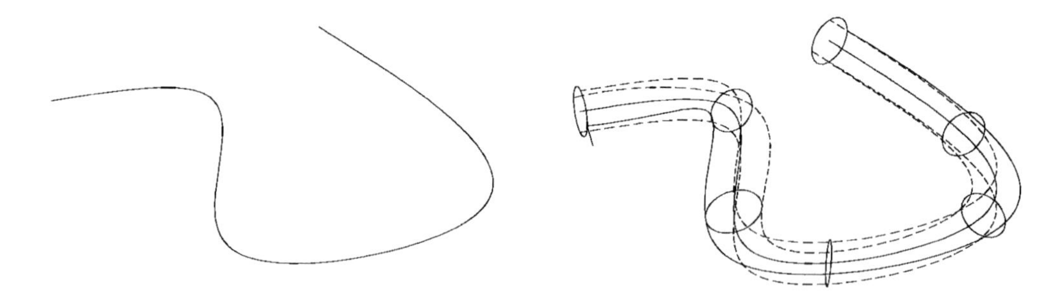

Figure 14-6 *The spline to be used for creating the tubular surface*

Figure 14-7 *Tubular surface created using the spline*

AMLOFTU Command

Toolbar:	Surface Modeling > Loft U Surface
Menu:	Surface > Create Surface > Loft U
Command:	AMLOFTU

This command is used to create a surface using a set of base curves representing the surface at that position. Minimum of two base curves are required for creating the surface using this command. You can even select non coplanar objects to be used as base curves in this command. Once you have selected all the base curves, the **Loft Surface** dialog box will be displayed as shown in Figure 14-8.

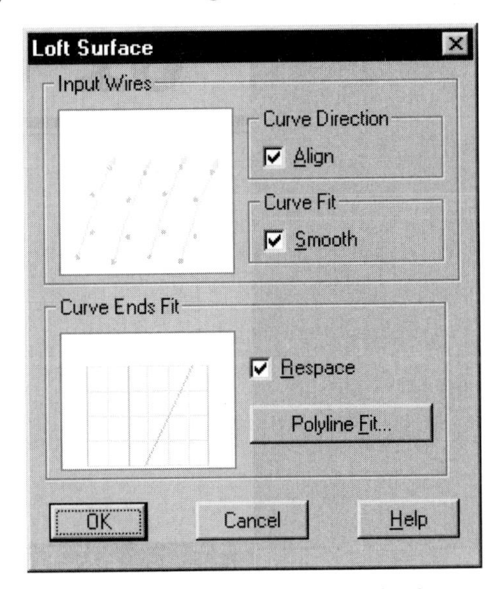

Figure 14-8 *Loft Surface dialog box*

Loft Surface Dialog Box Options

Input Wires Area

Align. This check box is selected to align directions of the selected base curves.

Smooth. This check box is selected to smoothen the lofted surface.

Curve Ends Fit Area

Respace. This check box is used to respace the base curves in case of poorly proportioned base curves, see Figures 14-9 and 14-10.

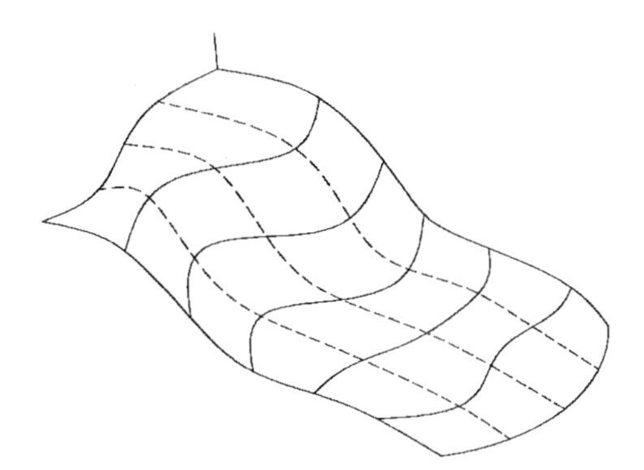

Figure 14-9 *Surface created without respacing*

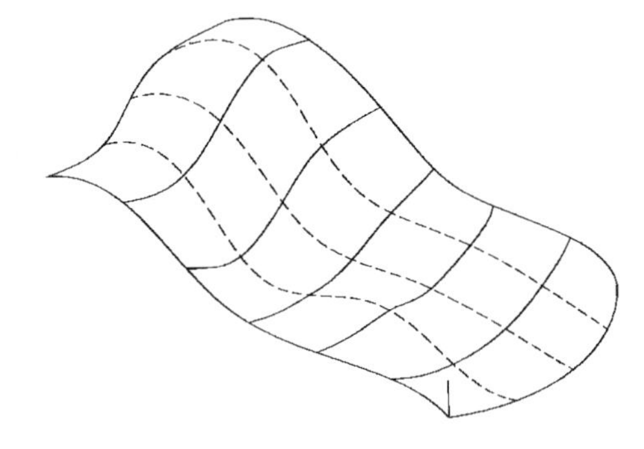

Figure 14-10 *Surface created with respacing*

Polyline Fit. This button is chosen to display the **Polyline Fit** dialog box for editing the polyline fit length and angle.

AMLOFTUV Command

Toolbar:	Surface Modeling > Loft U Surface > Loft UV Surface
Menu:	Surface > Create Surface > Loft UV
Command:	AMLOFTUV

This command is used to create a surface using two sets of base curves namely the U and V curves, see Figures 14-11 and 14-12. Each set of base curves should contain a minimum of two entities.

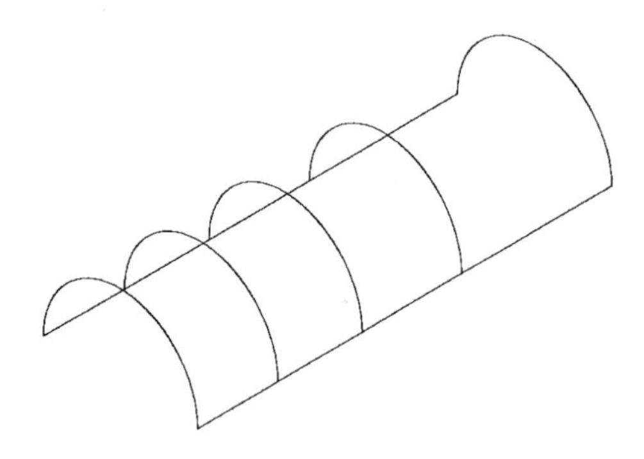

Figure 14-11 The U and V wires for creating the Loft UV surface

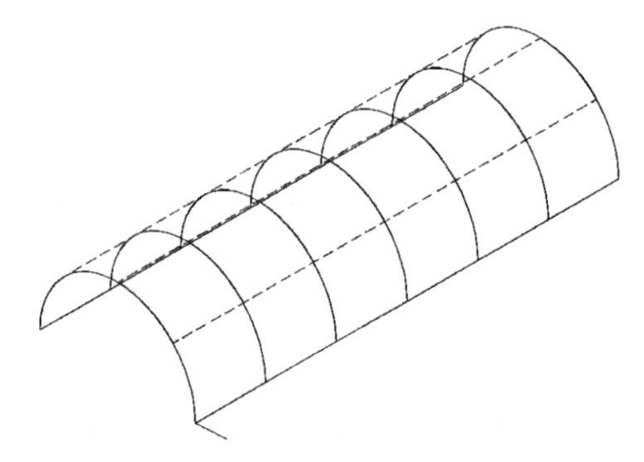

Figure 14-12 The Loft UV surface

AMPLANE Command

Toolbar:	Surface Modeling > Loft U Surface > Planar Surface
Menu:	Surface > Create Surface > Planar
Command:	AMPLANE

This command is used for creating the planar or planar trimmed surfaces. The options provided under this command are:

Plane

This option is used to first specify a plane in which the surface has to be created. The plane can be world XY, YZ, ZX planes or can be specified using Z axis option, a 2D entity, or view direction. You can even select the last plane as the new plane. After specifying the plane, you will be prompted to specify two points to draw a rectangle that will be the sketch for surface.

Wire

This option is used for selecting the existing sketches to be used as base curves for the surface. If you select more than one object, the resultant will be a trimmed planar surface as shown in Figures 14-13 and 14-14.

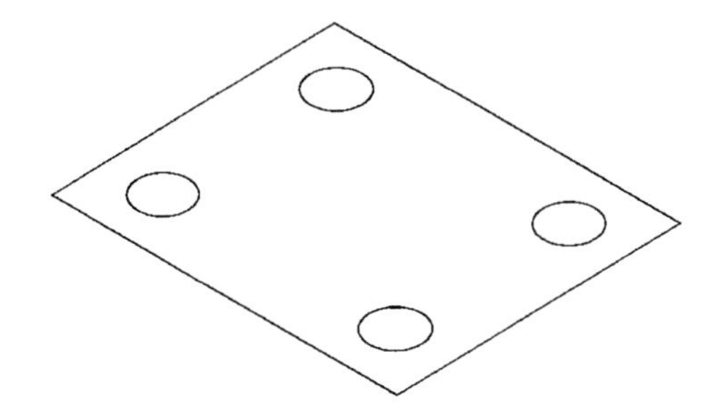

Figure 14-13 *The base curves to be used as wires*

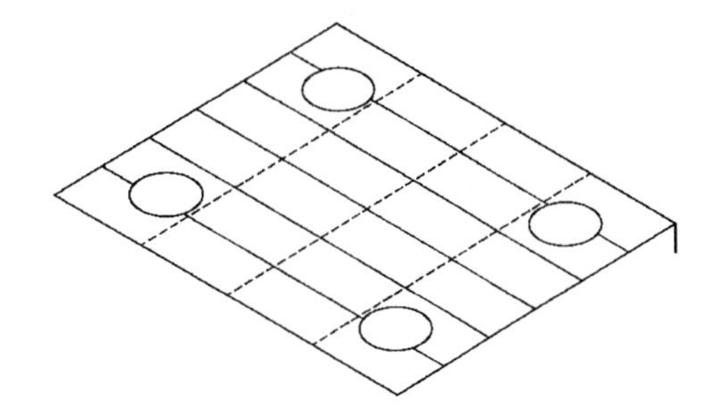

Figure 14-14 *Trimmed surface created using the selected wires*

AMRULE Command

Toolbar:	Surface Modeling > Loft U Surface > Ruled Surface
Menu:	Surface > Create Surface > Rule
Command:	AMRULE

This command is used to create a ruled surface between the two selected base curves. You cannot select more than two entities in this case.

AMBLEND Command

Toolbar:	Surface Modeling > Blended Surface
Menu:	Surface > Create Surface > Blend
Command:	AMBLEND

This command is used to create a surface between two, three or four wires or surfaces. You can also use this command to fill the empty spaces between the existing surfaces. Figure 14-15 shows the surfaces without blending and Figure 14-16 shows the surfaces after blending.

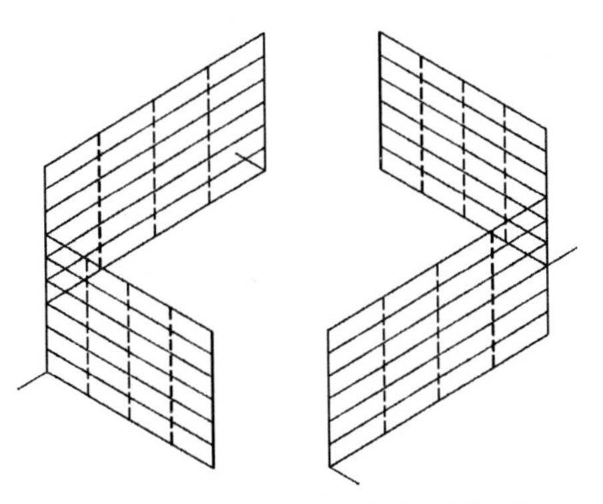

Figure 14-15 *Surfaces before blending*

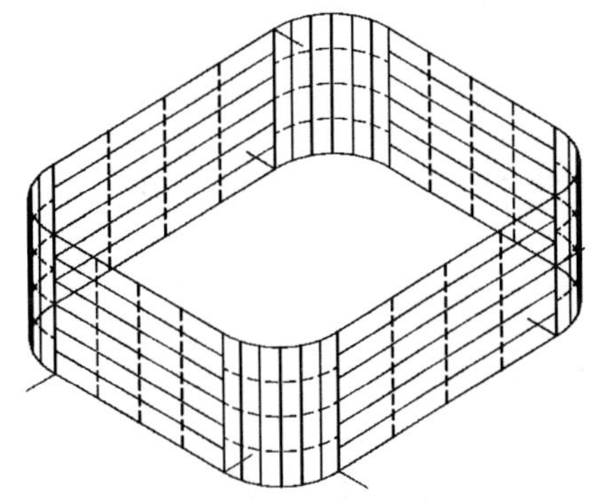

Figure 14-16 *Surfaces after blending*

AMPRIMSF Command

This command is used to create the standard primitive surfaces. Using this command you can create conical, cylindrical, spherical or torus shaped surfaces. You can also directly choose these buttons from the **AutoSurf Options** flyout in the **Surface Modeling** toolbar. These options are discussed below:

Creating Conical Surfaces

Toolbar:	Surface Modeling > AutoSurf Options > Cone Surface
Menu:	Surface > Create Primitives > Cone
Command:	AMPRIMSF

This option of the **AMPRIMSF** command is used to create conical surfaces. Using this option you can also define some top diameter thus creating partial cones. You can also specify the start angle and the included angle for the conical surface.

Creating Cylindrical Surfaces

Toolbar:	Surface Modeling > AutoSurf Options > Cylinder Surface
Menu:	Surface > Create Primitives > Cylinder
Command:	AMPRIMSF

This option of the **AMPRIMSF** command is used to create cylindrical surfaces. You can specify some start angle and included angle for the cylindrical surfaces.

Creating Spherical Surfaces

Toolbar:	Surface Modeling > AutoSurf Options > Sphere Surface
Menu:	Surface > Create Primitives > Sphere
Command:	AMPRIMSF

This option of the **AMPRIMSF** command is used to create spherical surfaces. You can specify some start angle and included angle for the cylindrical surfaces.

Creating Torus Surfaces

Toolbar:	Surface Modeling > AutoSurf Options > Torus Surface
Menu:	Surface > Create Primitives > Torus
Command:	AMPRIMSF

This option of the **AMPRIMSF** command is used to create the torus surfaces. You will be asked to specify the diameter which is the diameter of the torus and then you will be asked to specify the diameter of tube. You can also specify some start angles and included angle for the torus surfaces.

EDITING THE SURFACES

Once you have created various surfaces, you may need to edit them. The various editing operations available for editing the surfaces are discussed next.

AMBREAK Command

Toolbar: Surface Modeling > Lengthen Surface > Break Surface
Menu: Surface > Edit Surface > Break
Command: AMBREAK

This command is used to break the selected surface. The surface can be broken at the C1 tangency, in the U edge direction or in V edge direction. The options provided under this command are:

C1 tangents

This option is used to break the surface at the specified C1 tangency.

Flip

This option is used to toggle between breaking the surface in U and V edge directions.

Reposition

This option is used to reposition the break in the surface. The break is defined in the terms of percentage. However, you can also select a point using the **Specify break** option.

AMINTERSF Command

Toolbar: Surface Modeling > Flow Wires > Intersection Wires
Menu: Surface > Edit Surface > Intersect Trim
Command: AMINTERSF

This command is similar to the **TRIM** command and is used to trim the selected surfaces at their intersection point, see Figures 14-17 and 14-18. When you invoke this command, you will be prompted to select the first and the second surface.

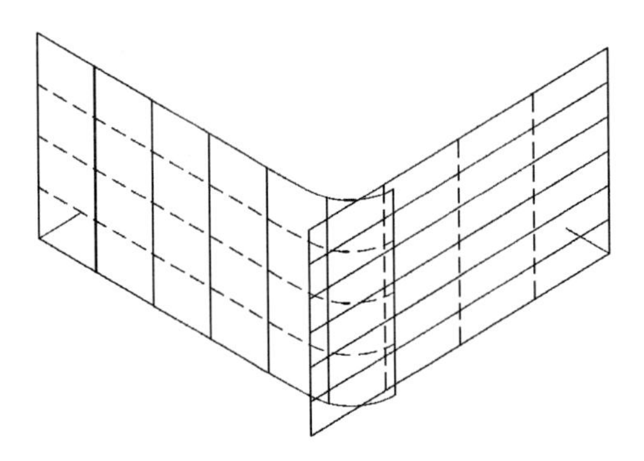

Figure 14-17 *Surfaces before intersection*

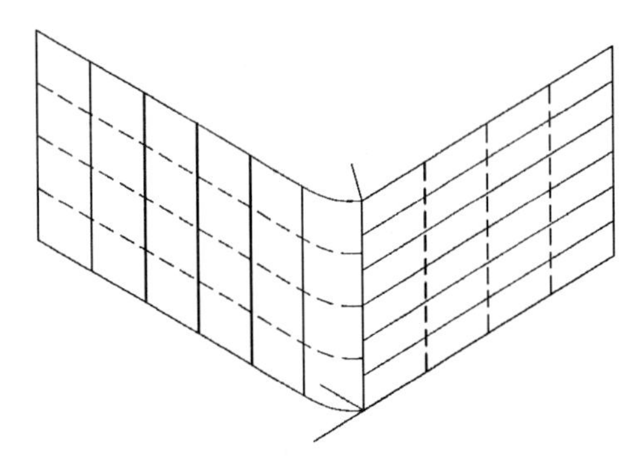

Figure 14-18 *Surfaces after intersection*

After you have selected both the surfaces, the **Surface Intersection** dialog box will be displayed as shown in Figure 14-19.

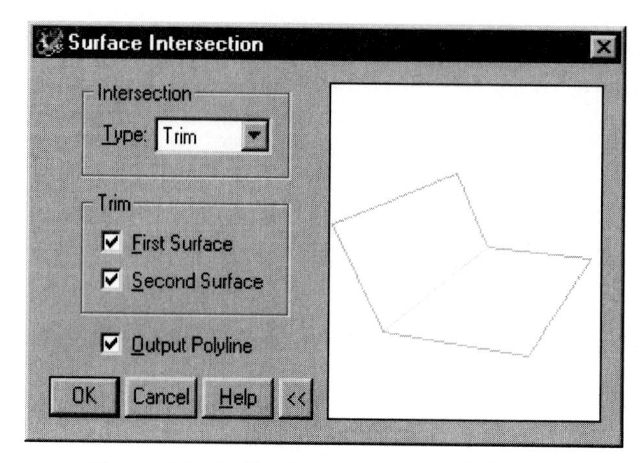

Figure 14-19 *Surface Intersection dialog box*

Surface Intersection Dialog Box Options

The options provided under this area are:

Intersection Area

This area displays the **Type** drop-down list. This drop-down list provides the option for the resultant intersection surface. The options under this list are:

Trim. This option is selected to trim the surfaces at the intersection. The portions of the surfaces from where you select them are retained.

Break. This option is used to break the selected surfaces at intersection.

Trim Area

First Surface. This check box is selected to trim the first surface at the intersection.

Second Surface. This option is selected to trim the second surface at the intersection.

Output Polyline

If this check box is selected then a polyline will be created along the intersection edge.

AMJOINSF Command

Toolbar:	Surface Modeling > Lengthen Surface > Join Surfaces
Menu:	Surface > Edit Surface > Join
Command:	AMJOINSF

This command is used to join the selected surfaces into a single continuous surface. You can also join the surfaces that are not matching using this command. It is very important here to mention that the trimmed surfaces cannot be joined using this command. Figure 14-20 shows the surfaces before joining and Figure 14-21 shows the surfaces after joining.

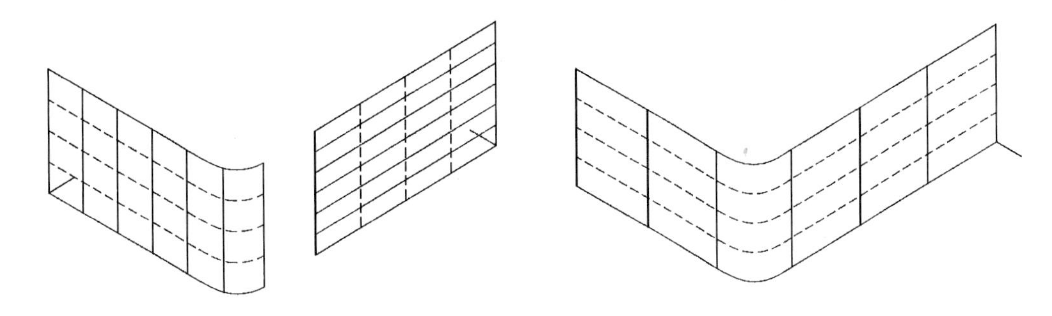

Figure 14-20 Surfaces before joining *Figure 14-21* Surfaces after joining

AMPROJECT Command

Toolbar:	Surface Modeling > Flow Wires > Projection Wires
Menu:	Surface > Edit Surface > Project Trim
Command:	AMPROJECT

This command is used to trim a selected surface by projecting the specified entity on to it. The entities allowed for projection are lines, polylines, arcs, circles, ellipses and splines. You can also use 3D splines for projecting. The portion from where you select the surface is retained. When you invoke this command, you will be asked to select the wires to project and the target surface. Once you have selected both of them, the **Project To Surface** dialog box will be displayed as shown in Figure 14-22.

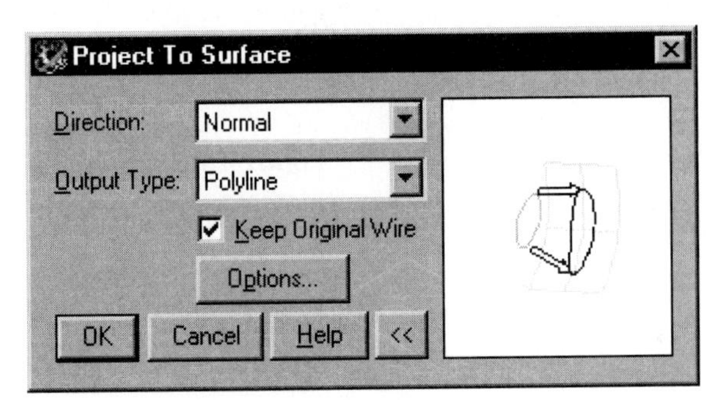

Figure 14-22 Project To Surface dialog box

Project To Surface Dialog Box Options

Direction

This drop-down list provides the projection direction options. The options provided under this drop-down list are as follows:

Normal. This option is used to project the wires in the direction perpendicular to the surface. If the surface is curved, the projection trim is increased or reduced depending upon the curvature of the surface, see Figures 14-23 and 14-24.

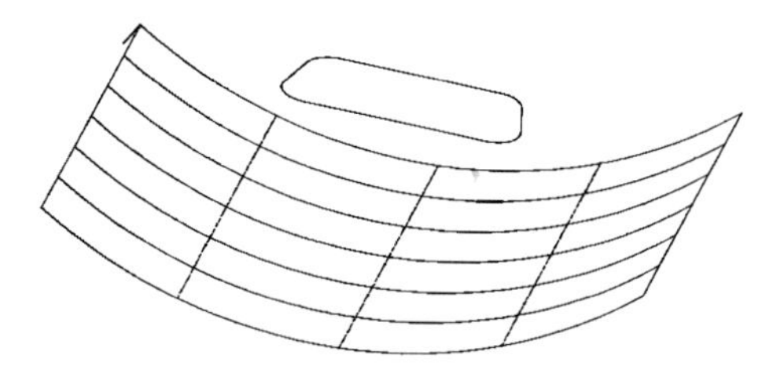

Figure 14-23 Surface before projecting the wire normal to the surface

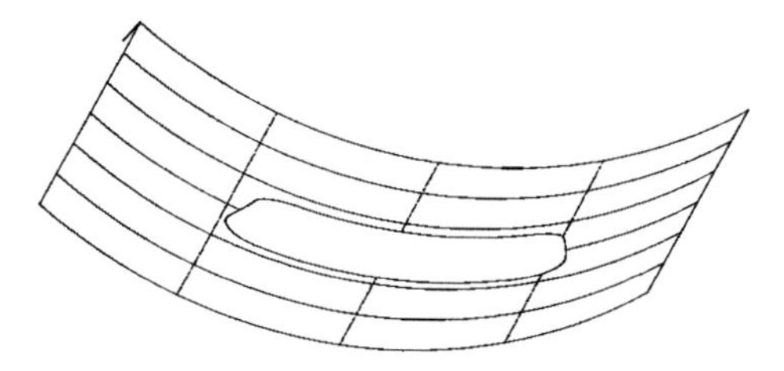

Figure 14-24 Surface after projecting the wire normal to the surface

UCS. This option is used to project the selected wires normal to the current UCS, see Figures 14-25 and 14-26.

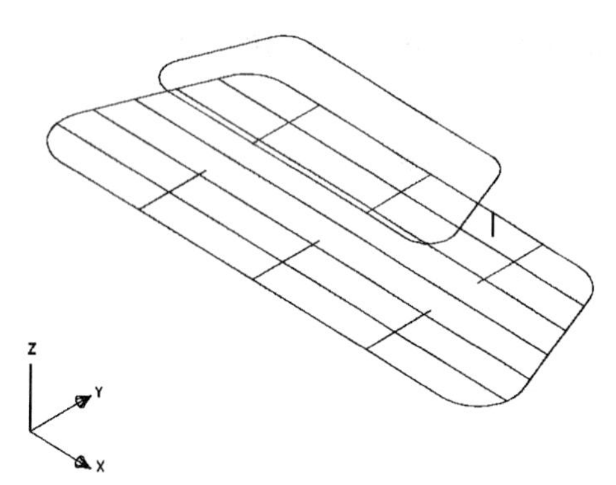

Figure 14-25 Surface before projecting the wire normal to the UCS

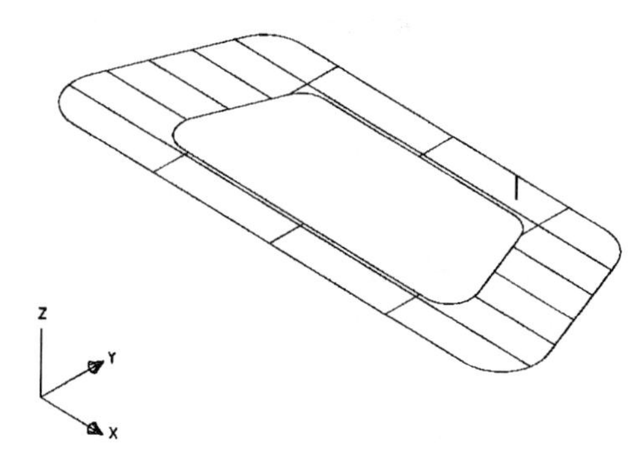

Figure 14-26 Surface after projecting the wire normal to the UCS

Vector. This option is used to specify the projection direction using the command line. The options that can be used to define the projection direction are view direction, a selected wire or X, Y, or Z axes.

View. This option is used to project the selected wire normal to the current view plane.

Output Type

This drop-down list provides the options related to the resultant output of the selected wire.

Augmented line. This option is used to get the resultant of the project trim in the form of a polyline with vectors normal to the surface, see Figure 14-27.

Polyline. This option is used to get the resultant of the project trim in the form of a polyline, see Figure 14-28.

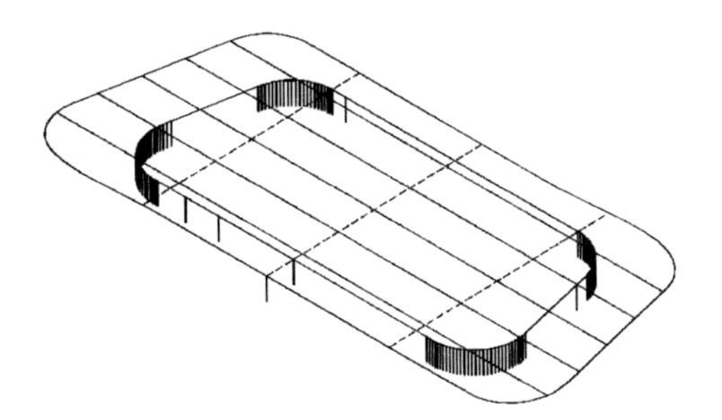

Figure 14-27 *Resultant of project trim in the form of augmented lines*

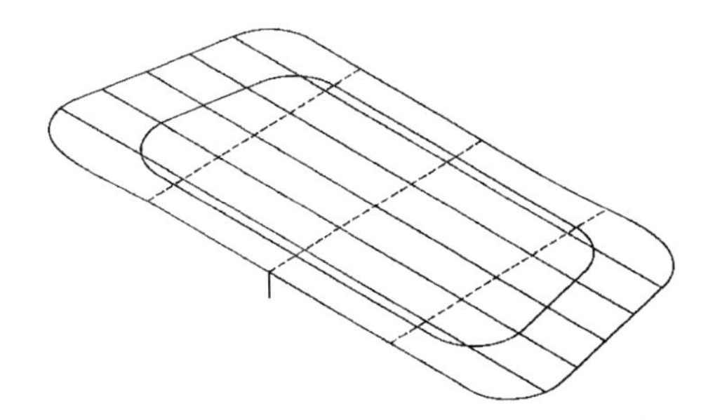

Figure 14-28 *Resultant of project trim in the form of polyline*

Trim surface. This option is used to trim the surface by projecting the selected wire.

Break Surface. This option is used to break the surface into two. The new surface will be the surface created upon projecting the wire on to the original surface.

Keep Original Wire

If this check box is selected then the original wire used for projection will be retained even after the projection is complete.

Options

This button is chosen to display the **Surface Projection Options** dialog box see Figure 14-29.

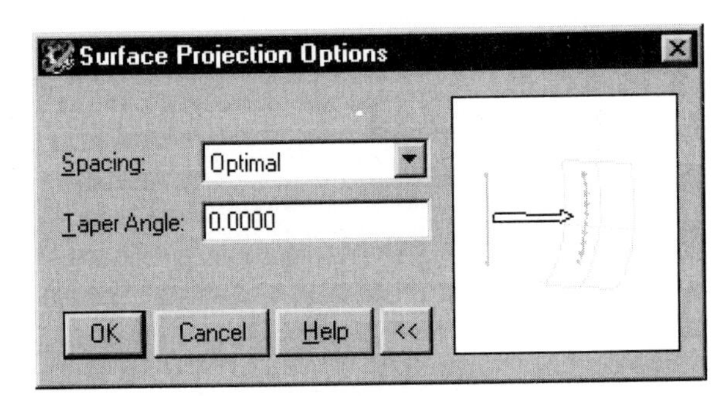

Figure 14-29 *Surface Projection Options dialog box*

Surface Projection Options Dialog Box Options

Spacing. The options under this drop-down are used to define the spacing of the points in the wires resultant of project trim.

Taper Angle. This option is used to define the draft angle for the project trim.

AMSURFPROP Command

Toolbar:	Surface Modeling > Surface Analysis > Surface Mass Properties
Menu:	Surface > Utilities > Mass Properties
Command:	AMSURFPROP

This command is used to calculate the mass properties of the selected surface. The options provided under this command are:

Density

The **Density** option is used to set the density for the selected surface.

Thickness

The **Thickness** option is used to set the thickness value for the selected surface.

tYpe

The **tYpe** option is used to specify the type of the surface under analysis. It can be a shelled surface or an enclosed model.

Calculate properties

This option is used to invoke the **AutoSurf Mass Properties** dialog box (Figure 14-30). This dialog box displays the mass properties of the selected surface.

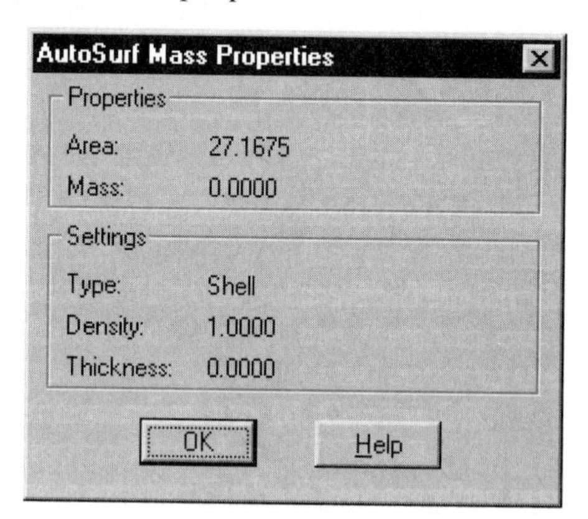

Figure 14-30 *AutoSurf Mass Properties dialog box*

TUTORIALS

Tutorial 1

In this tutorial you will create the tray shown in Figure 14-31a using the surfaces. The dimensions to be used are given in Figures 14-31b, 14-31c and 14-31d.

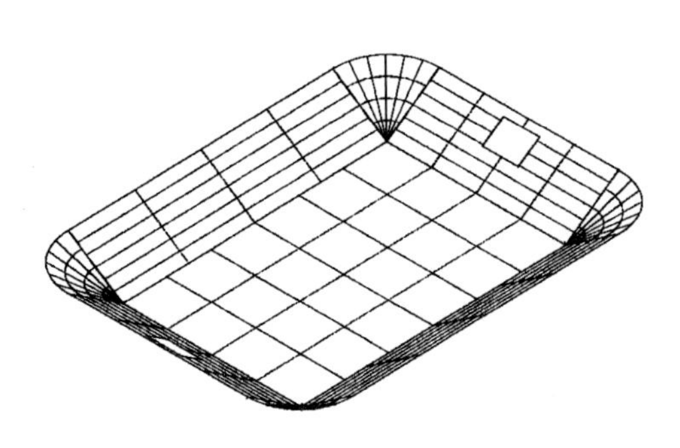

Figure 14-31a *Model for Tutorial 1*

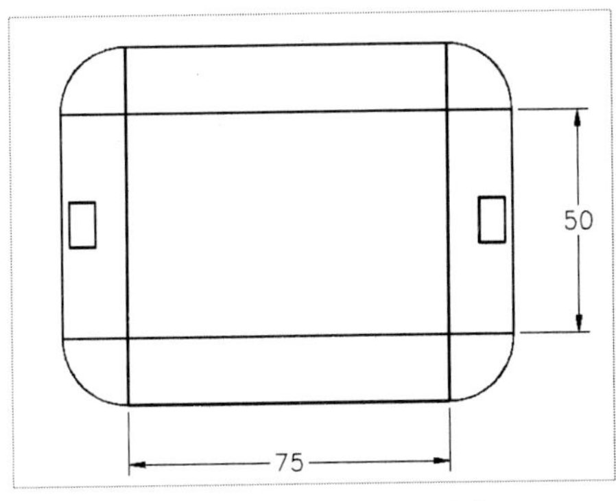

Figure 14-31b *Top view of the tray*

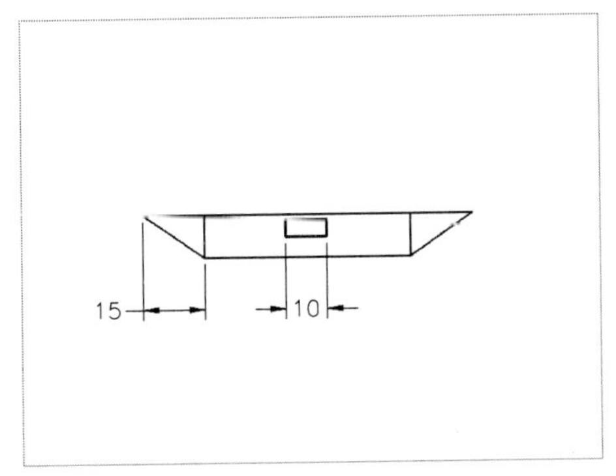

Figure 14-31c *Side view of the tray*

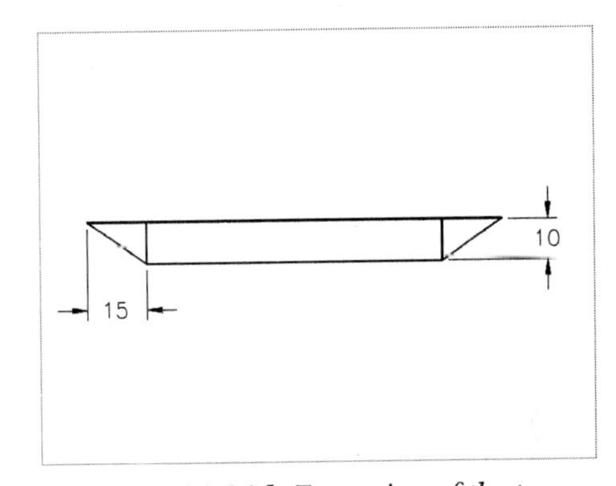

Figure 14-31d *Front view of the tray*

1. Create a rectangle that measures 50x75.

2. Choose the **Planar Trimmed Surface** button from the **Loft U Surface** flyout in the **Surface Modeling** toolbar. The prompt sequence is as follows:

   Specify first corner or [Plane/Wires]: _wire
   Select wires: *Select the rectangle.*
   Select wires: Enter

3. Draw a line from the lower end point of the right vertical edge. The end point of the line is defined using the coordinates @15,0,10.

4. Change the viewpoint and then choose the **Extruded Surface** button from the **Swept Surface** flyout in the **Surface Modeling** toolbar. The prompt sequence is as follows:

   Select wires to extrude: *Select the extreme right vertical edge of the surface.*
   Select wires to extrude: Enter

Define direction and length:
Specify start point or [Viewdir/Wire/X/Y/Z]: **W**
Select wire to define direction: *Select the line.*

Enter an option [Flip/Accept] <Accept>: *Make sure the direction is upwards.*

Enter taper angle <0>: Enter

5. Mirror this surface on to the other side of the base rectangular surface, see Figure 14-32.

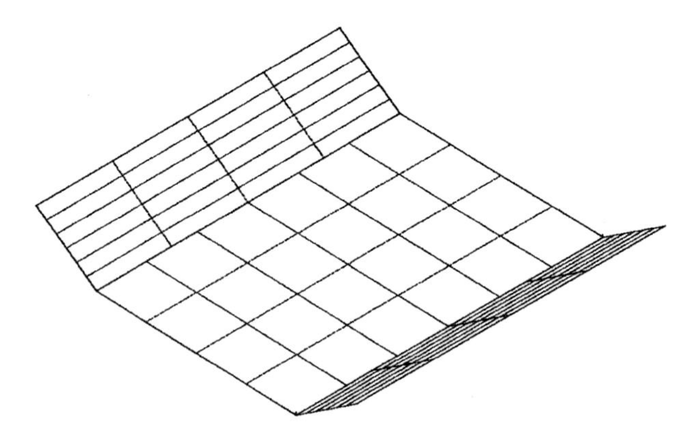

Figure 14-32 *Figure after mirroring the surface*

6. Draw a line from the left end point of the bottom horizontal edge of the base rectangular surface. The end point of the line is defined using the coordinates @0,-15,10.

7. Choose the **Extruded Surface** button from the **Swept Surface** flyout in the **Surface Modeling** toolbar. The prompt sequence is as follows:

Select wires to extrude: *Select the bottom horizontal edge of the base surface.*
Select wires to extrude: Enter

Define direction and length:
Specify start point or [Viewdir/Wire/X/Y/Z]: **W**
Select wire to define direction: *Select the line.*

Enter an option [Flip/Accept] <Accept>: *Make sure the direction is upwards.*

Enter taper angle <0>: Enter

8. Mirror this surface on the other side of the base surface, see Figure 14-33. For the ease of selecting the surfaces, they have been numbered as shown in Figure 14-33.

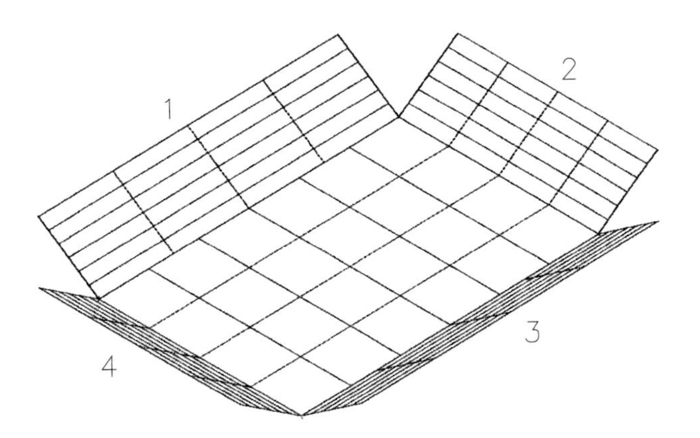

Figure 14-33 Surfaces on all four edges

9. Choose the **Blended Surface** button from the **Surface Modeling** toolbar. The prompt sequence is as follows:

Select first wire: *Select the extreme right edge of surface 1.*
Select second wire: *Select the extreme left edge of surface 2.*

Select third wire [Weights]: Enter

10. Similarly, create the blended surfaces at the remaining three corners.

11. Change the UCS position and then draw a rectangle of the specified dimensions as shown in Figure 14-34.

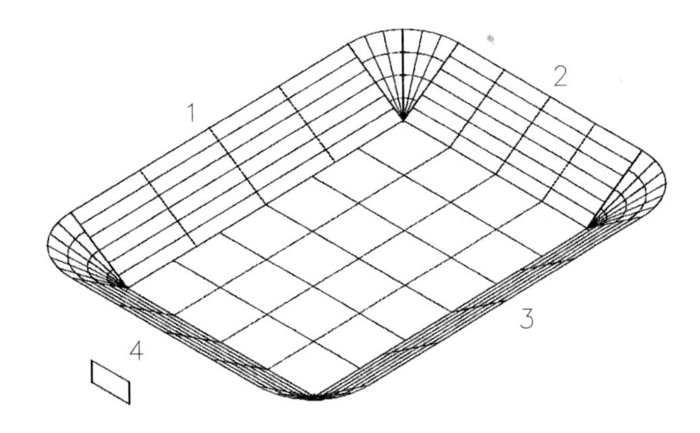

Figure 14-34 The rectangle for project trim

12. Choose the **Projection Wire** button from the **Flow Wires** flyout in the **Surface Modeling** toolbar. The prompt sequence is as follows:

Select wires to project: *Select the rectangle.*
Select wires to project: Enter

Select target surfaces: *Select surface 4.*
Select target surfaces: *Select surface 2.*
Select target surfaces: Enter *The **Project To Surface** dialog box will be displayed, see Figure 14-35.*

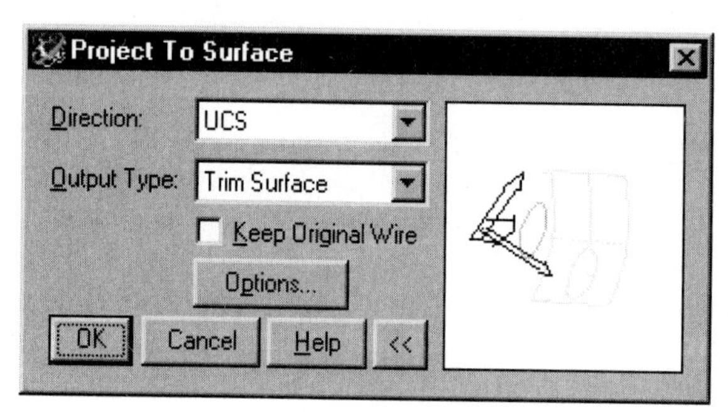

Figure 14-35 *Project To Surface dialog box*

13. Select **UCS** from the **Direction** drop-down list.

14. Select **Trim Surface** from the **Output Type** drop-down list.

15. Clear the **Keep Original Wire** check box. Choose **OK**. The final model of the tray should look similar to the one shown in Figure 14-36.

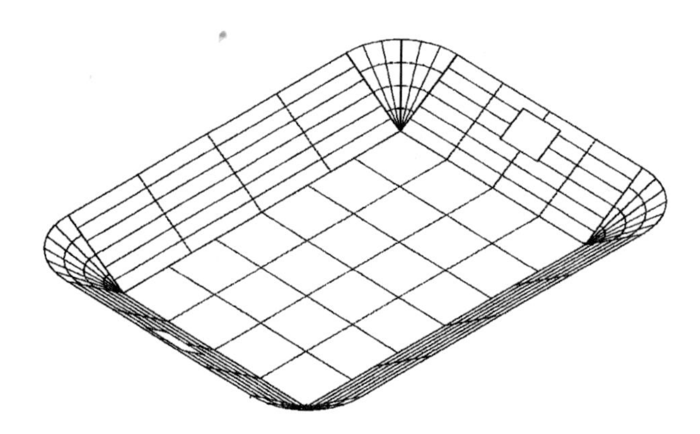

Figure 14-36 *Final tray model for Tutorial 1*

15. Save this drawing with the name given below:

    \MDT Tut\Ch-14**Tut1.dwg**

Review Questions

Answer the following questions.

1. Define surfaces.
2. Which command is used to create extruded surfaces?
3. What is the difference between Loft U and Loft UV surfaces?
4. If you select more than one object in the **Wire** option of the **AMPLANE** command, the resultant surface will be __________.
5. What are the various types of surfaces that can be created using the **AMPRIMSF** command?
6. Which command is used to join the existing surfaces?
7. What is the use of **AMPROJECT** command?
8. The mass properties of the surfaces can be calculated using the __________ command.

Chapter 15

Miscellaneous Commands

Learning Objectives

After completing this chapter, you will be able to:

- *Use the* ***AMREPLAY*** *command.*
- *Use the* ***AMPARTPROP*** *command.*
- *Use the* ***AMMAKEBASE*** *command.*
- *Use the* ***AMPARTSPLIT*** *command.*
- *Use the* ***AMFACESPLIT*** *command.*
- *Use the* ***AMTEXTSK*** *command.*
- *Use the* ***AMWHEREUSED*** *command.*
- *Use the* ***AMASSIGN*** *command.*
- *Use the* ***AMMANIPULATOR*** *command.*
- *Use the* ***AMHOLENOTE*** *command.*
- *Use the* ***AMTEMPLATE*** *command.*

Commands Covered

- ***AMREPLAY***
- ***AMPARTPROP***
- ***AMMAKEBASE***
- ***AMPARTSPLIT***
- ***AMFACESPLIT***
- ***AMTEXTSK***
- ***AMWHEREUSED***
- ***AMASSIGN***
- ***AMMANIPULATOR***
- ***AMHOLENOTE***
- ***AMTEMPLATE***

AMREPLAY COMMAND

Toolbar:	Part Modeling > Update Part > Feature Replay
Menu:	Part > Part > Replay
Context Menu:	Part > Replay
Command:	AMREPLAY

This command is used in the **Model** mode to display all the steps used in the creation of the selected component, starting from the first step, in the sequence in which they were used. You can truncate the steps where ever you want to lop off the remaining features. The options provided under this command are:

Display

This option is used to display the sketch along with all the geometric constraints applied on to it. This option is available when this command reaches the step where a sketch was created.

Exit

This option is used to exit the **AMREPLAY** command.

Next

This option is used to advance to the next step.

Size

This option is used to display the **Constraint Display Size** dialog box to modify the size of the constraint as they appear on the sketch.

Suppress

This option is used to suppress all the features created after the current step.

Truncate

This option is used to remove all the features created after the current step.

AMPARTPROP COMMAND

Toolbar:	Part Modeling > Options > Mass Properties
Menu:	Part > Part > Mass Properties
Context Menu:	Part > Mass Properties
Command:	AMPARTPROP

This command is used to calculate the mass properties of the selected component. You can specify some value of the density and calculate the mass properties based on that value. The calculated mass properties can also be written to a .MRP file. When you invoke this command, the **Part Mass Properties** dialog box will be displayed as shown in Figure 15-1. To write the mass properties to a .MPR file, choose the **File** button. The **Save As** dialog box will be displayed. You can specify the name of the .MPR file in this dialog box.

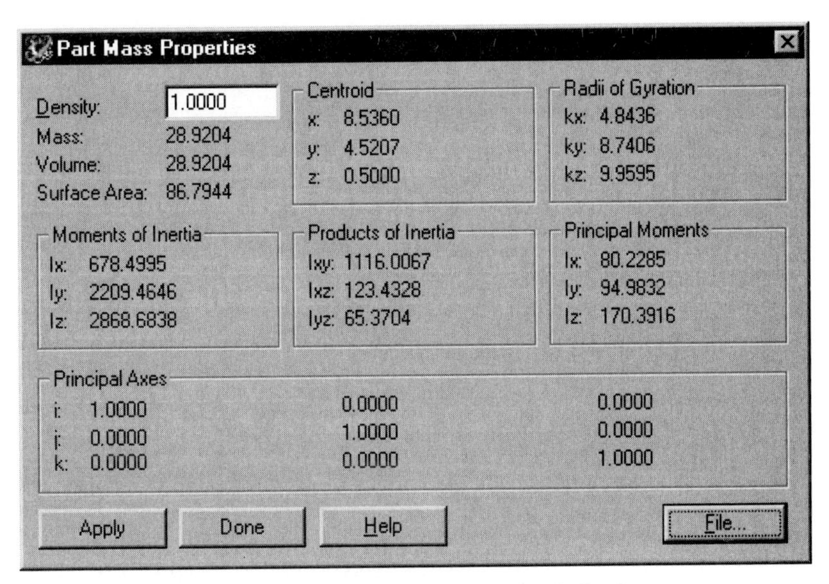

Figure 15-1 Part Mass Properties dialog box

AMMAKEBASE COMMAND

Toolbar:	Part Modeling > New Part > Make Base Feature
Menu:	Part > Part > Make Base Part
Command:	AMMAKEBASE

This command is used to convert the active part into the base feature. No information regarding the previous features remain in the memory of the drawing and therefore the new base feature cannot be edited or modified.

AMPARTSPLIT COMMAND

Toolbar:	Part Modeling > Placed Features-Hole > Part Split
Menu:	Part > Placed Features > Part Split
Context Menu:	Placed Features > Part Split
Command:	AMPARTSPLIT

This command is used to split the selected part in two. You can use a work plane, a sketch plane or a split line for splitting the part. The name of the new part is specified by you. Figure 15-2 shows a part before splitting and Figure 15-3 shows the same part after splitting using the work plane.

AMFACESPLIT COMMAND

Toolbar:	Part Modeling > Sketched Features-Extrude > Face Split
Menu:	Part > Sketched Features > Face Split
Context Menu:	Sketched & Worked Features > Face Split
Command:	AMFACESPLIT

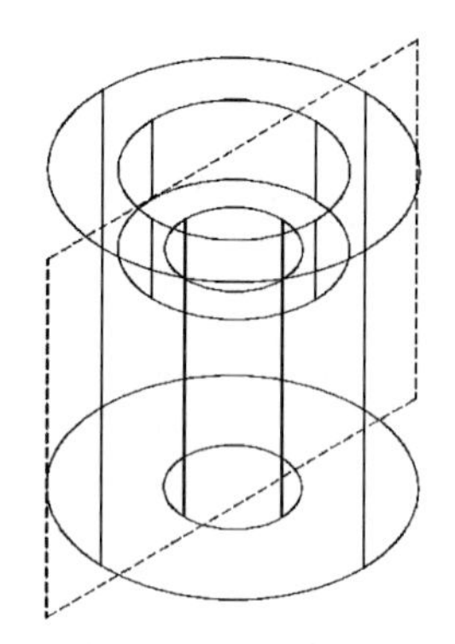

Figure 15-2 Part before splitting

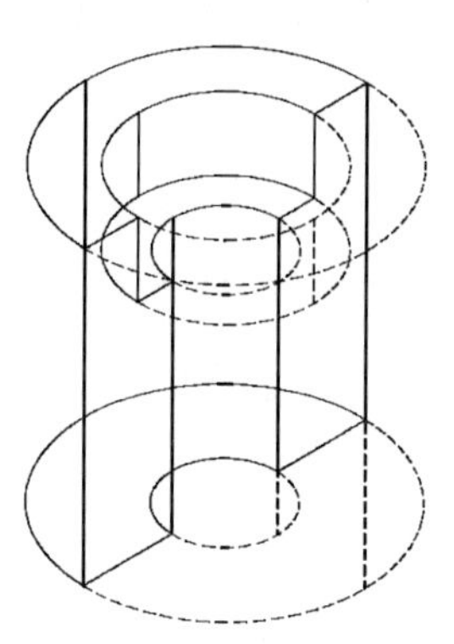

Figure 15-3 Parts after splitting using the work plane

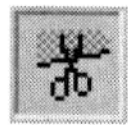

This command is used to split the specified faces using a selected plane or by projecting a split line. You can split all the faces or select the faces to split. The options provided under this command are:

Planar

This option is used to split the faces using an existing planar face or a work plane. When you select this option, you will be asked to select the faces to be split. After selecting the faces you will be asked to select the planar face or the work plane to split.

Project

This option is used to split the selected faces by projecting the split line. It is necessary that you have a split line created, before invoking this command, that will be used for projection. When you invoke this option, you will be asked to select the faces to be split. As soon as you select the faces, the split line will be projected on them and they will be split, see Figures 15-4 and 15-5.

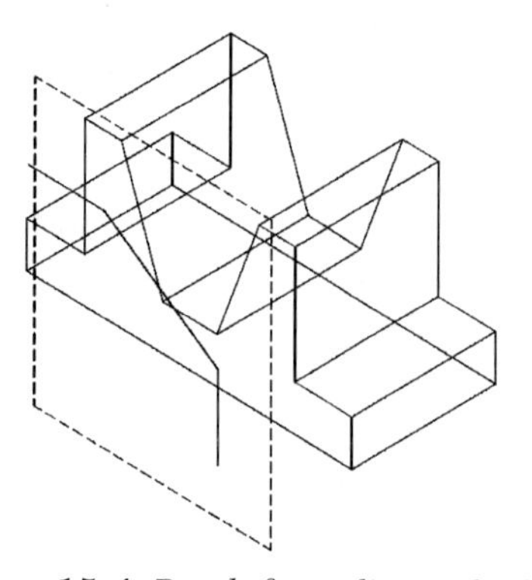

Figure 15-4 Part before splitting the faces

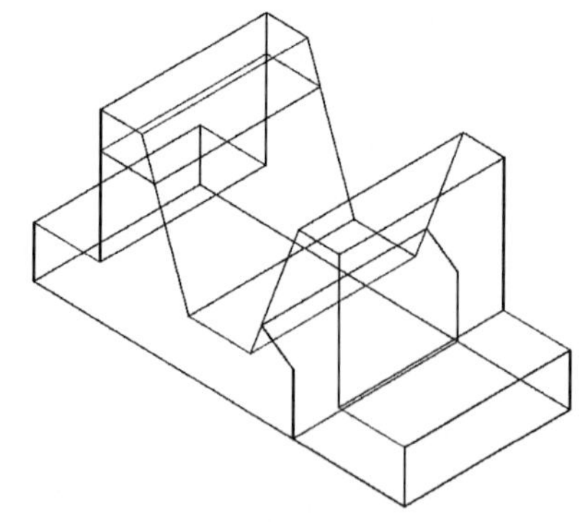

Figure 15-5 Part after splitting the faces by projecting the split line on all faces

AMTEXTSK COMMAND*

Toolbar:	Part Modeling > Profile a Sketch > Text Sketch
Menu:	Part > Sketch Solving > Text Sketch
Context Menu:	Sketch Solving > Text Sketch
Command:	AMTEXTSK

This command is used to create an embedded or an engraved text. When you invoke this command, the **Text Sketch** dialog box will be displayed, see Figure 15-6. Using this dialog box you can write the text that can be extruded, revolved or swept using the Mechanical Desktop commands.

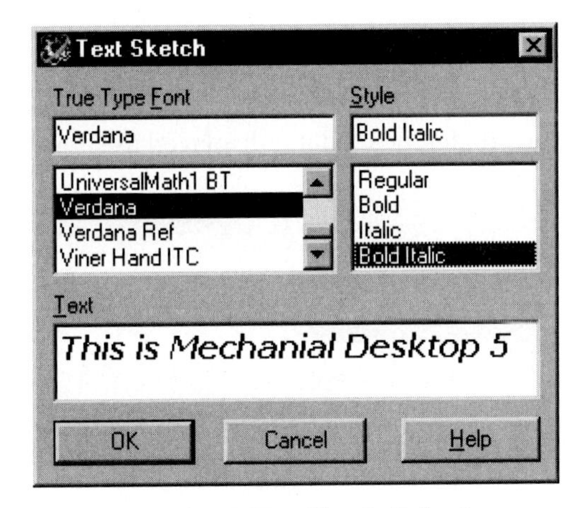

Figure 15-6 *Text Sketch dialog box*

Text Sketch Dialog Box Options

The options provided under this command are:

True Type Font Area

This area is used to select the font for the text. You can directly enter the name of the font in the edit box or select it from the drop-down list provided under this area.

Style Area

This area is used to select the style for the text. You can directly enter the style in the edit box or select it from the text box provided under this area.

Text

This box is used to type the text.

Once you have selected the font type and style and typed the text, choose **OK**. You will be prompted to specify the first corner and second corner for specifying the height of the text and the location of the text. You can also directly specify the height of the text and the rotation angle for the text using the **Height** and **Rotation** options.

Tip: *The text created using the **AMTEXTSK** command can be edited using the **AMEDITFEAT** command once it is converted into a feature using the Mechanical Desktop commands.*

AMWHEREUSED COMMAND

Toolbar: Assembly Modeling > Assembly Catalog > Where Used
Menu: Assembly > Assembly > Where Used
Context Menu: Assembly Menu > Assembly > Where Used
Command: AMWHEREUSED

This command is used to find the location of the selected designer part or component of assembly in the current drawing. You can also write the information regarding the selected component into a .LST file. When you invoke this command, the **Part/Subassembly Locations** dialog box will be displayed as shown in Figure 15-7.

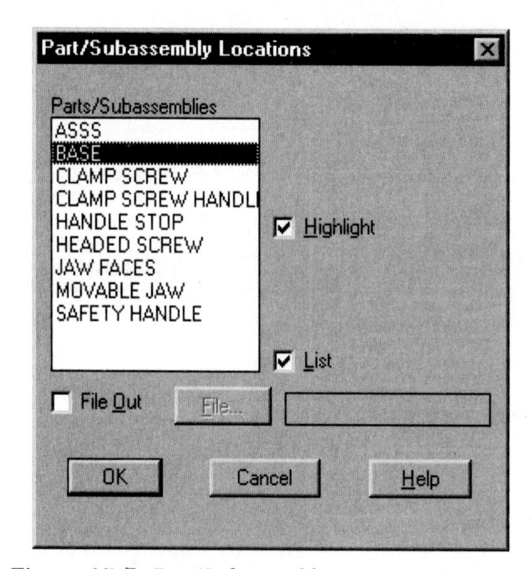

Figure 15-7 *Part/Subassembly Location dialog box*

Part/Subassembly Location Dialog Box Options

Part/Subassemblies Area

This area displays all the parts or the subassemblies available in the current drawing.

Highlight

If the **Highlight** check box is selected, the selected part or subassembly definition will be highlighted after you choose **OK**.

List

The **List** check box is selected to list the selected part or subassembly in the command line after you choose **OK**.

File Out

This check box is selected to write the information regarding the selected part or subassembly definition into a .LST file. You can specify the location of the file in the **File** edit box. You can also directly select the directory and specify the name of the file in the **Enter output file name** dialog box displayed by choosing the **File** button, see Figure 15-8.

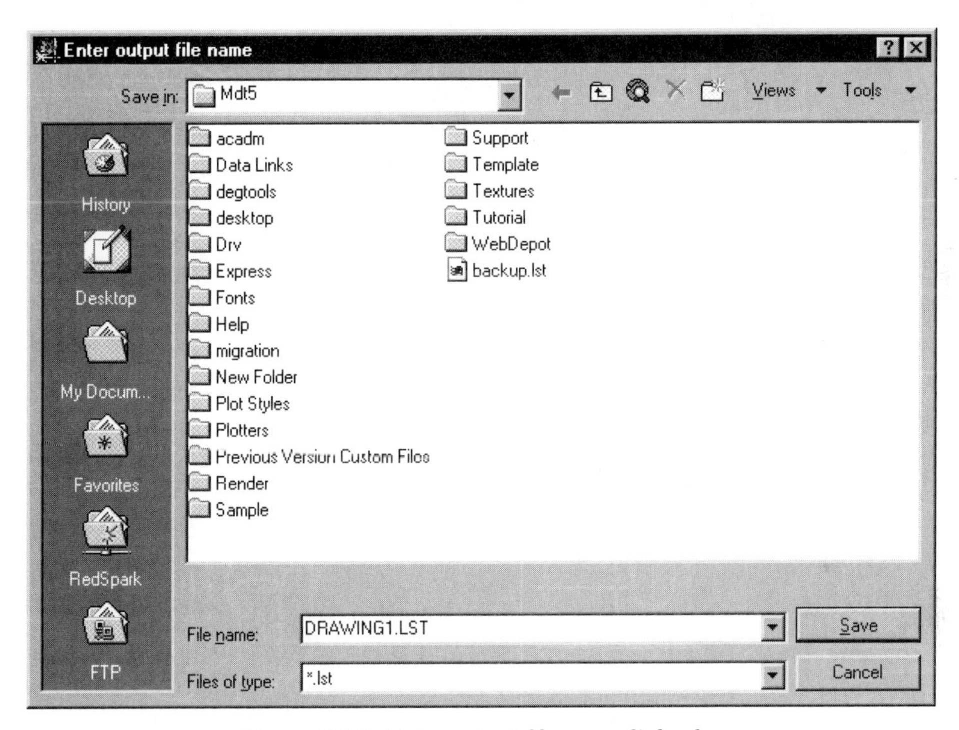

Figure 15-8 *Enter output file name dialog box*

AMASSIGN COMMAND

Toolbar:	Assembly Modeling > Assign Attributes
Menu:	Assembly > Assembly > Assign Attributes
Command:	AMASSIGN

This command is used to attach user defined information to the selected designer part or the subassembly. When you invoke this command, the **Assign Attributes** dialog box will be displayed as shown in Figure 15-9.

Assign Attributes Dialog Box Options

Part/Subassembly Definition Area

This area displays all the available parts or subassemblies in the current drawing. You have to select a part or subassembly from this area to attach the information.

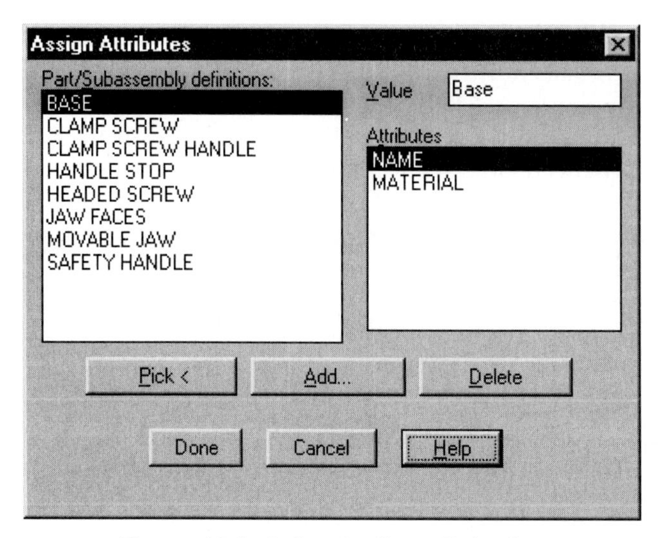

Figure 15-9 *Assign Attributes dialog box*

Pick

The **Pick** button is used to select the desired part or subassembly from the drawing area. When you choose this button, the **Assign Attributes** dialog box is temporarily closed and you will be asked to select the desired part or subassembly. The **Assign Attributes** dialog box will reappear on the screen after you have made the selection.

Add

The **Add** button is chosen to attach the information. You have to first select the part or the subassembly from the **Part/Subassembly definitions** area and then choose this button. The **Add new Attribute** dialog box will be displayed upon choosing this button, see Figure 15-10.

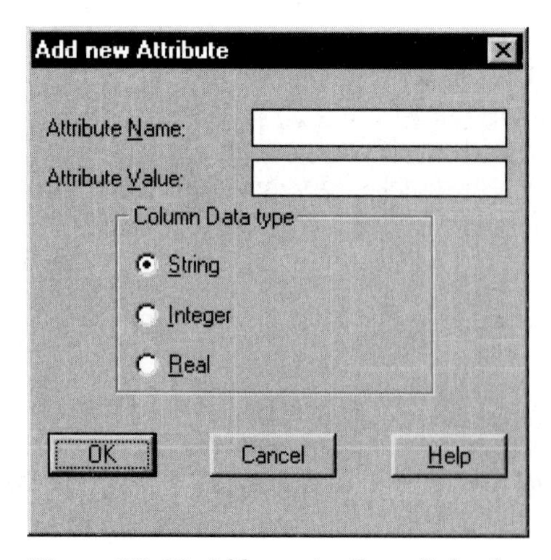

Figure 15-10 *Add new Attribute dialog box*

You have to enter the name of the new attribute in the **Attribute Name** edit box and the value in the **Attribute Value** dialog box. You can display the attributes as characters, whole numbers or in decimals using the radio buttons provided under the **Column Data type** area.

Delete

This button is chosen to delete the attributes assigned to the part or subassembly selected from the **Part/Subassembly definitions** area.

Value

This edit box displays the value of the attributes assigned to the part or subassembly selected from the **Part/Subassembly definitions** area.

Attributes

This area displays the names of the attributes assigned to the part or subassembly selected from the **Part/Subassembly definitions** area.

AMMANIPULATOR COMMAND

Toolbar:	Assembly Modeling > 3D Manipulator
Menu:	Assembly > 3D Manipulator
Context Menu:	Assembly Menu > 3D Manipulator
Command:	AMMANIPULATOR

This command is used for dynamically rotating or moving the selected object in the X, Y or Z direction or in the 3D space. This command is also used to create the instances of the selected object at the specified distance or angle. When you invoke this command, you will be asked to select the object. As soon as you select the object, the 3D Manipulator icon is displayed on the selected objects, see Figure 15-11. The axes of the 3D Manipulator icon have handles at each end that can be used for moving or rotating the selecting object.

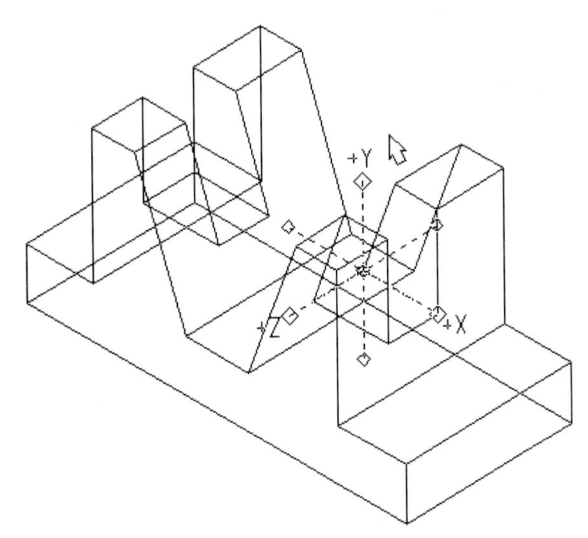

Figure 15-11 *Selected model displaying the 3D Manipulator icon*

Moving The Selected Object

To move the selected object, select the handle of the axis along which you want to move the object. Now move the cursor along the selected axis direction and the object will move in that direction. You can specify the exact movement value using the **Move** dialog box displayed by pressing ENTER after selecting the direction of movement, see Figure 15-12.

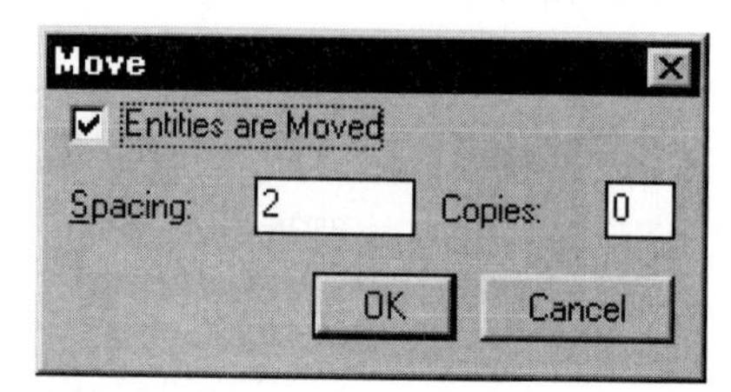

Figure 15-12 *Move dialog box for specifying the exact values*

You can also create the instances of the selected object using this dialog box. The number of instances can be specified in the **Copies** edit box. The specified number of instances of the selected object will be created at the distance specified in the **Spacing** edit box. If the **Entities are Moved** check box is cleared, the object will remain stationary and the 3D Manipulator icon will move by the specified distance.

Rotating The Selected Object

To rotate the selected object, select the handle of the axis about which you want to rotate the object. Now move the cursor about the selected axis direction and the object will rotate in that direction. You can specify the exact rotation angle using the **Rotate** dialog box displayed by pressing ENTER after selecting the direction of rotation, see Figure 15-13.

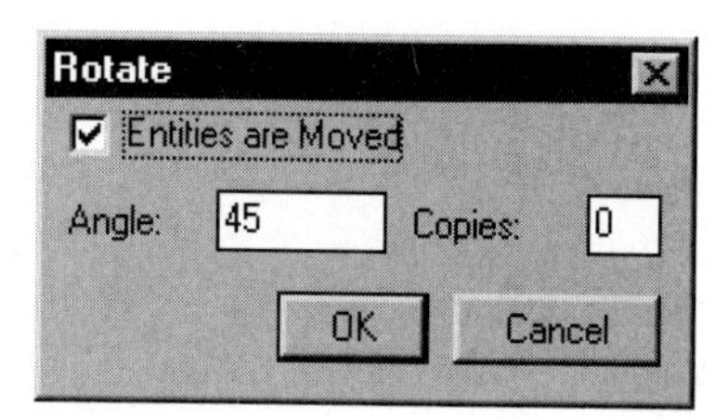

Figure 15-13 *Rotate dialog box for specifying the exact values*

You can also create the instances of the selected object using this dialog box. The number of instances can be specified in the **Copies** edit box. The specified number of instances of the selected object will be created at the angle specified in the **Angle** edit box. If the **Entities are Moved** check box is cleared, the object will remain stationary and the 3D Manipulator icon will rotate by the specified angle.

Moving The Selected Object Freely In The 3D Space

You can move the selected object freely in the 3D space by selecting the middle handle. The middle handle is also called the Free handle. By selecting this handle, you can move the selected object anywhere in the drawing window.

Selecting No Handle

If you click any where on the screen other than at the handles, the **Common Options** dialog box will be displayed as shown in Figure 15-14.

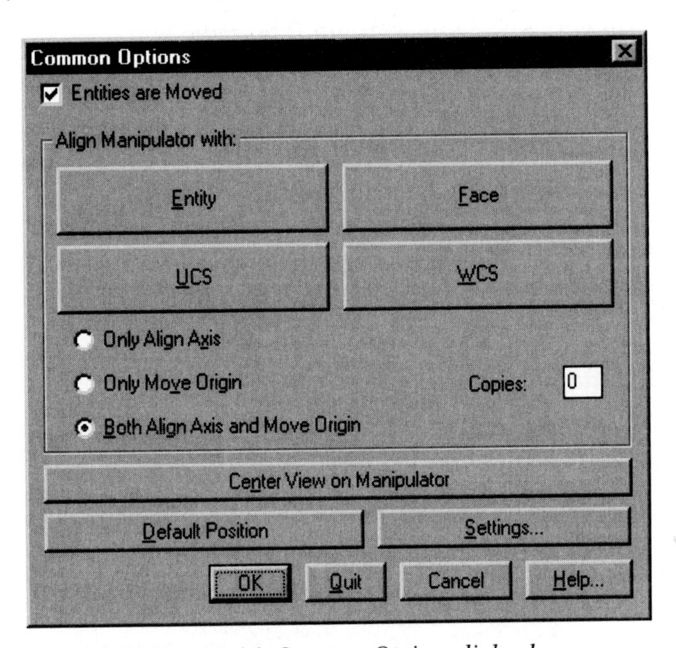

Figure 15-14 *Common Options dialog box*

Common Options Dialog Box Options

Entities are Moved

This check box is selected to move or rotate the selected object. If this check box is cleared, only the 3D Manipulator icon is moved or rotated.

Align Manipulator with Area

Entity. This button is used to align the 3D Manipulator icon with selected object.

Face. This button is chosen to align the 3D Manipulator icon with the specified face. When you choose this button, you will be prompted to select the face to align the icon.

UCS. This option is chosen to align the 3D Manipulator icon with the origin of the current UCS.

WCS. This button is chosen to align the 3D Manipulator icon with the origin of WCS.

Only Align Axis. This radio button is selected to align only the axis of the 3D Manipulator icon with the selected entity, face, UCS or WCS. In this case the origin will not move.

Only Move Origin. This radio button is selected to move the origin of the 3D Manipulator icon to the selected entity, face, UCS or WCS.

Both Align Axis and Move Origin. This radio button is chosen to align the axis of the 3D Manipulator icon with the selected entity, face, UCS or WCS as well as move the origin to the same point.

Copies. This edit box is used to specify the number of instances to be created of the selected object.

Center View on Manipulator

This option is used to reposition all the objects on the screen such that the 3D Manipulator icon is at the center of the screen.

Default Position

This option is used to move the 3D Manipulator icon to its default position.

Settings

This button is chosen to display the **Configuration** dialog box as shown in Figure 15-15.

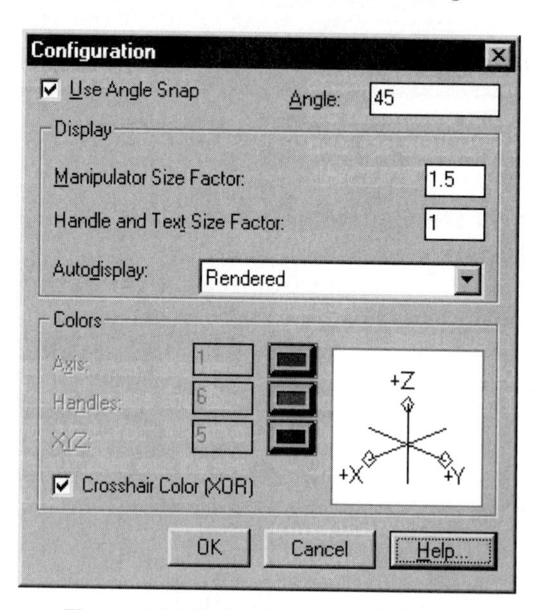

Figure 15-15 Configuration dialog box

Configuration Dialog Box Options

Use Angle Snap. This check box is selected so that the angle snap is used while rotating the selected object. The angle snap value is specified in the **Angle** edit box.

Manipulator Size Factor. This edit box is used to define the 3D Manipulator icon size factor.

Handle and Text Size Factor. This edit box is used to define the handle and text size.

Autodisplay. This drop-down provides the options related to the display of the selected part during this command. It is very important to mention here that the effect of these options will be viewed the next time you invoke this command.

Colors Area. Options provided under this area are used to define the color of the 3D Manipulator icon. You can specify its color same as that of the UCS icon (Crosshair). You can also define different colors for the axes, handles and text of the 3D Manipulator icon by clearing the **Crosshair Color (XOR)** check box.

AMHOLENOTE COMMAND

Toolbar: Drawing Layout > Power Dimensioning> Annotation > Hole Note
Menu: Annotation > Annotation > Hole Note
Context Menu: Annotate Menu > Annotation > Hole Note
Command: AMHOLENOTE

This command is used in the **Drawing** mode to automatically extract and display information related to the selected hole in the drawing views. The information that can be displayed is the diameter of hole, termination of the hole, type of hole and in case of counterbore and countersink holes, the counter diameter, depth and angle. There are some default standard templates stored in the memory of Mechanical Desktop and depending upon the type of hole selected, the related template is selected to display the information. You can also edit an existing holenote using this command. The options provided under this command are:

New

This option is used to extract and display new information related the hole. When you select this option, you will be asked to select the hole feature. The **Create Holenote** dialog box will be displayed when you select the hole, see Figure 15-16.

Template Name Area

This area displays the available holenote template related to the selected hole. You can select the template to be used.

Leader Justification

This area displays the options related to the positioning of the endpoint of the holenote leader.

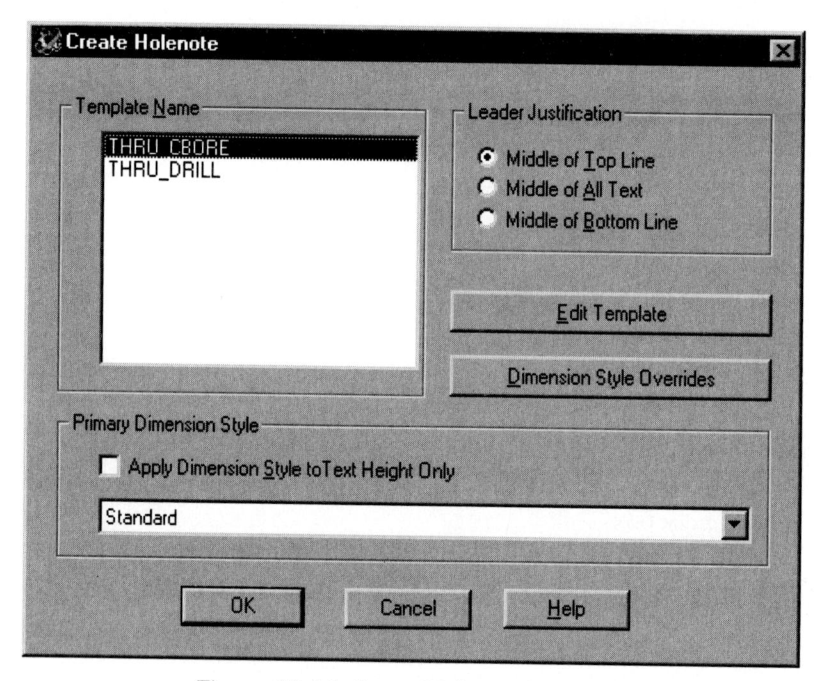

Figure 15-16 *Create Holenote dialog box*

Edit Template

This button is chosen to edit the template selected from the **Template Name** area. When you choose this button, the **Multiline Text Editor** will be displayed with the selected template. The templates in the **Multiline Text Editor** consists of the following variables:

AMDIA

This variable is related to the diameter of the hole. In case of the counterbore or countersink holes, this variable is related to the drill diameter.

AMDEP

This variable is related to the depth of the hole.

AMCDIA

This variable is related to the counter diameter of the counterbore or countersink hole.

AMCBDEP

This variable is related to the depth of the counterbore hole.

AMCSANG

This variable is related to the angle of the countersink hole.

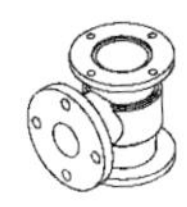

Note

The Multiline Text Editor will consider the variables only if they are preceded with %%. In case the variable is not preceded with %%, only the name of the variable will appear in the holenote.

Dimension Style Overrides

This button is chosen to display the **Holenote Dimension Style Overrides** dialog box, see Figure 15-17. The options provided in this dialog box are used to edit the precision, dimension style and measurement type options for the holenote.

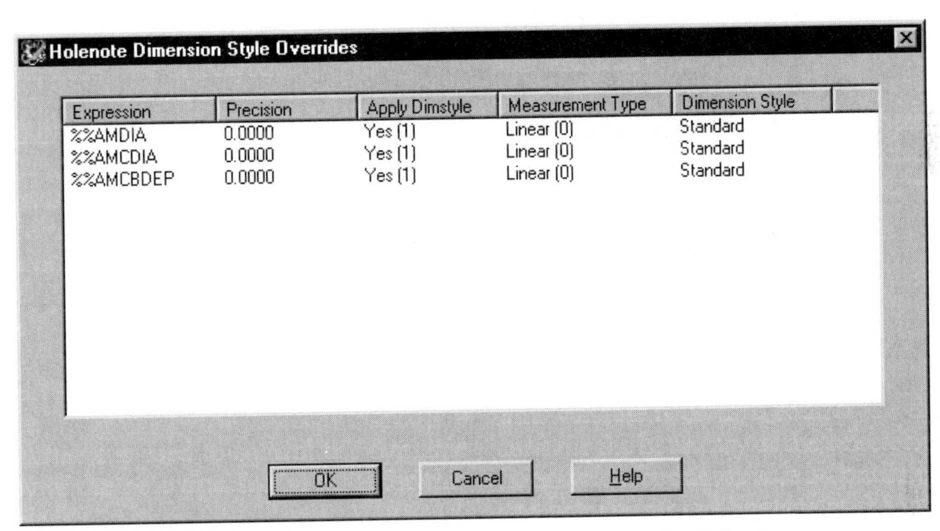

Figure 15-17 *Holenote Dimension Style Overrides dialog box*

Primary Dimension Style Area

Apply Dimension Style to Text Height Only. This check box is selected to apply the dimension style only to the text.

Apart from this check box, there is a drop-down list that displays all the available dimension styles to be applied to the holenote. You can select the desired dimension style from this drop-down list.

Edit

This option is used to edit the holenotes. When you invoke this option, you will be prompted to select the holenote to edit. When you select the holenote to edit, the **Edit Holenote** dialog box will be displayed. The options in this dialog box are similar to those in the **Create Holenote** dialog box.

AMTEMPLATE COMMAND

Toolbar:	Drawing Layout > Power Dimensioning> Annotation > Hole Note > Hole Note Template
Menu:	Annotation > Annotation > Hole Note Template
Context Menu:	Annotate Menu > Annotation > Hole Note Template
Command:	AMTEMPLATE

This command is used to create and modify new holenote template using the standard holenote templates. When you invoke this command, the **Hole Templates** dialog box will be displayed as shown in Figure 15-18.

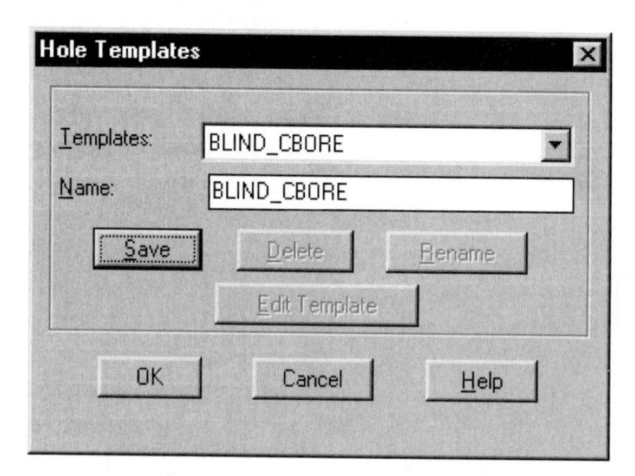

Figure 15-18 *Hole Templates dialog box*

To create a new holenote template, select the template using which the new template will be created from the **Templates** drop-down list. Enter the name of the new template in the **Name** edit box and then choose the **Save** button. All the other buttons in the dialog box will be activated. You can edit the new holenote template and add new variables to it using the **Multiline Text Editor** displayed by choosing the **Edit Template** button.

Chapter 16

Projects

Learning Objectives

After completing this chapter, you will be able to:

- *Create the components of the project assemblies using the commands of Mechanical Desktop.*
- *Assemble the components of the assemblies using the assembly constraints.*
- *Explode the assemblies in the **Scene** mode.*
- *Generate the drawing views of the assemblies and add the Bill Of Material to the drawing views.*

Tutorial 1

In this tutorial you will create and assemble all the components of the Pipe Vice assembly, the details of which are given in the following figures (assume the missing dimensions). Create the exploded assembly scenes. Generate the following drawing views of the assembly in the layout 1:

1. Top view.
2. Front view.
3. Left side view.
4. Isometric view of the front view.

Add BOMs and Balloons to the drawing views with JIS standard. The BOMs should have the following columns:

1. Item.
2. Quantity.
3. Name.

1. Open a new drawing. You will start with creating the Part 1.

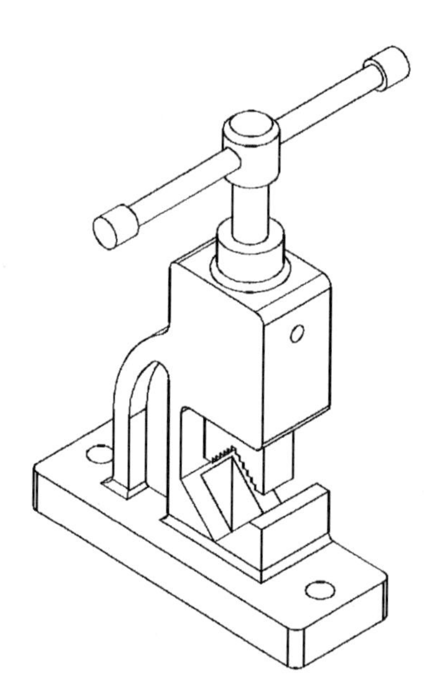

Figure 16-1 *Assembled view of the Pipe Vice assembly*

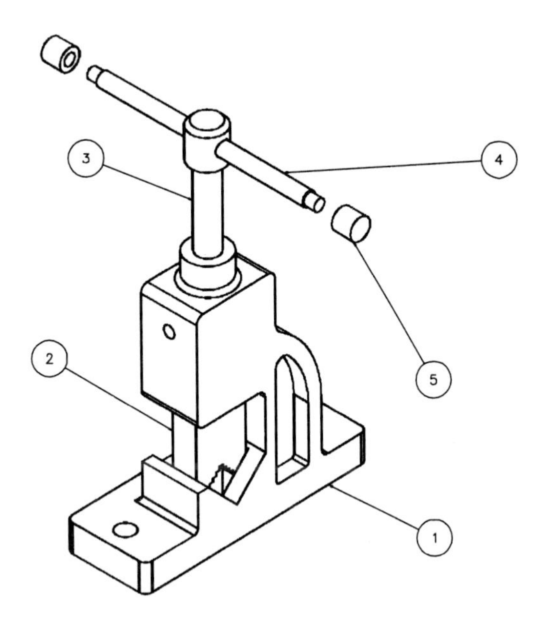

Figure 16-2 Exploded view of the Pipe Vice assembly displaying various components

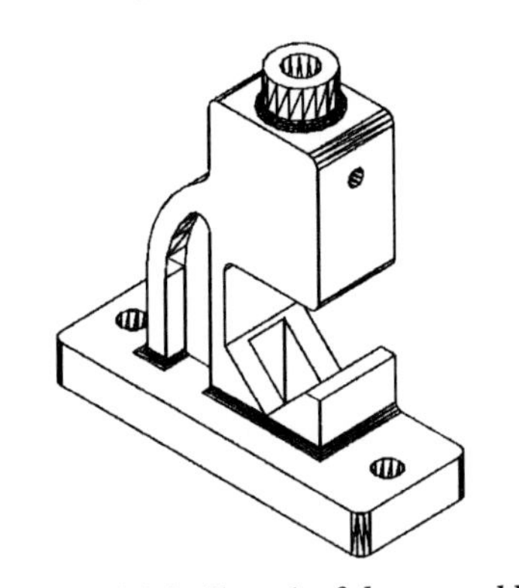

Figure 16-3 Part 1 of the assembly

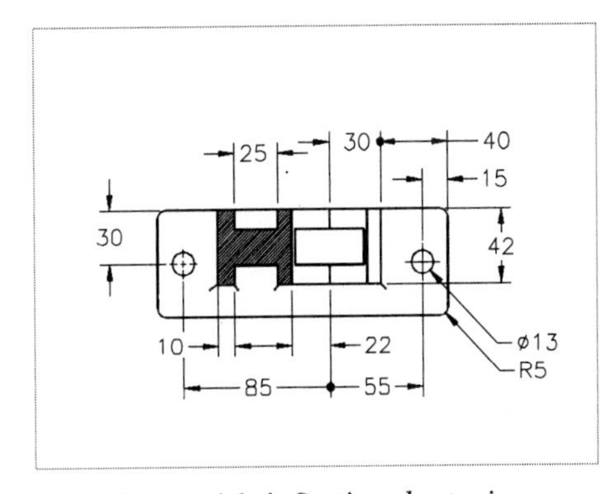

Figure 16-4 Sectioned top view

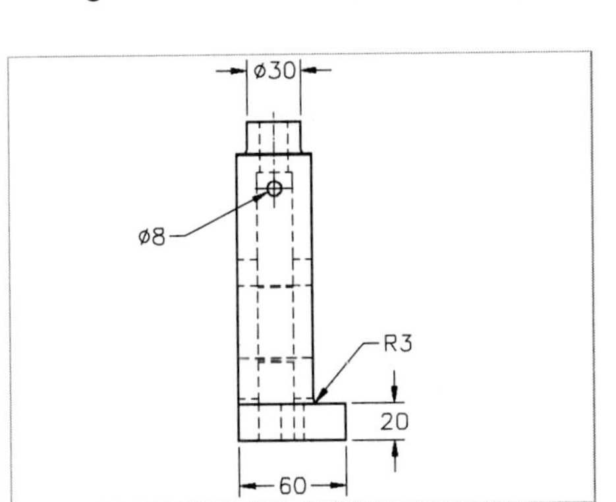

Figure 16-5 Side view

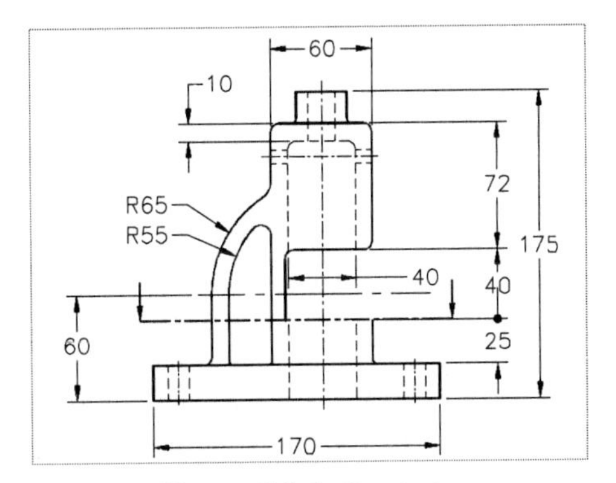

Figure 16-6 Front view

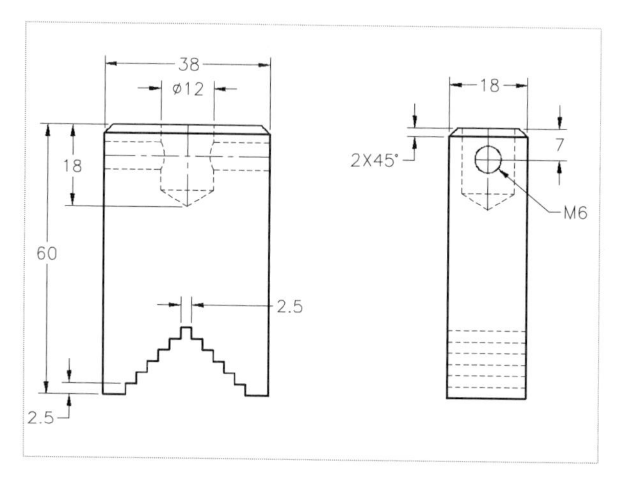

Figure 16-7 *Front and side view of Part 2*

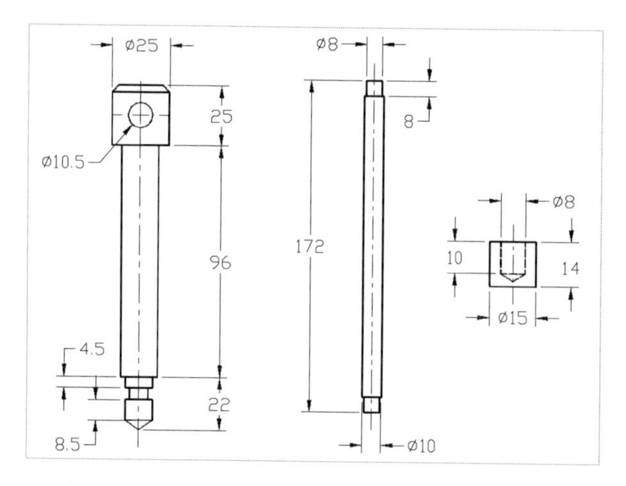

Figure 16-8 *Dimensions of various components*

2. Choose **Format** > **Drawing Limits** from the **Assist** menu. The prompt sequence is as follows:

   Reset Model space limits:
   Specify lower left corner or [ON/OFF] <0.0000,0.0000>: Enter
   Specify upper right corner <12.0000,9.0000>: **200,200**

3. Enter **FF** at the Command prompt. Choose **Format > Dimension Style** from the **Assist** menu to display the **Dimension Style Manager** dialog box. Choose the **Modify** button to display the **Modify Dimension Style: Standard** dialog box. Choose the **Fit** tab and set the value of the **Use overall scale of** spinner to **20**. Choose **OK** and then choose **Close**.

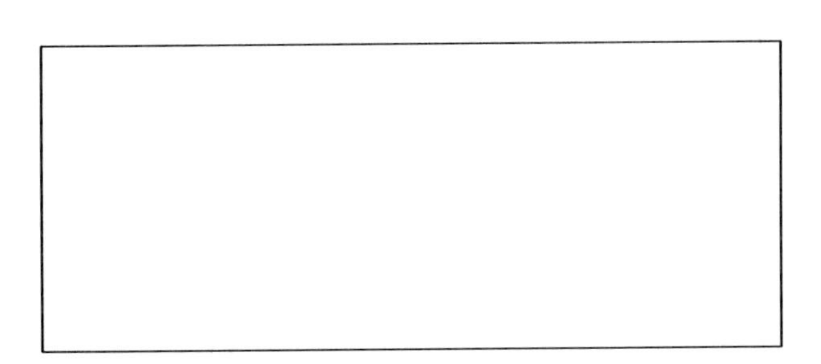

Figure 16-9 The rough sketch for the base feature

4. Create the rough sketch for the base feature as shown in Figure 16-9.

5. Choose the **Single Profile** button from the **Profile a Sketch** flyout in the **Part Modeling** toolbar.

6. Choose the **New Dimension** button from the **Power Dimensioning** flyout in the **Part Modeling** toolbar. The prompt sequence is as follows:

Select first object: *Select the lower horizontal line.*
Select second object or place dimension: *Place the dimension.*
Enter dimension value or [Undo/Hor/Ver/Align/Par/aNgle/Ord/Diameter/pLace] <default value>: **170**
Solved under constrained sketch requiring 1 dimension or constraint.
Select first object: *Select the right vertical line.*
Select second object or place dimension: *Place the dimension.*
Enter dimension value or [Undo/Hor/Ver/Align/Par/aNgle/Ord/Diameter/pLace] <default value>: **60**

Solved fully constrained sketch.
Select first object: *Press ESC.*

7. The sketch after applying the dimensions should look similar to the one shown in Figure 16-10. Enter **8** at the Command prompt.

8. Choose the **Sketched Feature-Extrude** button from the **Part Modeling** toolbar to display the **Extrusion** dialog box.

9. Set the value of the **Distance** spinner to **20**. Choose **OK**. The base feature should look similar to the one shown in Figure 16-11.

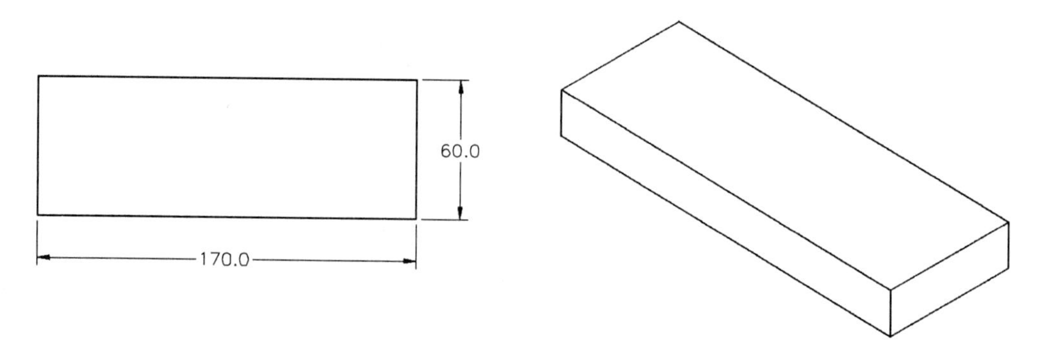

Figure 16-10 Fully dimensioned sketch for base

Figure 16-11 The base feature

10. Choose the **New Sketch Plane** button from the **Part Modeling** toolbar. The prompt sequence is as follows:

Select work plane, planar face or [worldXy/worldYz/worldZx/Ucs]: *Select the back face of the base feature.*
Enter an option [Next/Accept] <Accept>: Enter
Select edge to align X axis or [Flip/Rotate/Origin] <Accept>: *Align the X axis.*
Select edge to align X axis or [Flip/Rotate/Origin] <Accept>: Enter

11. Enter **9** at the Command prompt and then create the rough sketch for the next feature, see Figure 16-12.

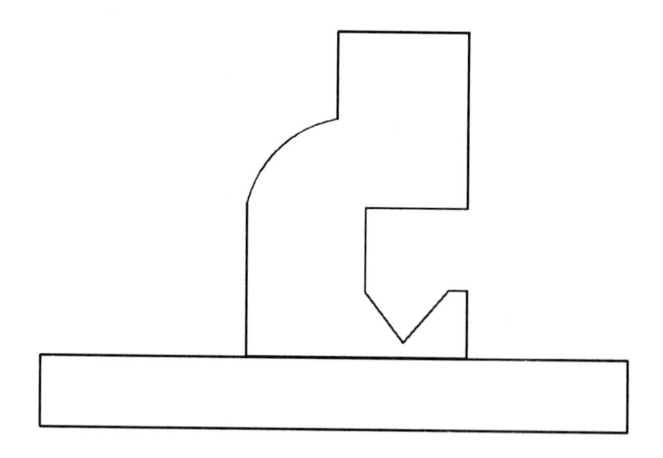

Figure 16-12 Rough sketch for next feature

12. Choose the **Single Profile** button from the **Profile a Sketch** flyout in the **Part Modeling** toolbar.

13. Choose the **New Dimension** button from the **Power Dimensioning** flyout in the **Part Modeling** toolbar and dimension the sketch. The sketch after dimensioning should look similar to the one shown in Figure 16-13.

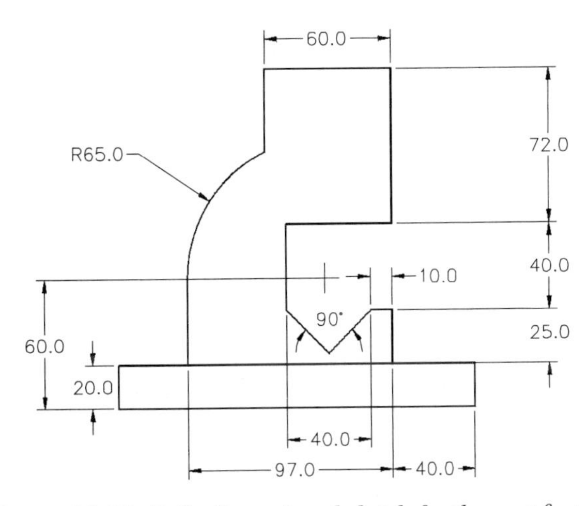

Figure 16-13 *Fully dimensioned sketch for the next feature*

14. Choose the **Sketched Feature-Extrude** button from the **Part Modeling** toolbar to display the **Extrusion** dialog box.

15. Choose **Join** from the **Operation** drop-down list.
16. Set the value of the distance spinner to **42**. Check the direction of the feature creation.
17. Select **Blind** from the **Type** drop-down list under the **Termination** area. Choose **OK**.
18. The model after creating this feature should look similar to the one shown in Figure 16-14.

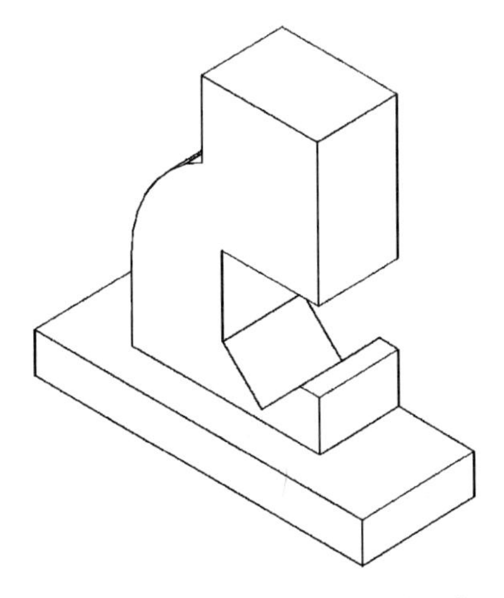

Figure 16-14 *Designer model after creating the next feature*

19. Create the next cut feature in the center of the new feature as well as the front and back face of the new feature, see Figure 16-15.

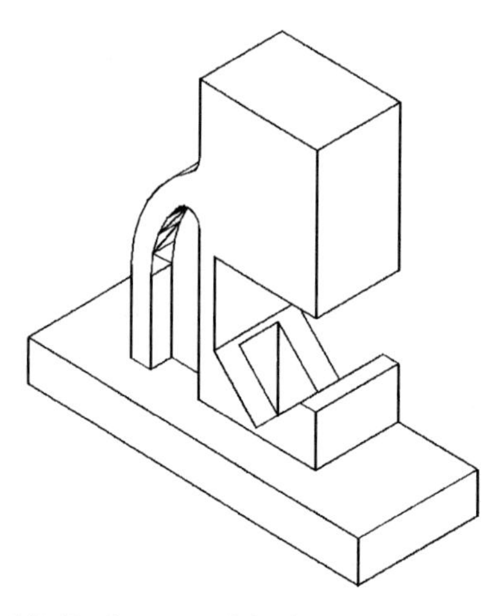

Figure 16-15 *Designer model after creating the cut features*

20. Create the next feature on the top face of the second feature and then create a through hole in this feature. Create the other remaining holes and the fillets. The final desired model for Part 1 should look similar to the one shown in Figure 16-16.

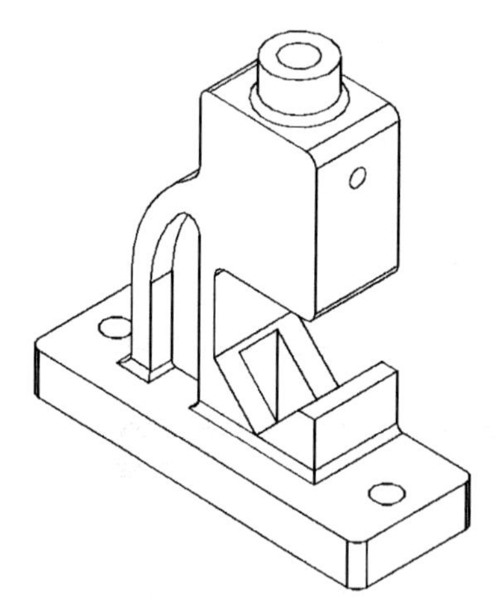

Figure 16-16 *Final designer model for Part 1*

21. Save the current drawing and close it. Open a new file to create Part 2.

22. Rotate the UCS through an angle of 90 degrees about X axis and then enter **9** at the Command prompt. Choose **Format > Drawing Limits** from the **Assist** menu to increase the limits of the drawing. The prompt sequence is as follows:

Specify lower left corner or [ON/OFF] <0.0000,0.0000>: Enter
Specify upper right corner <12.0000,9.0000>: **100,100**

23. Enter **FF** at the Command prompt. Choose **Format > Dimension Style** from the **Assist** menu to display the **Dimension Style Manager** dialog box. Choose the **Modify** button to display the **Modify Dimension Style: Standard** dialog box. Choose the **Fit** tab and then set the value of the **Use overall scale of** spinner to **10**. Choose **OK** and then choose **Close**.

24. Create the rough sketch for the base feature as shown in Figure 16-17.

***Figure 16-17** Rough sketch for the base feature*

25. Choose the **Profile a Sketch** button from the **Part Modeling** toolbar. The prompt sequence is as follows:

Select objects for sketch: *Select the sketch.*
Computing ...
Solved under constrained sketch requiring 20 dimensions or constraints.

26. Choose the **Equal Length** button from the **2D Constraints** toolbar and apply the equal length constraints on all the small lines except the bottom two horizontal lines.

27. Choose the **New Dimension** button from the **Power Dimensioning** flyout in the **Part Modeling** toolbar. The prompt sequence is as follows:

Select first object: *Select the left vertical line*

Select second object or place dimension: *Place the dimension.*
Enter dimension value or [Undo/Hor/Ver/Align/Par/aNgle/Ord/Diameter/pLace] <default value>: **60**
Solved under constrained sketch requiring 2 dimensions or constraints.

Select first object: *Select the upper horizontal line.*
Select second object or place dimension: *Place the dimension.*
Enter dimension value or [Undo/Hor/Ver/Align/Par/aNgle/Ord/Diameter/pLace] <default value>: **38**
Solved under constrained sketch requiring 1 dimension or constraint.
Select first object: *Select one of the smaller lines.*
Select second object or place dimension: *Place the dimension*
Enter dimension value or [Undo/Hor/Ver/Align/Par/aNgle/Ord/Diameter/pLace] <default value>: **2.5**

Solved fully constrained sketch.
Select first object: *Press ESC.*

The sketch after applying the dimensions should look similar to the one shown in Figure 16-18.

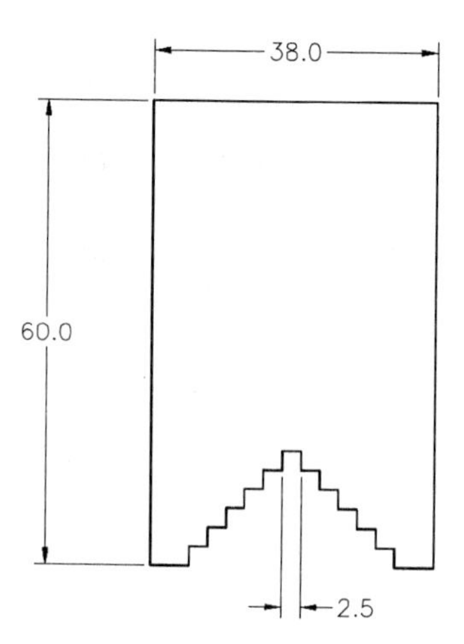

Figure 16-18 *Fully dimensioned sketch for the base feature*

28. Choose the **Sketched Feature-Extrude** button from the **Part Modeling** toolbar to display the **Extrusion** dialog box.

29. Set the value of the **Distance** spinner to **18**. Choose **OK**.

30. Enter **8** at the Command prompt. The base feature after creation should look similar to the one shown in Figure 16-19.

31. Choose the **Chamfer** button from the **Placed Feature-Hole** flyout in the **Part Modeling** toolbar to display the **Chamfer** dialog box.

32. Select **Distance and angle** from the **Operation** drop-down list.

33. Set the value of the **Distance 1** spinner to **2**.

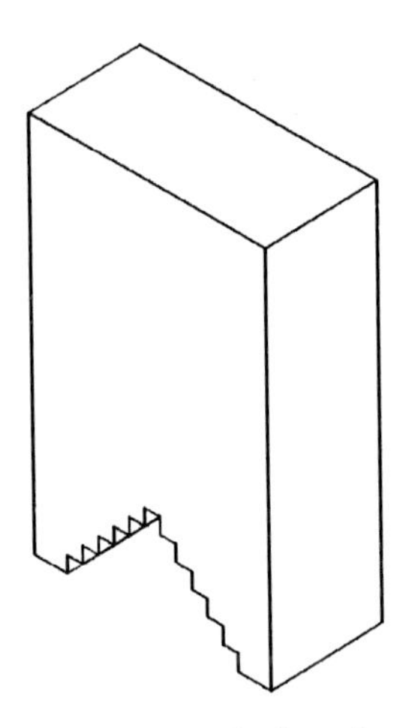

Figure 16-19 The base feature

34. Set the value of the **Angle** spinner to **45**. Choose **OK**. The prompt sequence is as follows:

    Select an edge or face to chamfer: *Select the top face of the base feature.*
    Enter an option [Next/Accept] <Accept>: Enter

35. Using the desktop browser, suppress the chamfer feature.

36. Choose the **Placed Feature-Hole** button from the **Part Modeling** toolbar to display the **Hole** dialog box.

37. Select **Blind** from the **Termination** drop-down list.

38. Select **2 Edges** from the **Placement** drop-down list.

39. Set the value of the **Dia** spinner under the **Drill Size** area to **12**.

40. Set the value of the **Depth** spinner to **18**.

41. Set the value of the **PT Angle** spinner to **118**. Choose **OK**. The prompt sequence is as follows:

    Select first edge: *Select the bigger edge on the top face of the model.*
    Select second edge: *Select the smaller edge on the top face of the model.*
    Computing ...
    Specify hole location: *Specify the location of the hole.*
    Enter distance from first edge (highlighted) <default value>: **9**
    Enter distance from second edge (highlighted) <default value>: **19**
    Computing ...
    Select first edge: Enter

42. Similarly, place the other hole on the base feature.

43. Unsuppress the chamfer feature. The final designer model should look similar to the one shown in Figure 16-20.

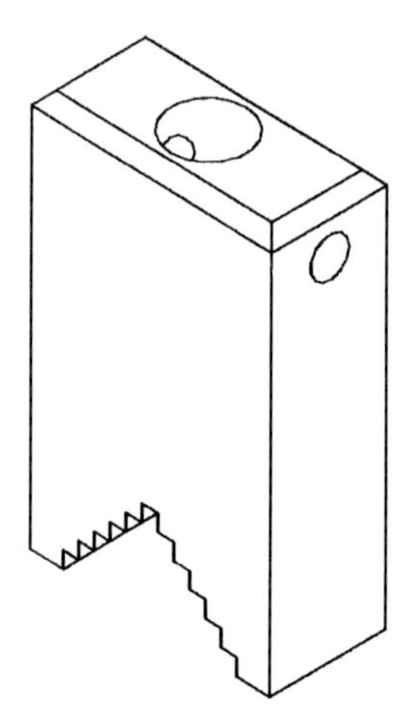

Figure 16-20 *Final designer model for Part 2*

44. Save this file and close it. Open a new file for creating Part 3.

45. Rotate the UCS about X axis through an angle of 0 degrees. Choose **Format > Drawing Limits** from the **Assist** menu to increase the limits of the drawing. The prompt sequence is as follows:

    Specify lower left corner or [ON/OFF] <0.0000,0.0000>: Enter
    Specify upper right corner <12.0000,9.0000>: **150,150**

46. Enter **9** and then **FF** at the Command prompt. Choose **Format > Dimension Style** from the **Assist** menu to display the **Dimension Style Manager** dialog box. Choose the **Modify** button to display the **Modify Dimension Style: Standard** dialog box. Choose the **Fit** tab and then set the value of the **Use overall scale of** spinner to **15**. Choose **OK** and then choose **Close**.

47. Create the rough sketch for the base feature..

48. Choose the **Profile a Sketch** button from the **Part Modeling** toolbar. The prompt sequence is as follows:

    Select objects for sketch: *Select the sketch.*
    Computing ...
    Solved under constrained sketch requiring 10 dimensions or constraints.

49. Choose the **New Dimension** button from the **Power Dimensioning** flyout in the **Part Modeling** toolbar. Add the dimensions to the sketch. The sketch after adding the dimensions should look similar to the one shown in Figure 16-21.

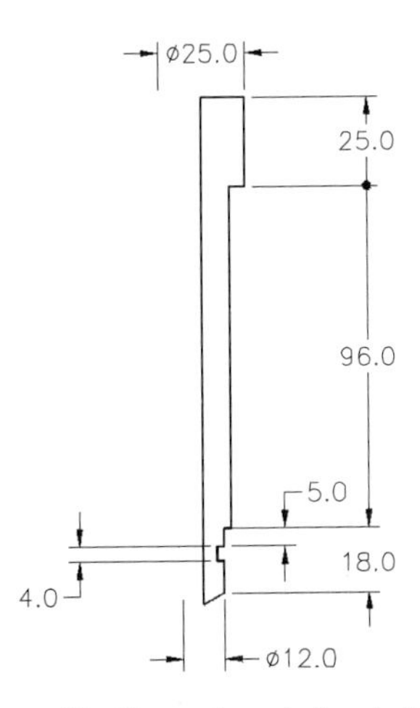

Figure 16-21 *Fully dimensioned sketch for base feature*

50. Choose the **Revolve** button from the **Sketched Feature-Extrude** flyout in the **Part Modeling** toolbar. The prompt sequence is as follows:

Select revolution axis: *Select the left vertical line as the revolution axis.*

51. In the **Revolution** dialog box, set the value of the **Angle** spinner to **360**.

52. Select **By Angle** from the **Type** drop-down list in the **Termination** area. Choose **OK**.

53. Choose **Chamfer** button from the **Placed Feature-Hole** flyout in the **Part Modeling** toolbar to display the **Chamfer** dialog box.

54. Select **Distance and angle** from the **Operation** drop-down list.

55. Set the value of the **Distance 1** spinner to **2**.

56. Set the value of the **Angle** spinner to **45**. Choose **OK**. The prompt sequence is as follows:

Select an edge or face to chamfer: *Select the top face of the base feature.*
Enter an option [Next/Accept] <Accept>: Enter

57. Now, create the hole feature. The final designer model for Part 3 should look similar to the one shown in Figure 16-22.

58. Save this drawing and close it. Similarly, create the remaining components of the assembly in different drawings.

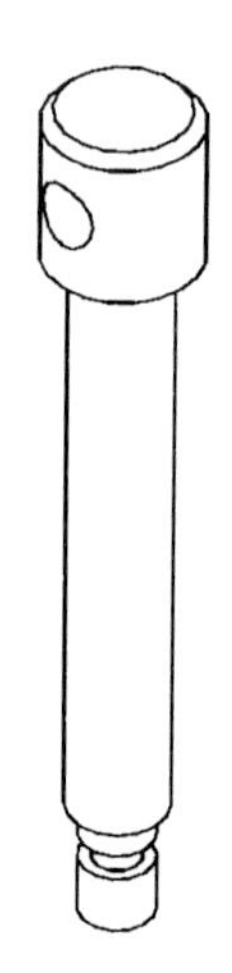

Figure 16-22 Designer model for Part 3

59. Once you have created all the components and saved them, open a new drawing for assembling the components.

60. Choose **Format > Drawing Limits** from the **Assist** menu to increase the limits of the drawing. The prompt sequence is as follows:

    Specify lower left corner or [ON/OFF] <0.0000,0.0000>: Enter
    Specify upper right corner <12.0000,9.0000>: **250,250**

61. Enter **8** and then **FF** at the Command prompt.

62. Choose the **Assembly Modeling** button from the **Desktop Main** toolbar.

63. Now, choose the **Assembly Catalog** button from the **Assembly Modeling** toolbar to display the **Assembly Catalog** toolbar.

64. Right-click under the **Directories** area to display the shortcut menu. Choose **Release All** to release all the current directories.

65. Again right-click under the directories area and choose **Add Directory** from the shortcut menu. Add the directory in which the parts were saved using the **Browse for Folder** dialog box.

66. Double-click on the Part 1 under the **Part and Subassembly Definition** area to attach the Part 1. Similarly, attach Part 2 also.

67. Choose the **Insert** button from the **3D Assembly Constraints** flyout in the **Assembly Modeling** toolbar. The prompt sequence is as follows:

Select first circular edge: *Select the circular edge of the hole on the top face of Part 2, see Figure 16-23.*
First set = Plane/Axis
Enter an option [Clear/Flip] <accEpt>: Enter
Select second circular edge: *Select the lower circular edge of the hole feature on the top face of Part 1, see Figure 16-23.*
Second set = Plane/Axis
Enter an option [Clear/Flip] <accEpt>: Enter
Enter offset <.0000>: **40**

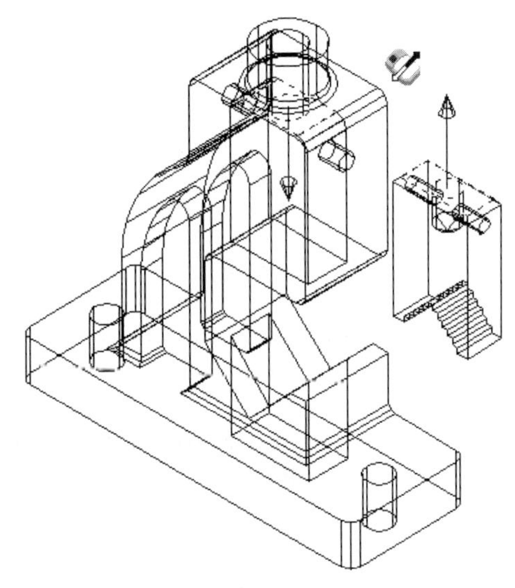

Figure 16-23 *Assembling Part 1 and Part 2*

68. Make Part 1 invisible using the desktop browser.

69. Attach Part 3 using the **Assembly Catalog** dialog box displayed upon choosing the **Assembly Catalog** button from the **Assembly Modeling** toolbar.

70. Choose the **Insert** button from the **3D Assembly Constraints** flyout in the **Assembly Modeling** toolbar. The prompt sequence is as follows:

Select first circular edge: *Select the circular edge of the hole on the bottom face of Part 3, see Figure 16-24.*
First set = Plane/Axis
Enter an option [Clear/Flip] <accEpt>: Enter
Select second circular edge: *Select the circular edge of the hole feature on the top face of Part 2, see Figure 16-24.*
Second set = Plane/Axis
Enter an option [Clear/Flip] <accEpt>: Enter
Enter offset <.0000>: **-14**

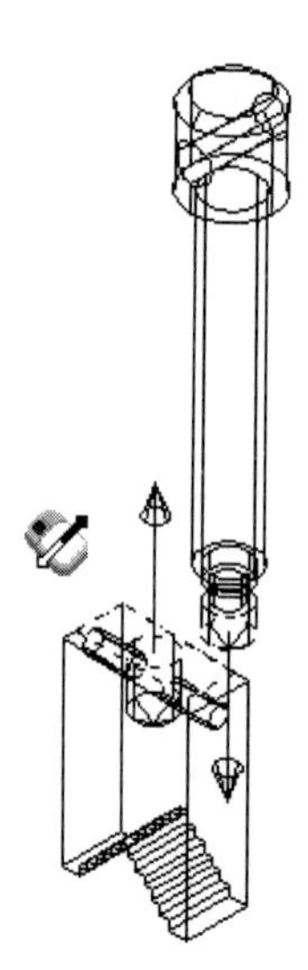

Figure 16-24 *Assembling Part 2 and Part 3*

71. Similarly, assemble the remaining components of the assembly. After assembling all the components, the assembly should look similar to the one shown in Figure 16-25.

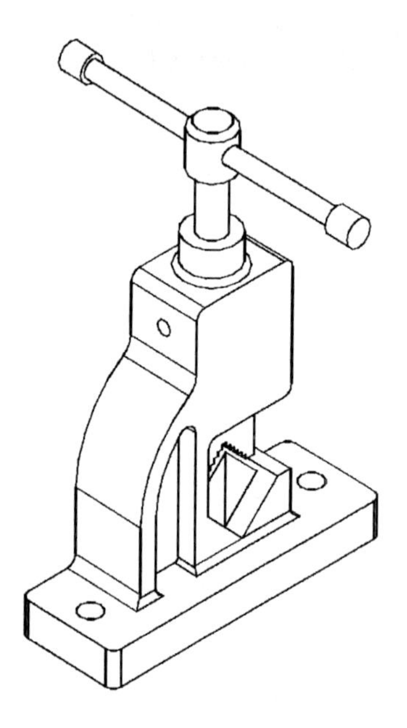

Figure 16-25 *The final Pipe Vise assembly*

72. Choose the **Scene** tab in the desktop browser to proceed to the scene mode.

73. Right-click in the desktop-browser to display the shortcut menu. Choose **New Scene** from the shortcut menu. The prompt sequence is as follows:

Enter new scene name of the active assembly () <SCENE1>: Enter

74. Similarly, create another scene with the name SCENE2.
75. Right-click on the scene 2 in the tree view of the desktop browser to display the shortcut menu. Choose **Explode Factor** to display the **Desktop Browser** dialog box.
76. Enter **25** in the **Explode Factor** edit box to explode the Pipe Vise assembly.
77. Choose the **Drawing** tab in the desktop browser to proceed to the drawing mode.
78. Choose the **New View** button from the **Drawing Layout** toolbar to display the **Create Drawing View** dialog box.

79. Select **Base** from the **View Type** drop-down list.
80. Select **Scene:SCENE3** from the **Data Set** drop-down list.
81. Set the value of the **Scale** spinner under the **Properties** area to **0.015** and choose **OK**. The prompt sequence is as follows:

Select planar face, work plane or [Ucs/View/worldXy/worldYz/worldZx]: *Select the front face of the assembly to generate the front view of the assembly.*
Enter an option [Next/Accept] <Accept>: Enter
Define X axis direction:
Select work axis, straight edge or [worldX/worldY/worldZ]: *Align the X axis of the view using the appropriate edge.*
Adjust orientation [Flip/Rotate] <Accept>: Enter
Regenerating layout.
Specify location of base view: *Place the drawing view.*
Specify location of base view: Enter

82. Enter **LTSCALE** at the Command prompt. The prompt sequence is as follows:

Enter new linetype scale factor <1>: **0.4**

83. Right-click on the base view in the tree view of the desktop browser to display the shortcut menu.
84. Choose **New View** to display the **Create Drawing View** dialog box.
85. Select **Ortho** from the **View Type** drop-down list.
86. Select the **Return to Dialog** check box. Choose **Apply**. The prompt sequence is as follows:

Specify location for orthogonal view: *Specify the placement point above the front view.*
Specify location for orthogonal view: Enter

87. Similarly, create the right side view and the isometric view of the front view.

88. Double-click on the top ortho view in the tree view of the desktop browser to display the **Edit Drawing View: Ortho - Scene** dialog box.

89. Clear the **Display Hidden Lines** check box under the **Hidden Lines** tab so that for clarity, the hidden lines are not displayed in this view. Choose **OK**.

90. The layout after generating all the required views should look similar to the one shown in Figure 16-26.

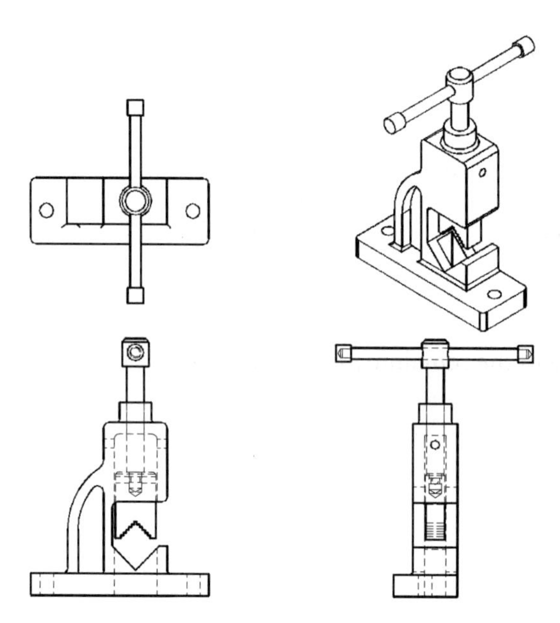

Figure 16-26 *The drawing views with hidden lines suppressed in the plan view*

91. Right-click on **Layout1** to display the shortcut menu and choose **Symbol/BOM Standards** to display the **Symbol Standard** dialog box.

92. Double-click on the **JIS** standard to display the other related options.

93. Double-click on the **Parts List** to display the **Parts List Properties for JIS** dialog box.

94. Keep only the **Item**, **Quantity** and the **Name** column in the **Part List** area and remove the remaining columns.

95. Select **Top Left** from the **Attach Point** drop-down list. Choose **Apply** and then choose **OK** in the **Parts List Properties for JIS** dialog box.

96. Double-click on the **Balloon** to display the **Balloon Properties for JIS** dialog box.

97. Select **3.5** from the **Text Size** drop-down list. Set the value of the **Balloon size factor** spinner to **2**. Choose **Apply** and then choose **OK**. Choose **OK** in the **Symbol Standard** dialog box.

98. Right-click on **Scene: SCENE 1** in the tree view of the desktop browser to display the shortcut menu and choose **BOM Database** to display the **BOM** dialog box.

99. Choose the **Insert parts list** button to display the **Parts list** dialog box. Enter **Bill Of Material** in the **Part List Name** edit box. Choose **Apply** and then choose **OK**. The prompt sequence is as follows:

    Bom table name is PIPE VISE.
    Specify location: *Specify the placement point of the BOM.*

100. Choose the **Ballooning** button. The prompt sequence is as follows:

    Select part/assembly or [auTo/autoAll/Collect/Manual/One/Renumber/rEorganize]: **O**
    Select pick object: *Select the point on any one of the component.*
    Select a start point of balloon: *Specify the start point of the balloon on the selected part.*
    Next Point: *Specify the next point.*
    Next Point: Enter

Similarly, add balloons to the remaining components.

101. Choose **OK** in the **BOM** dialog box. Figure 16-27 shows the layout with the drawing views, balloons and the BOM.

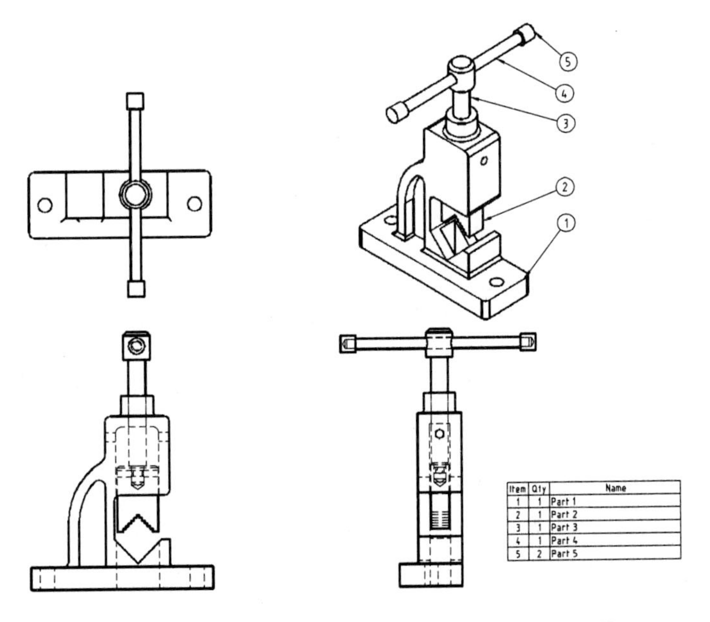

Figure 16-27 *The drawing views with BOM and balloons*

102. Save this file with the name given below:

\MDT Tut\Projects**Project1.dwg**.

PROJECTS

Project 1

In this project you will create and assemble all the components of the Tool Head Of Shaping Machine the details of which are given in the following figures (assume the missing dimensions). Create the exploded assembly scenes. Generate the following drawing views of the assembled drawing in the layout 1:

1. Top view.
2. Full sectioned front view.
3. Left side view.
4. Isometric view of the section view.

Generate the following views of the exploded assembly in the layout 2:

1. Top view.
2. Right half sectioned front view.
3. Right side view.
4. Isometric view of the section view. (Display the hatching in this view also.)

Make sure all the components in the assembly are sectioned with different hatch patterns. Add BOMs and Balloons to the drawing views with ISO and DIN standards respectively for layout 1 and layout 2. The BOMs should have the following columns:

1. Item.
2. Quantity.
3. Name.
4. Material.

Save the final drawing with the name given below:

\MDT Tut\Projects**Project2.dwg**.

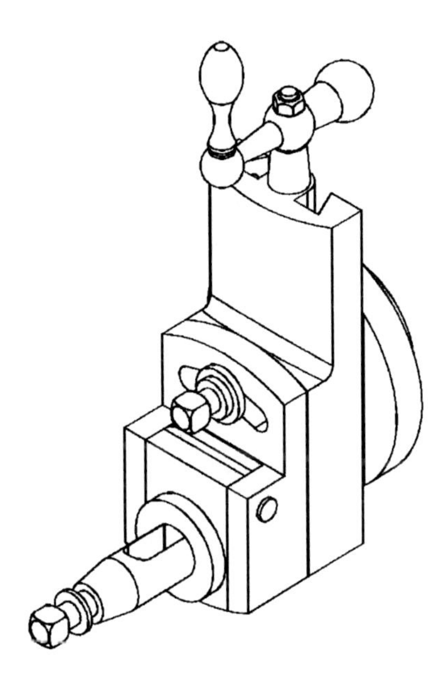

Figure 16-28 Assembled view of Tool Head Of Shaping Machine

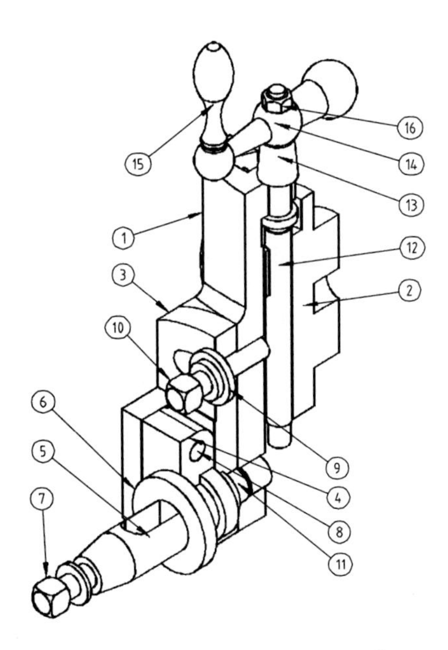

Figure 16-29 Various components in the assembly

Bill Of Material			
Item	Qty	Name	Material
1	1	VERTICAL SLIDE	CAST STEEL
2	1	BACK PLATE	CAST STEEL
3	1	SWIVEL PLATE	MILDSTEEL
4	1	DRAG PLATE	MILD STEEL
5	1	TOOL HOLDER	MILD STEEL
6	1	WASHER	MILD STEEL
7	1	TOOL FIXING SCREW	STEEL
8	1	PIVOT PIN	MILD STEEL
9	1	WASHER	MILD STEEL
10	1	CLAMPING SCREW	MILD STEEL
11	1	SWIVEL SCREW PIN	MILD STEEL
12	1	SCREW ROD	MILD STEEL
13	1	SPACER BUSH	MILD STEEL
14	1	HANDLE BAR	MILD STEEL
15	1	HANDLE	MILD STEEL
16	1	NUT M 10	MILD STEEL

Figure 16-30 *Bill of Material for Project 1*

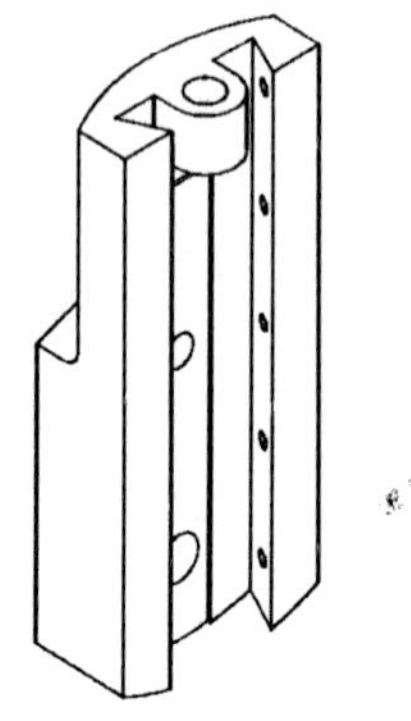

Figure 16-31 *3D view of Vertical Slide*

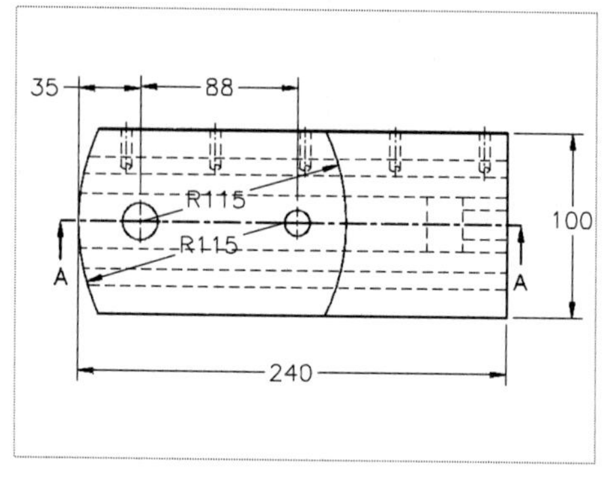

Figure 16-32 *Top view of Vertical Slide*

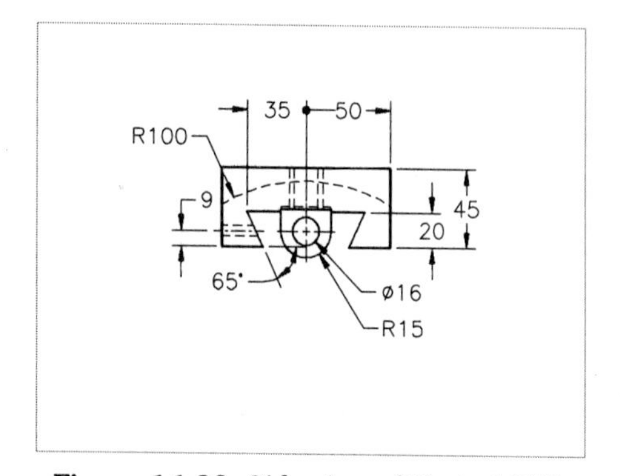

Figure 16-33 *Side view of Vertical Slide*

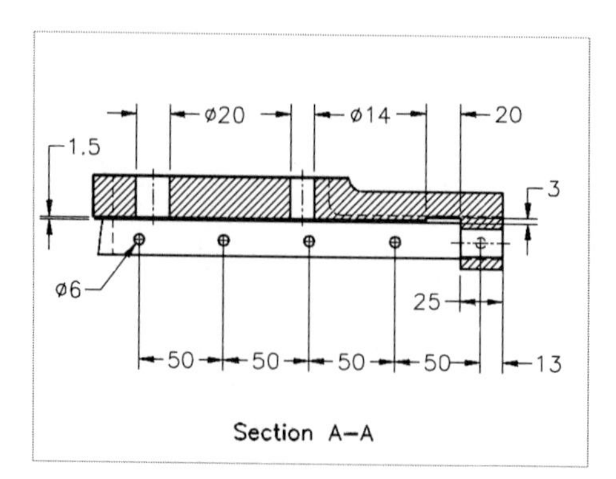

Figure 16-34 *Sectioned front view of Vertical Slide*

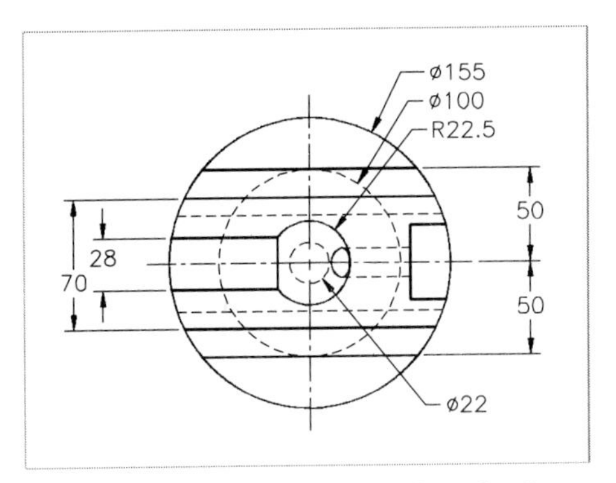

Figure 16-35 *Front view of Back Plate*

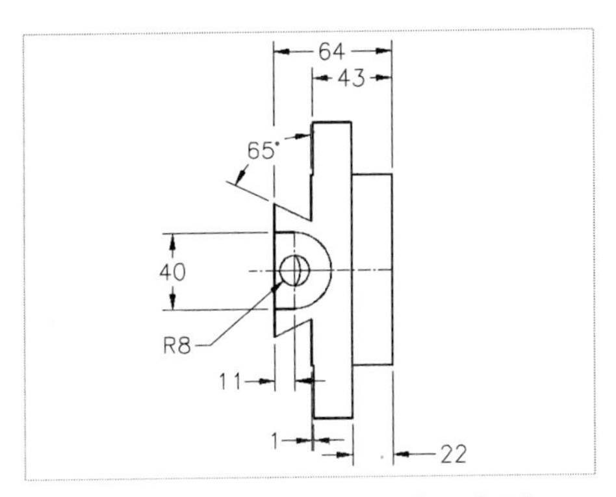

Figure 16-36 *Side view of Back Plate*

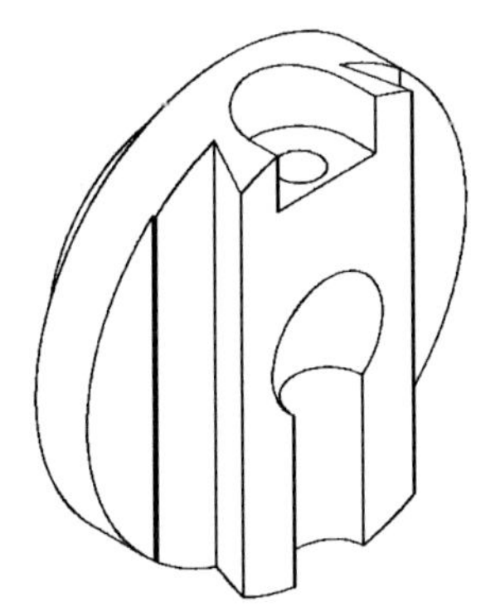

Figure 16-37 *3D view of Back Plate*

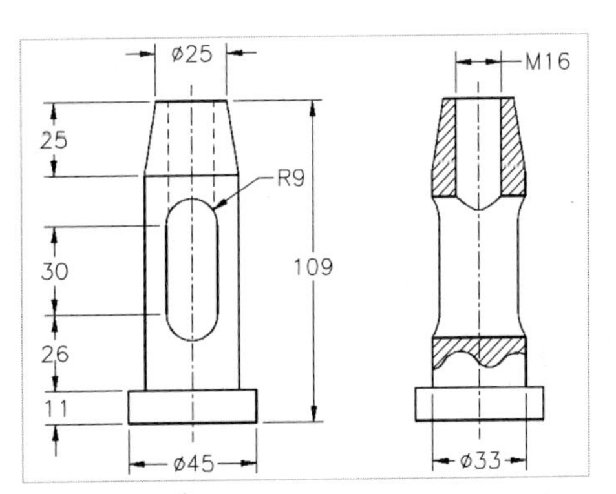

Figure 16-38 *Dimensions for Tool Holder*

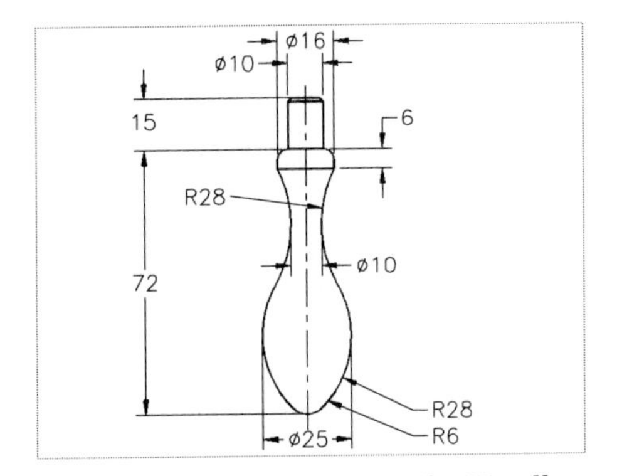

Figure 16-39 *Dimensions for Handle*

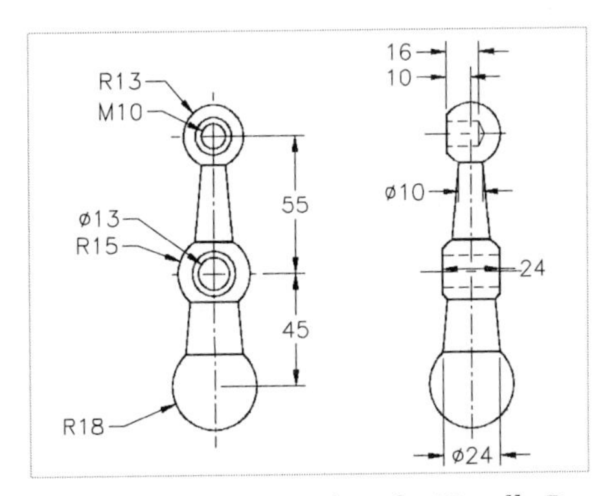

Figure 16-40 *Dimensions for Handle Bar*

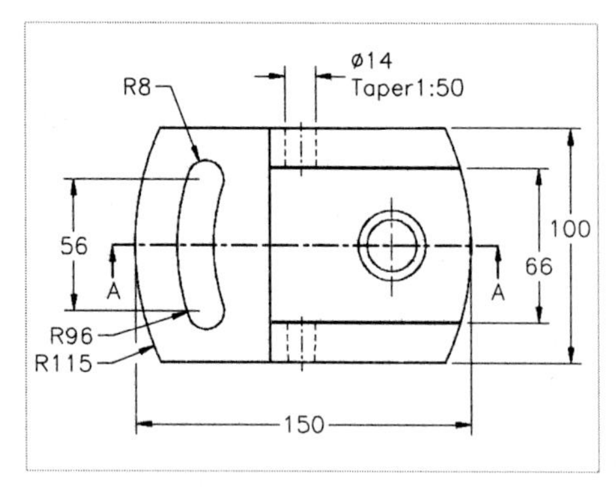

Figure 16-41 *Top view of Swivel Plate*

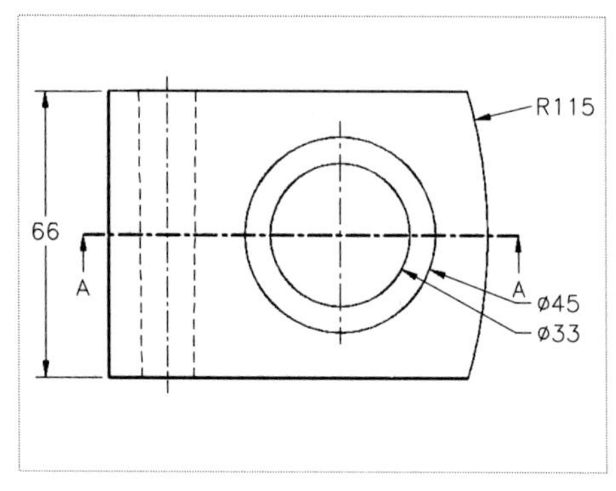

Figure 16-42 *Top view of Drag Plate*

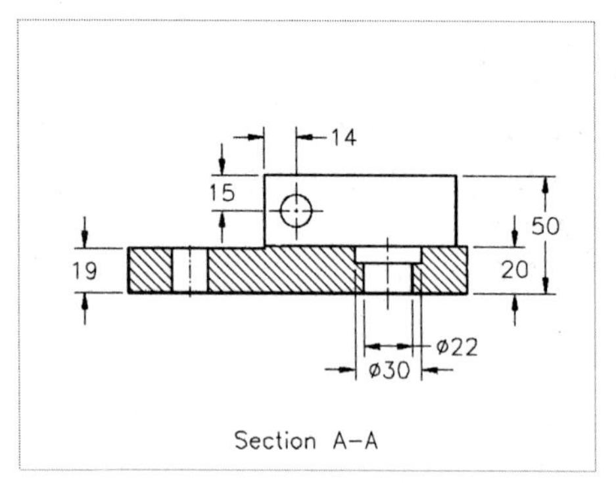

Figure 16-43 *Front view of Swivel Plate*

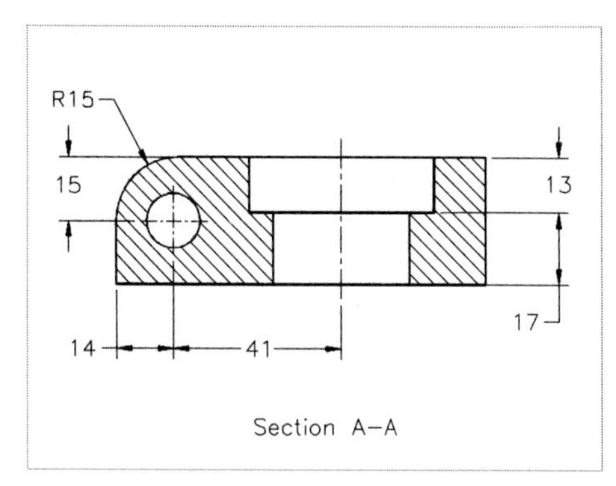

Figure 16-44 *Front view of Drag Plate*

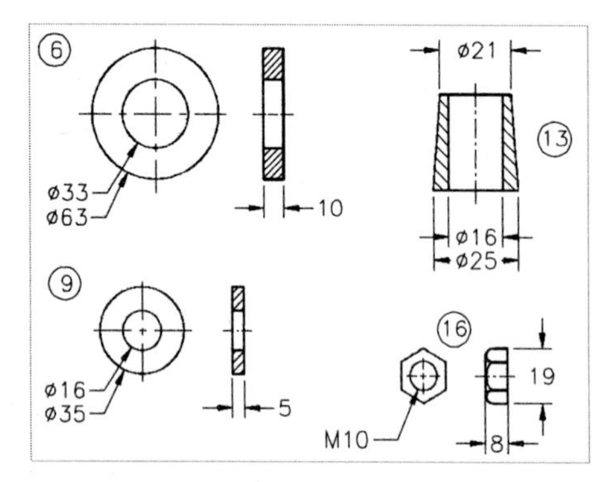

Figure 16-45 *Dimensions for Project 1*

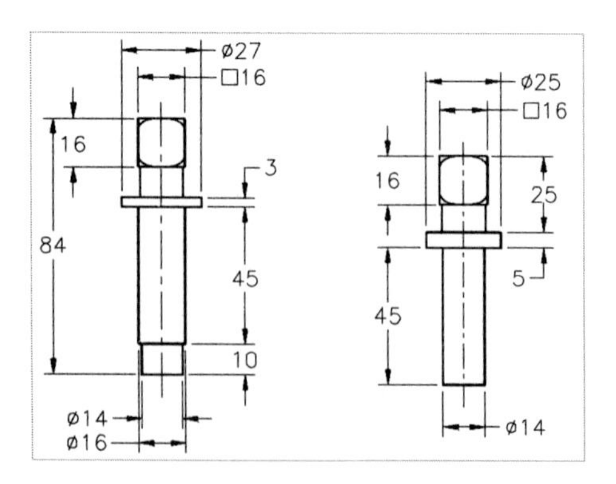

Figure 16-46 *Dimensions for Project 1*

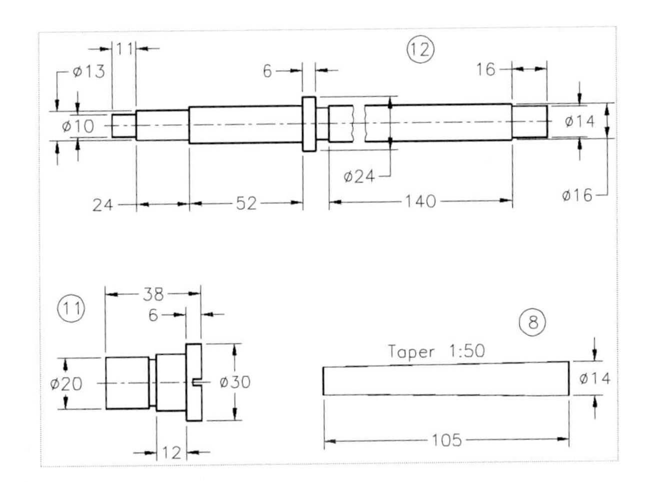

Figure 16-47 *Dimensions for Project 1*

Project 2

In this project you will create and assemble all the components of the Radial Engine Subassembly, the details of which are given in the following figures (assume the missing dimensions). Create the exploded assembly scenes. Generate the following drawing views of the assembly in the layout 1:

1. Top view.
2. Front view.
3. Left side view.
4. Isometric view of the front view.

Generate the following drawing views of the exploded assembly in the layout 2:

1. Top view.
2. Front view.
3. Right side view.
4. Isometric view of the section view.

Add BOMs and Balloons to the drawing views with JIS and GB standards respectively for layout 1 and layout 2. The BOMs should have the following columns:

1. Item.
2. Quantity.
3. Name.
4. Material.

Save this project with the name \MDT Tut\Projects**Project3.dwg**.

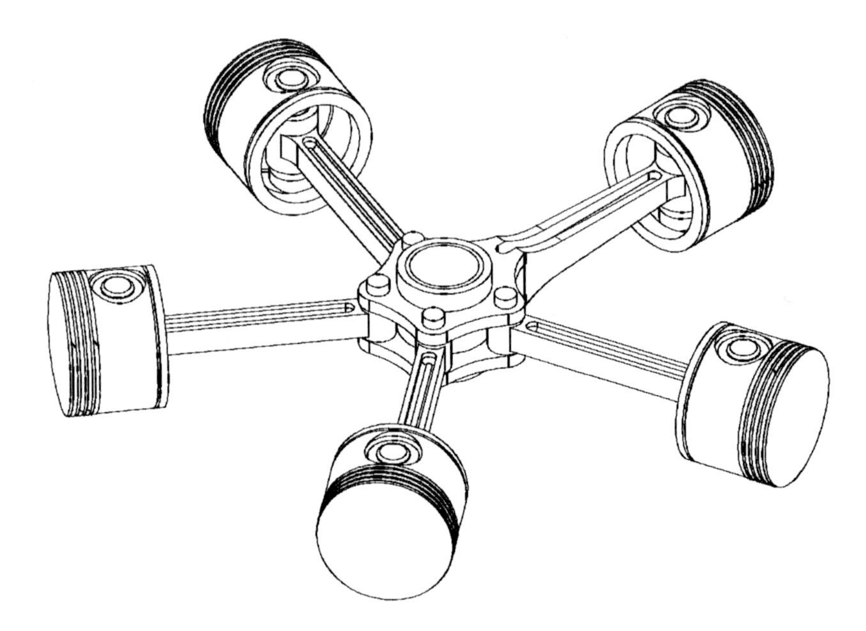

Figure 16-48 *Assembled view of Radial Engine Subassembly*

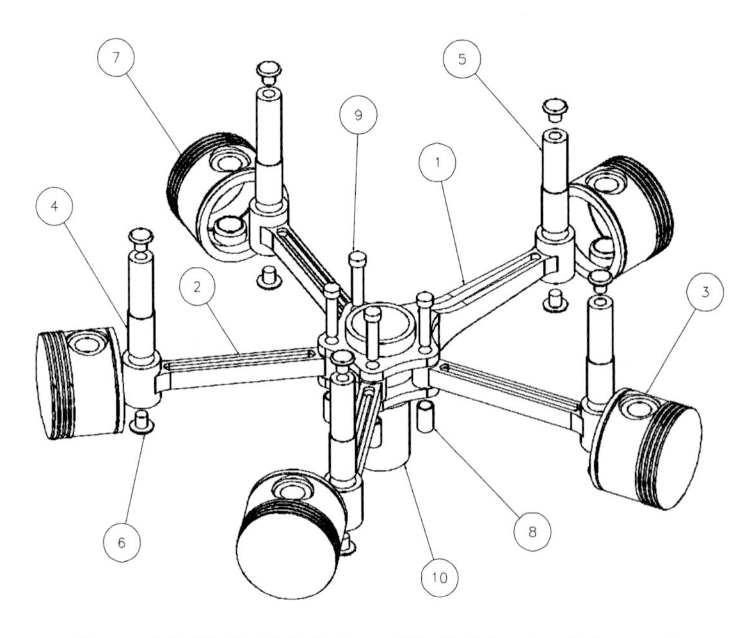

Figure 16-49 *Exploded view of Radial Engine Subassembly*

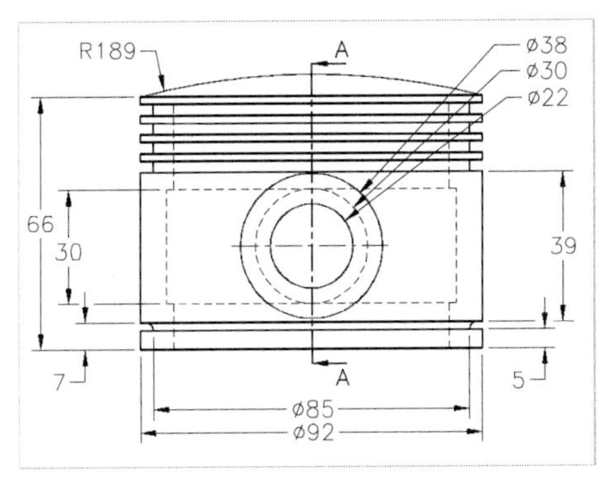

Figure 16-50 *Dimensions for Piston*

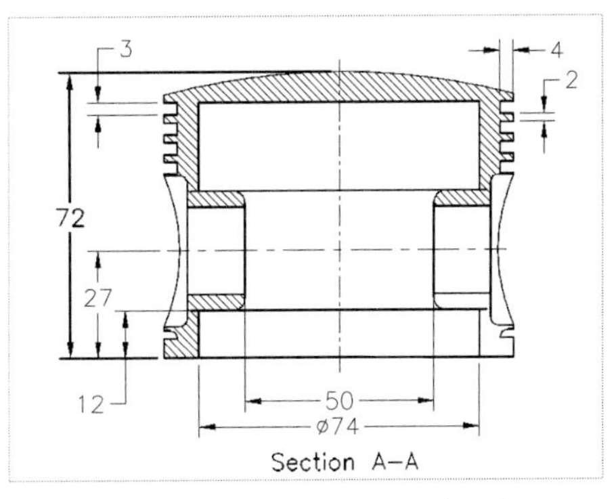

Figure 16-51 *Dimensions for Piston*

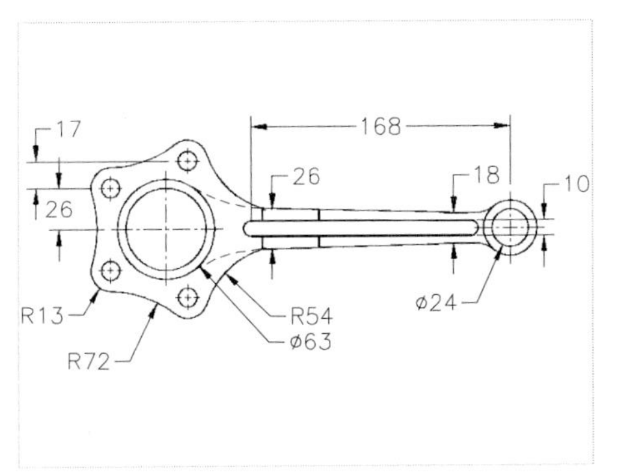

Figure 16-52 *Top view of Master Rod*

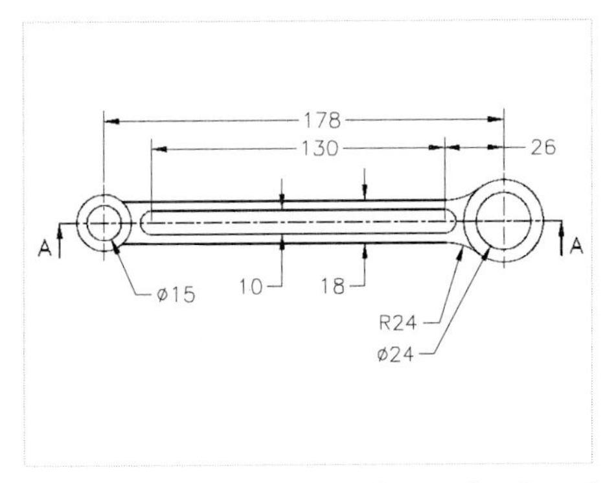

Figure 16-53 *Top view of Articulated Rod*

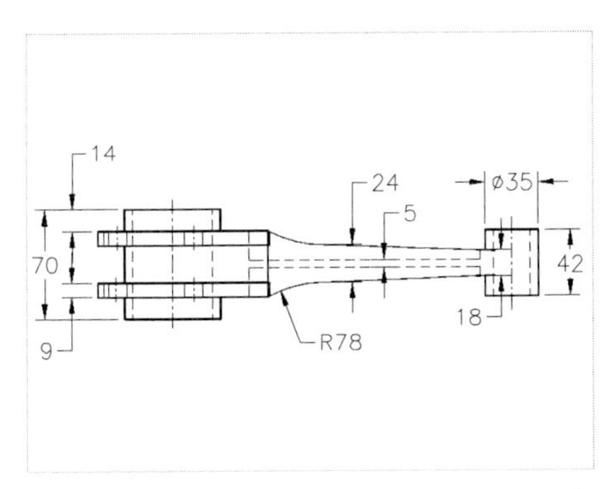

Figure 16-54 *Front view of Master Rod*

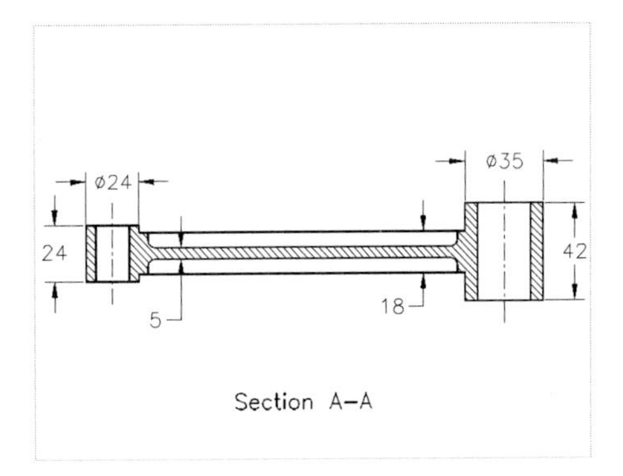

Figure 16-55 *Front view of Articulated Rod*

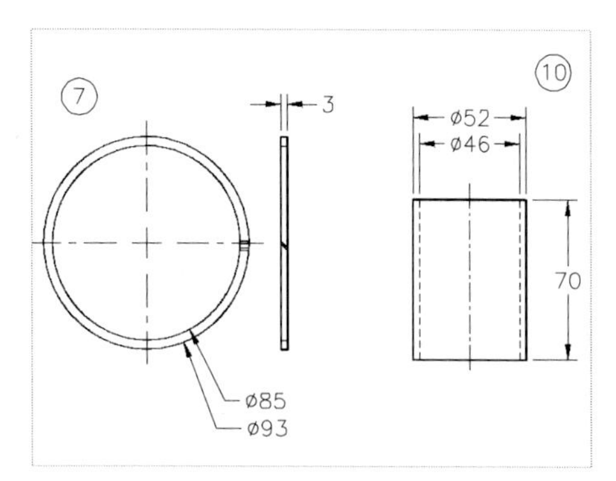

Figure 16-56 Dimensions for Project 2

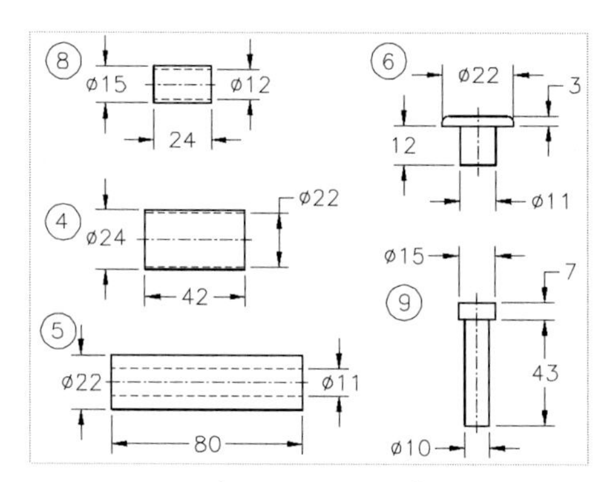

Figure 16-57 Dimensions for Project 2

Bill of Material			
Item	Qty	Name	Material
1	1	MASTER ROD	MCS
2	4	ARTICULATED ROD	MCS
3	5	PISTON	AL
4	5	ROD BUSH–UPPER	BABBIT
5	5	PISTON PIN	Ni–Cr STEEL
6	10	PISTON PIN PLUG	MS
7	20	PISTON RING	CS
8	4	ROD BUSH–LOWER	BABBIT
9	4	LINK PIN	HCS
10	1	MASTER ROD BEARING	Cd–Ag

Figure 16-58 Bill of Material for Project 2

Appendix-A

Toolbars

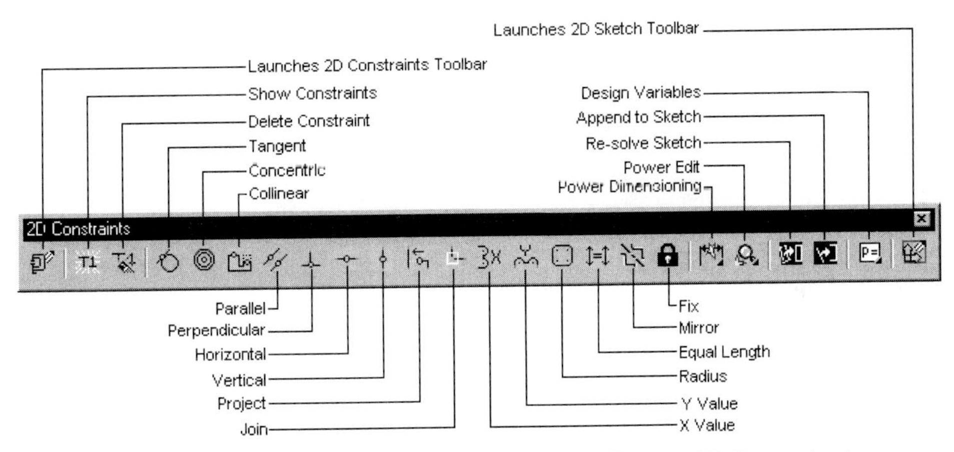

Figure A-1 *The 2D Constraint toolbar (View > Toolbars > 2D Constraints) or (View > Toolbars > Customize Toolbars > 2D Constraints)*

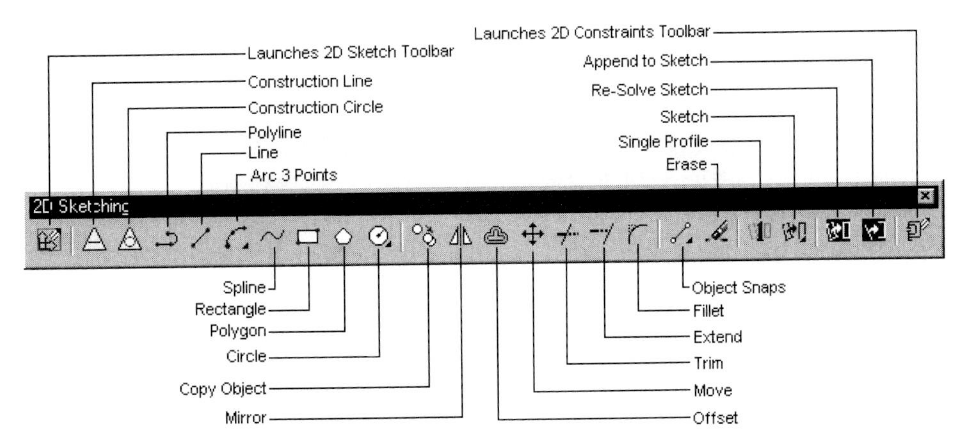

Figure A-2 *The 2D Sketch toolbar (View > Toolbars > 2D Sketching) or (View > Toolbars > Customize Toolbars > 2D Sketching)*

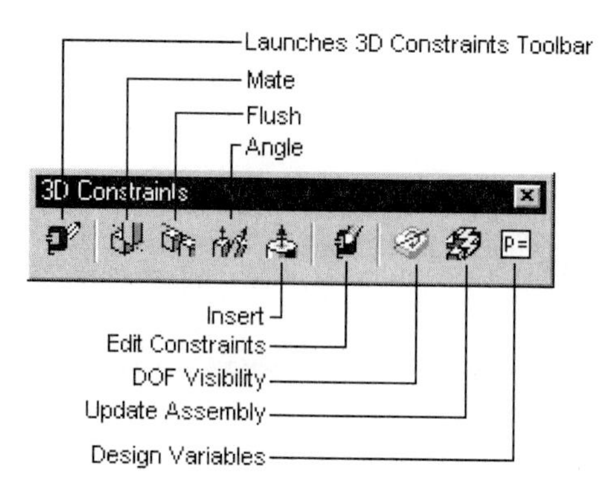

Figure A-3 *The 3D Constraints toolbar (View > Toolbars > 3D Constraints) or (View > Toolbars > Customize Toolbars > 3D Constraints)*

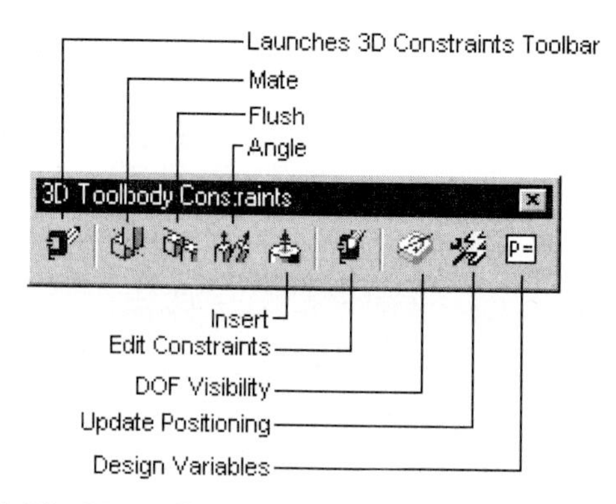

Figure A-4 *The 3D Toolbody Constraints toolbar (View > Toolbars > Customize Toolbars > 3D Toolbody Constraints)*

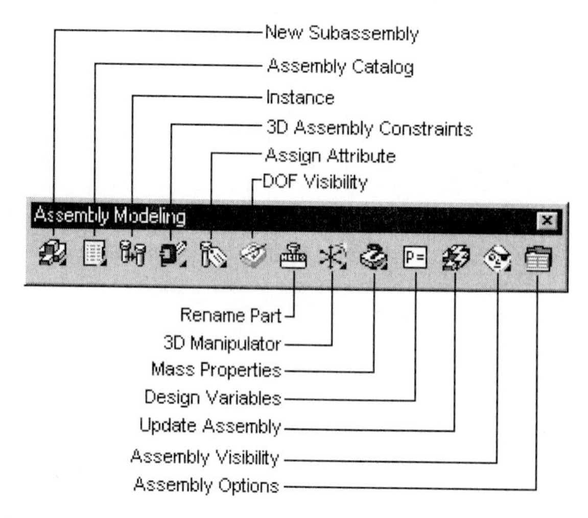

Figure A-5 *The Assembly Modeling toolbar (View > Toolbars > Assembly Modeling) or (View > Toolbars > Customize Toolbars > Assembly Modeling)*

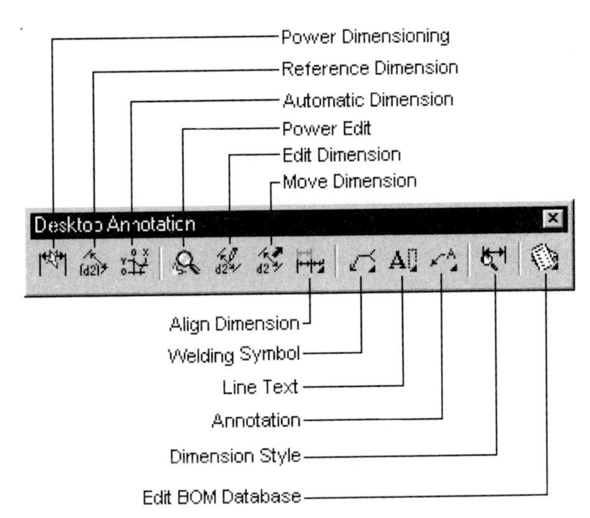

Figure A-6 *The Desktop Annotation toolbar (View > Toolbars > Desktop Annotation) or (View > Toolbars > Customize Toolbars > Desktop Annotation)*

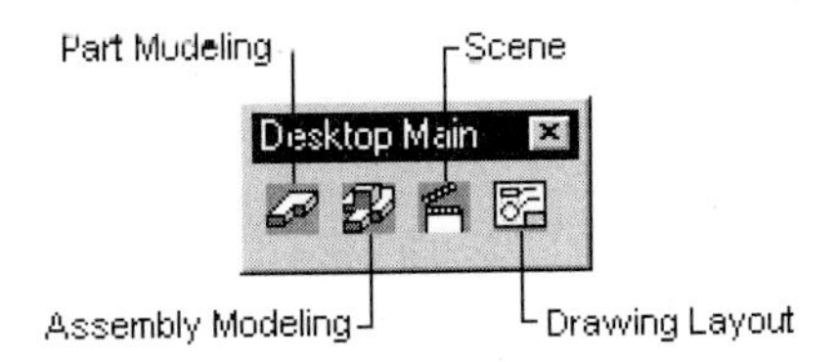

Figure A-7 *The Desktop Main toolbar (View > Toolbars > Customize Toolbars > Desktop Main)*

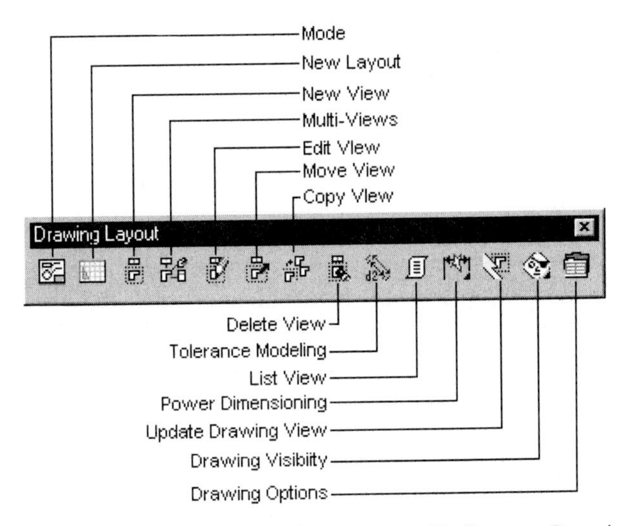

Figure A-8 *The Drawing Layout toolbar (View > Toolbars > Drawing Layout) or (View > Toolbars > Customize Toolbars > Drawing Layout)*

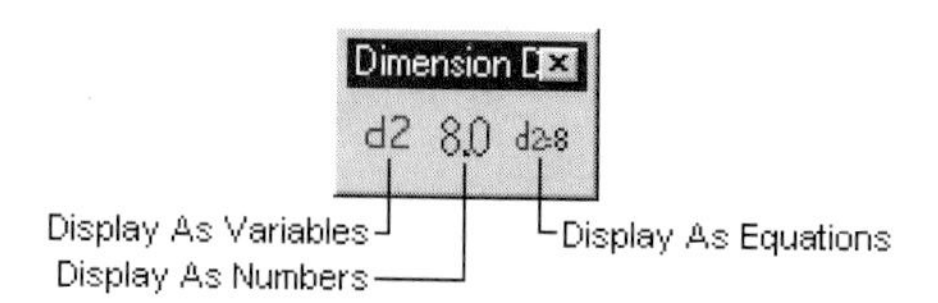

Figure A-9 *The Dimension Display toolbar (View > Toolbars > Customize Toolbars > Dimension Display)*

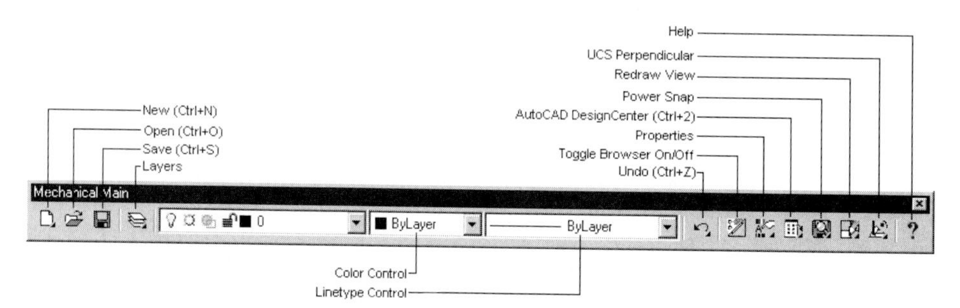

Figure A-10 *The Mechanical Main toolbar (View > Toolbars > Mechanical Main) or (View > Toolbars > Customize Toolbars > Mechanical Main)*

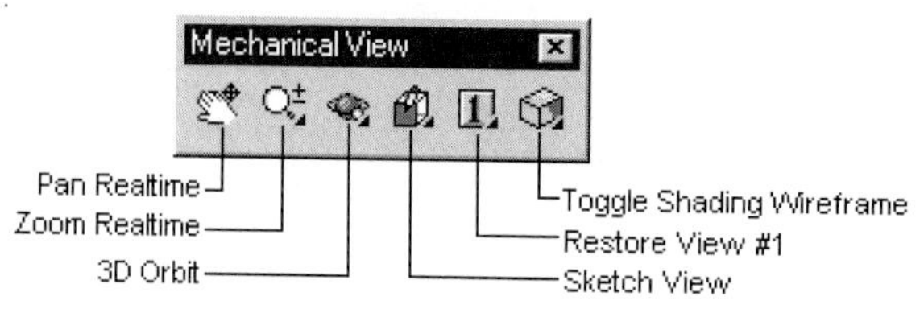

Figure A-11 *The Mechanical View toolbar (View > Toolbars > Mechanical View) or (View > Toolbars > Customize Toolbars > Mechanical View)*

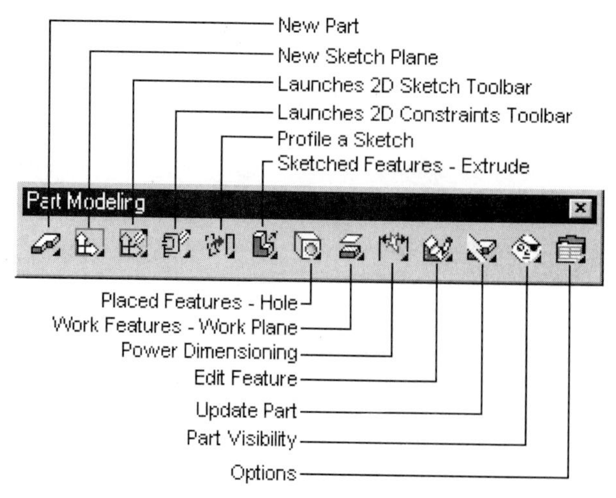

Figure A-12 *The Part Modeling toolbar (View > Toolbars > Part Modeling) or (View > Toolbars > Customize Toolbars > Part Modeling)*

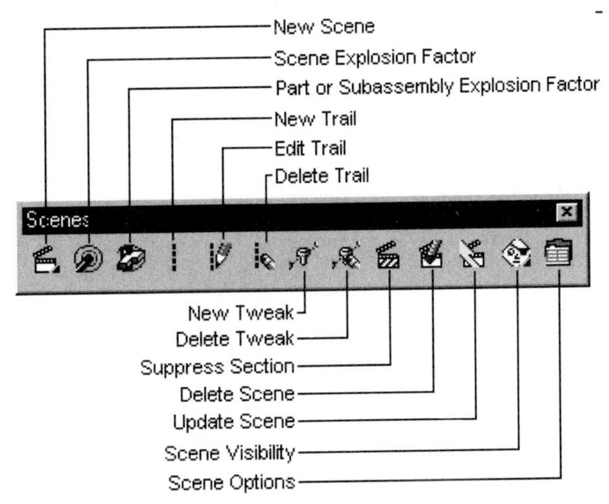

Figure A-13 *The Scene toolbar (View > Toolbars > Scene) or (View > Toolbars > Customize Toolbars > Scene)*

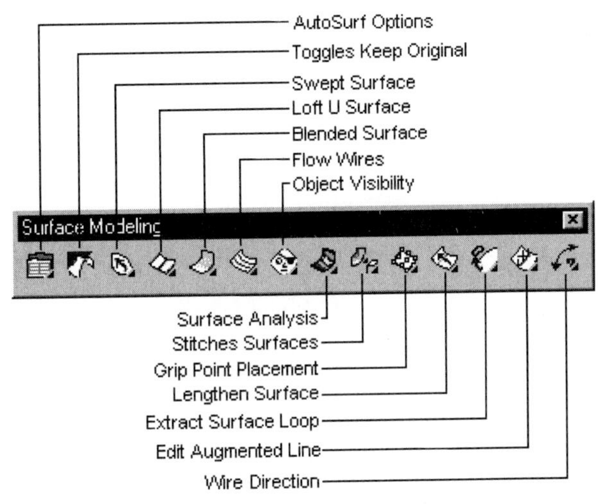

Figure A-14 *The Surface Modeling toolbar (View > Toolbars > Surface Modeling) or (View > Toolbars > Customize Toolbars > Surface Modeling)*

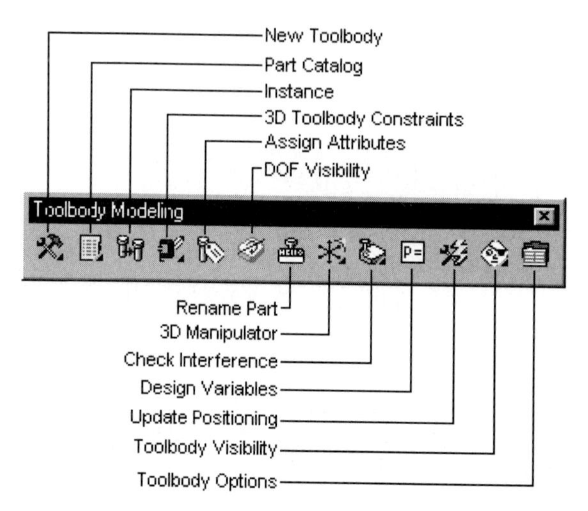

Figure A-15 *The Dimension Display toolbar (View > Toolbars > Customize Toolbars > Dimension Display)*

Appendix-B

System Variables

ACISOUTVER
Integer
This system variable is used to set the value for the version into which you create the ACIS file. Default value for this variable is 0.
0- Currently used version.
15- ACIS 1.5.
16- ACIS 1.6.
17- ACIS 1.7.
18- ACIS 1.8.
20- ACIS 2.0.
21- ACIS 2.1.
30- ACIS 3.0.
40- ACIS 4.0.
50- ACIS 5.0.

AMANNOTEPRESERVE
Integer
This variable is used to control the setting for the annotations that are not updated. Initial value for this variable is 1.
0- Annotations that are not updated are deleted.
1- Annotations that are not updated are saved into a group.

AMAUTOASSEMBLE
Integer
It is used to control the automatic updating of the assembly when a new constraint is applied to it. If it is turned off, the assembly is not automatically updated after the new constraint is applied to it. Initial value is 1.
0- Off.
1- On.

AMAUTOCL
This system variable is used to control the creation of the centerlines automatically when the views are generated. Initially this variable is on but you can make this variable off also.

AMAUTODIMARRANGE
This variable is used to automatically arrange the parametric dimensions. Initial value for this variable is 1.
0-Off.
1-On.

AMAUTOLOCKSCN
Integer
It is used to control the locking of the scenes.

If it is set to 1, the scenes can be locked. Initial value is 1.
0- Off.
1- On.

AMBLENDTOL
Real
It is used to avoid the creation of the CO breaks during the joining of the surfaces or splines in the surface modeling. Initial value is 0.01. The value for this variable ranges from 0.0001 to infinity.

AMCLCM
Real
It is used to control the length of the center line annotations. Initial value of this variable for the English drawings is 0.11 and for Metric drawings are 2.8. Any positive number can be set as the value for this variable.

AMCLFEAT
Integer
This variable is used to control the features for which the centerlines are automatically placed. Initial value for this variable is 5.

0-No feature
1-Holes
2-Fillets
3-Circular extrusions

AMCLGAP
Real
It is used to control the gap between the center line annotations. Its initial value for English drawings is 0.156 and for Metric drawings is 4. Any positive number can be set as the value for this variable.

AMCLOSHT
Real
It is used to control the overshoot of the center lines. Its initial value for English drawings is 0.11 and for Metric drawings is 2.8. Any positive number can be set as the value for this variable.

AMCLPAR
Integer
It sets the option for parametrically controlling the center line lengths during the updating. Initial value for this variable is 1.
0- Off.
1- On.

AMCLVIEW
Integer
This system variable is used to control the view for which the automatic centerlines are placed. Initial value for this variable is 63.
0-No view
1-Base view
2-Ortho view
3-Auxiliary view
8-Broken view
16-Detail view
32-Section view
64-Iso view

Note
You can get a combination of the views for which the centerlines will be placed automatically by adding their values. For example, if you want the centerlines to be created only for the ortho and the auxiliary view, set the value of the AMCLVIEW to 5.

AMCMDDIM
Integer
This variable is used to control whether the dimensions can also be selected when the sketches are edited. Its initial value is 1.
0- Off.
1- On.

AMCOMPSV
Integer
It is used to control whether the data in the Mechanical Desktop files are compressed while saving or not. Initial value for this variable is 1.
0- Off.
1- On.

AMCONDSPSZ

Integer

This variable is used to control the size of the constraints as they are displayed on the sketches. The initial value for this variable is 5. The value can vary from 1 to 19.

AMDWGCOLOR

Integer

This system variable is used to control whether the hatch patterns and the edges will use the color of the designer part or of the layer in which they are created. Its initial value is 0.

0- Visible edges, hidden edges and the hatch patterns will follow the colors of their layers.

1- Visible edges will follow the color of the designer model and the hidden edges and hatch patterns will follow the color of their layer.

2- Hidden edges follow the color of the designer model and the visible edges and hatch pattern follow the color of their layer.

3- Visible and hidden edges follow the color of the designer model and the hatch pattern follow the color of their layer.

4- Hatch pattern follow the color of the designer model and the visible edges and hidden edges follow the color of their layer.

5- Visible edges and hatch pattern follow the color of the designer model and the hidden edges follow the color of their layer.

6- Hidden edges and hatch patterns follow the color of the designer model and the visible edges follow the color of their layer.

7- Visible edges, hidden edges and the hatch patterns follow the color of the designer model.

AMDYNPAT

Integer

This variable is used to turn on or turn off the dynamic preview. Initial value for this variable is 1.

0-Off

1-On

AMEXTREFDIM

Integer

This variable is used to control the display of parametric dimensions on the external parts as the reference dimensions. Initially, this is turned off.

AMGRPREFIX

String

This variable is used to prefix names for the groups. Initial value for this variable is ASG and it can take any three characters as its value.

AMHATCHISO

Integer

This variable is used to turn on or off the automatic display of hatching in the isometric views.

0-Off

1-On

2-Hide obscured hatch

AMHIDEZERODIM

Integer

This variable is used to hide the parametric dimensions of zero length. Initially this variable is turned on.

AMHLCALC

Integer

This variable is used to set the options for the hidden line calculation in the drawing views. Initial value for this variable is 1.

0- Hidden lines are displayed.

1- Hidden lines are removed.

AMINSERTABS

Integer

This variable is used to control the insertion point of the external parts or subassemblies in the current drawing. Initial value for this variable is 0.

0- Insertion point will be the center of the geometry.

1- Insertion point will be the insertion point of the external file.

AMINTERPOLY

Integer

This variable is used to control whether or not a polyline is created at the intersection of two surfaces. Its initial value is 0.

0- Polyline is not created.

1- Polyline is not created.

AMJOINGAP

Real

This variable is used to control the gap between the two endpoints of the wires and surfaces in the surface modeling. Its initial value is 0.01 and it can vary from 0.0001 to infinity.

AMLINETHICK

Real

This variable is used to control the thickness of the lines used to calculate the thread annotation for the ISO and DIN standards. The initial value of this variable is 0.0275 and there is no limit for the value of this variable.

AMPAGELEN

Integer

This variable is used to control the number of lines sent to the text window before pausing. The initial value of this variable is 25 and it can have any value ranging from 0 to 1000.

AMPFITANG

Real

This variable is used to control the polyline fit angle. A spline or a surface created using the polyline at an angle less then this will be broken. Initial value for this variable is 150 and it can have any value ranging from 0 to 180.

AMPFITLEN

Real

This variable is used to set the value for finding the flat polyline segments. The new splines or surfaces created will lie flat against the polylines whose value is more then this value. Its initial value is infinity and it can vary from 0.0 to infinity.

AMPFITTOL

Real

This variable is used to control the tolerance for fitting a spline into a polyline. Initial value for this variable is 0.01 and it can vary from 0.0001 to infinity.

AMPROJOUTPUT

Integer

This variable is used to control the output of projection of a selected wire onto a specified surface. Initial value for this variable is 2.

0- Creates augmented lines.

1- Creates polyline.

2- Trims the surface.

AMPROJTYPE

Integer

This variable is used to control the view projection method for generating the drawing views. Initial value for this variable is 1.

0- First angle projection method.

1- Third angle projection method.

AMREUSEDIM

Integer

This variable is used to control the display of the parametric dimensions in the drawing views. Initial value for this variable is 1.

0- Display off.

1- Display on.

AMRULEMODE

Integer

This variable is used to control whether or not the constraints are automatically applied on the sketches. Initial value for this variable is 1.

0- Constraints are not applied automatically. (Off)

1- Constraints are applied automatically. (On)

AMSCENEUPDATE

Integer

This variable controls whether or not the

scenes are automatically updated. Initial value for this variable is 1.
0- Scenes are not updated automatically.
1- Scenes are automatically updated.

AMSECTIONDIM
Integer
This variable controls whether or not the parametric dimensions are displayed in the section views. Initial value for this variable is 0.
0- Dimensions are not displayed.
1- Dimensions are displayed.

AMSELDYNAMIC
Integer
This variable is used to control whether or not the selected entity is dynamically highlighted. Initial value for this variable is 1.
0- Off.
1- On.

AMSELTIPS
Integer
This variable is used to control the dynamic display of the help tips. Initial value for this variable is 1.
0- Off.
1- On.

AMSFDISPMODE
String
This variable is used to control the linetype for the display lines of the surfaces. Initial value for this variable is on.
On- ASULTYPE and ASVLTYPE are used.
Off- Default surface linetypes are used.

AMSFTOL
Real
This variable is used to control the tolerance of the surface accuracy. Initial value for this variable is 0.001 and it can have any value ranging from 0.0001 to infinity.

AMSINGLEHATLAY
Integer
This variable is used to place all the hatch patterns on to a single, already created, layer. Initially this variable is turned off.

AMSKANGTOL
Real
This variable is used to control the angle tolerance for the horizontal and vertical constraints. Initial value for this variable is 4 and it can have any value ranging from 0.001 to 10.

AMSKMODE
Integer
This variable is used to control whether the current sketch is considered as a precise sketch or a rough sketch. Initial value or this variable is 1.
0- Rough sketch
1- Precise sketch

AMSKSTYLE
String
This variable is used to control the linetype used for the boundary of the selected sketch. Initial value for this variable is Continuous line and you can set any linetype as the value for this variable.

AMTAPANNOTE
Integer
This variable is used to control the calculations of the hole notes in the views. Initially this variable is turned on.

AMTRUEPAT
Integer
This variable is used to control whether the preview of the patterned features is a true preview or a simplified representation. Initial value of this variable is 1.
0-Off.
1-On.

AMULINES
Integer
This variable is used to control the number of surface display lines in U direction. The initial

value for this variable is 5 and it can have any value ranging from 0 to 1000.

AMVANISH
Integer
This variable is used to control whether or not a vanish is displayed for the ANSI tapped holes. Initial value for this variable is 0.
0- Off.
1- On.

AMVECAUG
Real
This variable is used to control the vector length for the augmented lines. Initial value for this variable is 0.5 and it can have any value ranging from 0.0 to infinity.

AMVECSF
Real
This variable is used to control the vector length for the surfaces. Initial value for this variable is 0.5 and it can have any value ranging from 0.0 to infinity.

AMVIEWREFRESH
Integer
This variable is used to control the automatic updating of the drawing views. Initial value for this variable is 1.
0- Off.
1- On.

AMVIEWRESTORE
Integer
This variable is used to control whether or not the drawing views of the assembly are updated once the modifications are made in the assembly. Initial value for this variable is 1.
0- Off.
1- On.

AMVLINES
Real
This variable is used to control the number of surface display lines in V direction. Initial value for this variable is 5 and it can have any value ranging from 0 to 1000.

AMVPBORDER
Integer
This variable is used to control whether or not the viewport borders are displayed in the drawings. Initial value for this variable is 1.
0- Borders are displayed.
1- Borders are not displayed.

AMXASSEMBLE
Integer
This variable is used to control whether or not the external assembly constraints are updated automatically. The initial value for this variable is 0.
0- External assembly constraints are not updated.
1- External assembly constraints are updated.

Appendix-C

Commands

AM2DPATH
This command is used to solve (profile) the 2D entities and create 2D path for sweeping the profiles.

AM2SF
This command is used to convert the selected designer model, AutoCAD solid, 3D meshes, lines, arcs, circles and polylines with thickness into 3D AutoSurf surfaces.

AM3DPATH
This command is used to create 3D paths for sweeping the profiles. You can create helical path, spline path or pipe path using this command. You can also create a 3D path from an existing edge using this command.

AMABOUT
This command is used to display the information regarding the version and copyright of the Mechanical Desktop.

AMACISOUT
This command is used export the designer models or assemblies into an ACIS (.SAT) file. The version of the ACIS file is controlled using the AMACISOUTVER system variable.

AMACTIVATE
This command is used to activate a local or external designer parts, subassemblies, assemblies or the scenes.

AMADDCON
This command is used to apply the parametric constraints to a closed sketch, an open entity, cut line, split line or a break line.

AMADJUSTSF
This command is used to close the gap between the two selected surfaces by adjusting the edges of one or both the surfaces.

AMANALYZE
This command is used to analyze the selected surface. You can perform the analysis for curvature, draft or reflection lines using this command.

AMANGLE
This command is used to apply the angle assembly constraint to the selected local or external designer part or a subassembly.

AMANNOTE
This command is used to create or modify parametric annotations in the drawing views.

AMASSIGN
This command is used to create or modify the user-defined annotations to the selected designer part.

AMASSMPROP
This command is used to calculate the mass properties of the components of selected assembly. You can write the result of this command into a .MPR file.

AMATTACHSYM
This command is used to attach or display the symbols.

AMAUDIT
This command is used to find out whether or not all the external parts in the current drawing are updated.

AMAUGMENT
This command is used to create the augmented lines on the selected surface edge or the display line.

AMAUTODIM
This command is used to automatically dimension the drawing views in the layouts.

AMBALLOON
This command is used to add the parametric callouts to the part references in the drawing views in layouts.

AMBASICPLANES
This command is used to place the work planes at the top, front and side views of a specified point. A new part definition is created when you place the work planes using this command.

AMBEND
This command is used to bend an existing feature using an open profile.

AMBLDFEATNM
This command is used to list the features displayed in the desktop browser in the language of the last loaded version of Mechanical Desktop.

AMBLEND
This command is used to create a smooth surface between two, three or four wires.

AMBOM
This command is used to create or edit the BOM database to be used in the part lists in drawing views.

AMBREAK
This command is used to break the selected surface into two about the specified U or V edge.

AMBREAKLINE
This command is used to create the break lines to be used for generating the breakout section views in the layouts.

AMBROWSER
This command is used to control the display of the desktop browser.

AMBROWSERFONT
This command is used to control the type and size of the fonts used to list the features in the desktop browser.

AMCATALOG
This command is used to recall, attach or detach the external parts or subassemblies. You can localize the part or keep is as an external part in the drawing.

AMCENLINE
This command is used to create the parametric centerlines or center marks in the selected circles or arcs in the drawing views.

AMCHAMFER
This command is used to bevel the edges of

the designer model. The edges can be bevelled using one distance, two distances or one distance and one angle.

AMCHECKFIT
This command is used to check the minimum 3D distance or maximum 3D deviation between a surface and a wire or between two wires. This is done to check the product design and data for Coordinate Measuring Machines.

AMCOMBINE
This command is used to create complex designer models by applying the parametric boolean operations on them. The valid boolean operations that can be performed using this command are union, subtraction and intersection.

AMCOPYFEAT
This command is used copy the features within a designer model or from one designer model to the other. The new feature will be placed on the current sketch plane.

AMCOPYIN
This command is used to copy the external parts in the current drawing. However, this command localizes the parts placed.

AMCOPYOUT
This command is used to copy out the selected designer part into a new drawing file.

AMCOPYSCENE
This command is used to create a new scene by copying an existing scene.

AMCOPYSKETCH
This command is used to copy an unconsumed sketch or the sketch of a feature.

AMCOPYVIEW
This command is used to copy the selected drawing view to a new location in same layout or to a different layout.

AMCORNER
This command is used to create a blended surface at the corner of three filleting surfaces.

AMCUTLINE
This command is used to create the parametric offset or aligned cut lines for generating the offset or aligned section views in the layouts.

AMDATUMID
This command is used to create the datum identifiers in the form of frames. The drafting standards that can be used are ANSI, BSI, DIN, ISO, JIS, CSN, or GB.

AMDATUMTGT
This command is used to place the datum targets in the drawings. The drafting standards that can be used are ANSI, BSI, DIN, ISO, JIS, CSN, or GB.

AMDELCON
This command is used to delete the constraints applied to the profiled sketches.

AMDELETE
This command is used to delete the designer parts, instances or scenes.

AMDELFEAT
This command is used to delete the selected feature. If the selected feature has some dependent features, they will also be deleted.

AMDELTRAIL
This command is used to delete the trails from the exploded assemblies in the **Scene** mode.

AMDELTWEAKS
This command is used to delete the tweaks from the exploded assemblies in the **Scene** mode. The component moves back to its original position once the tweak is deleted.

AMDELVIEW
This command is used to delete the specified

drawing view from the layout. If the selected view has some dependent views then you will be asked whether you want to delete the dependent views or retain them.

AMDIMALIGN

This command is used to align the selected dimension in the drawing view with a specified base dimension.

AMDIMARRANGE

This command is used to arrange the dimensions in the drawing views or in the sketches in the **Model** tab.

AMDIMBREAK

This command is used to break the dimension line or the extension line in the drawing views. The associativity of the dimensions is not lost by breaking them using this command.

AMDIMDSP

This command is used to control the display mode of the dimensions. The dimensions can be displayed as parameters, equations or as numbers using this command.

AMDIMFORMAT

This command is used to modify the dimension format for the linear, angular, radial or diameter dimensions in the drawing views.

AMDIMINSERT

This command is used to break an existing dimension into two individual dimensions. The break point for the dimensions is specified by you.

AMDIMJOIN

This command is used to join two individual dimensions into a single dimension. This reduces the number of dimensions in the drawing views.

AMDIMMEDIT

This command is used to edit multiple dimensions in a single attempt.

AMDIRECTION

This command is used to display and reverse the direction of the selected surface. This is used in the NC machining or in the Coordinates Measuring Machines.

AMDISPSF

This command is used to control the display of the AutoSurf surfaces. You can change the surface normal size, the number of the U and V display lines and the surface patches using this command.

AMDIST

This command is used to calculate the minimum 3D distance between the selected components in the 3D space during the analysis of the assembly.

AMDWGDIMDSP

This command is used to control the display of the parametric dimensions in the drawing views. You can display the parametric dimensions in the drawing views as parameters, equations or numbers.

AMDWGVIEW

This command is used to create the 2D drawing views from the designer models or the assembly scenes. You can create orthographic, auxiliary, isometric, section, detail and broken views using this command.

AMEDGE

This command is used to untrim, copy and extract the edges of the surfaces. This command is also used to display the edge nodes at the surfaces and the faces.

AMEDGESYM

This command is used to generate a symbol that is used to define the current condition of the selected edge during the annotation of the drawing using the DIN standard.

AMEDIT
This command is used to edit the annotations like welding symbols, feature control systems, surface textures and datums.

AMEDITAUG
This command is used to resize, rotate, copy and modify the vectors on the augmented lines. You can also create augmented lines from lines and polylines using this command.

AMEDITCONST
This command is used to modify the distance and angle offset value of the assembly constraints. You can also delete the individual or all the assembly constraints applied to the assembly.

AMEDITFEAT
This command is used to edit the existing features on the designer models.

AMEDITSF
This command is used to modify the surfaces. Various modifications that can be performed using this command are lopping off the surfaces, reverse the direction of the surface normal, increasing the number of the surface grips, changing the density of the surface grips in the U and V directions and changing the distance between the surface grips.

AMEDITTRAIL
This command is used to edit the trails added to the exploded scenes of the assemblies in the **Scene** mode.

AMEDITVIEW
This command is used to modify the scale factor, associativity and the hidden lines display options of the drawing views created in the layouts.

AMEXTRUDE
This command is used to create 3D designer model from a profiled sketch by extruding it in the Z direction of the current sketch plane.

AMEXTRUDESF
This command is used to create the surface by extruding the lines, polylines, arcs, circles, ellipses or splines.

AMFACEDRAFT
This command is used to apply the draft angle to the faces of the designer model for its easy withdrawal from the molds during manufacturing.

AMFACESPLIT
This command is used to split the selected faces of the designer model using a planar face or by projecting a split line on to the selected faces.

AMFCFRAME
This command is used to create the feature control frames used to display the tolerances, orientation, position and so on. The standards that can be used are ANSI, BSI, CSN, DIN, GB, ISO, and JIS.

AMFEATID
This command is used to create the feature identification symbol to be used in the feature control frames. The standards that can be used are ANSI, BSI, CSN, DIN, GB, ISO, and JIS.

AMFILLET
This command is used to fillet the edges of the designer model. The fillet feature created on the designer model in parametric in nature and can be edited at any point of time.

AMFILLET3D
This command is used to fillet the selected wires in the current drawing. The wires that can be used for this command are lines, coplanar polylines, arcs, elliptical arcs, circles, ellipses and splines.

AMFILLETSF
This command is used to fillet any two selected

surfaces. The fillet surface thus created will be an independent surface.

AMFITSLIST
This command is used to copy the existing fits and their dimension values in a list that can be copied in the drawing. This command is also used to update the fits list when the fits are modified.

AMFITSPLINE
This command is used to fit the spline into a selected polyline, arc, circle or an ellipse and at the same time retaining the properties of the original entity. The spline can be fitted either using the tolerance value or the control points.

AMFLOW
This command is used to create the 3D polylines or the augmented lines to be displayed on the surfaces in the U and/or V directions or on the C1 tangencies.

AMFLUSH
This command is used to apply the flush assembly constraint to the components for placing them parallel to each other.

AMHOLE
This command is used to place a parametric drilled, counterbore or countersink hole on the selected planar face of the designer model.

AMHOLENOTE
This command is used in the **Drawing** mode to automatically extract and display the information related to the selected hole in the drawing view.

AMIDFIN
This command is used to convert a printed circuit board data in the ID format (Intermediate Data) into a Mechanical Desktop entity.

AMINSERT
This command is used to apply the insert assembly constraint to the two selected components with circular edges or faces. This constraint allows both the selected circular faces or edges to share same central axis and at the same time allows them to make their faces coplanar. You can also specify some offset value for the selected faces.

AMINTERFERE
This command is used to analyze the components of the selected assembly for interference. You have an option of creating the interference solid that can be used for the analysis of the extra material. This command can also be used for the combined designer parts.

AMINTERSF
This command is used to intersect the selected surfaces at their intersection. You can also create a polyline at the intersection of the selected surfaces.

AMJOIN3D
This command is used to join the selected lines, polylines, ellipses, arcs, circles, splines or augmented lines into a single polyline, spline or augmented line. The new entity will have the properties of the first object selected to join.

AMJOINSF
This command is used to join the selected surfaces. You can join two or more surface using this command.

AMLAYER
This command is used for setting the layer properties in the drawing. When you invoke this command, the **Layer Control** dialog box will be displayed. You can set the layer properties and control the layer groups using this dialog box.

AMLENGTHEN
This command is used to increase or decrease

the length of the selected surface by a specified distance or percentage.

AMLIGHT
This command is used to control the ambient and the direct light that falls on the designer model. When you invoke this command, the lights window appears next to the desktop browser. Using this window you can modify the ambient and the direct light.

AMLIGHTDIR
This command is used to change the target of the current light source that falls on the designer model.

AMLISTASSM
This command is used in the **Model** mode to display the information related to the selected designer parts or subassemblies. The information is displayed in the AutoCAD Text Window.

AMLISTPART
This command is used to list the selected part or all the parts in the **Model** mode.

AMLISTVIEW
This command is used to display the information related to the selected view in the **Drawing** mode. The information is displayed in the AutoCAD Text Window.

AMLOCKSCENE
This command is used in the **Scene** mode to lock or unlock the selected scene. You can not modify the explosion factor and tweaks of the locked scene.

AMLOFT
This command is used to create the complex designer models by blending together two or more dissimilar sketches of faces of the designer models.

AMLOFTU
This command is used to create the surfaces using a single set of base wires.

AMLOFTUV
This command is used to create a surface using two sets of base wires. The two sets of wires represent the surface in the U direction and in the V direction.

AMMAKEBASE
This command is used to convert the selected designer model comprising of any number of features into a single base feature. No data related to the previous features remain in the memory of Mechanical Desktop and therefore, you cannot edit the base feature.

AMMATE
This command is used to apply the mate assembly constraints to the selected components of the assembly.

AMMIGRATEB
This command is used to convert the data of the previous release of Genius software so that they can be used in the Mechanical Desktop 4.

AMMIRROR
This command is used to create a new designer part by mirroring an existing designer part. You have to specify the name of the new part.

AMMODDIM
This command is used to modify the parametric dimensions of the sketches in the **Model** mode or on the drawing views in the **Drawing** mode.

AMMODE
This command is used to toggle between the **Model** or the **Drawing** mode.

AMMOVEDIM
This command is used to move the selected dimension within the view or from one view to the other.

AMMOVEVIEW
This command is used to move the selected

view to a new location within the current layout or to another layout. If the selected view has some dependent views then they will also move along with the parent view.

AMMANIPULATOR
This command is used to dynamically rotate or move the selected object in the X, Y or Z axis or in the 3D space. You can also create the instances of the selected object at a specified angle or distance using this command.

AMNEW
This command is used to create a new designer part, subassembly or a scene. This command is also used to convert an AutoCAD solid into a designer model.

AMNOTE
This command is used to create an associative note for the parts, holes or the undercuts in the drawing views.

AMOFFSET3D
This command is used to offset the selected 3D polyline.

AMOFFSETSF
This command is used to create a new surface by offsetting the selected surface.

AMOPTIONS
This command is used to set preferences for the designer parts, assemblies, surfaces, scenes, drawing views and annotations in the current drawing.

AMPARDIM
This command is used to add the parametric dimensions to the profiled sketch. All the dimensions added using this command are displayed during the editing of the designer model.

AMPARTEDGE
This command is used to create a line, arc or a combination of both on the selected edge or face of an existing designer model. The new line or arc can be used for creating a new sketch.

AMPARTLINE
This command is used to create a 3D polyline representing the parting on the selected surface.

AMPARTLIST
This command is used to create a part list in the **Drawing** mode that displays the information related to all the part of the assembly. The information is used from the BOM database.

AMPARTPROP
This command is used to calculate the mass properties of the active designer model. You can also write the result of this command into a .MPR file.

AMPARTREF
This command is used to add the attributes to the nonparametric parts or a specified point. The attributes thus added can also be displayed in the Bill Of Material.

AMPARTREFEDIT
This command is used to edit the attributes added to the nonparametric parts or points using the **AMPARTREF** command.

AMPARTSPLIT
This command is used to split the active designer part into two using a planar face, work plane or a split line.

AMPATTERN
This command is used to create the rectangular, polar or axial patterns.

AMPATTERNDEF
This command is used in the **Model** mode to assign different hatch pattern to all the components of the assembly to be used in the

section views in **Drawing** mode. This is done to easily identify various components in the section views.

AMPLANE

This command is used to create planar surfaces. The surfaces can be created by defining two corners of a window enclosing the surface or select a closed wire. If you select more than one wire, the trimmed surfaces are created depending upon the boundaries of the selected objects.

AMPOWERDIM

This command is used to apply the parametric dimensions or nonparametric dimensions to the selected entities in the **Model** mode or in the **Drawing** mode. When applied to an unprofiled sketch in the **Model** mode or to the entities in the drawing views in the **Drawing** mode, these dimensions will be nonparametric in nature and when applied to the profiled sketch in the **Model** mode, these dimensions will be parametric in nature.

AMPOWEREDIT

This command is used to edit the parametric, nonparametric or reference dimensions in the **Model** or the **Drawing** mode. You can also edit the entities like Part Lists, Balloons, Datum Identifiers, Welding Symbols and so on using this command.

AMPOWERSNAP

This command is used to create four personalized snap settings in addition to the current snap settings. When you invoke this command, the **Power Snap Settings** dialog box will be displayed.

AMPRIMSF

This command is used to create the primitive surfaces like cone, sphere, cylinder or torus.

AMPROFILE

This command is used to solve a closed entity and apply the geometric constraints to it. You can also profile an open entity using this command provided the gap or the opening is very small.

AMPROJECT

This command is used to project the selected wire on the specified surface and output can be obtained in the form of a trimmed surface, polyline at the intersection area or an augmented line at the intersection area.

AMPROJECT2PLN

This command is used to project a selected 2D planar entity or a 3D planar face on the specified work plane, sketch plane or a 3D planar face.

AMPSNAP1

This command is used to load the power snap 1 settings set using the **AMPOWERSNAP** command.

AMPSNAP2

This command is used to load the power snap 2 settings set using the **AMPOWERSNAP** command.

AMPSNAP3

This command is used to load the power snap 3 settings set using the **AMPOWERSNAP** command.

AMPSNAP4

This command is used to load the power snap 4 settings set using the **AMPOWERSNAP** command.

AMRECOVER

This command is used to check the Mechanical Desktop file for the errors and correct the errors, if any, in the selected file.

AMREFDIM

This command is used to add reference dimensions to the entities in the drawing views.

AMREFINE3D
This command is used to refine the selected line or polyline that is used for fitting the surfaces by changing the density of the points on the lines or polylines.

AMREFINESF
This command is used to refine the selected surface by changing the number of U and V patches.

AMREFRESH
This command is used to refresh all the external components in the current drawing if they have been modified since their last reference in the current drawing or the execution of the last **AMREFRESH** command.

AMRENAME
This command is used to rename the selected designer part, instance, scene or the drawing view.

AMREORDFEAT
This command is used to change the order of the feature used in the creation of the designer model.

AMREPLACE
This command is used to replace the external component referenced in the current drawing with other referenced components.

AMREPLACEDGE
This command is used to replace the edge of a selected surface with that of a specified surface. This command not only replaces the edge but also trims the surface at the apparent intersection and blends the transition between the two surfaces.

AMREPLAY
This command is used to display all the steps used in the creation of the selected designer part in the sequence in which they were used. You can truncate the steps at any point of time or suppress the further steps using this command.

AMRESTRUCTURE
This command is used to move the selected designer part or subassembly within the assemblies.

AMREVOLVE
This command is used to create a designer model or a new revolved feature by revolving a profiled sketch about a line included in the sketch, a work axis or an edge of the existing designer model.

AMREVOLVESF
This command is used to create a revolved surface by revolving the base curve about a selected axis.

AMRIB
This command is used to convert an open profile into a rib feature.

AMROTCENTER
This command is used to find out a new center of rotation for all the visible objects.

AMRSOLVESK
This command is used to resolve the selected sketch and display the number of constraints that are remaining to fully constraining the sketch. This command is also used to solve the sketch in case you have added an additional entity to the already profiled sketch.

AMRULE
This command is used to create a surface using two base wires.

AMSCALE
This command is used to scale the selected surface or wires individually or uniformly in the direction of their axes.

AMSECTION
This command is used to cut one or more section in a selected set of surfaces.

AMSELROT
This command is used to specify a user define rotation point for all the visible objects.

AMSHELL
This command is used to create hollow designer model by defining some wall thickness and removing the remaining material. You can define uniform wall thickness for the designer model or multiple wall thickness.

AMSHOWACT
This command is used to highlight the active designer part or the current sketch plane in the current file.

AMSHOWCON
This command is used to display all the constraints that are applied on the selected profiled sketch in the **Model** mode.

AMSHOWINST
This command is used to highlight all the instances of the selected designer model in the desktop browser.

AMSHOWSKETCH
This command is used to highlight all the geometries on the current sketch plane.

AMSKPLN
This command is used to create a new sketch plane on the specified planar face for creating the features in other planes.

AMSOLCUT
This command is used to cut an AutoCAD solid using the AutoSurf surface that is extending beyond all the faces of the AutoCAD solid.

AMSPLINE
This command is used to create a spline that is tangent to a specified surface, designer part or an AutoCAD solid.

AMSPLINEDIT
This command is used to edit the control point or the fit point spline.

AMSPLITLINE
This command is used to create a parametric split line that is used to split the faces or the parts.

AMSTITCH
This command is used to stitch the selected surfaces or create a blended surface to fill the gap between the selected surfaces.

AMSTLOUT
This command is used to copy out the selected solids, regions, designer parts or subassemblies in the .STL format to be used for Stereo Lithography.

AMSTYLEI
This command is used to copy only the dimension style from a specified drawing without actually inserting that particular drawing.

AMSUPPRCOLOR
This command is used to control the color of the dimensions and the Degree Of Freedom symbols on the suppressed features.

AMSUPPRESS
This command is used when you do not want some components to be sectioned in the section view created using the scenes.

AMSUPPRESSFEAT
This command is used to suppress the selected feature on the active designer model. This command also allows you to suppress the features with the help of a spreadsheet.

AMSURFCUT
This command is used to cut a selected

designer model using an AutoSurf surface extending beyond all the faces of the designer model.

AMSURFPROP
This command is used to calculate the mass properties of the selected surface to be used during the analysis of the surface model characteristics.

AMSURFSYM
This command is used in the **Drawing** mode to add the surface texture symbols to the drawing view entities. The standards that can be used for the surface texture symbols are ANSI, BSI, CSN, DIN, GB, ISO, and JIS.

AMSWEEP
This command is used to create a complex designer model by sweeping a profiled sketch about a parametric path created using the **AM2DPATH** or the **AM3DPATH** commands.

AMSWEEPSF
This command is used to create a complex surface by sweeping one or more base curves about one or more rails or one augmented line and one rail.

AMSYMLEADER
This command is used to attach or detach a leader from a selected symbol in the drawing mode.

AMSYMSTD
This command is used to select a drafting standard for the BOMs, Balloons, or other symbols used in the drawing views.

AMTEMPLATE
This command is used to modify an existing or create a new holenote template. The new templates are created using the existing standard holenote templates.

AMTEXTSK
This command is used to used to create an embedded or an engraved text on the designer part. The text material written using this command can be extruded, revolved or swept using the Mechanical Desktop commands.

AMTHICKEN
This command is used to add thickness to the selected AutoSURF surface. The surface is thus converted into a solid part.

AMTOLCONDITION
This command is used in the **Drawing** mode to control the tolerances used by the parametric dimensions.

AMTRAIL
This command is used to add the parametric lines to the exploded or tweaked assemblies displaying the path and the direction of the mating components.

AMTUBE
This command is used to create tubular surfaces along a specified wire that acts as the axis for the tubular surface.

AMTWEAK
This command is used to redefine the position of the assembled components in the **Scene** mode by moving or rotating them.

AMUCSFACE
This command is used to align the UCS to a specified face.

AMUCSPERP
This command is used to define a UCS perpendicular to the selected line. You can also define a sketch plane on the new UCS using this command.

AMUNSPLINE
This command is used to break the selected spline into a polylines fit into a specific tolerance or into fit points.

AMUNSUPPRESSFEAT
This command is used to unsuppress the features that were manually suppressed using the **AMSUPPRESSFEAT** command.

AMUPDATE
This command is used to view the effect of any kind of modifications made to the designer part, assembly, scene or drawing view.

AMVARS
This command is used to create the design variables that can be a numeric value, a parameter or an equation to manage the designer models in the drawing. You can create two types of design variables using this command; the Active Part and the Global design variables.

AMVIEW
This command is used to control the display of the designer models in the **Model** mode. You can rotate the view orientation using this command.

AMVIEWOUT
This command is used to export the drawing views of the designer model created in the **Drawing** mode into a new drawing file in the form of 2D AutoCAD entities.

AMVISIBLE
This command is used to control the visibility of the parts, assemblies, scenes, drawings, work features and other geometric objects.

AMVRMLOUT
This command is used to export the selected object in the form of Virtual Reality Modeling Language so that it can be displayed on the Web pages.

AMWELDSYM
This command is used to create the welding symbols to be added to the drawing views in the **Drawing** mode. The standards that can be used for the welding symbols are ANSI, BSI, CSN, DIN, GB, ISO, and JIS.

AMWHEREUSED
This command is used to find out the location of the selected designer model or the component of the assembly in the current drawing.

AMWORKAXIS
This command is used to create a parametric axis for the circular or cylindrical features in the active designer model. You can also sketch the work axis on the current sketch plane.

AMWORKPLN
This command is used to place a parametric work plane, with the help of the modifiers, for creating the features on the other faces of the designer model.

AMWORKPT
This command is used to place a parametric point at the specified location on the current sketch plane.

AMXFACTOR
This command is used to specify the explosion factor for the scene or individually for different parts of the assembly.

AMXREFCONVERT
This command is used to remove the O symbol from the layers, linetypes, blocks, dimension styles, text style and so on of the external references in the current file.

AMZOOM
This command is used to zoom in on to a specified designer part or subassembly.

Index

B

C